Remote sensing of vegeta

G:\vege-indices-wk\

SPOT

Remote sensing of vegetation

Principles, techniques, and applications

Hamlyn G. Jones

Honorary Fellow,

College of Life Sciences,

University of Dundee at SCRI,

Invergowrie, Dundee, Scotland

Robin A. Vaughan

Honorary Fellow,

Centre for Remote Sensing and

Environmental Monitoring,

University of Dundee,

Scotland

OXFORD
UNIVERSITY PRESS

Great Clarendon Street, Oxford OX2 6DP

Oxford University Press is a department of the University of Oxford.
It furthers the University's objective of excellence in research, scholarship,
and education by publishing worldwide in

Oxford New York

Auckland Cape Town Dar es Salaam Hong Kong Karachi
Kuala Lumpur Madrid Melbourne Mexico City Nairobi
New Delhi Shanghai Taipei Toronto

With offices in

Argentina Austria Brazil Chile Czech Republic France Greece
Guatemala Hungary Italy Japan Poland Portugal Singapore
South Korea Switzerland Thailand Turkey Ukraine Vietnam

Oxford is a registered trade mark of Oxford University Press
in the UK and in certain other countries

Published in the United States
by Oxford University Press Inc., New York

First published 2010

British Library Cataloguing in Publication Data
Data available

Library of Congress Cataloging in Publication Data
Data available

Typeset by MPS Limited, A Macmillan Company
Printed in Great Britain on acid-free paper by
CPI Antony Rowe, Chippenham, Wiltshire

ISBN 978–0–19–920779–4

1 3 5 7 9 10 8 6 4 2

Preface

There are already many textbooks on remote sensing, many of them admirable in their own way; so why another? Our prime motivation was our feeling that many of the current texts do not appear adequately to reflect the increasing interest in remote sensing (RS) of vegetation and the rapid development of new tools for such studies. As in many areas of research, a substantial proportion of new work is presented at conferences and in the subsequent volumes of their proceedings. These, however, are not widely available to most students and researchers. Even papers published in the primary literature are dispersed over a number of different journals, and may not be easily digestible, especially at undergraduate level. It is the role of a textbook to distil the available essential information and to put it in context, as well as to direct the reader to further sources of information on the subject.

Although there are many texts that cover the basics of radiation physics, image manipulation and analysis, and also many specialist texts on plant and ecosystem functioning, we felt that it was important to present an integrated approach to bring information relating to remote sensing of vegetation together in one place. Our interpretation of remote sensing for the purposes of this book has been broad, including not only airborne and satellite sensing, but also close-range 'in-field' sensing (sometimes referred to as 'proximal sensing') and even small-scale or laboratory 'remote' sensing as that is widely used for calibration and validation studies. Wherever possible we have treated results and applications at all scales, though for some topics it is more convenient to separate them. Separation is particularly appropriate where measurements at a particular scale have specific properties only available at that scale; for example close-range 'in-field' imaging provides access to within and between leaf variability that can provide specific information (e.g. for extraction of canopy structure from shaded/unshaded areas in an image, Section 8.6.3), while the larger spatial scale of most satellite images has particular value in derivation of spatially averaged quantities for management of large areas of land or for political purposes.

Remote sensing (RS) is not really a specialist subject in its own right, but one that brings together a wide range of disciplines, including physics, maths, and computing as well as the environmental sciences and biology. The aim of this book is to provide a rigorous, yet fairly simple, grounding in the relevant basic physics and plant physiology to allow the reader to choose from and critically assess the plethora of new techniques that are becoming available for the remote study of vegetation (and other surfaces). We have throughout kept the mathematical derivations to a level accessible to most upper undergraduates, emphasizing the meaning rather than the mathematics. This is an approach that we have used in our teaching and that one of us has previously adopted with some success in a previous text (Jones, 1992).

Although our main objective has been to describe the application of remote sensing to the study of plants and vegetation canopies, for completeness, and for the convenience of the reader, we have included both a fairly comprehensive introduction to much of the basic radiation physics, image analysis, and remote sensing technology, and an outline of the physiological basis of key aspects of plant functioning that might be amenable to remote sensing. Even though there are good introductory texts in each of these fields, an understanding of these aspects is so fundamental to the effective use of remote sensing data for vegetation studies that we have felt it critical to bring it all together into one place. We also strongly believe that in order to be able to obtain the most benefit from a subject, it is essential for the user to have a thorough understanding of the principles involved, rather than to just apply techniques and procedures by rote. With the increasing availability of free satellite data, and indeed the pressures for more data to be made available free to users, there have been concerns expressed (Mather, 2008) that this is likely to lead to the proliferation of unconsidered or indeed erroneous results. Correct interpretation of RS data to provide useful information requires a good understanding of the ways in which these data are obtained and especially of the inherent limitations.

The book is aimed at senior undergraduates and at new postgraduates to help them appreciate the ways in which remote sensing can be used for the study and monitoring of vegetation. Throughout, we have tried to guide the reader to further detailed information that appears in textbooks and published papers that are relevant to that particular aspect, but at the end of each chapter there is a 'further reading' section that suggests suitable texts for particular aspects of that chapter. Some chapters include boxes, in which are collected relevant definitions or explanations so as not to break up the text. There are a number of appendices, listing, for example, the symbols used in the text, abbreviations and acronyms, some current remote sensing systems, etc. Also included, where appropriate, are some questions to help the reader appreciate some of the quantitative material, and their answers.

The first part of the book covers fundamental principles of remote sensing, biological properties and detecting systems. Then follow several chapters that consider how we can extract useful information from the, mainly spectral, remote sensing data. Finally, after considering possible ways in which errors may creep into the interpretation, we look at a few applications in more depth in order to illustrate how the principles we have been discussing may be applied to the study of vegetation properties. Although we include information on a number of current observing systems (e.g. Appendix 3), as the technology is rapidly changing, we have not tried to be exhaustive but rather to give examples of the most common types. Because of the generality of the principles described we hope that the resulting book will be applicable in fields beyond purely vegetation studies.

Hamlyn Jones and Robin Vaughan
April 2010
Dundee, Scotland

Acknowledgements

We are very grateful to a number of people who have read and commented on drafts of various parts of this book. These include especially: Clement Atzberger, Raffaele Casa, Graeme Horgan, John Raven, Andrew Skidmore, and Meredith Williams, together with a number of anonymous referees.

Particular thanks also to Clement Atzberger for providing Fig. 7.7; Claus Buschmann for data for Fig. 3.5; Raffaele Casa for the data for Fig. 8.4, and for images in Figs 8.8(b), 8.12, 8.21, and Plates 8.2, 8.3; Jürg Schopfer for Fig. 8.8; F.C. Bosveld for data in Fig 9.2; Eyal Rotenberg for analysing the data in Fig 9.2; Ashley Wheaton for part of Fig. 6.7/part of Plate 6.2 and for Fig. 8.1/Plate 8.1. We also thank Kevin Watts, J. Humphrey, M. Griffiths, C. Quine, D. Ray, and the Forestry Commission for Figs 11.8, 11.9, and Plates 11.5 and 11.6.

We are also grateful to the following for permission to reproduce figures and other copyright material: American Geophysical Union – Journal of Geophysical Research (Fig 8.16); Cambridge University Press (Figs 2.10, 2.11, 3.19, 3.22, and part of Fig. 4.2); (Elsevier Ltd. – Agricultural and Forest Meteorology (Fig. 3.16), Elsevier Ltd. – Remote Sensing of Environment (Fig 8.20, Fig. 8.2); Elsevier Ltd. – International Journal of Applied Earth Observation and Geoinformation (Fig. 7.8); Elsevier Ltd. – Scientia Horticulturae (Fig 11.3/Plate 11.2); Elsevier Ltd. – Biosystems Engineering (Fig. 11.3/Plate 11.2); European Space Agency (Fig. 6.20/Plate 6.5; and part of Fig. 8.6); Journal of Experimental Botany (Fig. 3.7); NASA (image of the globe in Fig. 8.6 – downloaded from http://visibleEarth.nasa.gov/); NEODAAS and University of Dundee (Fig. 6.1/Plate 6.1); NERC ARSF for permission to use images obtained during project MC/04/07 (Figs 6.4, 6.8, 6.13, 6.14, 6.15); Springer – Oecologia (8.14); Springer – Photosynthetica (Fig 11.5/Plate 11.4); USGS for permission to use Landsat images (Fig. 6.12/Plate 6.4 and Landsat data for Fig. 7.2).

We are grateful to LOPEX (Fig. 3.3), ASTER spectral library (courtesy of the Jet Propulsion Laboratory, California Institute of Technology, Pasadena, California) (Fig. 3.10 and Table 3.2) and the USGS spectral library (Fig. 3.12) for spectral data.

H. G. Jones and R. A. Vaughan

Table of contents

Symbols

Where possible we have tried to use the most common symbols for quantities, but as different scientific specialisms often employ different symbols for a particular quantity, in an interdisciplinary text such as this we have inevitably had to make decisions as to which to use here. We have also attempted to minimize the number of alternative meanings for any individual symbol, but it has been inevitable that some symbols are used for several quantities. In a few cases we have resorted to the use of non-standard multiple letter 'symbols' where these are commonly applied in the literature. Nevertheless, in general, we hope that ambiguities have been avoided. We have not attempted to list all the different subscripted versions of the symbols used, nor have we listed all the constants used.

a	angular separation between two sources (microwaves)
a_i	absorption coefficient ($\Delta \boldsymbol{I}/\boldsymbol{I}$ per unit thickness (of an infinitely thin) layer of absorber; m^{-1}); specific absorption coefficient = a/concentration (m^2/kg, or for water often expressed as cm^{-1})
A	ampere (SI unit of electric current)
A	albedo
A	amplitude (peak–peak) of surface temperature change (N.B. some authors use A for half this value)
A	area (m^2)
AI	angle index (e.g. AI_λ)
B	slope MVI (Section 6.3.5)
$\boldsymbol{B}_i$	radiosity, or total flux density leaving the ith discrete surface (W m^{-2})
$\boldsymbol{B}$	magnetic vector of an electromagnetic wave
c	speed of light (2.99792458×10^8 ms^{-1}) [from NIST calibration tables]
c	volumetric concentration m^3 m^{-3} or %
c	mass concentration = density (ρ = kg m^{-3})
c	cost per unit of sampling
$\hat{c}_i$	molar concentration (= n_i/V; moles of the quantity i per unit volume; mol m^{-3})
c_{ij}	coefficients in the $i \times j$th cell of a kernel for convolution
c_p	specific heat capacity of air (at constant pressure) (J kg^{-1} K^{-1}); $\boldsymbol{c}_s$ heat capacity of surface material
c_{area}	heat capacity per unit area (J m^{-2})
$\hat{c}_p$	molar specific heat of air (= 29.3 J mol^{-1} K^{-1})
$\boldsymbol{C}$	sensible heat flux (J m^{-2} s^{-1})[N.B. many authors use $\boldsymbol{H}$]
C	coulomb
C	total cost (of sampling)
$\boldsymbol{C}_{24h}$	24-h integral sensible heat flux (J m^{-2} s^{-1})
C_{chl}	concentration of chlorophyll; C_w concentration of water
$CWSI$	crop water stress index, dimensionless = $(T_s - T_{wet})/(T_{dry} - T_{wet})$
d	diameter
dB	decibel (logarithmic unit of power)

D_a	humidity deficit of air (kPa); D_a humidity deficit of air (mol mol^{-1})
$\boldsymbol{D}$	diffusion coefficient or diffusivity (m^2 s^{-1}) e.g. thermal diffusivity
D	day of the year
D	Simpson's diversity index (some authors use (1–D))
D_{ab}	Euclidean distance between a and b
$\boldsymbol{D}_H$	thermal diffusivity (some authors use κ)(m^2 s^{-1}; $= K/(\rho_s c_s)$)
DVI	difference vegetation index
e	eccentricity of the earth's orbit (= 0.0167)
e	partial vapour pressure of water (Pa) (inside a leaf = e_{leaf} assumed to equal e_{sat} or the saturation vapour pressure of water (Pa) at leaf temperature; or $e_{sat(Ta)}$
etr	electron transport rate
eV	electron volt ($= 6.24150974 \times 10^{-18}$ J)
E	energy (J)
E	echoes from lidar (ΣE_{veg}, ΣE_{total}, respectively refer to the sum of echoes from vegetation and of total echoes for a footprint)
$\boldsymbol{E}$	electric vector of an electromagnetic wave
$\boldsymbol{E}$	evaporation or transpiration rate (mol m^{-2} s^{-1}) or (kg m^{-2} s^{-1}); $\boldsymbol{E}_o$ reference evapotranspiration
EWT	equivalent water thickness (m)
f	focal length of lens
f	frequency (of radiation) (Hz; cycle s^{-1})
$f(\ldots)$	mathematical function of (...).
f	weighting factors (f_{iso}, f_{vol} and f_{geo}) for kernel-driven models
f	fractions of sunlit and shaded ground (g) and canopy (c) (f_g, f_c, $f_{g\text{-}sh}$, $f_{c\text{-}sh}$)
$f_{veg(\theta)}$	apparent fractional cover of vegetation apparent at a given view zenith angle (θ)
f_{veg}	true fractional cover of vegetation viewed at nadir
$\boldsymbol{F}$	radiant flux (W)
$F(\theta)$	canopy gap fraction at zenith angle (θ)
F	fluorescence (where F_v is the variable fluorescence, F_m is maximal fluorescence, F_o is the basal level of fluorescence with low light, F'_m is the quenched maximal fluorescence, F_s is the steady-state fluorescence, and F'_o is the quenched F_o, F_{UV} is that stimulated by UV light, and F_B is that stimulated by blue light, while other subscripts are used to indicate the fluorescence wavelength)
F	force (N; F_c centripetal force; F_a gravitational attraction)
g	conductance (m s^{-1}) subscripts include: g_W conductance to water vapour transfer; g_a boundary layer conductance; g_H conductance to heat transfer; g_{HR} parallel conductance to heat and radiative transfer
$\hat{g}$	molar conductance (mol m^{-2} s^{-1}); subscripts as for g
G	gain of antenna
G	total conductance of system ($= g_1 + g_2$ for conductances in series) (mol m^{-2} s^{-1} or m s^{-1})
$\boldsymbol{G}$	heat flux into the soil (W m^{-2}) (see also $\boldsymbol{S}$ for flux into storage in general)
G	gravitational constant (6.67300×10^{-11} m^3 kg^{-1} s^{-2})
$G(\theta)$	g-function expressing the projection of unit area of leaf in the direction θ

GNDVI	green normalized difference vegetation index ($GNDVI = (\rho_{NIR} - \rho_G)/(\rho_{NIR} + \rho_G)$)
h	Planck's constant (6.6261×10^{-34} J s)
h	the lag in a semivariogram
H	height above a surface; or *scale height* (the equivalent depth of the atmosphere)
H	Shannon–Weaver diversity index
H_0	null hypothesis
H_1	alternative hypothesis
HSAI	hemispheric surface area index (defined as half the actual surface area of leaves per unit ground area)
Hz	hertz (= 1 cycle s^{-1})
I	electrical current (A)
I	incident radiant flux density (W m^{-2})
$\boldsymbol{I}_{PAR}$	incident radiant flux density of photosynthetically active radiation (400–700 nm; W m^{-2})
$\boldsymbol{I}_S$	incident radiant flux density of total solar radiation (W m^{-2})
$\boldsymbol{I}_0$	incident flux density; at top of atmosphere = $\boldsymbol{I}_{toa}$; at top of atmosphere normal to the solar beam = solar constant = $\boldsymbol{I}_{pS}$
$\boldsymbol{I}^0, \boldsymbol{I}^1, \boldsymbol{I}^M$	unscattered, singly scattered, and multiply scattered radiation
$\boldsymbol{I}_S(D)$	extraterrestrial solar irradiance on day of the year D (W m^{-2})
J	joule
$\boldsymbol{J}$	mass flux (kg m^{-2} s^{-1})
J	source function for radiation
$\hat{\boldsymbol{J}} = \boldsymbol{J}_i^M$	molar flux (mol m^{-2} s^{-1})
k	Boltzmann constant (1.3807×10^{-23} J K^{-1})
k	extinction coefficient
k	radiation transfer kernels (k_{vol} and k_{geo})
k_D, k_F, k_P	rate constants for de-excitation through thermal dissipation as heat, fluorescence, and energy transfer to photosynthesis
K	kelvin (SI unit of temperature)
K	thermal conductivity (W m^{-1} K^{-1})
K_c	crop coefficient for evaporation calculation
l	correlation length (soil surface roughness parameter)
L	leaf area index (version for use in equations; see also *LAI*)
$\boldsymbol{L}$	radiance (W m^{-1} sr^{-1}); $\boldsymbol{L}_\lambda$ spectral radiance (W m^{-2} sr^{-1} μm)
LAI	leaf area index or leaf area per unit ground area (note in projected leaf area – m^2 m^{-2}); (see also L, and *HLAI*)
LUE	light use efficiency (see ε_V)
LWI	Leaf water index (most commonly $LWI = \rho_{1300}/\rho_{1450}$)
LWC	leaf water content (kg m^{-2})
m	optical airmass (defined as the ratio of the mass of atmosphere traversed per unit cross-sectional area of the actual solar beam)
m_s	mass of body (kg)
M	radiant excitance, radiant emittance (W m^{-2})
$\boldsymbol{M}$	metabolic heat output (W m^{-2})

M	merit function
$\mathrm{M_i}$	molecular mass or mass of one mole of substance *i*, (kg)
MPa	megapascals
MPDT	microwave polarization difference temperature
M_e	mass of earth (= 5.9736×10^{24} kg)
n_i	number of moles of *i*; number of replicates
n	refractive index
N	newton
N	near infrared reflectance (= ρ_N)
N	number of layers
NDTI	normalized difference temperature index
NDVI	normalized difference vegetation index
NPI	normalized polarization index (a normalized *MPDT*)
NRCS	normalized radar cross-section (= σ^o)
p	proportion
p_i	partial pressure of a component gas (Pa)
p	recollision probability
$\hat{p}$	average recollision probability
P	probability
Ppt	precipitation (mm)
P	pressure, atmospheric pressure, total gas pressure (Pa)
P_o	atmospheric pressure at sea level (Pa)
P	thermal inertia or thermal admittance (J m^{-2} K^{-1} $s^{-1/2}$)
P	power; P_t power leaving antenna, P_s power reaching the target, P_r power reaching detector
Pa	pascal (= 10^{-5} bar)
P_n, P_g	net and gross photosynthesis (mol m^{-2} s^{-1})
r	radius (m; e.g. of earth)
r	resistance (m s^{-1}); (includes r_W resistance to water vapour transfer; r_a boundary layer resistance; r_H resistance to heat transfer; r_{HR} parallel resistance to heat and radiative transfer)
ρ_s	hemispherical reflectance at surface
$\hat{r}$	molar resistance (m^2 s mol^{-1}) (subscripts as for *r*)
R	red reflectance (= ρ_R)
$R(\theta_i, \varphi_i; \theta_r, \varphi_r)$	*BRF* or bidirectional reflectance factor where $\boldsymbol{\theta}_i$ and $\boldsymbol{\varphi}_i$ refer to illumination zenith and azimuth angles, $\boldsymbol{\theta}_r$ and $\boldsymbol{\varphi}_r$ to view zenith and azimuth angles
R	radius of earth's orbit around the sun
R	radar range; $\boldsymbol{R}_r$ range resolution for SLAR radar
R	total resistance (e.g. for series = r_1+r_2)
R	electrical resistance (ohm)
$\boldsymbol{R}$	radiative flux density (W m^{-2}) and includes: $\boldsymbol{R}_n$ net radiation flux absorbed by canopy (W m^{-2}); $\boldsymbol{R}_{ni}$ isothermal net radiation; $\boldsymbol{R}_L$ longwave radiation ($\boldsymbol{R}_{Ld}$, $\boldsymbol{R}_{Lu}$); $\boldsymbol{R}_S$ shortwave (<4 μm) radiation
$\mathfrak{R}$	gas constant (8.3143 J mol^{-1} K^{-1})
RVI	ratio vegetation index

RWC	relative water content defined as *RWC* = (fresh mass – dry mass)/(turgid mass – dry mass) (dimensionless, often expressed as %)
s	sample standard deviation
s	the slope of a curve relating saturation water vapour pressure and temperature (at $(T_a - T_s)/2$); $\hat{s}$ the slope expressed as saturation water vapour mole fraction
S	heat flux (W m^{-2}) into storage (see ***G*** as a special case)
t	time
T	absolute temperature (K) (T_a, air; T_s, surface, T_ℓ or T_{leaf} leaf temperature, T_e equilibrium temperature for specific environmental conditions (at time *t*), T_{kin} for kinetic temperature, T_B apparent brightness temperature = $\varepsilon^{1/4} T_{kin}$
T	period of satellite/planet
u	wind velocity (m s^{-1})
V	volume m^{-3}
V	voltage
VI	vegetation index
VI^*	scaled vegetation index
w_i	mixing ratio
W	volume water content; W_o = initial amount
$\bar{x}$	sample mean value of *x*
x_i	mole fraction (mol mol^{-1})
x	Campbell's ellipsoidal parameter (the ratio of the horizontal semi-axis (*b*) to the vertical semi-axis (*a*))
z	*z*-test statistic for comparing means of two large samples
z	depth
Z	damping depth (m)
α	probability level at which we accept that an alternative hypothesis is true
α	absorptivity (dimensionless)
α	Priestley–Taylor constant relating ***E*** to net radiation (dimensionless; $\cong$1.26)
α	woody area index to plant area index
β	elevation of beam above the horizontal (e.g. solar elevation)
β	probability level we choose at which the null hypothesis is accepted when it is really false
γ	attenuation coefficient; γ_a fraction of radiation absorbed per unit thickness; γ_s fraction of radiation backscattered per unit thickness
γ	psychrometer constant (P_{cp} /0.622 λ; P_a K^{-1})
$\hat{\gamma}$	thermodynamic psychrometer constant ($\hat{c}_p /\lambda$; K^{-1})
$\hat{\gamma}^*$	modified psychrometer constant allowing for different conductances for heat and water vapour ($\hat{c}_p \hat{g}_H / \lambda \hat{g}_w$)
Γ	$\boldsymbol{G}/\boldsymbol{R}_n$ – the fraction of the incident net radiation energy that goes into the soil (Γ' is the partitioning at the soil surface, Γ'' the fraction of above canopy $\boldsymbol{R}_n$ that reaches the soil surface)
δT	temperature difference (= $T_s - T_a$)
ϵ	dielectric constant (ratio of the electric permittivity of a medium to the electric permittivity of free space) (dimensionless)
ε	emissivity (dimensionless); ε_λ spectral emissivity (dimensionless)
ε	s/γ

ε	efficiency (e.g. of photosynthesis; ε_V – vegetation conversion efficiency in terms of absorbed total solar radiation ($g\ MJ^{-1}$); ε_V' vegetation efficiency in terms of PAR– see also *LUE*); ε_V^* – conversion efficiency reduced by environmental stress.
η	dynamic viscosity (kg m^{-1} s^{-1})
θ_i	zenith angle; angle between normal to a surface and an incident beam; θ_v angle of view
κ	kappa coefficient of agreement
λ	wavelength (m, though note that we often use μm for convenience)
λ	latent heat of vaporization; λ (J mol^{-1}) (which for water = 44 172 J mol^{-1})
λ_m	the peak wavelength of the Planck distribution
λ_{RE}	position of red edge inflection point
λ_0	clumping index
μ	thermal admittance (see also thermal inertia, *P*)
μ	population mean
υ	wave number (cm^{-1}) (note that wave numbers in cm^{-1} are commonly used for historical reasons, even though strictly they should be m^{-1} according to SI; note also that in some fields wave number refers to what might be termed a circular wave number defined as $2\pi/\lambda$, rather than $1/\lambda$); conversions given below.
ρ	reflectivity (dimensionless); ρ_λ spectral reflectance; ρ_h, reflectance of deep horizontal leaved canopy; ρ reflection coefficient
ρ	density (kg m^{-3}) (see also concentration, *c*)
$\hat{\rho}_i$	molar density of a gas (n_i/V) = 44.6 mol m^{-3} at standard temperature and pressure
σ	Stefan–Boltzmann constant (5.6703×10^{-8}; W m^{-2} K^{-4})
σ	standard deviation of a population
$\sigma_{M(\lambda)}$	Mie scattering
$\sigma_{R(\lambda)}$	Rayleigh scattering
σ	microwave backscattering cross-section (dB)
σ^o	normalized microwave backscatter coefficient – *NRCS* (dB), normalized radar cross-section.
τ	(radar) pulse length
τ	transmissivity (dimensionless)
τ'_λ	optical depth or optical thickness at wavelength (λ) (= kx)
$\boldsymbol{\tau}$	shearing stress (= momentum flux) (kg m^{-1} s^{-2})
ϕ	evaporative fraction
ϕ	phase
φ	azimuth angle; φ_i for incident beam; φ_v for viewing beam
Φ_F	quantum yield of fluorescence (dimensionless)
Φ_{PSII}	quantum yield of *PSII*
ψ	water potential (MPa); ψ_π osmotic potential (MPa); ψ_p pressure potential or turgor; ψ_g gravitational potential
ω	angular frequency
Ω	solid angle (sr)
$\boldsymbol{\Omega}$	decoupling coefficient

Abbreviations and acronyms

AATSR Advanced Along Track Scanning Radiometer
AgIIS Agricultural Irrigation Information System
ALA Average Leaf Angle
ALI Advanced Land Imager on EO-1 Satellite
ALOS Advanced Land Observation Satellite (NASA)
ALS Airborne Laser Scanning
AMI Active Microwave Instrument
ANN Artificial Neural Net
ANOVA Analysis Of Variance
Aqua A NASA EOS satellite
ARVI Atmospherically Resistant Vegetation Index
ASAR Advanced Synthetic Aperture Radar (on Envisat)
ASM Angular Second Moment Texture Measure
ASTER Advancer Spaceborne Thermal Emission and Reflection Radiometer
ATCOR Atmospheric Correction Models (http:\\www.ReSE.ch)
ATI Apparent Thermal Inertia
ATM Airborne Thematic Mapper
ATP Adenosine Tri-Phosphate (an energy-containing molecule)
ATSR Along Track Scanning Radiometer
AVHRR Advanced Very High Resolution Radiometer
AVIRIS Airborne Visible and Infrared Imaging Spectrometer
AVNIR Advanced Visible and Near-Infrared Radiometer
BAI Burned-Area Index
BIRD Biospectral Infrared Detection Satellite
BRDF Bidirectional Reflectance Distribution Function
BRF Bidirectional Reflectance Factor (dimensionless)
CART Classification And Regression-Tree Approaches
CASI Compact Airborne Spectral Imager
CCD Charge Coupled Device
CEOS Committee on Earth Observation Systems
CHRIS Compact High Resolution Spectrometer
CI Confidence Interval
CON Contrast Texture Measure
CryoSat European Satellite For Measuring Ice Mass Balance
CZCS Coastal Zone Colour Scanner
DDV Dense Dark Vegetation
DN Digital Number
DEM Digital Elevation Model
DIFN Diffuse Non-Interference
DMC Disaster Monitoring Constellation
DN Digital Number (subscripted according to wavelength used)
DOS Dark Object Subtraction
DT Dark Target
DTM Digital Terrain Model
DVI Difference Vegetation Index
EMR Electromagnetic Radiation
ENT Entropy Texture Measure
Envisat European Space Agency's Earth Observation Satellite
EO Earth Observation
EOS Earth Observing System (NASA)
ERS-1, -2 Earth Resources Satellites, 1 and 2 (European)
ERTS Earth Resources Technology Satellite (later called Landsat)
ESA European Space Agency
ET Evapotranspiration
ETM Enhanced Thematic Mapper
ETM+ Enhanced Thematic Mapper Plus (on Landsat 7)
EVI MODIS Enhanced Vegetation Index
EWT Equivalent Water Thickness
fAPAR Fraction of Absorbed Photosynthetically Active Radiation
FCC False-Colour Composite
FDRI Fire Danger Rating Index
FER Fluorescence Emission Ratio

FLD — Fraunhofer Line Depth
FLEX — Fluorescence Explorer (proposed ESA mission)
FMC — Fuel Moisture Content
FOV — Field Of View
FPI — Fire Prediction Index
FWHM — Full Width Half-Maximum
GA — Genetic Algorithm
GAC — Global Area Coverage
GCP — Ground Control Point
GCM — General Circulation Model
GEOSS — Global Earth Observing System of Systems
GIFOV — Ground Instantaneous Field of View
GIR — Geographic Image Retrieval (or object retrieval)
GIS — Geographic Information System
GLAS — Geoscience Laser Altimeter System
GMES — Global Monitoring for Environment and Security programme (European Union)
GMS — Geosynchronous Meteorological Satellite (Japan)
GNDVI — Green Normalized Difference Vegetation Index
GOES — Geostationary Operational Environmental Satellite (NOAA)
GOMS — Geostationary Operational Meteorological Satellite (Russia)
GPS — Global Positioning System
GRACE — Gravity Recovery and Climate Experiment
GRD — Ground Resolution Cell
GSD — Ground Sampling Distance
HBW — Half Band Width
HCMM — Heat Capacity Mapping Mission
HOME — Height Of Mean Energy
HRG — High-Resolution Geometric Sensor on Spot-5
HRV — High-Resolution Visible Scanner on Spots 1–4
HSAI — Hemisurface Area Index
HSRS — Hot Spot Recognition Sensor
ICESat — Ice Cloud and Land Elevation Satellite
IFOV — Instantaneous Field Of View
IHS — Intensity Hue and Saturation (system of colours)
InSAR — Interferometric SAR
IPCC — Intergovernmental Panel on Climate Change
IR — InfraRed
IRS — Indian Remote Sensing satellite
ISS — Inertial Stabilizing System
ISCCP — International Satellite Cloud Climatology Project
JERS — Japanese Earth Resources Satellite
JPL — NASA's Jet Propulsion Laboratory
KOMPSAT — Korean Earth Observing Satellite
LAC — Local Area Coverage
LAD — Leaf-Angle Distribution (fraction of the total leaf area in each of a set of defined angle classes)
LAI — Leaf-Area Index
Landsat-1, 2, ... — Series of NASA/NOAA Land Observation Satellites
LEO — Low-Earth Orbit (satellite)
Lidar — Light Detection And Ranging
LIFT — Laser-Induced Fluorescence Transients
LISS — Linear Self-Scanning Radiometer
LUT — Look-Up Table
LW — Longwave radiation (TIR; 4–15 μm)
LWC — Leaf Water Content
LWI — Leaf Water Index
MERIS — Medium-Resolution Imaging Spectrometer
Meteor — Soviet/Russian series of meteorological satellites
Meteosat — European series of geostationary meteorological satellites
MIR — Mid-InfraRed (1–4 μm)
MNF — Minimum Noise Fraction Transform
MISR — Multiangle Imaging Spectroradiometer
MODIS — Moderate-Resolution Imaging Spectrometer
MODTRAN — Moderate Spectral Resolution Atmospheric Transmittance (algorithm for atmospheric correction, based on original US Air Force code)
MSG — Meteosat Second Generation
MSS — Multispectral Scanner on Early Landsats
MTF — Modulation Transfer Function
MW — Microwave

MVC Maximum Value Composite
NASA National Aeronautics and Space Administration (USA)
NBARS Nadir Bidirectional Distribution Function Adjusted Reflectance
NBR Normalized Burn Ratio
NCE Net Carbon Exchange (see also *NEE*, *NPP*, etc.; kg or mol of CO_2, or of C or of dry matter m^{-2})
NDVI Normalized Difference Vegetation Index
NEE Net Ecosystem Exchange – Consider Units
NERC The UK Natural Environment Research Council
NHI Normalized Heading Index
NIR Near-InfraRed (700 nm – 1 μm)
NOAA National Oceanic and Atmospheric Administration (USA)
NOAA-1, 2, . . NOAA (series of polar orbiting meteorological satellites)
NPOESS National Polar-Orbiting Operational Environmental Satellite System
NPP Net Primary Productivity (kg ha^{-1} (day or year)$^{-1}$)
NPQ Non-Photochemical Quenching
NRCS Normalized Radar Cross-Section
OLI Operational Land Imager
PAR Photosynthetically Active Radiation
PIT Pseudo-Invariant Target
Pixel Picture Element
POES Polar-Orbiting Environmental Satellite
PPA Parks and Protected Areas
PRI Photochemical Reflectance Index
PRI Precision Radar Image
PSI, PSII Photosystems 1 and 2
Radar Radio Detection And Ranging
Radarsat Canadian Radar Satellite
RAR Real Aperture Radar
REP Red-Edge Position (= λ_{RE})
RGB Red, Green, Blue (system of colours)
RGI Relative Greenness Index
RMSE Root Mean Square Error
ROS Reactive Oxygen Species
RS Remote Sensing
RTC Radiative Transfer Code
RTE Radiative Transfer Equation
RTM Radiative Transfer Models
Rubisco Ribulose bis-phosphate carboxylase-oxygenase
RVI Ratio Vegetation Index
RWC Relative Water Content
SAR Synthetic Aperture Radar
SAVI Soil-Adjusted Vegetation Index
SE Standard Error Of The Mean
Seasat NASA Proof-of-Concept Satellite For Microwave Ocean Remote Sensing
SDD Stress Degree Day
SeaWiFS Sea-Viewing Wide Field of View Sensor
SI le Système International de Unités
SI Stress Index (e.g. SI_{CWSI}, SI_{gs})
SIF Solar-Induced Fluorescence
SIR-A, -B, -C Shuttle Imaging Radar
SLAR Side-Looking Airborne Radar
SLR Side-Looking Radar
SMOS Soil Moisture and Ocean Salinity (satellite)
SNR Signal-to-Noise Ratio
SOM Self-Ordering Map (a neural network-based approach to classification)
SPOT Satellite Pour l'Observation de la Terre (French satellite)
SST Sea Surface Temperature
SVM Support Vector Machine
SVR Support Vector Regression
SW ShortWave or Solar Radiation (c.300 nm – 4 μm)
SWIR ShortWave InfraRed (1–4 μm)
Terra A NASA EOS Satellite
TES Temperature Emissivity Separation method
TIR Thermal InfraRed (4–15 μm)
TIRS Thermal InfraRed Scanner
TM Thematic Mapper on Later Landsats
TOVS TIROS Operational Vertical Sounder on the TIROS satellites
UAV Unmanned Aerial Vehicle
UV Ultraviolet radiation (UV-A, 315–400 nm; UV-B, 280-315 nm; UV-C, <280 nm)

VHRR Very High Resolution Radiometer

VI Vegetation Index

VIFIS Variable Interference Filter Imaging Spectrometer

VISSR Visible and Infrared Spin-Scan Radiometer

WDRVI Wide Dynamic Range Vegetation Index

WiFS Wide Field of View Sensor

X-ray X-ray waveband

X-SAR SAR flown on Space Shuttle

γ-ray Gamma-Ray

1 Introduction

Plants are crucially dependent on their physical environment for growth, survival, and reproduction; in order to understand these responses we need tools both for quantification of the environment and for the study of plants and their functioning. Indeed, there has been an increasing recognition in recent years that studies as diverse as the future improvement in crop yields and the improved monitoring and management of natural ecosystems, for example in response to climate change, requires a good understanding of the mechanisms underlying plant response to the environment. Similarly the prediction of climate change and climate-change impacts is critically dependent on understanding the role of vegetation in controlling changes in the atmospheric CO_2 concentration and the terrestrial energy balance. Many of the methods used for the necessary investigation of plant functioning, whether they involve harvest for traditional growth analysis, or biochemical studies on tissue extracts, are destructive; otherwise they often rely on measurements of processes at rather small scales (such as the gas-exchange of single leaves) which are difficult to scale up to plant, field, or regional scales.

Remote-sensing techniques, in the sense of the gathering of information by a device separated from the target, are increasingly providing an important component of the battery of technologies available for the study of vegetation systems and their functioning. This is in spite of the fact that many applications only provide indirect estimates of the biophysical variables of interest. Particular advantages of remote sensing for vegetation studies include the fact that it is non-contact and non-destructive and the ease with which observations are readily extrapolated to the larger scales that are of particular interest in ecological and climate studies. Even at the plant scale, 'remotely' sensed imagery can allow rapid sampling of large numbers of plants as required, for example, in plant breeding and the identification of varieties with specific physiological characteristics.

1.1 History

To the present generation of students, brought up as they are in the ubiquitous presence of satellite technology – television, mobile phones, satnavs, GPS, Google Earth, etc. – it may seem surprising that much remote-sensing technology is surprisingly recent: the first observation of the earth from space occurred only in 1972. It was in 1960 that reference was first made to the term remote sensing (RS), meaning then merely the observation and measurement of an object without touching it. Since then, it has taken on a more discipline-oriented meaning relating specifically to earth observation (EO). An outline of the main milestones in the development of remote sensing is provided in Appendix 2, while details of some current and recent satellite and airborne sensors are summarized in Appendix 3.

In spite of the relatively recent development of satellite remote sensing, aerial photography had already been extensively used over the previous 40 years, mainly for mapping, but increasingly for monitoring the earth's surface, and the sciences of photogrammetry and photointerpretation had been extensively developed. In the 1950s, Robert Colwell used aerial colour infrared photography to identify small-grain cereal crops and diseases, as well as to study other problems in plant sciences (Colwell, 1956), but many of the basic principles of this work had been established even earlier. Around the same time hemispherical photography was being developed as a tool for the study of canopy structure (Evans and Coombe, 1959). Aerial thermal scanners were employed in the early 1970s to study the development of crop water stress (Bartholic *et al.*, 1972). At the leaf scale, although colour photography has been used for many years (e.g. for disease diagnosis), imaging of physiological processes such as energy transfer started only in the 1970s with some early thermal imaging of leaves (Clark and Wigley, 1975). Nevertheless, it is only since the late 1990s that other imaging techniques, such as chlorophyll fluorescence imaging, have been fully utilized for the study of plant function.

The launch of Sputnik in 1957 signalled the beginning of the satellite era, and the development of electronic detectors opened up a whole new range of possibilities for imaging the earth's surface. In the following five decades, sensors positioned on space platforms, aircraft, balloons, ships, and ground platforms collected data for the study and management of renewable and non-renewable resources and as input for the monitoring of natural and man-made disasters. The parallel development of computer technology enabled the handling and analysis of more complex datasets to be undertaken and opened up the whole field of remote sensing to a large new group of users.

Already in the late 1960s various meteorological institutions had started to incorporate satellite data into their research and weather forecasts, based upon images produced by low-earth orbit (LEO) satellites. The US TIROS/NOAA series was launched in 1960, soon followed by the Soviet Meteor series. The first geostationary satellites (placed at 36 000 km above the Equator) started to supply data in 1966 (ATS) followed by GOES (from 1975) and Meteosat (from 1977). Images from both types of satellite showed the earth's cloud systems in panoramic detail.

The first successful photographs of the earth's surface from space were obtained from a Mercury satellite in 1965, followed by more and more from the Gemini and Apollo series. However, a new dimension was added to existing programmes when the first US Earth Resources Technology Satellite (ERTS-1) was placed in a near-polar orbit on 23rd July 1972. This carried electronic detectors, rather than the earlier photographic cameras, and, compared to the approximate 1-km resolution for the NOAA weather satellites, the 80-m resolution of the ERTS (later called Landsat) sensors, together with the demonstrated success of multispectral scanning, caused a revolution in the whole field of remote sensing. Over the years, Landsat and SPOT have produced much valuable data that has added greatly to our understanding of vegetation and its processes.

Other examples of remote sensing include photography, industrial quality control, security surveillance, monitoring in hazardous environments, such as in nuclear reactors, meteorology, atmospheric science, planetary studies from unmanned probes, and non-invasive procedures in medicine. Earth observation – the gathering of information about the surface of the earth and events on it – may be regarded as a subset of this much broader field of remote sensing.

1.2 Interpretation of the term 'remote sensing'

Although remote sensing as a field tends today to be associated with the use and interpretation of optical imagery (from satellites and aircraft), its potential in vegetation studies is much greater. For example, the recent developments in the use of microwaves and Lidar, as well as enhancements in the optical region, such as the increasing availability of higher spectral and spatial resolutions, have greatly extended the precision with which we can remotely monitor vegetation characteristics and function. Developments in analytical techniques such as data assimilation and the use of machine learning and object-oriented classification procedures are also substantially increasing the power of the technology.

Another important extension to the traditional view of remote sensing is the use of satellites for earth observation, and one that we adopt in this book, is the increasing recognition that other scales of 'remote sensing' are equally useful. Some of these other types of remote sensing had been in use long before satellite data became available. Examples include not only photography, but also the use of hand-held and static detectors, balloons, aircraft, and so on. In this book, we recognize the increasing use of both airborne and especially 'in-field' or 'proximal' remote sensing as key tools in their own right, rather than just for validation of larger scale observations. Throughout this book we therefore use the term remote sensing in its broadest sense – to encompass technologies ranging from classical satellite RS, though airborne sensing, right down to close-range remote sensing of individual leaves and canopies. Although we generally omit techniques only applicable in the laboratory, these are mentioned where they are relevant for calibration or validation purposes. Nevertheless close-range remote sensing provides a particularly powerful tool for the diagnosis and monitoring of plant stresses and will be discussed in some detail in Chapter 11.

Classical satellite remote sensing has been, and will continue to be, a key tool in the studies of vegetation functioning in relation to climate and climate change because of its suitability for integration over large areas and regions and even to the global scale. Remote sensing has therefore been a key contributor to major integrated studies of land-surface functioning in different ecosystems (such as FIFE (Strebel *et al.*, 1998), BOREAS (http://daac.ornl.gov/BOREAS/boreas_home_page.html), ABRACOS (Gash *et al.*, 1996), etc.) that have aimed at understanding the contribution of different ecosystems to the global carbon and energy balances. The large spatial scale imagery obtained by meteorological and passive microwave sensors and others with greater than 1 km pixels is particularly appropriate for the study of functioning at these larger scales, providing data that are hard to obtain from ground-based technologies and thus allowing efficient scaling-up of leaf and canopy-scale measurements.

1.3 Plant physiology and remote sensing

In spite of the fundamental similarity between many of the processes being studied in the fields of plant environmental physiology on the one hand and in remote sensing of vegetation on the other, these fields have tended to develop in parallel in recent years. This is in spite of the fact that there is much commonality in the principles underlying the various sensing techniques. For example both crop physiologists and remote sensing specialists have considered radiation transfer in plant canopies, often for the same purposes such as the estimation of leaf area, but the physiologists have looked at the system from below and emphasized canopy transmission, while in remote sensing the perspective is from above the canopy and an interest in reflection, and often at rather larger scales than used by the physiologists. Both these approaches to radiation-transfer modelling in canopies, however, were based on work dating back to the 1930s (Kubelka and Munk, 1931). It seems useful to bring together these contrasting approaches and to treat in one place all the relevant canopy processes at all these different scales. It is worth noting, however, that although the principles may be similar at different scales, there are often substantial scaling issues involved in transferring between scales; these will be addressed in Chapter 10, although they do recur at various stages in the book.

The application of remote imaging to the study of vegetation and plant canopies, was given a great stimulus with the development of vegetation indices (see Chapter 7) and their use with Landsat and the early AVHRR sensors after 1978.

1.4 Some important considerations for the future

1.4.1 Continuity

Historically, there appears to have been very little coordination between satellite missions and programmes. Remote sensing has been carried out on an *ad hoc* basis, with little apparent consistency. Many sensors and programmes have been developed often with considerable duplication and often seemingly in competition with one another (e.g. both ICESat and CryoSat were designed to measure the mass balance of polar ice sheets, amongst other things, although the latter failed to achieve orbit). In spite of there being many multispectral instruments around, there are still important gaps and areas not being adequately covered, such as the measurement of geopotential or even surface temperatures. There is also considerable lack of continuity, with many satellites being only for research of limited duration (e.g. GRACE). Even Landsat was not conceived as an operational system and it is only now, with the proposed Landsat Data Continuity Mission (Landsat 8: http://ldcm.nasa.gov/), that it is hoped to have it as a truly operational system. The announcement that Envisat will continue till 2013, and that Sentinels 1, 3, and 5 (components of the European Space Agency's GMES system) will continue to provide imaging radar, altimetry, and optical data well into the future, will ensure some form of continuity from when ERS-1 was launched in 1991. But even so, among the many satellite missions that are launched annually, there are still relatively few that are components of long-term operational systems providing adequate back-up or redundancy.

As we shall see, the need for continuity is paramount for many ecological and long-term studies. Continual changes in sensors severely limit the ability to determine long-term ecological trends. There is also a critical need to integrate national space programmes to avoid excessive duplication and competition and to make use of all available resources to provide most efficiently the information that is needed by users. This need for integration remains a major challenge as there is still a tendency for countries to operate independently in setting their objectives and priorities. Nevertheless, there are some promising moves that are improving international coordination. In particular at the First Earth Observation Summit in Washington, D.C. in July 2003 an *ad hoc* Group on Earth Observation was established. This group has led to the setting up of the Global Earth Observation System of Systems (GEOSS; http://Earthobservations.org/) that aims to connect the producers of environmental data and decision-support tools with the end users of these products. Importantly, it involves many countries and organizations with the aim of coordinating the disparate activities relating to earth observation and the provision of EO information for use in activities as diverse as management of energy resources, responding to climate change, managing ecosystems, and promoting sustainable agriculture. It is hoped that the synergies and improved efficiencies that such international coordination brings will be a very powerful driver in the improved availability of data for all vegetation-based studies in the future.

1.4.2 Data availability

There has been an enormous growth in the availability of remotely sensed data over the past two or three decades, and a fall in the real cost. Also, the cost and difficulty of analysing such data have decreased immeasurably. In the early 1980s, there were only a few large mainframe computers capable of analysing such digital data. The development of desktop computers prompted a rush of simple image processing software often developed in-house. Now, very sophisticated processing can be done on a laptop. Some of these key developments are listed in Appendix 2; further information may be found in many texts (e.g. Campbell, 2007; Lillesand *et al.*, 2007).

A critical factor in the take up of remote sensing will be the increasing availability of free satellite data, and also of basic data products such as vegetation indices and digital terrain models. This has been particularly stimulated by the decision by NASA to open up much of their data archive, as well as all new data from critical systems such as Landsat and MODIS, to free access. This has led to an explosion in the downloading of satellite imagery.

There is much more to remote sensing, however, than just the collection of data, or even the making of maps; it is the conversion of that into useful information (though what may be considered useful will differ between users). Although there are currently 35 standard MODIS data products (http://modis.gsfc.nasa.gov/data/dataprod/index.php) this number is greatly expanded when one takes account of the many composite products available (e.g. https://lpdaac.usgs.gov/lpdaac/products/modis_products_table); the challenge is to choose the most appropriate. In spite of the ready availability of satellite data, or even because of it, there will probably remain an increasing need for more advanced 'value-added' products targeted to specific user needs. Examples would include the particular opportunities in precision agriculture, where there is a need for time-critical information on crop growth, disease, or water status to allow for timely management intervention, and in ecology where the needs of policy makers and managers for land-use mapping for regulatory monitoring, reserve management, and so on will continue to grow.

Unfortunately with the increasing availability of pre-processed data there is less understanding among the user community about the potential errors and caveats that must be considered when using the data. A key objective of this text, therefore, is to provide an introduction to the more important principles of remote sensing to enable users to make optimal use of standard products such as *NDVI* or *EVI*, and to decide when and what other information might be needed for their own applications. Many people from researchers to agronomists and farm managers are interested in the capacity of remote sensing to answer questions of importance to them. Not only is there an enormous range of data/image sources available but modern remote-sensing software provides a potentially bewildering range of tools with 'ready-made' solutions for vegetation analysis. For example ENVI (one of the leading softwares available) at the time of writing (2009) provides a built-in ability to calculate 27 different vegetation indices. Users need to be able to decide which products and which analysis tools are most appropriate for their situation, even when they may get a specialist to provide the necessary analysis or higher-level product.

1.5 The structure of this book

There are several ways in which we could have assembled in this text the diverse range of disciplines involved in remote sensing. The book is structured so that we start by introducing in Chapters 2 and 3 the fundamental principles underlying much of remote sensing. Chapter 2 covers the basics of radiation physics, in which we discuss the nature of electromagnetic radiation, from its sources to the detector, considering the complexities of its interactions, on the way, with the atmosphere. Interactions with the target vegetation and soil are outlined in Chapter 3, while Chapter 4 provides essential background information for non-biology specialists on basic plant physiology and plant–environment interactions. The next two chapters provide a general introduction to remote-sensing technology and the processing of data. Though the details are common to any remote-sensing study they are included here as they are critical to the vegetation-specific discussion in the remainder of the book. In Chapter 5 we therefore describe the workings of detectors such as scanners and the various platforms, aircraft, satellites, etc., on which they are mounted. We mention a number of the more common systems but, because these are usually only short lived and are being replaced fairly frequently by other systems that may or may not be similar, we have not tried to be exhaustive but have given examples of the most common generic types (see also Appendix 3 for details of some of those that are more commonly used for vegetation work). Chapter 6 explains how images, especially optical ones, are corrected for imperfections and analysed. We do not go into the theory of image processing, but merely give an idea of what the manipulations do and when and why you need to do them so that the reader is in a position to decide on the best course of action when faced with interpreting a dataset. Chapter 7 concentrates on the uses that can be made of spectral information in optical remotely sensed imagery, introducing both the concept of vegetation indices and the principles of image classification. The next two chapters outline some of the vegetation-specific information that can be derived from remote imagery, with Chapter 8 concentrating on the potential of multiangular imagery, and Chapter 9 addressing the use of remote sensing to estimate heat and mass fluxes to and from vegetated surfaces. Important questions relating to assessment of accuracy and the validation of remote-sensing data are covered in Chapter 10.

It would not be appropriate to have a book about a technique such as remote sensing without demonstrating that it has a use, so the final chapter (Chapter 11) describes a representative selection of applications. These are chosen to indicate how remote sensing enables key vegetation properties to be derived as an aid to landscape management and precision agriculture. Throughout we concentrate on outlining the principles underlying the main approaches used, referring to a number of representative examples to illustrate the ways in which different remote-sensing approaches can be combined with each other and with supplementary information to answer questions relating to vegetation structure and function. The choice of examples is not exhaustive; they are merely indicative of the range of applications. We have largely omitted studies of global CO_2 balance and climate change, and also have not addressed in detail many of the high-level satellite products that are increasingly available free to users. Rather, we concentrate on evaluating the basis underlying the various techniques available so that users can make rational decisions on their choice.

Websites

GEOSS: **http://Earthobservations.org/**

Landsat Data Continuity Mission: **http://ldcm.nasa.gov/**

BOREAS data site: **http://daac.ornl.gov/BOREAS/boreas_home_page.html**

MODIS products: **http://modis.gsfc.nasa.gov/data/dataprod/index.php; https://lpdaac.usgs.gov/lpdaac/products/modis_products_table**

2 Basics of radiation physics for remote sensing of vegetation

2.1 Introduction

The natural way to observe vegetation, and hence assess its properties, is by looking at it. Light, usually from the sun, is reflected from the plant and enters our eyes where it interacts with the retina and a signal is then sent to the brain where the information is analysed. Exactly what we see depends on the conditions under which we observe, and on the sensitivity of our eyes. Whether it is sunny or cloudy, wet or dry, might affect the appearance of the vegetation as it is the spectral distribution of the reflected light that our eyes detect. We then rely on the brain to interpret these signals in the context of the situation, relying on experience and ancillary information in order to 'know' what it is that we see, what type of plant it is, whether it is healthy or diseased, or whether it is stressed. Similar principles apply in remote sensing. The detector, whether on a satellite, aeroplane, or in the hand, merely records data. The user requires information. It is the job of the interpreter to add this value, and in order to do that effectively there is a need to understand the significance and reliability of the data for the particular purpose in hand.

In the above example, the information arrives at the eye as visible light, a form of electromagnetic radiation, simply because that is the signal to which the human eye is sensitive. Artificial detectors can be made to be sensitive to other wavelengths of electromagnetic radiation (EMR), such as ultraviolet (UV), infrared (IR) or microwaves (MW). Each of these contains different types of information because of the ways in which that particular wavelength interacts with surface materials or the intervening atmosphere.

In fact, the basis of nearly all remote sensing is electromagnetic radiation (though acoustic waves (sound) are sometimes used, especially under water). It is therefore important to understand the properties of that radiation and how it interacts with matter, not just with the target, but with the atmosphere through which it travelled and with the detector used to record it. It is only by understanding the physical principles behind the techniques of remote sensing that one is in a position to make full use of the data for making meaningful measurements and to interpret them correctly.

We shall see in this chapter how the amount and nature of the radiation detected depends not only on the properties of the object itself but also on the properties of the source of the radiation and on the properties of the medium through which it travels. We shall see also how to describe the nature of the radiation and its interactions and how to quantify them.

2.2 Radiation characteristics

Electromagnetic radiation is a form of energy, ranging from low-energy radio waves to high-energy γ-rays. Other forms of energy exist, such as chemical, mechanical, sound, nuclear, etc., and, although the total amount of energy must always stay constant in any process (law of conservation of energy), each type can be converted one into another. For example, chemical energy can be converted into electrical energy in a battery, or into mechanical energy in an engine, or into light in a glow worm. It is the energy that the radiation contains that determines the way in which it interacts with matter. High-energy X-rays interact with body tissues, often removing electrons from molecules, whereas low-energy radio waves pass straight through causing no damage. Different wavelengths may be absorbed by a substance depending on the physical or chemical properties of that substance.

2.2.1 Electromagnetic radiation

Electromagnetic radiation consists of time-varying electric and magnetic fields that travel in the form of a wave at the speed of light. Let us examine what is meant by this statement.

We are all familiar with magnetic fields. The earth has a magnetic field that affects a compass needle. The compass needle is itself a bar magnet that lines up in the direction of the field. Two bar magnets exert a force on each other that is stronger the closer the magnets are. Without going into the philosophical question as to *what* a magnetic field is, we see that we can quantify it by means of a *vector*, a line with magnitude and direction. In a similar way we can talk about an electric field, which affects electric charges rather than magnets.

A time-varying field is one whose magnitude and/or direction changes with time. In an electromagnetic wave, the magnitude of the electric and magnetic vectors changes regularly from a maximum value in one direction to a maximum value in the opposite direction, passing through zero at the midpoint.

Electromagnetic radiation is composed of both a time-varying magnetic ($\boldsymbol{B}$) and a time-varying electric ($\boldsymbol{E}$) field that oscillate in phase and at right angles to each other, and they travel in the direction perpendicular to the two vectors (see Fig. 2.1). It can be seen that, as the fields travel along and change in magnitude and direction, the tips of the vectors representing the fields draw out a sinusoidal waveform. Hence, we can talk about an *electromagnetic wave*. It should be noted that once the wave has been formed (*launched into space*) it will continue to travel directly from the source, and does not require a medium in which to travel (unlike a sound wave that cannot travel in a vacuum). This is why light can travel from the sun to the earth. Another property of such a wave is that it travels with a characteristic velocity, called the *speed of light*, which is constant in a vacuum, with a value of 2.998×10^8 m s^{-1}. An electromagnetic wave travels more slowly than this in any medium denser than a vacuum due to the interaction of the electric field with the electric charges in the medium. This leads to a number of important properties, such as refraction and dispersion, which we will come across later. Note that no material body can ever travel faster than the speed of light.

This description of electromagnetic radiation as a wave is an example of a *model* in which a physical phenomenon is described by analogy with something with which we are familiar and that enables us to parameterize it in terms of known variables. Later, we will see that we can also view electromagnetic radiation as a stream of particles, which is a particularly useful description when the radiation interacts with matter. The concept of modelling is quite common in many branches of science, and other types of models will be discussed later.

We can now define *wavelength*, λ, measured in metres, as the distance between adjacent wave crests and *frequency*, f, measured in cycles (oscillations) per second or Hertz (Hz), as the number of waves that pass a given point in one second. It follows that the shorter the wavelength the higher is the frequency (see eqn (2.1)). An alternative measure that is sometimes used is the *wave number*, which is the reciprocal of the wavelength,[1] with units cm^{-1}.

Note that the wave illustrated in Fig. 2.1 actually illustrates *plane-polarized radiation* with the plane of the electric field being parallel to the y-axis. The polarization, however, can be in any plane, or where there is no preferential plane for the electromagnetic wave it is said to be *randomly polarized* or *unpolarized*. If instead of being constrained to a fixed plane the electric field vector rotates in the xy-plane it is circularly polarized.

2.2.2 Electromagnetic spectrum

In nature, electromagnetic waves of all frequencies and wavelengths can exist, and it is these parameters that distinguish the different types, such as visible light, X-rays, and radio waves. This continuum is referred to

[1] Strictly the SI unit is m^{-1}, though units of cm^{-1} are generally used. An alternative definition of wave number, $2\pi/\lambda$, is sometimes employed.

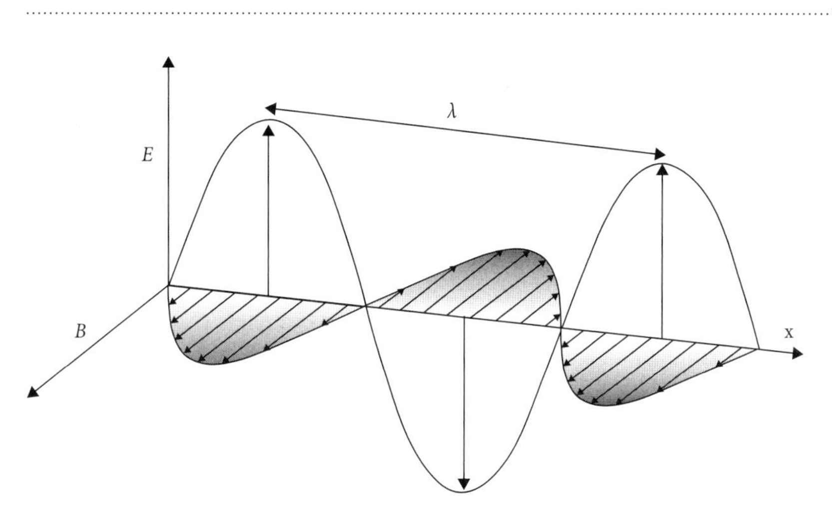

Fig. 2.1. Electric (E) and magnetic (B) vectors of an electromagnetic wave of wavelength λ viewed at a given instant. E and B are mutually at right angles and both are at right angles to the direction of propagation, x.

as the *electromagnetic spectrum* (see Fig. 2.2). Because the speed of all electromagnetic waves, c, is constant and independent of wavelength, we can relate the frequency, f and wavelength, λ by

$$\lambda f = c \quad \text{or} \quad \lambda = c/f. \tag{2.1}$$

This inverse relationship is shown in Fig. 2.2 where the longwave radio waves (hundreds of metres) have very low frequencies (tens of kHertz), whereas the very short-wavelength γ-rays (10^{-11} m) have very high frequencies (10^{20} Hz). Visible light is intermediate, green light having a wavelength of around 500 nm and a frequency of about 6×10^{14} Hz.

The spectrum is usually divided up into regions, as shown in Fig. 2.2. The visible (VIS) region, to which the human eye responds, is only a very small portion that approximately corresponds with the photosynthetically active region (PAR). This is flanked by the short-wavelength (higher-frequency) ultraviolet (UV) region and the even more energetic X-rays and γ-rays. On the longer-wavelength side is the infrared that is conveniently partitioned into the near infrared (NIR from 700 nm to 1 μm), the mid-infrared (MIR 1–4 μm), the thermal infrared (TIR from 4–15 μm; often split into the shortwave thermal, <8 μm, and the longwave thermal, >8 μm) and the far infrared

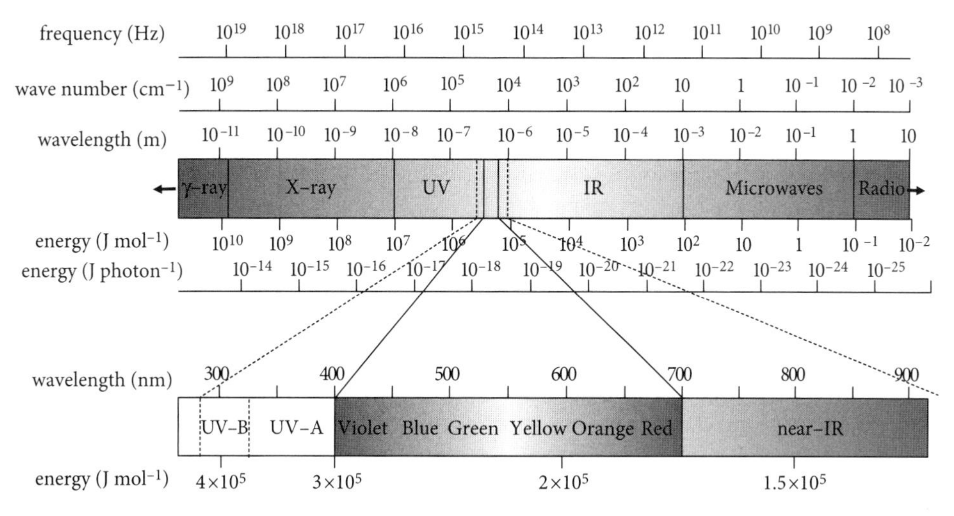

Fig. 2.2. The electromagnetic spectrum; note that energy is given in J mol^{-1} (to convert to J photon^{-1} it is necessary to divide by Avogadro's number (6.022×10^{23})) so a photon of red light at 700 nm contains 2.84×10^{-19} J.

n–100 μm). This then merges through the hertz region (THz, 100 μm to 3 mm, or 3 to 0.1 THz) with the radiowave region that itself is often taken to include microwaves (3 mm to 1 m). There are no sharp boundaries between these arbitrary regions, one type merging into its neighbour, but we shall see that each region has different uses in remote sensing, and usually requires different types of detectors. In the remainder of the text, we use the term *optical* when we refer to the UV–MIR range, and use the term *thermal* when we refer to the TIR. The region between ultraviolet and short radio waves is that most useful for environmental remote sensing, but the wavelengths that are of primary concern in environmental plant physiology lie between about 300 nm and 15 μm.

2.2.3 Electromagnetic energy

Electromagnetic waves, consisting as they do of force fields, carry energy. The energy content at a given wavelength λ, E_λ, is proportional to the frequency of the radiation and inversely proportional to wavelength, as given by

$$E_\lambda = hf = hc/\lambda, \tag{2.2}$$

where h is a fundamental physical quantity called Planck's constant, which has a value of 6.63×10^{-34} J s. It is sometimes more convenient to think about the energy of the wave in terms of a small quantity or *quantum*. These quanta behave a bit like particles, in that their effect is localized rather than being distributed along a wave, and have the name *photons*. Simple substitution demonstrates that a γ-ray photon has an energy of about 10^{-14} J whereas a radio wave carries only about 10^{-30} J. So, we usually refer to high- or low-energy photons,[2] and it is the photon energy that determines the behaviour of the radiation and how it interacts with matter. It is often more convenient, for example when we are relating absorbed radiation to photosynthesis, to express the energy in terms of the energy in a mole of photons (i.e. Avogadro's number of photons $= 6.022 \times 10^{23}$) as shown in Fig. 2.2. Although it may possibly be more intuitive to think about energy in terms of frequency, in remote sensing wavelength is often a more useful quantity with shortwave radiation being more energetic than longwave radiation (eqn (2.2)). This variation in energy has a number of important consequences when it comes to remote sensing, as it affects such factors as the sensitivity of the detectors used in different spectral regions and the interactions of different wavelengths with the atmosphere and surface of the earth.

2.2.4 Sources of radiation

Electromagnetic energy can be generated by a number of mechanisms. These include changes in the energy levels of electrons in atoms (line spectra), decay of radioactive materials (e.g. γ-rays), the acceleration of electrical charges, for example in the aerial of a transmitter, and the thermal motion of atoms and molecules. All matter above the absolute zero of temperature (0 K or −273.16 °C) emits radiation. This is due to the internal motion of the constituent atoms and molecules, which becomes more energetic as the temperature increases. Each source of radiation emits a characteristic spectrum of different wavelengths and intensities, and, importantly, because the same internal processes are involved, it can also absorb that same range of radiation. It is usual to consider the behaviour of an ideal source – called a *black body*. A black body is a body that is a perfect absorber and emitter of radiation. It is defined as a hypothetical object or material that absorbs all radiation incident upon it and emits the maximum amount of radiation possible at that temperature. This is a useful concept (or model) that enables us to formulate some useful laws and equations.

The first of these is the Stefan–Boltzmann law relating the total amount of energy, E, emitted per unit time (power) per unit area of a body to its absolute temperature, T, as

$$E = \sigma T^4, \tag{2.3}$$

where σ is the *Stefan–Boltzmann constant* that has a value 5.67×10^{-8} W m^{-2} K^{-4}. E therefore has units of J m^{-2} s^{-1} or W m^{-2} and is known as the emitted *flux density* or the *radiant exitance*. Any real body emits less energy than this, and so, for the general case, we use Kirchhoff's law

$$E = \varepsilon \sigma T^4, \tag{2.4}$$

where ε is called the *emissivity* of the body, which is a measure of the efficiency of emission and can take

[2] The energy of an individual photon is very small, and it is often convenient to use a unit more in keeping with the atomic scale, the *electron volt*, or eV, which is equal to 1.602×10^{-19} J. On this scale, γ-rays have energies about 10^5 eV and radio waves about 10^{-8} eV. Photons in the visible part of the spectrum have energies of about 1 eV.

values from 1, for a black body, to 0, for a *white body* that does not emit or absorb any energy. For most real bodies ε takes values between these extremes. Where emissivity is independent of wavelength these are termed *grey bodies*, and where it does depend on wavelength we refer to *selective radiators*. Note that the spectral emissivity (ε_λ) for any body at a wavelength, λ, is equal to the efficiency of radiation absorption at that wavelength.

The energy distribution for emission from a true black body ($\varepsilon = 1$ at all wavelengths) is given by *Planck's distribution law*:

$$E_\lambda\,(\mathrm{d}\lambda) = \frac{2hc^2}{\lambda^5(e^{hc/\lambda kT} - 1)}\,\mathrm{d}\lambda, \qquad (2.5)$$

where $E_\lambda\,(\mathrm{d}\lambda)$ is the spectral radiant exitance (power emitted per unit area in a wavelength range $\mathrm{d}\lambda$), h is Planck's constant, c is the speed of light, and k is Boltzmann's constant. Note that eqn (2.3) is obtained by integrating this expression over all wavelengths. Examples of this distribution are shown in Fig. 2.3 for black bodies at 5800 K (the approximate effective temperature of the sun) and 300 K (approximately equivalent to the temperature of the earth). Conventionally for meteorological purposes we refer to all radiation originating from the sun (c. 0.25 μm < λ < c. 4 μm) as *shortwave radiation* (SW), while we refer to the thermal radiation emitted from bodies at terrestrial temperatures (c. 4 μm < λ < 15 μm) as *longwave* (LW) or *thermal radiation* (TIR). For a real object, ε varies with wavelength and so the spectral emittance curves would not be as smooth or regular as for a black body (see the spectral curve for the sun in Fig. 2.6).

The peak wavelength (λ_m μm) of the Planck distribution is an inverse function of the temperature, K, and is given approximately by *Wien's displacement law*

$$\lambda_m = 2897.769/T. \qquad (2.6)$$

Simple substitution shows that for a body at 5800 K λ_m is 499.5 nm, which is in the visible region in which about 44% of the total radiation from the sun falls. For a body at the mean temperature of the earth, λ_m would be 9.65 μm, which is in the thermal part of the infrared. This simple law shows why the sun emits visible light and the earth emits invisible thermal infrared radiation. It should be noted that, not only does the peak wavelength decrease with increasing temperature (frequency increases), but the total amount of energy (intensity) increases as the fourth power of the temperature (eqn (2.4)).

2.2.5 Radiometric terms and definitions

The main radiometric terms that will be encountered in this book are summarized in Table 2.1 while the SI base units and conversions between units may be found in Appendix 1. *Radiant energy*, E, is the energy (i.e. capacity to do work) that is contained in an electromagnetic wave. The SI unit of measurement of energy is the joule (J). The rate of transfer of energy, or the rate at which work is performed, is known as power (with units of watts; W = J s^{-1}). The rate of transfer of radiant energy, such as from the sun to the earth, is called the *radiant flux*, ***F***, measured in watts, and the amount of radiant flux incident upon, or emitted from, a unit area

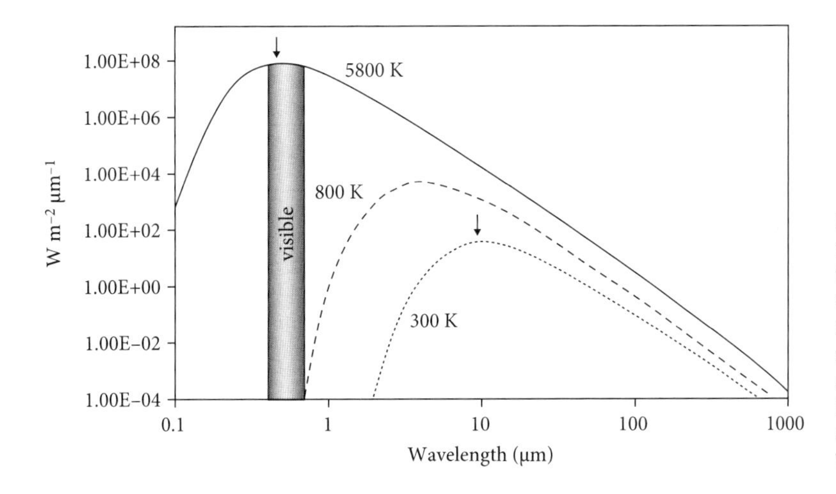

Fig. 2.3. Spectral distribution of radiation emitted from black bodies at temperatures of 5800 K, approximately equal to that of the sun, 800 K and 300 K typical of the earth's surface showing the region of visible radiation and the large shift in peak emissions between shortwave solar radiation and thermal longwave as predicted by Wien's law.

Table 2.1 Terminology for radiometric quantities. All of the terms can be wavelength dependent, and this can be indicated by using the word 'spectral' in front of the term and by using the subscript λ after the symbol (e.g. Jones, 1992).

Term	Unit	Definition
Radiant energy (E)	joule (J) = newton m (N m)	Total energy radiated in all directions
Radiant flux ($\mathbf{F}$)	joules per second = watt (W) = J s^{-1}	Radiant energy emitted or absorbed per unit time
Radiant flux density ($\mathbf{R}$)	watts per square metre (W m^{-2})	Net radiant flux through a surface per unit area
Irradiance ($\mathbf{I}$)	watts per square metre (W m^{-2})	Radiant flux incident on unit area
(Radiant) emittance	See radiant exitance	
Radiance ($\mathbf{L}$)	watts per square metre per steradian (W m^{-2} sr^{-1})	Radiant flux per unit area of surface per unit solid angle
Radiant intensity	watts per steradian (W sr^{-1})	Radiant flux from a source into unit solid angle
Radiant exitance ($\mathbf{M}$)	watts per square metre (W m^{-2})	Total energy radiated in all directions in unit time per unit area (for isotropic radiation emitting into a hemisphere $\mathbf{M} = \pi\mathbf{L}$)
Spectral irradiance ($\mathbf{I}_\lambda$)	watts per square metre per micrometre (W m^{-2} μm^{-1}) (N.B. the strict SI unit is W m^{-2} m^{-1})	Irradiance per unit wavelength

of a surface is the *radiant flux density* ($\mathbf{R}$), measured in watts per square metre (W m^{-2}). If the radiant energy is incident on a horizontal surface, the alternative term *irradiance* ($\mathbf{I}$) is often used. Similarly, if the radiant energy is leaving the surface, the term *radiant exitance* ($\mathbf{M}$) or *radiant emittance* may be used. Both are measured in W m^{-2}. The radiant flux density emitted from a unit area of an emitter into a unit solid angle is the *radiance* ($\mathbf{L}$), measured in watts per square metre per steradian (W m^{-2} sr^{-1}). All the above expressions refer to the total amount of energy, regardless of its wavelength. In fact the energy emitted by any natural source is wavelength dependent, so when referring to the radiance or irradiance at a specific wavelength the terms *spectral radiance* ($\mathbf{L}_\lambda$) or *spectral irradiance* ($\mathbf{I}_\lambda$) are used (measured in W m^{-2} sr^{-1} μm^{-1} or W m^{-2} μm^{-1}).

Lambert's cosine law

The irradiance at a surface depends on the angle between the surface and an incoming radiant beam according to *Lambert's cosine law*:

$$\mathbf{I} = \mathbf{I}_0 \cos\theta = \mathbf{I}_0 \sin\beta, \tag{2.7}$$

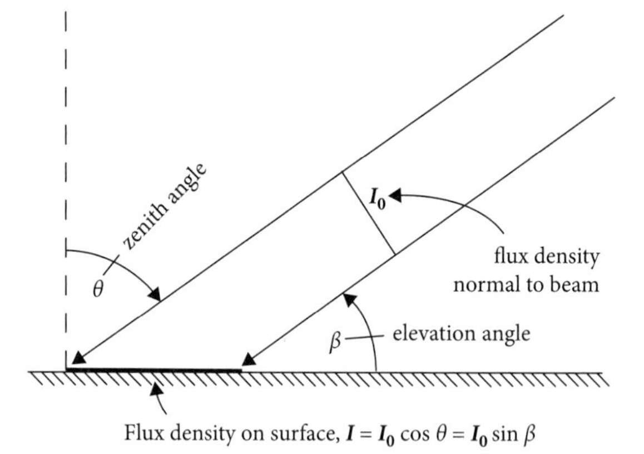

Fig. 2.4. An illustration of Lambert's cosine law showing how the irradiance on a surface varies with beam elevation (or zenith angle).

where $\mathbf{I}_0$ is the incident flux density normal to the beam, θ is the angle between the beam and the normal to the surface (= the *zenith angle*) and β is the complement of θ and is known as the beam *elevation*. It follows that the larger the angle between the beam and the surface, the lower the irradiance; this arises because the same beam is spread out over a larger area as the beam moves away from the normal (Fig. 2.4).

2.3 Interaction of radiation with matter

2.3.1 General principles

There are a number of ways in which radiation can interact with matter depending on the wavelength of the radiation and on the nature of the matter. In the optical region, and in particular at shorter wavelengths, specific wavelengths can cause electronic transitions in atoms and molecules when the energy of the quantum involved is exactly equal to the separation of electronic energy states. These wavelengths are either emitted if the atoms or molecules are 'excited' in some way (e.g. by heating them or by passing an electrical discharge through them), or absorbed, since the same electronic transitions are involved. It follows that at any wavelength, λ, the spectral emissivity (ε_λ) equals the *spectral absorptivity* (α_λ). These electronic transitions are the origin of atomic *line spectra* and provide the basis of spectroscopy as a technique for studying and analysing atoms and molecules. Larger molecules and dense gases interact over a wider range of energies, leading to *band spectra* and absorption bands, respectively, but still characteristic of the molecules concerned. Specific chemicals in solids and liquids also absorb characteristic wavelengths relating to the wider range of possible energy transitions; the combination of all absorption processes in the visible account for the perceived colour of an object.

In the thermal region, the interaction with solids is largely by means of molecular vibrations and rotations. Unlike electronic transitions that occur at specific wavelengths, the absorption and emission of low-energy radiation occurs over broader spectral ranges giving rise to continuous spectra. As pointed out above the emissivity of a surface determines the efficiency with which the interactions (absorption and reflection) occur.

Interactions at optical and thermal wavelengths take place mainly through the electric field vector in electromagnetic radiation that causes electronic and vibrational transitions in the material. At microwave wavelengths and longer, however, it is the magnetic field vector that interacts with the magnetic dipoles in the material. Water is a highly polar liquid that allows the shorter visible wavelengths (i.e. blue and green), to pass through with little attenuation, red is substantially attenuated, and microwaves are absorbed very strongly by the first few molecules encountered at the surface. Microwaves can penetrate several wavelengths into dry soil or sand, but this *penetration depth* is highly sensitive to the presence of water. In the case of microwaves it is the *dielectric constant* of the material that determines the strength of the interaction. We will return to this later in the chapter (Section 2.5).

When radiation interacts with a material all the energy is conserved such that the total amount of energy dissipated by reflection, transmission, and absorption equals the incident energy. Therefore, we can write:

$$E_{I(\lambda)} = E_{R(\lambda)} + E_{A(\lambda)} + E_{T(\lambda)}, \tag{2.8}$$

where $E_{I(\lambda)}$ = incident energy, $E_{R(\lambda)}$ = reflected energy, $E_{A(\lambda)}$ = absorbed energy, $E_{T(\lambda)}$ = transmitted energy. Dividing eqn (2.8) through by $E_{I(\lambda)}$ gives

$$E_{R(\lambda)}/E_{I(\lambda)} + E_{A(\lambda)}/E_{I(\lambda)}) + E_{T(\lambda)}/E_{I(\lambda)} = 1, \tag{2.9}$$

which can be written as

$$\rho + \alpha + \tau = 1, \tag{2.10}$$

where the first three terms, respectively, define the *reflectance* (ρ), *absorptance* (α), and *transmittance* (τ) of the material. When referring to the reflection, absorption, or transmission by a body we sometimes use the term coefficient, as in *reflection coefficient*, to describe the proportion dissipated by that mechanism. The proportions of incident energy that are reflected, absorbed, or transmitted depend upon the nature of the surface material and its condition. It is the component that is reflected that is detected in optical remote sensing (excluding thermal) and it is the variation in this that enables us to distinguish between different surface materials and especially to study the vegetation. It is important to note that these three parameters are also wavelength dependent, and so the relative amounts of energy reflected, absorbed, or transmitted vary with wavelength. We should strictly speak about *spectral reflectance*, ρ_λ, and so on. A summary of many of the terms used for description of electromagnetic radiation and its interaction with matter is provided in Box 2.1.

2.3.2 Propagation of radiation – interaction with the atmosphere

Meaningful information about a surface target is contained in the physical properties of the radiation

BOX 2.1 Key terms associated with electromagnetic radiation

Word endings: Materials or bodies are distinguished using 'er' or 'or' as a suffix, e.g. emitter/radiator, absorber, reflector, and transmitter. The processes associated with these carry the suffix 'ion', thus emission, absorption, reflection, and transmission. The suffix 'ivity' refers to their properties, emissivity, absorptivity, reflectivity, and transmissivity, and 'ance' defines the behavioural characteristics of a specific body, taking account of its size and shape, emittance, absorptance, reflectance, and transmittance.

Absorption: The retention of electromagnetic energy by a substance or by a body.

Coherence: Two waves are coherent if their relative phase difference remains constant over time. The observation of interference requires that the two waves be coherent. Speckle in a radar image is due to interference between the coherent echoes from the individual scatterers within a resolution cell. Light, except when generated by a laser, is usually incoherent.

Dispersion: The spreading out of the different wavelength components in a beam of light. The amount by which a beam of radiation is bent in refraction depends upon its wavelength, hence different coloured light emerges from a prism at different angles, accounting for the observed spectrum. It also accounts for the formation of a rainbow.

Diffraction: The change in direction of an electromagnetic wave when part of the wavefront is obstructed by an obstacle of some kind. Diffraction is most apparent when the wavelength is of similar magnitude to the object. The usual manifestation of this effect is the spreading out of a beam of light or microwaves when it passes through an aperture (a lens or antenna) that is of the same order of magnitude as the wavelength. It is responsible for the fundamental limit to the resolution of remote-sensing detectors.

Doppler effect: The shift in frequency that occurs when there is relative motion between the transmitter and receiver. This effect is used to improve the spatial resolution of radar systems (SAR) and synthetic aperture altimeters.

Interference: The superposition of two or more waves that results in a new wave pattern. Constructive interference occurs when the two waves are in phase, destructive interference occurs when the two waves are exactly out of phase. Interferometry combines the returns of two coherent signals that have travelled different distances and hence have a relative phase difference, using two receivers that are separated in space or time. SAR interferometry can be used for topographic measurements in a way similar to the use of stereo-photography.

Polarization: The plane in which the electric vector in an electromagnetic wave oscillates in space. Natural radiation is usually unpolarized, but active systems can generate polarized signals. Polarimetric radar makes use of the fact that different types of vegetation may have characteristic responses to waves of different polarizations.

Reflection: The redirection of a beam of radiation when it encounters a boundary. Reflection from a smooth surface produces *specular reflection* in which all the energy is concentrated in a beam symmetrically oriented with respect to the normal to the surface as is the incident beam (e.g. reflection in a mirror). The scattering of radiation when it encounters a rough surface can be considered in terms of the specular reflection of different portions of the beam from many randomly oriented facets: this is commonly termed *diffuse reflection*.

Refraction: A change in direction of a beam of electromagnetic radiation at a boundary between two materials having different refractive indices. It is refraction at the interface between glass and air that causes a prism to bend light and for a lens to focus it.

Scattering: The redirection of electromagnetic energy, without absorption, when it encounters an object. The amount of scattering depends upon the size of the object with respect to the wavelength of the radiation. Atmospheric scattering is wavelength dependent with shorter wavelengths being more scattered compared to longer wavelengths.

leaving that surface target (both reflected and emitted). But because both source and detector are 'remote' from the target, the characteristics of the radiation detected by the sensor are affected not only by its reflection/emission from the target but also by any interactions with the intervening atmosphere.

Two processes have to be considered. The first is the attenuation of solar radiation in the atmosphere both on its way to the surface and, after reflection, back to the detector. This leads to an overall decrease in the amount of radiation intercepted by the sensor as compared to the hypothetical (atmosphere-free) observation. The second atmospheric effect is the addition of the fraction of the incident sunlight that is scattered by the atmospheric constituents back into the field of view of the sensor (e.g. by molecules and aerosols) without having interacted with the earth's surface. This atmospheric scattering, (or 'reflection') is independent of the brightness of the land surface and would even be measured over a completely black target.

Therefore, the signal detected contains information not only about the surface observed but also about the atmosphere, and it is necessary to correct for, or at least to understand, these atmospheric effects if one is to make sensible deductions, particularly quantitatively, about the surface being remotely sensed. Although the atmosphere can itself be the subject for remote sensing, from the point of view of earth surface observations, the atmosphere's contribution can be regarded as noise that has to be removed in order to isolate the useful signal from the surface target. The quantification and removal of the effects of the atmosphere is usually referred to as *atmospheric correction* and will be described in more detail in Section 6.2.2.

Of the shortwave radiation (<4 μm) that enters the atmosphere from the sun, a substantial proportion is absorbed and scattered by aerosols and gases or reflected (from clouds) before it reaches the earth's surface. Radiation transmission through the atmosphere depends on the concentration of scattering and absorbing particles in the atmosphere, their optical properties, and upon the optical path traversed by the radiation. Absorption and scattering processes attenuate the radiation as it passes through the atmosphere; for parallel monochromatic radiation the amount of attenuation ($d\boldsymbol{I}_\lambda$) for an infinitesimal path length (dx) is proportional to the incoming irradiance ($\boldsymbol{I}_{0\lambda}$), d$x$ and what is known as an extinction coefficient (k_λ) according to

$$d\boldsymbol{I}_\lambda = \boldsymbol{I}_{0\lambda}.k_\lambda.dx. \tag{2.11}$$

This equation can be integrated over a finite path-length x to give the following well-known expression of exponential decay

$$\boldsymbol{I}_\lambda = \boldsymbol{I}_{0\lambda} \exp(-k_\lambda x), \tag{2.12}$$

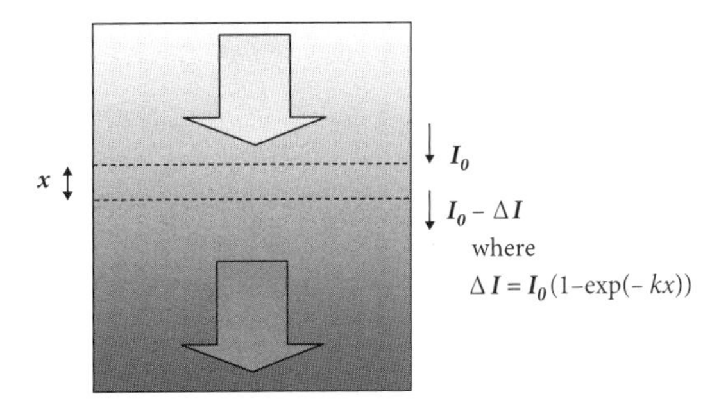

Fig. 2.5. Attenuation of parallel monochromatic radiation in an absorber (either a gas or a solution).

which is often referred to as the *Beer–Lambert Law* or *Beer's Law* (see Fig. 2.5). For vertical transmission through the atmosphere (i.e. the solar zenith angle = 0), the term $k_\lambda x$ is called the *optical depth* or *optical thickness* (τ'_λ).[3] Equation (2.12) can be written to define the transmittance ($\tau = \boldsymbol{I}_\lambda / \boldsymbol{I}_{0\lambda}$) as

$$\boldsymbol{\tau} = \exp(-k_\lambda x) = \exp(-\tau'_\lambda). \tag{2.13}$$

In order to estimate the attenuation when the radiation comes from other zenith angles we introduce the concept of *optical air mass* (m), defined as the ratio of the mass of atmosphere traversed per unit cross-sectional area of the actual solar beam to that traversed at a site at sea level if the sun were overhead. In this case we can rewrite eqn (2.13) as

$$\boldsymbol{I}_\lambda = \boldsymbol{I}_{0\lambda} \exp(-\tau'_\lambda m). \tag{2.14}$$

The value of m in this equation decreases with elevation above sea level (z) in proportion to the atmospheric pressure, P_z, and increases with increasing solar zenith angle, θ, according to Lambert's cosine law (eqn (2.7)) in proportion to $1/\cos\theta$, giving

$$m = (P_z/P_0)/\cos\theta = (P_z/P_0)\sec\theta, \tag{2.15}$$

where P_z is the atmospheric pressure at elevation z and P_0 is the atmospheric pressure at sea level (see Box 2.2 for estimation of P_z/P_0).

The optical air mass corrects for the geometric variation in the path through the atmosphere, while the transmittance term accounts for the concentration of the important absorbers and scatterers in the

[3] Unfortunately, optical depth and atmospheric transmittance are often both given the symbol τ, though transmittance (I/I_0) equals the negative exponential of optical depth, so here we distinguish optical depth with a prime (τ').

BOX 2.2 Composition of the atmosphere

The earth's atmosphere is a mixture of gases, of which O_2 and N_2 are the main constituents, representing 99% of the atmosphere, and solid/liquid particles (aerosols and clouds). The gases O_2 and N_2 are reasonably uniformly distributed around the world. Water vapour, carbon dioxide, ozone, methane, aerosols, etc. comprise the remaining 1%. The distribution of these minor constituents is highly variable, depending on creation and destruction mechanisms which occur at specific locations. For example water vapour is confined to the troposphere (below 15 km in height) and to low latitudes and mainly over oceans, and carbon dioxide tends to be well distributed vertically but concentrated in regions of human activity. Turbulent mixing prevents the heavier molecules from being concentrated close to the earth's surface, and global mean density shows an approximately exponential decrease with altitude, halving in value for every 5 km increase in height. Atmospheric pressure decreases to only 10% of that at ground level at about 15 km, and only 1% of the atmospheric mass occurs above 32 km.

If the temperature is assumed to be roughly constant, a simple approximate expression for the pressure P_z at an elevation z can be derived:

$$P_z/P_0 = \exp(-z/H),$$

where P_0 is the atmospheric pressure at sea level and H is called the *scale height* and has an approximate value of 8 km at T=273 K. The scale height is the thickness that the atmosphere would have if the density were constant. In reality, the temperature of the atmosphere is not constant, but, since it only varies by about a factor of two, whereas the pressure changes by six orders of magnitude, then this approximation holds quite well.

atmosphere: these include the various atoms, molecules, dusts, and aerosols present. The total optical depth is the sum of the optical depths of its components. On the clearest days atmospheric transmissivity approaches 0.75 so that when the sun is overhead, about 75% of the radiation reaches the surface giving a total shortwave irradiance of c. 1000 W m^{-2}.

Not only do absorption and scattering reduce the overall intensity of the radiation at the ground, but the amount of this attenuation is a function both of the wavelength of the radiation and of the concentration, size, and properties of the gases and aerosols encountered. It is particularly important to recognize the impact that these absorption and scattering processes have on the spectral distribution of the transmitted radiation. The concentration of such particles in the atmosphere is not constant. The concentrations of atmospheric gases (oxygen and nitrogen) vary exponentially with height, being most dense at ground level, with 90% of the atmospheric molecules existing below 10 km. The density of water vapour, aerosols, etc. that are present mainly in the lower atmosphere, varies also in the horizontal direction, sometimes over quite small distances. The reflected solar radiation received at a satellite at a height of about 800 km has therefore made a double transit of the entire column of the atmosphere, and even radiation detected by a hand-held receiver at ground level has passed through it once. This clearly demonstrates the importance of a sound radiometric and atmospheric correction of the analysed imagery. Only a proper radiometric correction of the imagery allows a quantitative interpretation of data acquired from satellite or airborne sensors. We will come back to this issue later (Section 6.2). Fortunately, only a fraction of the total radiation is strongly attenuated by atmospheric scattering and absorption processes. Large parts of the spectrum are only weakly attenuated – those in specific *'atmospheric windows'* as shown in Figs. 2.6 and 2.7.

The impact of some of the more important atmospheric constituents including ozone, water vapour, and carbon dioxide on the transmission of solar radiation to the earth's surface is illustrated in Fig. 2.6 for clear-sky conditions. This shows very clearly the strong absorption bands due to CO_2 and H_2O vapour in the infrared wavelengths, as well as the very strong absorption due to ozone (mainly in the UV). The US National Renewable Energy Laboratory has developed a suite of tools that are useful for modelling atmospheric transmission of solar radiation (http://www.nrel.gov/rredc/; Bird and Riordan, 1984; Gueymard, 1995; Gueymard, 2001).

Absorption – As indicated in Figs 2.6 and 2.7, certain molecules, such as water vapour, carbon dioxide, and ozone, are particularly good absorbers of solar radiation. Some of the energy from the sunlight may

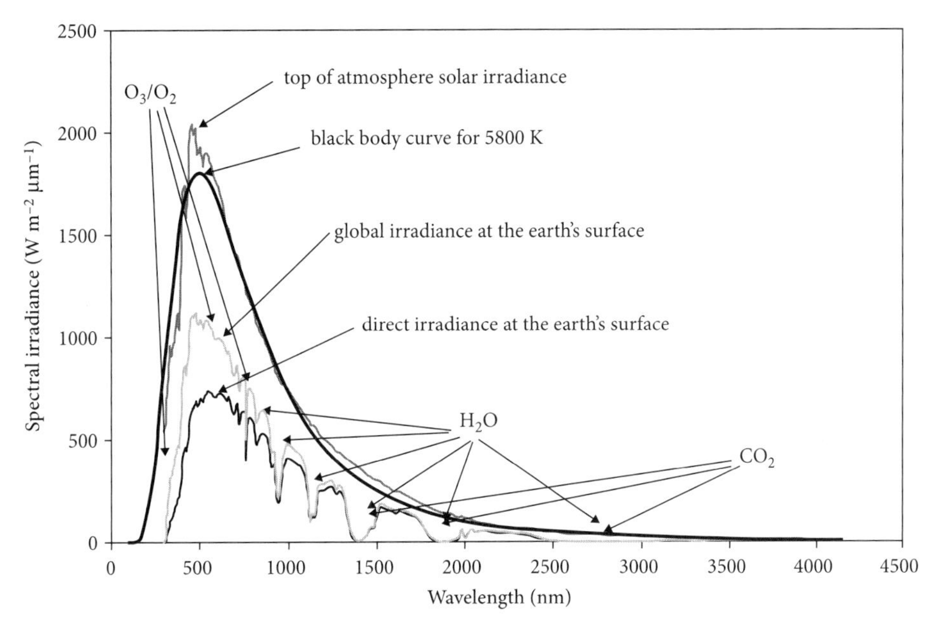

Fig. 2.6. Illustration of the solar spectrum expressed in terms of energy per unit wavelength at the top of the earth's atmosphere as compared with the Planck distribution (black body curve) for an emitter at 5800 K, together with the corresponding total and direct radiation received on a horizontal surface for typical clear-sky conditions on March 1st at Golden, Colorado, calculated using the SPCTRAL2 and SMARTS2 models (Bird and Riordan, 1984; Gueymard, 1995). Also shown are the main absorption bands due to O_3, O_2, water vapour, and carbon dioxide. The narrow absorption lines apparent in the solar spectrum (not visible at this scale) are the so-called Fraunhofer lines associated with absorption in the sun's photosphere by specific elements such as sodium, hydrogen, and helium.

be converted into heat energy and re-emitted at these longer wavelengths. Ozone absorbs strongly in the ultraviolet region, and hence helps to protect us from the harmful radiation that can cause skin cancer and damage to the eye's retina. This is the reason for the concern about the depletion of stratospheric ozone by chlorofluorocarbons leading to the so-called *ozone hole*. Also, some atoms and molecules may selectively absorb radiation at certain wavelengths and re-emit it at the same wavelength but in all directions. Hence, the intensity of these spectral components in the direct beam will be considerably reduced, thus altering the spectral distribution of the sunlight arriving at ground level (see Fig. 2.6).

A large part of the energy in the solar beam, specifically the component in the visible region (from about 400 to 700 nm), passes through the atmosphere with relatively little attenuation. Scattering losses increase both with increasing solar zenith angle and with decreasing wavelength (see next section); furthermore O_2 and O_3 are also particularly strong absorbers of shorter-wavelength UV radiation. In the near infrared, the mid-infrared, and thermal infrared regions, the main absorbers are water vapour and carbon dioxide. There are, however, some useful windows in the near infrared, at about 1.0, 1.2, and 1.6 µm and another in the mid-infrared between 3.5 and 4.1 µm. In the thermal infrared, there is a broad window between about 8.0 and 12.5 µm, except for an oxygen absorption band at about 9 µm. The wavebands in which different sensors record (see Chapter 5 and Appendix 3) have been deliberately chosen to be in the regions with least atmospheric attenuation, i.e. in the atmospheric windows.

Scattering – Radiation travelling through the atmosphere may be scattered by interaction with the particles it encounters. The amount of scattering depends on the size of the scattering particles in relation to the wavelength. Hence, a water droplet may look 'big' to shortwave visible radiation, and hence produce considerable scattering, whereas it looks negligibly 'small' to longer-wave microwave radiation that passes straight through it. This is why, for example, it is possible to use

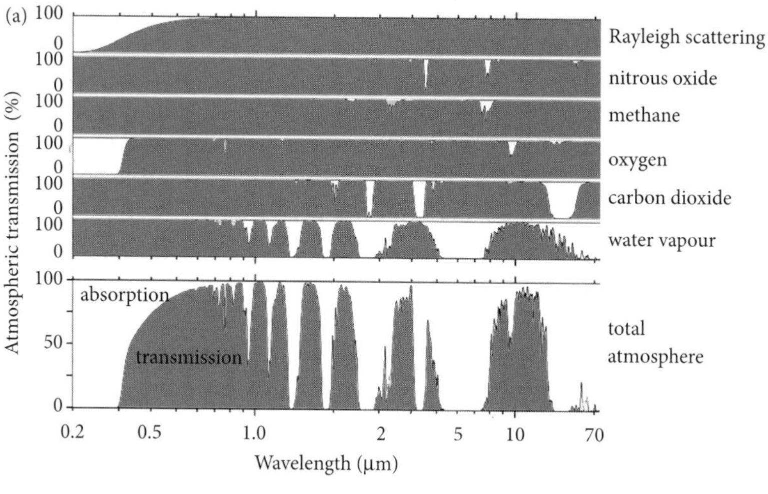

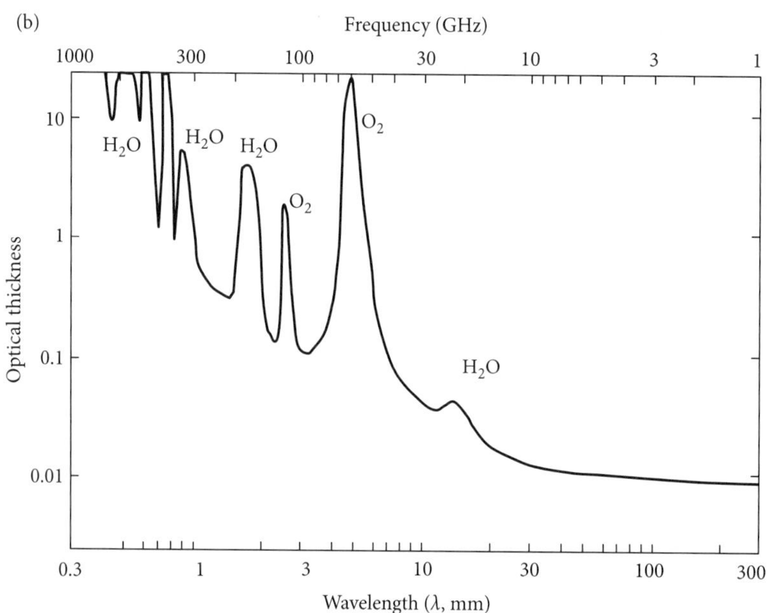

Fig. 2.7. (a) Atmospheric absorption and scattering for sunlight at nadir showing the elimination of short wavelengths by Rayleigh scattering and by absorption by O_3 together with the increasing absorption in the infrared by various atmospheric components including especially CO_2 and H_2O. The dark areas represent atmospheric windows. (based on data from http://www.globalwarmingart.com/wiki/Greenhouse-Gases-Gallery) for further data see the HITRAN website (http://www.cfa.harvard.edu/HITRAN/). (b) Atmospheric transmission in the microwave/radio region expressed as total zenith optical thickness (transformed from Rees, 2001).

microwaves for remote sensing even when it is cloudy, which is not possible using visible light. Three common types of scattering can be considered. The first is *Rayleigh* scattering, which occurs when the diameter of the scattering particles is much smaller than the wavelength of the radiation. This is the case for scattering of optical wavelengths by the smaller molecules in the atmosphere such as oxygen and nitrogen. Rayleigh scattering ($\sigma_{R(\lambda)}$) is strongly wavelength dependent, varying inversely as approximately the fourth power of λ, hence blue light is scattered about eight times as much as is red light and ultraviolet by about sixteen times as much (Fig. 2.7).

It is this phenomenon that makes the sky look blue (because when viewing the sky from below you see mainly the scattered radiation) and for sunsets to look red (because the direct solar radiation is strongly depleted in the blue when the sun is low). It is also the primary cause of *haze* in an image, resulting in a loss of contrast. In photography this is usually corrected for by using a yellow-coloured filter that prevents scattered blue light from entering the camera.

The second type is *Mie* scattering ($\sigma_{M(\lambda)}$). This occurs predominantly in the forward direction when the particle diameters are of the same order of magnitude as the wavelength. In the optical region, water vapour and dust particles are the main causes of this type of scattering. The wavelength dependence of Mie scattering can be expressed as follows:

$$\sigma_{M(\lambda)} = \beta\lambda^{-\alpha}, \tag{2.16}$$

where β is proportional to the concentration of aerosols, and α is inversely proportional to the size of the aerosols. This type of scattering is still wavelength dependent, but not as strongly as Rayleigh. For very small particles, α approaches 4 leading to a Rayleigh-type scattering behaviour. Most aerosols can be characterized by an α of 1.3.

The third is *non-selective scattering*. This occurs when the diameter of the particles is several times larger than the wavelengths involved. As the name implies, all wavelengths within that range are scattered by the same amount. An example of this is the scattering of white light by suspended water droplets, which is why a cloud looks white and why car headlights are strongly scattered by mist and fog.

Two major consequences of atmospheric scattering are firstly that optical remote sensing is not much use under cloudy conditions, and secondly that it is better to avoid using the shorter wavelengths of the spectrum. On the other hand, the longer the wavelength used in the analysis, the lower the signal-to-noise ratio. Hence, there is a trade-off between atmospheric attenuation and detector performance.

2.3.3 Interactions with target surfaces

Transmission – For the purposes of remote sensing the surface of the earth may be regarded as opaque to radiation, so that transmission can be ignored and all incident radiation is therefore either absorbed or reflected. Nevertheless, at the smaller scale, when considering plant canopies or water bodies, substantial (spectrally dependent) transmission may occur in the surface layers giving rise to characteristic scattering and absorption profiles that are useful in diagnosing surface characteristics. Clear water transmits short-wave blue and green light quite well, but is completely opaque to infrared light that it absorbs strongly. Some plant leaves, on the other hand, are almost opaque to visible light but transmit infrared. A detailed treatment of these effects and their use in remote sensing will be presented in Chapter 3. Microwaves can pass through dry soil and sand up to several wavelengths thick; this can be useful for detecting subsurface phenomena under ideal conditions. In general, however, remote sensing relies on the reflected/emitted components of radiation.

Absorption – We saw above that the earth's surface absorbs a large proportion of the incident sunlight; this is either converted into heat raising the temperature of the body, used in evapotranspiration, or, in the case of plants, a fraction (up to about 5%) is converted into metabolic energy in the process of photosynthesis (Chapter 4). This 'primary production' underlies most life on earth. The nature of the surface material determines which wavelengths are absorbed most strongly and hence which wavelengths are reflected, and thus determines its perceived colour. For example, plant pigments (such as chlorophylls) are very good absorbers of red and blue wavelengths. The green part of the spectrum, on the other hand, is less strongly absorbed. Hence, if plants are illuminated by (white) sunlight they will preferentially absorb red and blue wavelengths; the green part of the incident light is less absorbed, and is thus mostly reflected, leading to the green appearance of vegetation (See Box 2.3). All colours can be analysed in terms of the proportion of three primary colours (red, green, and blue, RGB) that enables us to characterize and hopefully identify the object. For most remote-sensing purposes, however, we need to go beyond the familiar RGB terminology and look in more detail at the spectral distribution of the reflected radiation (see Chapter 7).

Reflection – The solar radiation reaching the earth's surface that is not absorbed by the target is reflected back to space (eqn (2.9)). As absorption is wavelength dependent, it follows that the same must be true for the reflected part. Hence, the same mechanisms have to be considered as described in the previous paragraph.

The most important parameter influencing the directional distribution of the reflected electromagnetic radiation is the roughness of that surface in relation to the wavelength. A surface can be considered *smooth* for a certain wavelength if the irregularities are smaller than one eighth of the wavelength (*Rayleigh's criterion*), otherwise it is *rough* (see Section 2.5). A smooth surface reflects a ray of light at an angle that is equal to the angle of incidence (*Snell's Law*), as is done by a mirror (Fig. 2.8(a)). This is called *specular* reflection. For a rough surface, however, each facet on the surface presents a different angle to the incident

BOX 2.3 Colour theory

It was Isaac Newton who, back in the 1660s, showed that white light can be split up into a spectrum of wavelengths or 'colours' using a prism. He also showed that white light can be analysed in terms of the three primary colours, red, green and blue. Any one of these cannot be made up of combinations of the other two, but any other spectral colour can be generated by adding the appropriate proportions of the primaries. For example, yellow can be formed by combining red and green, magenta from equal proportions of red and blue, and grey from reduced (but still equal) proportions of all three. Digital colour displays, such as a computer monitor, can generate palettes of millions of shades by using combinations of 256 levels (8-bit) of each of the three primaries.

The colour of an object is defined by the colour of the light that it reflects. So an object illuminated with white light appears 'blue' if it absorbs both red and green and reflects what is left, namely blue. But if that object were illuminated with red light, it would appear black, because there is no blue component in the red light.

The above are examples of 'colour addition'. If an object absorbs blue light, it reflects both red and green. The addition of these by the eye or a detector gives the appearance of yellow. Hence, we can say that yellow is obtained by subtracting blue from white light. Blue and yellow are referred to as *complementary* colours. Other pairs of complementary colours are red and cyan (a greeny blue), and green and magenta (a reddish blue). It can be seen that white light can be formed by adding pairs of complementary colours. A yellow photographic filter is sometimes referred to as a *minus blue* filter.

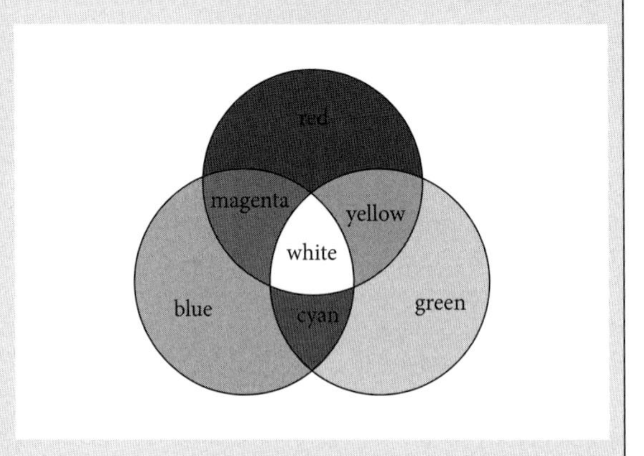

(See Plate 2.1)

The eye perceives colour because there are three different types of receptors in the retina, each stimulated by one of the primary colours, and the brain adds the appropriate quantities together to synthesize the concept of colour (colour addition). A photographic colour slide (transparency) registers colour because it consists of three layers, each sensitive to one of the primary colours, and when white light passes through, the eye will receive various quantities of the three primaries. On the other hand, the colour printing process (and colour filters) makes use of colour subtraction. The printer lays down successive layers of the secondary (complementary) colours, yellow, cyan and magenta, each of which will reflect its complementary colour, and the eye will add them. If successive layers of red ink and blue ink were laid down, all three primaries would be absorbed and there would be no reflected light, the surface would appear black.

beam and thus reflects at different angles, with the net result that the light emerges into a cone of angles (*diffuse reflection*). This effect is also known as *scattering*, analogous to the scattering effect in the atmosphere, where reflection say from a cloud is due to large angle scattering (*backscattering*). A perfectly diffuse (Lambertian) reflector would scatter energy equally in all directions such that it appears equally bright when observed from any angle (*Lambertian scattering*; Fig. 2.8(b)). Most natural surfaces, with the exception of smooth water and some leaf surfaces, are rough at optical wavelengths, whereas they may be smooth at microwave wavelengths. The anisotropic (non-Lambertian) scattering of most natural targets (Fig. 2.8(c)) leads to variation of the reflection coefficient with changing sun and view angles. Although the scattered signal is often much stronger in the forward direction, with vegetation canopies this is not necessarily true: we shall see how we can make use of this so-called *bidirectional* effect in the remote sensing of natural surfaces in Chapter 8.

When referring to reflection of solar radiation, we can define a *shortwave reflection coefficient* (ρ).[4]

[4] The reflectivity ρ includes a directional component. When referring to reflection from land surfaces, an alternative term, *albedo*, is commonly used. This encompasses a whole hemisphere and takes account of all the incoming and outgoing radiation at all angles.

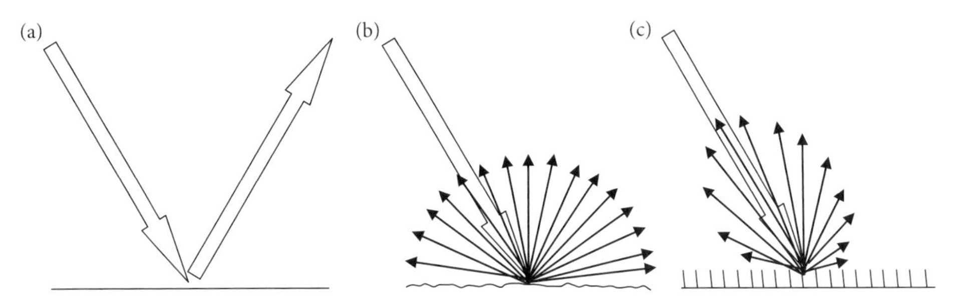

Fig. 2.8. Illustration of (a) specular (forward) scattering, (b) 'Lambertian' or diffuse reflection, and (c) typical asymmetric scattering from a plant canopy showing the enhanced reflection near the 'hotspot' (i.e. with the sun directly behind the observer).

As absorption is wavelength dependent, it follows that the same must be true for reflection.

The *reflectance spectrum* of a surface over a range of wavelengths can be measured in either a laboratory or in the field using a spectrometer. Many remote sensing detectors however, measure the response in only a few, relatively broad, wavebands (hyperspectral sensors – see Section 5.3.3 – are an exception). With broadband sensors the output from, for example, the green channel would represent the integrated and weighted radiance received within the range of wavelengths over which it records. This would be represented as a *digital number*, say DN_g. Similarly, DN_r and DN_{ir} would represent the output in the red and infrared channels. We can define a *spectral signature* for a particular surface or object in terms of its brightness/reflectance over a range of wavelengths, for example by the above three digital numbers. Strictly speaking, we should refer to the *spectral response pattern* of a surface, because the particular signature may vary over time (as vegetation develops or senesces) or over space (the vegetation may not have the same vigour over the whole field if soil condition or moisture content varies). The spectral signature of an object provides valuable diagnostic information and lies at the heart of much optical remote sensing. Vegetation, for example, absorbs red light, but reflects near infrared light very strongly. The ratio or difference of these reflectivities is very sensitive to plant species and plant health, and can be used as a good indicator of both of these. We will return to the use of spectral signatures to classify land surfaces and to retrieve canopy biophysical characteristics in Chapter 7.

A further feature of reflection is that when radiation interacts with a boundary the amounts of reflection and transmission depend on the angle of incidence and the angle of polarization. Where the electric field plane is parallel to that containing the incident and reflected rays it is known as parallel-polarized (or, when referring to microwave radiation the corresponding term is *vertical polarization*) and when perpendicular to the plane it is known as perpendicularly (*horizontally*) polarized. Because of polarization-dependent differences in the angular dependence of reflection it is usual for randomly polarized incident radiation to be partially polarized on reflection so that one can generalize that the degree of polarization is changed on reflection. This effect is particularly useful in microwave remote sensing (see Chapter 5).

Fluorescence – Sometimes radiation is absorbed at one wavelength and immediately re-emitted at a longer (less energetic) wavelength. This phenomenon is called *fluorescence* and is reasonably common in some minerals that absorb ultraviolet radiation and emit visible radiation. Vegetation also fluoresces following absorption by the plant's photosynthetic pigments; as we shall see later (Chapter 4), the magnitude of this fluorescence can be used to indicate the activity of the photosynthetic systems.

Interaction with vegetation – A detailed description of radiation interaction with vegetation is provided in Chapters 3, 7, and 8; here it is just relevant to introduce some of the key features. The interaction of radiation with plants and vegetation is complicated by the structure of the plant leaves and other tissues, by their chemical composition, and by their distribution and arrangement in space. The most obvious spectral feature of vegetation, in addition to its green appearance in the visible, is the comparison between a strong absorption in the visible region and a relatively high reflectance in the near infrared; this characteristic is crucial to much remote sensing of vegetation. The spatial arrangement and density of leaves and branches in the canopy space leads to a strong dependence of reflection on both the angle of illumination and the sensor angle; this behaviour is developed in Chapter 8 where it is shown that the angular dependence of the reflected signal provides a key tool for identification and for inferring biophysical properties of the vegetation canopy.

2.4 Thermal radiation

The earth has a mean temperature of around 300 K and so has its peak emission in the thermal infrared part of the spectrum (see Fig. 2.3) at wavelengths between about 4 and 50 μm with the majority of this emission in the atmospheric window (about 8–14 μm). Detection of this emitted radiation can be used to derive the effective temperature of objects on the earth's surface. In geology, the spectral variability of the emitted radiation is in addition used to identify rock types and minerals. The apparent radiative temperature of a surface depends upon a number of factors:

Emissivity – Since natural objects are not black bodies, they absorb and emit radiation less efficiently at a given temperature than Planck's Law predicts (eqn (2.5)). This means that the apparent *brightness temperature* (T_B) measured by a detector that records the radiation emitted is less than the true *kinetic temperature* (that measured with a thermometer), T_{kin}. It follows from Kirchhoff's law (eqn (2.4)) that these temperatures are related by the expression

$$T_B = \varepsilon^{1/4} T_{kin}. \quad (2.17)$$

It is therefore essential to know the emissivity of a body if its real temperature is to be estimated from the emitted radiation. The emissivity of different surfaces is difficult to measure exactly and varies with wavelength, although it is fairly constant in the range of wavelengths usually used for remote sensing, namely 8–14 μm. Vegetation has an emissivity ranging from 0.96 to 0.99 (although the emissivity of individual leaves may be as low as 0.91 or 0.92; see Section 3.2), and soil emissivity can be as low as 0.89. Water has an emissivity of 0.99, which is fairly constant over this range of wavelengths, and also varies little with temperature. Thermal measurements over water bodies (e.g. sea-surface temperatures) are thus usually much more successful than over land due to the greater homogeneity of the emissivity over large areas.

There are a number of properties of materials that contribute to their surface temperature at any time. The first is the *thermal capacity* (or *specific heat capacity*, c_p measured in J kg^{-1} K^{-1}), which is a measure of the amount of heat energy required to raise the temperature of a body by one degree, and hence of its ability to store heat. An object with a high thermal capacity will not heat up as much as one with a low thermal capacity when subjected to the same amount of radiation. The second parameter is the *thermal conductivity*, K (W m^{-1} K^{-1}), which is a measure of the rate at which heat can pass through a material. The energy incident on a body with a high conductivity is transported more rapidly into the bulk of the material than would be the case for a body of low conductivity, so the surface heats up more slowly (as heat is dissipated in a larger volume). In general, wet soil will diffuse heat downwards more rapidly than will dry soil. It is frequently more convenient to use *thermal diffusivity* ($D_H = K/\rho c_p$, m^2 s^{-1}) rather than thermal conductivity to allow an easy comparison with the units used for mass transport (see, e.g., Jones, 1992). The physical basis underlying the rate of change of temperature of a surface will be developed in Chapter 4, here we simply note that it depends on both heat capacity of the material and the rate at which heat is conducted away from the surface and can be described by what is known as the *thermal inertia* P (J m^{-2} K^{-1} s$^{-1/2}$)

$$P = (K\rho c_p)^{1/2} = \rho c_p (D_H)^{1/2}. \quad (2.18)$$

This measures the thermal response, or inertia, to a change in temperature. Bodies with a high thermal inertia have a smaller diurnal temperature range than do bodies with low thermal inertia. The temperature of a water body does not change markedly between day and night, whereas the land will usually heat up rapidly after dawn, but cool rapidly after sunset. The thermal inertia of vegetation can show quite complex behaviour. Indeed, the temperature of vegetation is actively controlled by the plant itself through evaporation. Further details and derivations for a range of vegetated and unvegetated surfaces can be found in Chapter 4.

2.5 Microwave radiation

The microwave wavelengths commonly used in remote sensing range from about 4 mm to 1 m; shorter-wavelength microwaves tend to be attenuated too much by the atmosphere. Sometimes frequencies are used when discussing microwaves, rather than wavelengths, so a conversion, together with the conventional terms for different regions of the microwave spectrum is provided in Table 2.2. It should be noted that there is no universal agreement as to the boundaries between, or indeed the letters used for, the different wavelength ranges. The letter codes were initially selected arbitrarily for military security.

All materials potentially emit radiation as a function of their temperature according to the Planck function (eqn (2.5)) with the peak intensity and the wavelength at which this occurs depending on the temperature of the body. Although the conventional diagrammatic presentation of this function on a linear scale suggests that the upper limit of terrestrial radiation emission is in the thermal infrared, it is apparent from a logarithmic plot that there is a tail of emission right into the microwave region (Figs 2.3 and 2.9). The amount of this microwave radiation received at the earth's surface from the sun is very small and more than two orders of magnitude smaller than that emitted from the earth. Although weak, however, the microwave emission from the earth and from the molecules in the atmosphere is large enough to be used in passive remote sensing.[5]

The intensity of the radiation emitted by a body is sometimes considered as a surface *brightness*. At the long-wavelength (low-frequency) end of the microwave spectrum, where $hc \ll \lambda kT$, the Planck function (eqn (2.5)) reduces to

$$E_\lambda \approx (2k/\lambda^2)T. \tag{2.19}$$

This is referred to as the *Rayleigh–Jeans limit*, and produces a linear relationship with the physical temperature. For $T = 280$ K this condition holds for $\lambda \gg 50$ μm, so the approximation holds true for microwave and

Table 2.2 Microwave and terahertz frequency nomenclature.

Band	Frequency (GHz)	Wavelength	Comments
P-band	0.3–1	30–100 cm	Penetrates cloud, haze, dust, etc.
L- band	1–2	15–30 cm	Deeper penetration of canopies and soil; for forests gives information on wood biomass
S-band	2–4	7.5–15 cm	Useful for rain radars
C-band	4–8	3.75–7.5 cm	Senses water in top 2–3 cm of soil; interacts with the volume of forest canopies and gives information on leaf biomass
X-band	8–12.5	2.5–3.75 cm	Sensitive to rain; interacts with surface of canopies giving information on surface roughness
Ku-band	12.5–18	1.7–2.5 cm	Also senses the top layers of canopies
Q-band	35–70	4–8 mm	Senses the top layers of canopies, but significant atmospheric attenuation
THz	100–3000 (0.1–3 THz)	0.1–3 mm	Strongly absorbed by water but not non-polar material, so useful for sensing water in plant leaves (close range)

[5] In order for the emission to peak at the wavelength of microwaves, a black body would need to be at about 1 K.

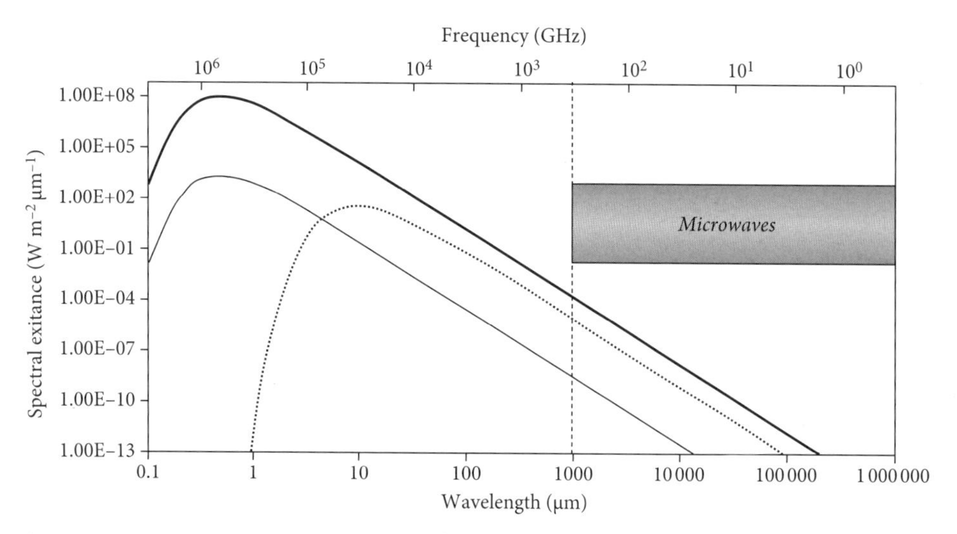

Fig. 2.9. Extension of Fig. 2.3 into the microwave region showing spectral radiant exitance for solar radiation at the sun's surface (thick solid line), and for terrestrial radiation at 300 K (dotted line). This illustrates the long tail into the microwave of radiation emission from the earth, which can be used for passive microwave sensing, even though it has very low energy content (more than 15 orders of magnitude less at 10 mm than for sunlight at 500 nm). Also shown is the radiant exitance for solar radiation at the top of the atmosphere illustrating the fact that terrestrial radiation is more than two orders of magnitude greater than solar radiation reaching the earth's atmosphere at wavelengths greater than 10 μm.

radio frequencies for objects at typical terrestrial temperatures. This implies that we can now refer to the intensity of the emitted radiation in terms of a *microwave brightness temperature*, T_B. This is the temperature of a black body that would emit the same radiance at the wavelength being considered. Since most surfaces are less than perfect emitters (i.e. their emissivity is less than one), eqn (2.4) indicates that T_B will normally be less than the true thermodynamic temperature of the body. At sufficiently long wavelengths, where the Rayleigh–Jeans approximation holds, we find that $T_B = \varepsilon T$ so that the true thermodynamic or kinetic temperature may be obtained as T_B/ε.

Although microwave radiation has properties similar to those of any other portion of the electromagnetic spectrum, being of longer wavelengths than those used in optical remote sensing, it has substantially less energy per photon (2×10^{-25} – 2×10^{-22} J) than thermal (1.4×10^{-20} – 5×10^{-20} J) or visible (2.8×10^{-19} – 5×10^{-19} J) radiation. This low-energy content and the corresponding differences in detection technology have led to microwave remote sensing developing as a subject in its own right in parallel with optical remote sensing, not as a development from it (Woodhouse, 2006).

A second key feature of microwave radiation, again resulting from its long wavelength (mm – m), is that it is not substantially scattered by the atmospheric gases or by microscopic (10 nm – 10 μm) aerosol components of the atmosphere, while there is only limited scattering occurring at wavelengths longer than about 2 cm (frequencies of less than about 15 GHz) even by quite thick layers of larger aerosols such as fog and cloud (droplets up to perhaps 50 μm). Water droplets in rain, however, may be of the order of 0.1–5 mm in radius (approaching the wavelength of the shorter microwaves) so that rain can lead to substantial microwave scattering[6] of high-frequency microwaves. The scattering coefficient also tends to increase with rain rate, and so the microwave backscatter from rain cells can be used as an indicator of rainfall distribution and intensity.

In the microwave region of the electromagnetic spectrum, atmospheric absorption is mainly due to water vapour and oxygen, but there are a series of intervening atmospheric windows that can be used even if the weather conditions are bad (Fig. 2.7). Also, because microwave emission does not rely on solar illumination, radar can be used in overcast conditions and at night. This can be very useful, particularly in tropical regions, where cloud cover usually persists, and at high latitudes.

6. The scattering of microwaves by such particles (Rayleigh scattering) increases rapidly with frequency (Section 2.3.2). Echoes from an individual drop of rain are proportional to d^6/λ^4.

Optical remote sensing measures the radiation reflected, scattered, transmitted, or absorbed by the first few layers of leaves and stems in dense canopies, thus providing information on leaf biochemistry and plant structure but giving almost no information about the understorey or surface soil characteristics below the canopy. Active microwaves, on the other hand, can penetrate to varying depths depending on wavelength and incident angle. Microwaves respond to features in the plant structure that have dimensions of the order of a few centimetres, and the backscattered signal can be useful for identifying spatial distributions, plant structure, leaf orientation, and canopy and soil water content as well as for distinguishing vegetation type and measuring biomass.

Because of their very different wavelengths, microwaves interact with surface features in a different way from optical waves. In the optical region, electromagnetic radiation is rarely specularly reflected, instead the wavelengths that are not absorbed are scattered. Microwaves, on the other hand, interact with the structure of the material. Whether a wave is specularly reflected or scattered depends on the roughness of the surface in comparison to its wavelength. A working criterion as to whether a surface is rough or smooth (*Rayleigh's criterion*) is that a surface can be considered 'rough', and hence act as a diffuse reflector, if the root-mean-square height of the surface variations exceeds one eighth of the wavelength divided by the cosine of the local incident angle.[7] A shingle beach would therefore look smooth to say P-band microwaves (30–100 cm) but rough to X-band (3 cm). Microwaves tend to be specularly reflected from a smooth water surface, resulting in no signal being scattered back to the detector, hence water areas look 'dark' in a radar image. The rougher the surface, the brighter it would appear. The roughness and orientation of plant structures, whose sizes tend to be of the same order of magnitude as micro-wavelengths, will therefore affect the appearance in the image. Short wavelengths tend not to penetrate dense canopies because of the multiple scattering that takes place.

Another factor that is important in microwave sensing is the *polarization* of the microwaves (i.e. the plane in which the electric field vector in the electromagnetic wave oscillates in space). Natural emissions of radiation are usually unpolarized, or rather consist of waves having all directions of polarization. Active systems, however, can generate waves having only one particular orientation. Laser light is polarized, as are the microwaves used in active systems. The way in which a polarized microwave interacts with a surface may depend on the angle of polarization because of the orientation or roughness of the structure and on its dielectric constant, and hence moisture content. Polarimetry uses this directional factor in order to help discriminate between different types of targets. A polarimetric radar system sends out microwaves with a particular polarization (horizontally or vertically) and detects the horizontal or vertical component in the returned signal. A polarized wave incident on a *depolarizing* target, such as a full forest canopy, will return a signal containing polarizations other than that incident. For example, if the incident wave is horizontally polarized (H), the return may contain a vertically polarized component (V) that can be detected using a polarimetric radar. This particular example would be referred to as HV. Other combinations such as HH, VV, or VH are possible. Because of the different polarimetric responses of vegetation, it may be possible to distinguish different types in this way.

2.6 Diffraction and interference

When a beam of light passes through an aperture and is allowed to strike a screen, a bright patch, similar in shape to the hole, is seen. An example of this is sunlight shining through a window. If the aperture is small, however, on close inspection, the edges of the light patch are seen not to be sharp, but diffuse. Similarly, if the shadow of an object is closely examined, again the edges are not sharp and light appears to have 'leaked' round the object into the geometrical shadow. This effect is caused by a combination of *diffraction* and *interference* of the electromagnetic waves. Both diffraction and interference are fundamental properties of any type of wave and their effect can often be seen in images of water waves when they pass round rocks or through a gap between islands. Both effects

[7] This is to do with whether or not there is a significant phase change between waves reflected from the top and the bottom extremes of the surface undulations.

have important consequences in remote sensing, limiting resolution and causing spurious noise in an image, as we shall see in Chapters 4 and 5 in particular.

2.6.1 Diffraction

Waves on open water when viewed from above appear to be straight (*plane waves*) at right angles to their direction of travel. After passing through a small gap, however, the waves take on a circular shape as though emanating from a point at the centre of the gap, and their direction of travel is radial to the new wavefront (Fig. 2.10). This change of direction when part of the wavefront is obstructed is characteristic of all waves and is called *diffraction*. Its effect is always present, but becomes more apparent as the wavelength approaches the size of the object or aperture. When a beam of microwaves is emitted from a circular antenna (a *dish* aerial) say on a satellite, it spreads out in a cone and the diameter of the circular area illuminated on the ground below[8] (the *footprint*) becomes larger the greater the distance the beam has travelled. It can be shown that the spreading of the beam (in terms of θ, the semi-angle of the cone) is related to the wavelength λ and the diameter of the antenna, d, by the expression

$$\theta \approx \lambda/2d. \tag{2.20}$$

Hence, the diameter D of the footprint for a satellite at height H will be given by

$$D = 2H\,\theta \approx 2H\,\lambda/d. \tag{2.21}$$

For a microwave system on a satellite, with an antenna of diameter 1 m using microwaves of wavelength about 6 cm flying at a height of about 800 km, D would be about 50 km, and even from a height of 5 km it would still produce a footprint of about 300 m. At optical wavelengths, say 5×10^{-7} m, using a lens with a diameter of a few mm from a platform at a height of 1 km, the footprint would be less than half a metre. Finer detail will therefore be observed by using Lidar rather than radar (Section 5.6). Diffraction at the first lens of any optical instrument will always set the lower limit to the spatial resolution of the system. A point source will be imaged on the film as an extended spot, and whether or not two adjacent point sources will be distinguishable on the photograph will depend on how large their spots are. A 35-mm camera on a satellite at 800 km would just be able to distinguish two points which are separated by a distance of about 40 m.

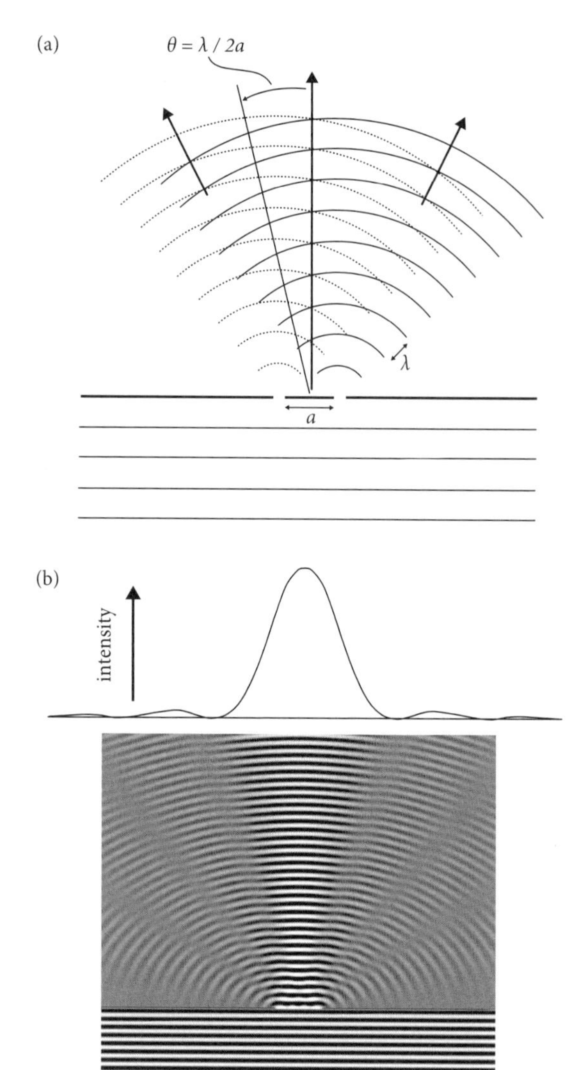

Fig. 2.10. (a) Illustration of interference between two slits, with the arrows indicating lines of constructive interference, with the areas of destructive interference between them. (b) Diagram showing the combined diffraction and interference at a slit with a width 4 × the wavelength, showing spreading out of the wavefront as a result of diffraction. Also shown is the interference between the waves from different parts of the slit leading to clear maxima and minima. As the slit gets narrower, the central maximum would get broader, until for a very narrow slit < λ the central maximum would extend over all angles and there would be no apparent interference pattern.

[8] Actually, all points on the wavefront at the aperture act as secondary sources emitting waves that are in phase with one another. Only in the straight-on direction do these waves travel the same distance and hence arrive in phase. For all other directions there will be phase differences between them, resulting in a reduction in intensity. Thus, the intensity will be a maximum at the centre of the diffracted beam, and will drop off as one moves radially outwards from the centre.

The diffraction-limited resolution of the OSA scanner on IKONOS, whose primary mirror has a diameter of 70 cm, would be less than half a centimetre, but in practice the spatial resolution of the instrument itself, which is about 1 m, is governed by the size of the IFOV that itself is defined by the extent of the detecting elements.

2.6.2 Interference

When two waves come together, the *law of superposition* says that the amplitude at that point will be the algebraic sum of the amplitudes of the two waves acting independently. If the two waves are in phase (i.e. their crests and troughs coincide) then the amplitude of the resulting wave will be the sum of the two (*constructive interference*). If the trough of one coincides with the crest of the other, then they will cancel out and the resultant will give zero amplitude (*destructive interference*). If there is some other phase difference, then the resulting amplitude will be somewhere between these extremes. This is the basis of *interference*. Interference between two waves will *always* occur, but whether or not it is observable depends on two factors – whether the waves are *coherent* and on their relative amplitudes. Coherent waves have a constant phase relationship with one another, incoherent waves do not.[9] If the waves are not coherent, then the amplitude at the point will fluctuate between maximum and zero at a rate determined by the difference between the frequencies of the two waves. If the waves differ greatly in amplitude, then the most intense one will swamp the other and any modulation will not be noticeable.

If you consider light from two coherent sources,[10] their phase difference is determined by the difference in pathlength travelled by the two waves when they come together (Fig. 2.10).

In the direction in which their paths are exactly equal, there will be maximum intensity. In directions in which their paths differ by an integral number of wavelengths there will also be maximum intensity (but see footnote 8 – these secondary maxima will not be as intense as the first one). Between these maxima there will be directions in which the paths differ by half a wavelength, or one and a half wavelengths, etc., meaning that no energy travels in these directions (Fig. 2.10). The angular separation between maxima is given by

$$\theta = \lambda / a, \qquad (2.22)$$

where a is the separation of the sources. For the case of two sources, there is a gradual gradation from maximum to minimum. But a similar effect occurs when there are more than two sources. A *diffraction grating* consists of many very closely spaced slits, light from adjacent ones interfering as above. But now, when the resultant of all waves is considered, the condition for exact phase equality is much more stringent, and the fall-off from maximum intensity is very rapid. When viewed on a screen, the pattern of light consists of very sharp bright lines separated by almost zero intensity (the *diffraction pattern*). Note that perhaps this should be called an *interference grating* even though diffraction is necessary in order that the rays from the slits spread out and overlap. This is also called a *transmission grating* because the light actually goes through the slits and the interference take place on the far side. In a *reflection grating*, the light is reflected off a number of closely spaced grooves etched on a surface, and interference takes place between rays reflected at different angles. This is now a scattering effect, not a diffraction effect, but the net result is similar. If the wave is not monochromatic, the angles of maxima for each wave are different, and the interference pattern would now consist of individual spectral lines (or continuous spectra if white light is used).

We see from the above discussion that diffraction and interference are often inextricably linked.[11] Diffraction causes the bending of rays. Interference takes place if two rays are bent so as to come together. Interference does not need diffraction if we consider two separate waves. Diffraction does not necessarily result in interference.

9. Natural light sources emit radiation in very short bursts due to electronic rearrangements within the atoms, hence a beam of natural light consists of many short wave pulses, each lasting about 10^{-8} s. Even though these wavelets do interfere with one another, no pattern will be observed since the maxima and minima will be constantly interchanging at this rate.

10. In the classic case of *Young's slits*, which is described in physics textbooks, these would be two fine closely spaced slits. Diffraction at each of these slits spreads the waves and enables them to overlap. If diffraction did not take place, each wave would continue straight on and produce its own geometric image of its slit on the screen.

11. Diffraction is actually caused by interference. The secondary waves emitted by each point on the wavefront (for example the portion in the slit) travel in all forward directions. In some directions they reinforce, in others they cancel. The final wavefront is the envelope of the individual wavefronts, most of the energy going in the forward direction, but increasingly smaller amounts away from that direction. On either side of the central maximum, there will be regions of zero intensity where the waves cancel. This gives rise to *diffraction fringes* (see Fig. 2.10), or in the case of a circular aperture, *diffraction rings*.

2.6.3 Bragg scattering

We saw in Section 2.6 that Rayleigh's criterion of roughness of a surface (a surface is rough if the variation in height exceeds one eighth of the wavelength) explains why microwaves are scattered rather than being specularly reflected. The coherence of the waves reflected from different elements of the surface structure causes interference effects, combining constructively in some directions and destructively in others. For a smooth surface, the energy is redistributed into a narrow beam in the forward direction. For a surface with random roughness (i.e. no periodicity), the energy is redistributed into a wide range of directions (Woodhouse, 2006). If the roughness is periodic, however, even if it is smooth by Rayleigh's criterion (such as microwaves interacting with sea surface waves having small millimetre or centimetre wavelengths), coherent scattering now scatters the waves into a number of discrete directions (this is similar to the behaviour of a reflection grating). This is due to the fact that the reflecting facets are now regularly distributed rather than being random. This is referred to as *Bragg scattering* and is particularly relevant when considering the interaction of microwaves with the periodic structures in soils and vegetation.

2.7 The radiation environment

We have seen how electromagnetic radiation would interact with materials on the surface of the earth under ideal conditions. In the real world, however, the illumination of a surface, and also the radiation emitted from a surface, may vary with time of day, season, wavelength and atmospheric conditions. The spectral radiance reflected from a surface depends on the incident radiation (itself depending on factors such as the orientation of the sun (*solar azimuth*), the height of the sun in the sky (measured in terms of the *solar zenith angle*), the cloudiness and proportions of direct and diffuse sunlight, on the direction in which the detector is pointing with respect to the vertical (*look angle*), and on the condition of the surface. As we shall see in Chapters 9 and 11, a basic knowledge of radiation climatology is required for incorporation into the many models where we utilize remote-sensing data to derive information about vegetation growth and performance. In particular we need to be able to convert instantaneous measurements (often obtained on the rather atypical clear days that are needed for satellite remote sensing in the optical wavelengths) into the integrated measures such as photosynthesis, evaporation, or growth that are usually of primary concern to plant biologists, agronomists, and ecologists. In this section, therefore, we outline the basics of radiation variation in natural environments.

2.7.1 Shortwave radiation

The shortwave radiation received at the remote-sensing detector contains components of radiation scattered from the atmosphere into the field of view of the sensor as well as that reflected from the surface of the earth. Both of these will be spectrally modified by interactions with the component atoms, molecules, or aerosols in the atmosphere, but it is only the component reflected from the surface that carries the information that we require about the surface. The illumination may not be uniform due to elevation, slope, and aspect of the surface and shadows and cloud shadow may obscure parts of an image even on bright days.

Solar constant

The amount of solar radiation received at the top of the atmosphere (the irradiance on a horizontal surface at the top of the atmosphere – $\boldsymbol{I}_{\text{toa}}$; W m^{-2}) varies throughout the year depending on the location, season, and time of day. The value of $\boldsymbol{I}_{\text{toa}}$ is given, according to Lambert's cosine law, by

$$\boldsymbol{I}_{\text{toa}} = \boldsymbol{I}_{\text{S}(D)} \cos\theta = \boldsymbol{I}_{\text{S}(D)} \sin\beta, \tag{2.23}$$

where $\boldsymbol{I}_{\text{S}(D)}$ is the extraterrestrial solar irradiance normal to the solar beam on a particular day of the year (D), θ is the solar zenith angle, and β is the solar elevation. The solar zenith angle is a simple function of latitude, time of day, and time of year and can be calculated using readily available equations (e.g. http://susdesign.com/). The extraterrestrial irradiance itself depends primarily on the *solar constant* ($\boldsymbol{I}_{\text{pS}}$), defined as the integral over wavelength of the extraterrestrial solar irradiance normal to the solar beam at the earth's mean distance from the sun, so $\boldsymbol{I}_{\text{S}(D)}$

is modified slightly by the seasonal variation in the earth–sun distance. The seasonal fluctuations due to changes in the earth's orbit constitute just over 3% of the radiation averaged over one year and can be well approximated by

$$\boldsymbol{I}_{S(D)} = \boldsymbol{I}_{pS}\left[1 + e.\cos\frac{2\pi}{365}(D-3)\right], \tag{2.24}$$

where $e = 0.0167$ is the eccentricity of the elliptical earth orbit. Variations in solar constant need to be taken into account in the radiometric correction of remotely sensed images (Section 6.2). Recent satellite measurements give a value for the solar constant of about 1366 W m^{-2} (ASTM, 2000) but with a weak seasonal variability.

The variability of solar radiation can have important consequences for the quality of remotely sensed images, and also for the sensitivity of the detected signal. Indeed, the SeaWiFS scanner is switched off in winter above about 54° latitude because the surface leaving radiance from water bodies is too weak to detect.

The most important contributions to the solar variability are the 27-day rotation cycle of the sun and the 11-year sunspot activity cycle. The first is responsible for variations of several per cent in the UV but less than 1% for wavelengths above 0.25 μm. The second factor modulates the solar emission by noticeable changes in the outer reaches of the sun's atmosphere. The wavelength of the shortwave radiation depends on the altitude in the solar atmosphere from where it is emitted. Wavelengths longer than about 0.3 μm originate from the sun's photosphere, while wavelengths of $0.05 < \lambda < 0.3$ μm originate from higher in the base of the chromosphere. The peak-to-peak variation in the solar flux between the extremes of solar activity is about 10% for $\lambda = 0.16$ μm but falls to less than 1% for $\lambda \sim 0.3$ μm. Therefore, the temporal variations in the shortwave spectrum are most noticeable in the shortwave UV region, contributing only about 0.1% to the solar constant, but because effectively all the radiation in the UV-C region (< 0.28 μm) is filtered out in the earth's atmosphere, this is not very important for vegetation studies. The non-homogeneity of the sun's temperature also accounts for the distortion of the sun's spectrum away from that which would be expected from a black body (Fig. 2.3). Most of the radiation is emitted from the photosphere that is indeed at about 6000 K, but the chromosphere is at only about 4500–5000 K, which produces a slight shoulder on the long-wavelength side of the spectrum.

Solar radiation at the earth's surface

As we have seen (Fig. 2.6) the spectrum of shortwave radiation reaching the earth's surface ranges roughly from 0.3 to 2.4 μm, as both tails of the solar spectrum have been trimmed, respectively, by scattering and absorption by ozone in the UV and by absorption by water vapour and carbon dioxide in the IR.

Although about 44% of the energy in the solar spectrum at the top of the atmosphere is in the visible, the exact value at the surface depends on solar angle and atmospheric conditions and especially the proportion of diffuse radiation. The diffuse radiation tends to be enriched in longer wavelengths. A useful generalization, which we shall use for calculations of photosynthesis, is to assume that 50% of the energy in solar radiation reaching the earth's surface is in the range of wavelengths below 700 nm and 50% in the infrared (e.g. Monteith and Unsworth, 2008). More detailed estimates and measurements of the spectral dependence of solar radiation on atmospheric conditions are available from the Solar Energy Research Institute, Golden, Colorado (http://www.nrel.gov/solar_radiation/data.html).

A useful simplification that will be used throughout this book is to distinguish between the parallel radiation in the direct solar beam (*direct radiation*, $\boldsymbol{I}_{dir}$) and the *diffuse radiation* ($\boldsymbol{I}_{diff}$) that includes all shortwave radiation reflected and scattered from the sky, clouds, and even canopy components. The sum of the direct and diffuse components is commonly termed *global radiation*. Even on a clear day diffuse radiation contributes between 10 and 30% of the total solar irradiance. In parts of the world where there is extensive cloud cover for up to two thirds of the time, the diffuse radiation may contribute on average between 50 and 100% of the shortwave radiation, depending on the season and time of day. In drier climates, the sunshine duration may reach 90% of the possible maximum amount, and the proportion of diffuse radiation will therefore be correspondingly lower. On a cloud-free day, most of the diffuse radiation comes from a region near the sun, as a result of forward scattering. On a very overcast day, the sky may appear to be equally bright in all directions (called a *uniform overcast sky*) so the radiance in any direction would be given by $\boldsymbol{I}_{diff}/\pi$.

Under most other conditions, however, the sky tends to be brighter near the zenith or from the direction of the sun and other approximations such as a *standard overcast sky* are required (Monteith and Unsworth, 2008). An approximate relationship between the proportion of diffuse radiation and the irradiance expressed as a fraction of the irradiance at the top of the atmosphere is given in Fig. 2.11 that takes account both of the effects of changing solar elevation (and hence of the optical air mass, m) and of absorption and scattering by clouds. Knowledge of the fraction of diffuse radiation is important both for modelling photosynthesis and when we start modelling radiation transfer in plant canopies (Chapters 3 and 8).

As one might expect, the shortwave irradiance at the earth's surface varies both diurnally and seasonally, with the proportion of diffuse radiation varying correspondingly (Fig. 2.12).

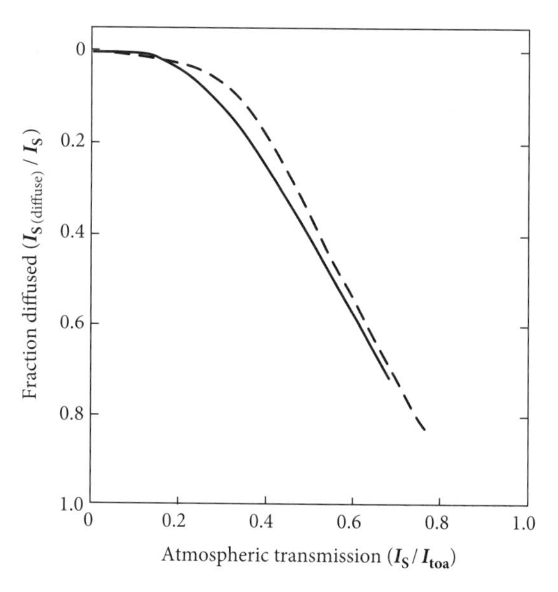

Fig. 2.11. Fraction of diffuse radiation as a function of atmospheric transmission expressed as the ratio between global shortwave irradiance at the surface and the irradiance at the top of the atmosphere. The solid line represents daily data and the dashed line hourly data (after Jones, 1992).

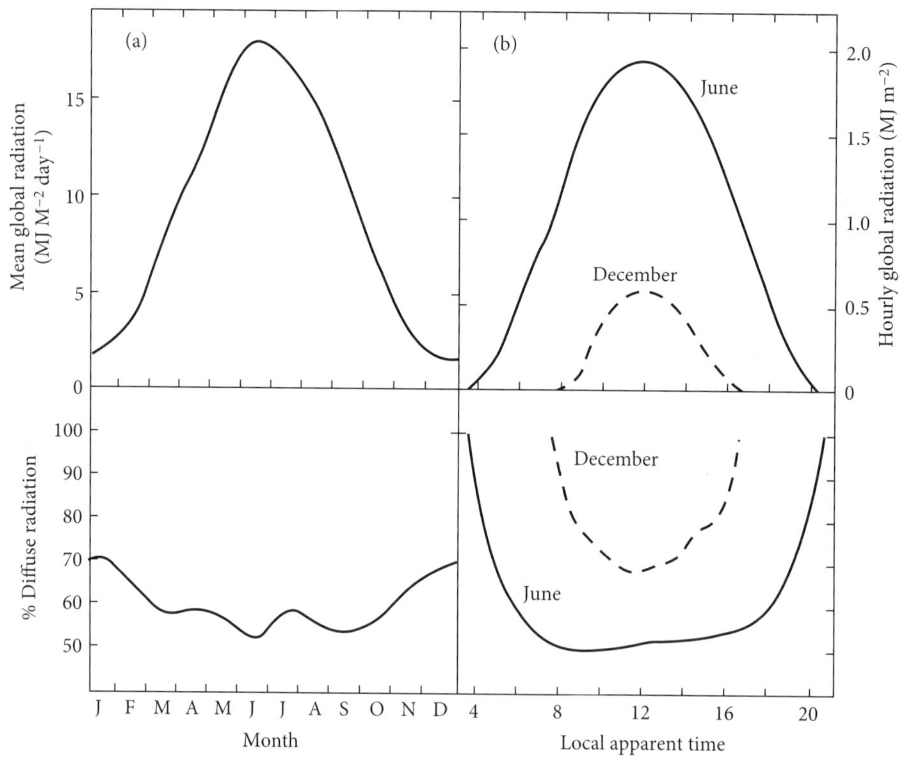

Fig. 2.12. Typical (a) seasonal and (b) diurnal variation in incident global shortwave irradiance (I_S) for central England (Rothamsted Experimental Station), together with the corresponding percentage diffuse radiation (from Jones, 1992).

2.7.2 Longwave radiation

We have seen that there are two contributions to the longwave radiation environment at the earth's surface, namely from the sky, due to absorption of shortwave radiation and its reradiation at thermal wavelengths (*downwelling*), and from the earth itself (*upwelling*). Both components vary strongly with time of year, time of day, and with region.

Downwelling longwave radiation is emitted by atmospheric gases, especially water vapour and carbon dioxide as a function of their temperature. However, since the atmosphere is not a perfect emitter (emissivity somewhat less that 1), its apparent radiative temperature is approximately 20 °C lower than its actual temperature. With a clear sky the apparent sky temperature viewed from the surface is about −40 °C, but much higher apparent temperatures are recorded from cloud. Clouds are not only very effective emitters, but they also reflect thermal radiation back to the surface of the earth and prevent it being radiated into space.

The upwelling radiation is the main component detected by a remote-sensing instrument operating in the thermal waveband. This will contain a contribution from the atmosphere as well as from the surface of the earth. The major loss of heat from the ground is by radiation, although for global energy-balance studies it is also necessary to account for the energy transfer by turbulent transfer (convection in the air due to heating of the lower atmosphere) and by evapotranspiration (from moisture in the soil, water bodies, and vegetation).

Figures 2.13(a) and (b) illustrate typical diurnal trends in the longwave radiation fluxes measured at the surface. The mean longwave fluxes show some diurnal variation and a slightly larger seasonal variation as temperatures change. Though both the upward and downward fluxes are substantial, the difference between them (equal to the *net longwave radiation*, $\boldsymbol{R}_{\mathrm{Ln}}$) remains fairly constant at a net loss of between 0 and 100 W m^{-2} for this site, which is probably representative of many temperate sites. The range of the longwave fluxes from the sunniest to cloudiest days is also shown in these graphs, indicating the magnitude of the variation expected.

2.7.3 Radiation and the global energy budget

Most of the energy that fuels the earth system (atmosphere, biosphere, and geosphere) comes from the sun (apart from a very small geothermal component), and it is the delicate balance between the incoming and outgoing radiation (the earth's *radiation budget*), and the distribution of this energy within the earth–atmosphere system, which ensures that surface temperatures have been maintained reasonably constant, give or take a few degrees, over the centuries. Components of the atmosphere, as well as the surface of the

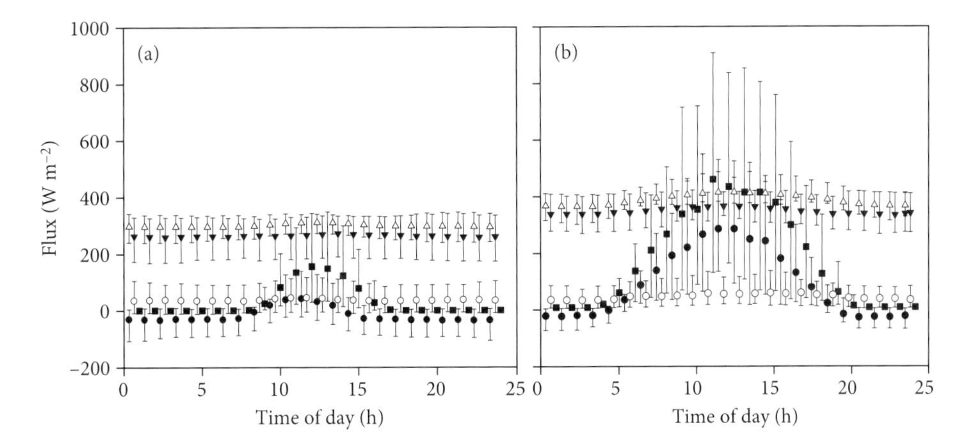

Fig. 2.13. Diurnal trends in shortwave radiation ($\boldsymbol{I}_S$; ■), net radiation ($\boldsymbol{R}_n$; ●), downwelling longwave ($\boldsymbol{R}_{\mathrm{Ld}}$; ▼), upwelling ($\boldsymbol{R}_{\mathrm{Lu}}$; Δ), and net longwave radiation ($\boldsymbol{R}_{\mathrm{Ln}}$; ○) for grassland in the Netherlands (CABAUW experiment as described by Beljaars and Bosveld, 1997) for (a) January 1987 and (b) June 1987. The figure shows the mean values and ranges of the monthly data.

earth, reradiate absorbed shortwave (solar) radiation at longer wavelengths. In a steady state, the emitted radiation must balance the absorbed radiation. Therefore if one assumes a globe with no atmosphere and an average global surface reflectance of 0.3, 239 W m^{-2} of the incident 342 W m^{-2} (Fig. 2.14) would be absorbed by the earth. Substitution into eqn (2.4) gives a surface temperature of −18 °C. In practice, however, although some of the emitted thermal radiation is radiated back into outer space, much is prevented from escaping by the so-called greenhouse gases and reradiated back to the surface in the 'greenhouse effect', leading to surface warming (Fig. 2.14). The most significant greenhouse gases (especially water vapour, but also carbon dioxide and methane) together ensure that the earth's surface is maintained approximately 33 °C warmer than it would be without any atmosphere.

Any anthropogenic or natural increase in carbon dioxide or methane concentrations would lead to more thermal energy being trapped, thus tending to lead to 'global warming'. The actual changes in surface temperature expected in response to a given radiative forcing, however, are very dependent on the *gains* in a complex series of feedback processes. These include those involving changes in vegetation cover, the hydrological cycle, and clouds. For example, increases in cloud cover can increase both the reflection of incident solar radiation (leading to cooling) and the trapping of terrestrial thermal radiation (leading to warming). The balance between these effects depends on cloud type and distribution (Solomon *et al.*, 2007) which further depend on other hydrological interactions, with, for example, increases in surface temperatures tending to increase the evaporation of water, and hence cloud cover. The science behind our current understanding of climate change and anthropogenic impacts has been reviewed by the Intergovernmental Panel on Climate Change (Solomon *et al.*, 2007). Over longer timescales especially, another significant factor in the earth's climate is variation in solar activity and proximity to the earth that can change the amount of solar radiation incident on the earth.

Figure 2.13 shows in addition to the longwave fluxes typical diurnal and seasonal variation in global shortwave irradiance and the net radiation ($\boldsymbol{R}_n = \boldsymbol{R}_{Ln} + \boldsymbol{I}_S$). At night there tends to be a net loss of energy from the earth's surface while there is a net gain by the surface during daylight hours from absorbed sunlight. The longwave radiation constitutes a major component of $\boldsymbol{R}_n$ in winter, though may be only 10–20% in summer.

The net contribution of these different radiative fluxes to the global energy balance is summarized in Fig. 2.14, which illustrates the magnitudes of the various energy fluxes including also the turbulent transport of sensible heat to the atmosphere and the transfer of heat associated with evaporation. As the earth's temperature remains essentially constant over time, the

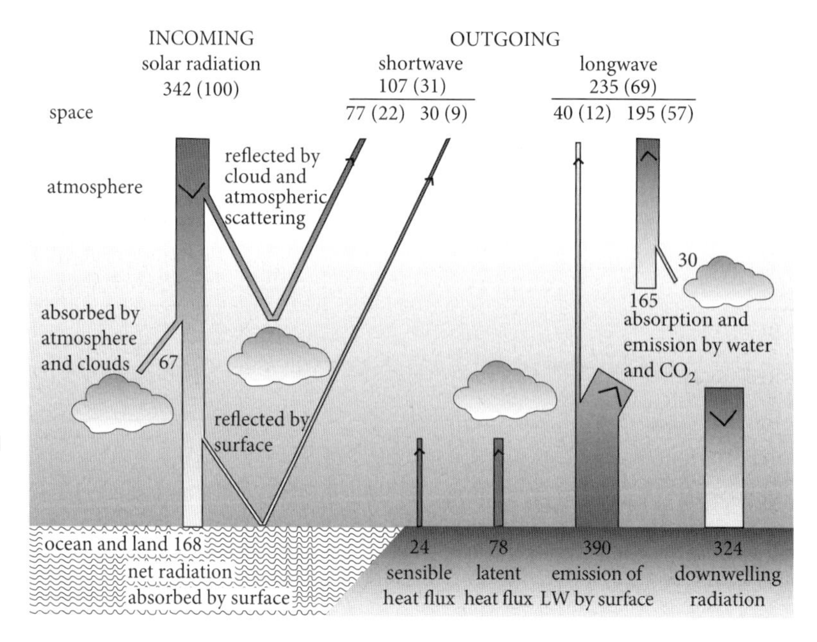

Fig. 2.14. Estimated global annual mean energy budget of the earth (averaged over land and water surfaces), with the magnitudes of the fluxes in W m^{-2} (values in parentheses %) (data from Le Treut *et al.*, 2007).

total amount of radiation entering the earth system is exactly balanced by the outgoing reflected shortwave and emitted thermal radiation. It is notable that the thermal radiation exchanges between the earth's surface and the atmosphere are among the largest fluxes, but that the net thermal flux is relatively small as the upwelling and downwelling fluxes largely balance out. As indicated above, a very large proportion of the emitted thermal radiation is absorbed by the atmospheric greenhouse gases (primarily water vapour and CO_2). The most important effect of this radiation absorption and subsequent re-emission is to redistribute energy and hence temperature between the different components of the earth–atmosphere system.

This absorption of thermal radiation by the atmosphere raises real problems when we attempt to measure surface temperature using thermal infrared. As can be seen from Fig. 2.14 most of the radiated energy is absorbed by the atmosphere and reradiated, one third to outer space and two thirds back to the ground. Although less than ten per cent of the thermal radiation emitted from the ground is transmitted through the atmosphere, it is fortunate for remote-sensing purposes that most of this transmission occurs in the atmospheric window in more or less the centre of the earth's spectrum, between about 8 and 14 μm (see Fig. 2.7). Most detectors on satellites and aircraft operate in this window. Because most thermal remote sensing is quantitative, i.e. measuring temperatures, it is necessary to correct for atmospheric effects (Section 5.2) even when using data collected through the atmospheric windows.

It is useful to be able to estimate the relative intensities of the radiation emitted by bodies at different wavelengths; this can be done using eqn (2.5). Some examples are shown in Table 2.3 for bodies at 300 K (the temperature of the earth) and at 6000 K (the effective temperature of the sun), with the latter reduced by a factor $(r/R)^2$ (where r is the radius of the earth and R is the radius of the earth's orbit around the sun) to account for the distance the radiation has travelled from the sun to the earth. No account has been taken of atmospheric effects or variations in emissivities. It can be seen that there is negligible earth radiation in the visible part of the spectrum, but that emitted radiation and reflected solar radiation in the mid-infrared are of similar magnitude, and in the thermal to microwave regions the earth's emitted radiation predominates. This indicates that measurements of surface temperatures using mid-infrared channels should only be carried out at night when there is no solar radiation present. Even at 12 μm there will be some solar contamination during daylight hours. This component may be reduced by interactions with the atmosphere, but lower surface emissivities may also reduce the emitted component. Even so, the solar contamination is usually ignored, particularly if only relative temperatures are required.

Table 2.3 Estimated relative intensities (in units of $8\pi hc$ m^{-5}) of radiation emitted from the earth and that received from the sun at the top of the atmosphere (adapted from Cracknell and Hayes, 2007), together with the ratio between the emitted terrestrial and received solar radiation. Some estimates of the ratio between emitted terrestrial and reflected solar radiation for vegetation are also shown.

Wavelength	Emitted intensity from the earth at 300 K	Received intensity of solar radiation (6000 K) at the earth (top of atmosphere)	Ratio of terrestrial to solar
Blue – 0.4 μm	7.7×10^{-20}	6.1×10^{24}	1.3×10^{-44}
Red – 0.7 μm	2.4×10^{0}	5.1×10^{24}	4.7×10^{-25}
Infrared – 3.5 μm	1.6×10^{21}	4.7×10^{22}	3.4×10^{-2} (2.4×10^{0})[a]
Thermal infrared – 12 μm	7.5×10^{22}	4.5×10^{20}	1.7×10^{2} (1.2×10^{4})[a]
Microwave – 3 cm	2.6×10^{10}	1.3×10^{7}	2.0×10^{3} (1.4×10^{5})[a]

[a] estimate (at the surface) of the ratio between reflected solar and emitted terrestrial radiation from vegetation assuming an atmospheric transmittance of 0.7, and a vegetation emissivity of 0.98.

2.7.4 Microwave radiation

The thermodynamic temperatures within the lower and middle atmosphere are well above 200 K, high enough to produce significant microwave emission. At frequencies between about 10 and 300 GHz, a number of important atmospheric constituents have rotational and vibrational spectral lines, and at higher frequencies absorption and emission from water vapour acts as a continuous background signal. Measurement of the atmospheric microwave signal forms the basis of *atmospheric sounding* for measuring distribution and concentration of the various atmospheric components. Since emitted radiation is measured with a passive system, there is no requirement for an external source and so sounding measurements can be made at any time of day or night allowing global coverage on a daily basis.

As we saw above, the amount of microwave energy arriving at the earth's surface from the sun is negligible compared with the amount emitted by the earth itself. This means that surface temperatures measured with a microwave radiometer will not be contaminated by a reflected solar contribution, as they are in the mid-infrared range. Cloud cover and aerosols do not have much impact (Jones, 1992) below about 20 GHz, which is also an advantage for longwave microwave sensors over thermal infrared for measuring surface temperatures. Even though the microwave emission from the earth's surface is very weak, it is very sensitive to changes in roughness, salinity, and water content that are not possible to measure at other wavelengths. Passive sensing, as we saw in Section 2.5, is also useful for large-scale mapping of the surface brightness temperatures.

Further reading

Most introductory textbooks in remote sensing cover the physical principles to a greater or lesser extent. Especially recommended texts include *Introduction to Remote Sensing by* Cracknell and Hayes (2007), the very comprehensive *Remote Sensing and Image Interpretation* by Lillesand, Kiefer, and Chipman (2007), *Introduction to Environmental Remote Sensing* by Barrett and Curtis (1999), and *Introduction to Remote Sensing* by Campbell (2007). A particularly good text that gives further details on the underlying principles of radiation physics is *Physical Principles of Remote Sensing* by W.G. Rees (2001). Of particular relevance to the microwave section is the excellent *Introduction to Microwave Remote Sensing* by I.H. Woodhouse (Woodhouse, 2006). Much useful information on radiation transfer and solar radiation may be found in the listed websites.

Websites

http://aeronet.gsfc.nasa.gov/ – NASA Aeronet site has useful data on atmospheric properties and optical depth.

http://www.cfa.harvard.edu/HITRAN/ – Home page for the HITRAN database.

http://www.globalwarmingart.com/wiki/Greenhouse_Gases_Gallery

http://jwocky.gsfc.nasa.gov/ozone/ozone_v8.html NASA total ozone information from TOMS.

http://www.nrel.gov/rredc/ Solar *radiation* information from US National Renewable Energy Laboratory, Golden, CO, USA.

http://www.nrel.gov/rredc/smarts/ Simple model of the Atmospheric Radiative Transfer of Sunshine (SMARTS).

http://rredc.nrel.gov/solar/models/spectral/SPCTRAL2/ Bird Simple Spectral Model: SPCTRAL2 this excel spreadsheet computes clear sky spectral beam, hemispherical diffuse and total irradiance on a prescribed receiver plane at a given time and place.

http://www.nrel.gov/solar_radiation/data.html – A calculator to estimate diffuse and direct spectral irradiance.

http://susdesign.com/ – Useful calculator for calculation of solar angles.

Sample problems

2.1 Estimate the most suitable wavelength ranges for detecting (a) forest fires, and (b) vegetation temperatures. In what parts of the electromagnetic spectrum do these wavebands fall?

2.2 (a) The element of an electric fire glows a dull red. Estimate its radiant temperature. (b) The radiant temperature of the filament of a light bulb is 7000 K. What is the value of λ_{max}?

2.3 Name the parts of the electromagnetic spectrum in which waves of the following wavelengths fall:

(a) 350 nm, (b) 3.5 μm, (c) 35 mm. In each case calculate their frequencies and the energies carried by the photons of electromagnetic radiation at these wavelengths.

2.4 Estimate by how much a temperature increase of 4 °C would increase the total radiation from the earth's surface. How does this compare with the effect of a 2% increase in surface emissivity?

2.5 Calculate the optical air mass for a surface at 1000 m altitude when the solar zenith angle is 30°. What is the irradiance on a horizontal surface, relative to that above the atmosphere, assuming that the transmittance of the atmosphere is 0.7 (for vertical transmission to sea level)? What is the equivalent optical thickness?

3 Radiative properties of vegetation, soils, and water

In remote sensing applications we sense the radiation emanating from the surface, which in the optical region represents for the most part reflected radiation. There are rare cases in the visible where we make use of radiation emission by natural surfaces (fluorescence or phosphorescence), though emitted radiation is more widely used in the longwave infrared and microwave regions. In order to infer the properties of a land surface from the remotely sensed (reflected) signal, it is necessary to understand how vegetation, soil, and water interact with the incoming radiation to generate the reflected signal. The nature of the reflection, in terms of its intensity, its spectral properties, and its spatial or angular properties, contributes information about the surface being studied. In this chapter we describe the basic principles underlying the observed radiative properties of different vegetated and non-vegetated surfaces, starting with a discussion of the radiative properties of the individual components of a vegetated surface including leaves, soils, and water and then extending this to a simple treatment of canopies.

As will be discussed in more detail in Chapter 6, the reflected signal is usually recorded as a simple digital number relating to the intensity of the radiation in a particular wavelength range at the sensor. This can then be converted to a (spectral) radiance *at the sensor* (W m^{-2} sr^{-1}). In order to estimate the corresponding reflectance *at the surface*, one needs to correct for atmospheric absorption and the scattering of electromagnetic radiation travelling between the sun, the surface, and the sensor (Fig. 3.1; Section 6.2.2). This requires information on the incident flux density at the surface, on the transmission between surface and sensor, and on the atmospheric path radiance (scattered radiation that adds to the detected radiation at the sensor). Perhaps the most straightforward way of deriving the reflectance is by use of a reference standard of known reflectance. For in-field sensing we frequently use a reflectance standard such as barium sulphate or more usually Spectralon® (a sintered polytetrafluoroethylene (PTFE) material that shows nearly Lambertian behaviour and that has a high and known reflectance (>99%) from 400–1500 nm). In this case, the reflectance for any sample pixel can be simply obtained as the observed sensor value for that pixel multiplied by the known reflectance of the reference surface and divided by the observed sensor value for the reference. The calibration of satellite or airborne radiances to reflectance is more difficult, though surfaces of known spectral properties such as deserts or snowfields can usefully be applied; methods will be discussed further in Section 6.2.2.

It is important to be clear at this stage that variations in observed radiance across an image, especially at high spatial resolution, may not correspond with actual variations in reflectance because the incident irradiance may be varying over the image. For example, the lower reflected radiance observed in shaded areas does not imply a lower reflectance; although it is sometimes (confusingly) reported as such, it is simply a result of the lower incident irradiance on these areas. Similarly topographic effects that affect the angle between the surface normal and the sun can have comparable effects.

3.1 Optical region

Historically, much of the development of remote sensing has been based on studies of the passive reflection of radiation in the visible, or at least in the broader optical-reflective region; this is still true today, even though use of the thermal region and microwaves is increasing. For the purposes of this chapter we will treat all wavelengths from the ultraviolet through the visible and near to mid-infrared (i.e. c. 250–3000 nm) as falling in the optical region.

3.1.1 Leaf radiative properties

Central to the development of sensitive techniques for inferring the structure, composition, and functioning of plant canopies is an understanding of the radiative properties of the canopy as a whole and of its components, whether leaves, stems, soil, or water. The interaction of radiation with plant leaves, and hence the magnitudes of spectral reflectance (ρ_λ), spectral absorptance (α_λ) and spectral transmission (τ_λ), depends not only on the wavelength but also on a range of structural and chemical characteristics such as chemical composition, leaf age, leaf thickness, leaf structure, leaf water content, and so on. Examples of the diversity of plant leaf structure are illustrated in Fig. 3.2. Although many higher plant leaves are flat (Fig. 3.2(a)), with the chlorophyll-containing photosynthetic tissue being optimally spread out, some, especially the conifers (Fig. 3.2(d)), have needle shaped or cylindrical leaves. Most leaves have an epidermis covered in a thick layer or cuticle (often waxy) that prevents water loss, with gas exchange occurring primarily through the stomatal pores (Fig. 3.2(b)). Some species have leaves covered with a thick tomentum of hairs on one or both surfaces (see Fig. 4.1). The lower plants such as mosses and liverworts (Fig. 3.2(c)) have less well developed structures for control of gas exchange. Each of the structural characters such as tissue density, presence, absence, or distribution of waxes, hairs, or air spaces, and differences in pigment composition all affect the radiative properties of the leaves of individual species. It follows that the characteristic combinations of these characters in individual species gives obvious scope for distinguishing them on the basis of reflected radiation. At the same time, plant growth and development, as well as specific stresses, will change the structural and chemical characteristics of the plants, which can thus be inferred from the recorded spectral signature.

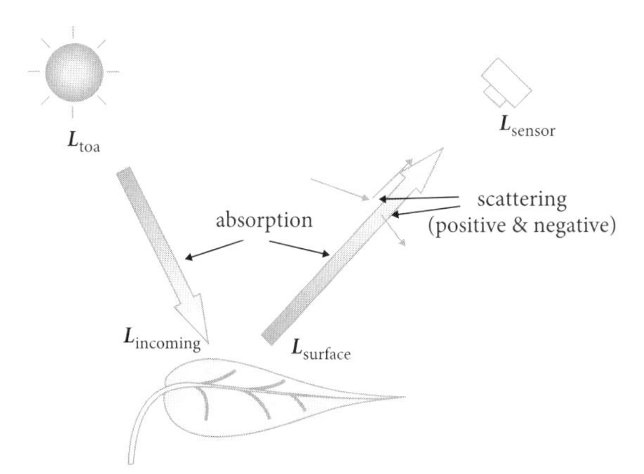

Fig. 3.1. Illustration of the relation between radiance received at a sensor (L_{sensor}), the radiance reflected from a surface ($L_{surface}$) and the incoming radiances at the top of the atmosphere (L_{toa}) and at the surface ($L_{incoming}$). The value of L_{sensor} depends on absorption and scattering processes in the atmosphere affecting both the incoming solar radiation and the reflected radiation, as well as on the reflectance of the surface.

Not only does the interaction of radiation with leaves depend on the angle of incidence of radiation with respect to the leaf, but also a further complication is that the arrangement of leaves to form a canopy also substantially modifies the canopy reflection, as compared with reflection from the individual components. This is a result of radiation scattering and secondary and tertiary interactions between the leaves at different levels in the canopy, as well as between leaves and the underlying soil. As a result, the scaling up of information from single leaves and other canopy components to derive the properties of whole canopies and landscapes is complex and will be developed in more detail in Chapter 8, where the whole topic of directional reflectance will be developed.

The main features of the spectral properties of plant leaves are illustrated in Fig. 3.3 for grape vine leaves. All leaves absorb a large proportion of incident radiation in the visible wavelengths (shorter than about 700 nm, though with a dip in the green), and absorb relatively little radiation in the infrared, except in the water-absorption bands. Conversely the reflection and transmission of incident radiation are low in the visible

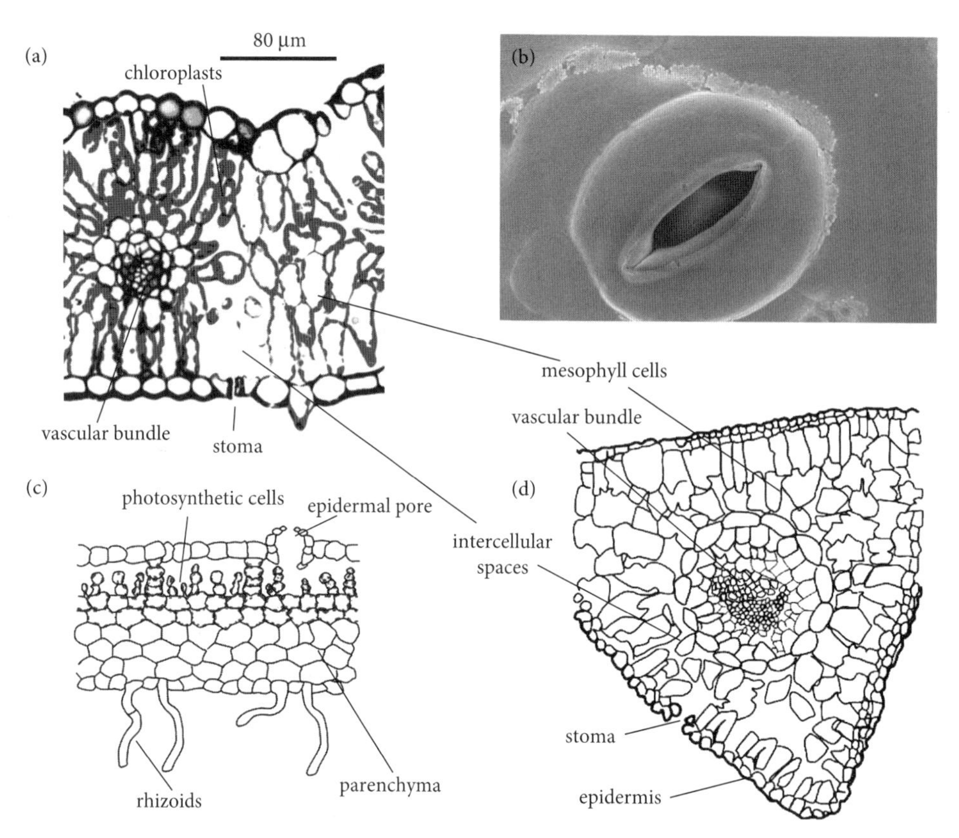

Fig. 3.2. Diversity in plant leaves: (a) Transverse section of a wheat leaf (*Triticum urartu* Tum.) showing the dense photosynthetic mesophyll cells containing chloroplasts, where the chlorophyll is, the mesophyll tissue with its airspaces and the vascular bundle, and the position of stomatal pores in the epidermis, (b) scanning electron micrograph of a single stoma from tobacco, (c) drawing of a cross-section of a liverwort leaf (*Marchantia*) showing airspaces surrounded by photosynthetic cells near the upper surface and rhizoids on the lower surface that acquire water and nutrients from the substrate, and (d) a cross-section of a typical pine needle showing the concentration of photosynthetic cells near the needle surface.

and high in the infrared. It is also worth noting that there is a tendency, at least for thin leaves, for regions of high reflectance to correspond with regions of high transmittance. Though the pattern of spectral properties illustrated here is common for most leaves, the precise reflectance and transmittance can vary substantially between different species and as a result of environmental stresses.

Measurement of leaf radiative properties

For laboratory studies of leaf radiative properties it is usual to perform the measurement on a single leaf using a spectrophotometer with an attached integrating sphere (as in Fig. 3.4). Reflectance can also be assessed for infinite stacks of leaves; this is all that can be conveniently achieved for conifer needles (Fig. 3.4(b)). When used as illustrated in Fig. 3.4 with a collimated beam of illuminating light this instrument allows measurement of what are known as directional-hemispherical (see Chapter 8) values for transmittance (τ_λ) or reflectance (ρ_λ). In any case it is important to define the angular properties of the illumination and the detector (Section 8.1.3).

Dependence on chemical/biochemical composition

The most important determinant of the spectral properties of a leaf, particularly in the visible wavelengths, is its chemical composition, of which components such as the photosynthetic pigments (especially chlorophyll) have the most influence. Other leaf components include pigments such as carotenoids (which include the yellow pigment carotene and also the xanthophylls that

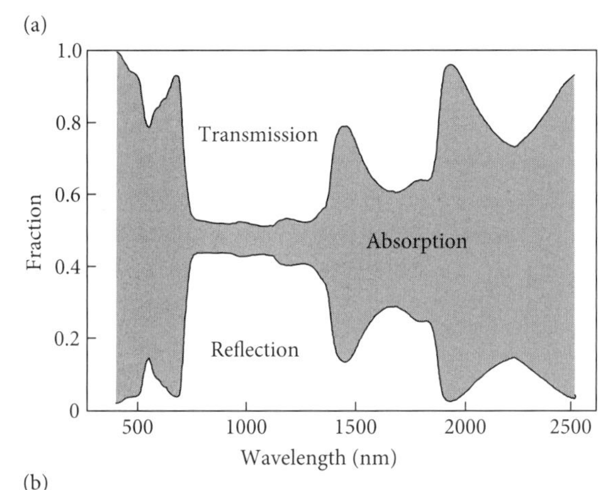

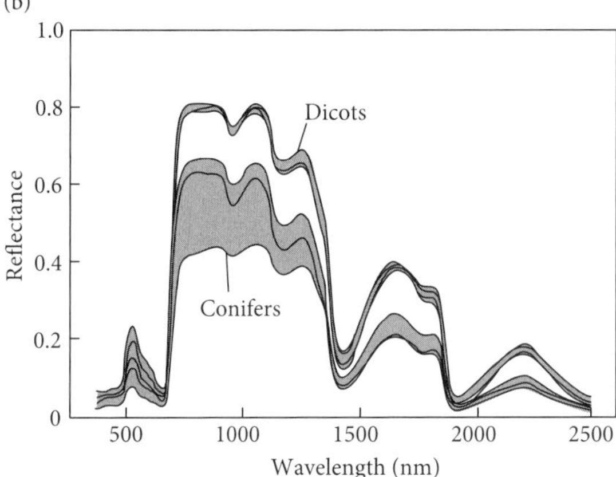

Fig. 3.3. (a) Typical patterns of radiation absorption, transmission, and reflection for plant leaves, illustrated for vine leaves (*Vitis vinifera* L.). Data were obtained from single leaves using an integrating sphere as illustrated in Fig. 3.4; (b) typical ranges of reflectance for optically thick stacks of either dicotyledonous leaves (lettuce, potato, poplar, and grape vine) and conifer needles (*Pinus contorta*, *Pseudotsuga menziesii*, and *Picea sitchensis*) (data from the LOPEX93 experiment (Hosgood *et al.*, 2005; http://ies.jrc.ec.europa.eu/index.php?page=data-portals)).

are important in photosynthesis (see Chapter 4)) and flavonoids (including anthocyanins) that contribute to autumn leaf colour. Absorption spectra for several of the most important leaf components when isolated in acetone are illustrated in Fig. 3.5, where Figs. 3.5(a) and (c) show the spectra for different pigments all normalized to the same area under the curve, and Figs. 3.5(b) and (d) show the relative absorption due to the different pigments in proportion to their actual concentration in a typical leaf. The dominant pigments, the chlorophylls, account for almost all the absorption in the red, and much of that in the blue. The carotenoids (including the yellow pigments lutein and β-carotene) and the xanthophylls (violaxanthin and zeaxanthin) extend absorption into the blue–green. Other structural components of leaves including lignins, proteins, and so on, also have characteristic spectral signatures that contribute to the net result, especially in the shortwave infrared. The difference between the total absorptance of an intact leaf and the sum of absorption by the different pigments, as shown in Fig. 3.5, is partly caused by the presence of other absorbers in the leaf (proteins, etc.), partly by structural effects (see below), and also partly by changes in pigment characteristics caused by removal from their natural pigment–protein complexes in the leaf and assay in an organic solvent.

Although the spectral properties in the visible of healthy leaves are dominated by chlorophyll, as leaves senesce during autumn or as a result of environmental stress, chlorophyll concentration decreases and the effects of pigments that absorb primarily in the blue, and/or green, such as carotene and the xanthophylls (giving orange and yellow colours) and anthocyanins

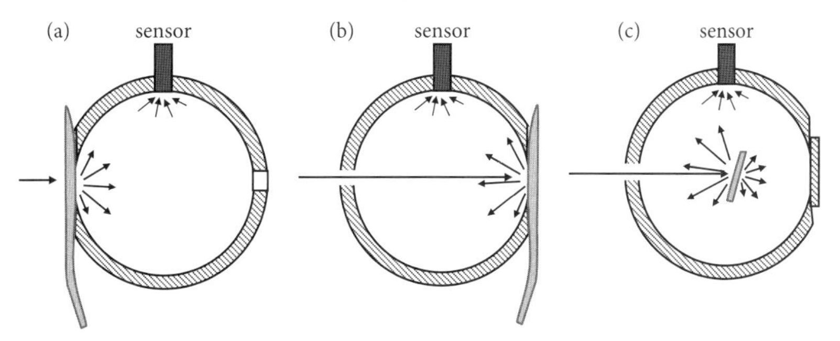

Fig. 3.4. Use of an integrating sphere for the measurement of leaf optical properties. The inside of the sphere is covered with a highly reflective surface (e.g. $BaSO_4$) so that all scattered radiation within the sphere eventually is collected by the detector. (a) This arrangement is used for measurement of directional-hemispherical leaf transmittance (τ_λ), (b) for measurement of directional-hemispherical leaf reflectance (ρ_λ), and (c) for measurement of directional leaf absorptance (α_λ); alternatively α_λ may be obtained from (a) and (b) by use of the relationship $\alpha_\lambda = 1 - \rho_\lambda - \tau_\lambda$.

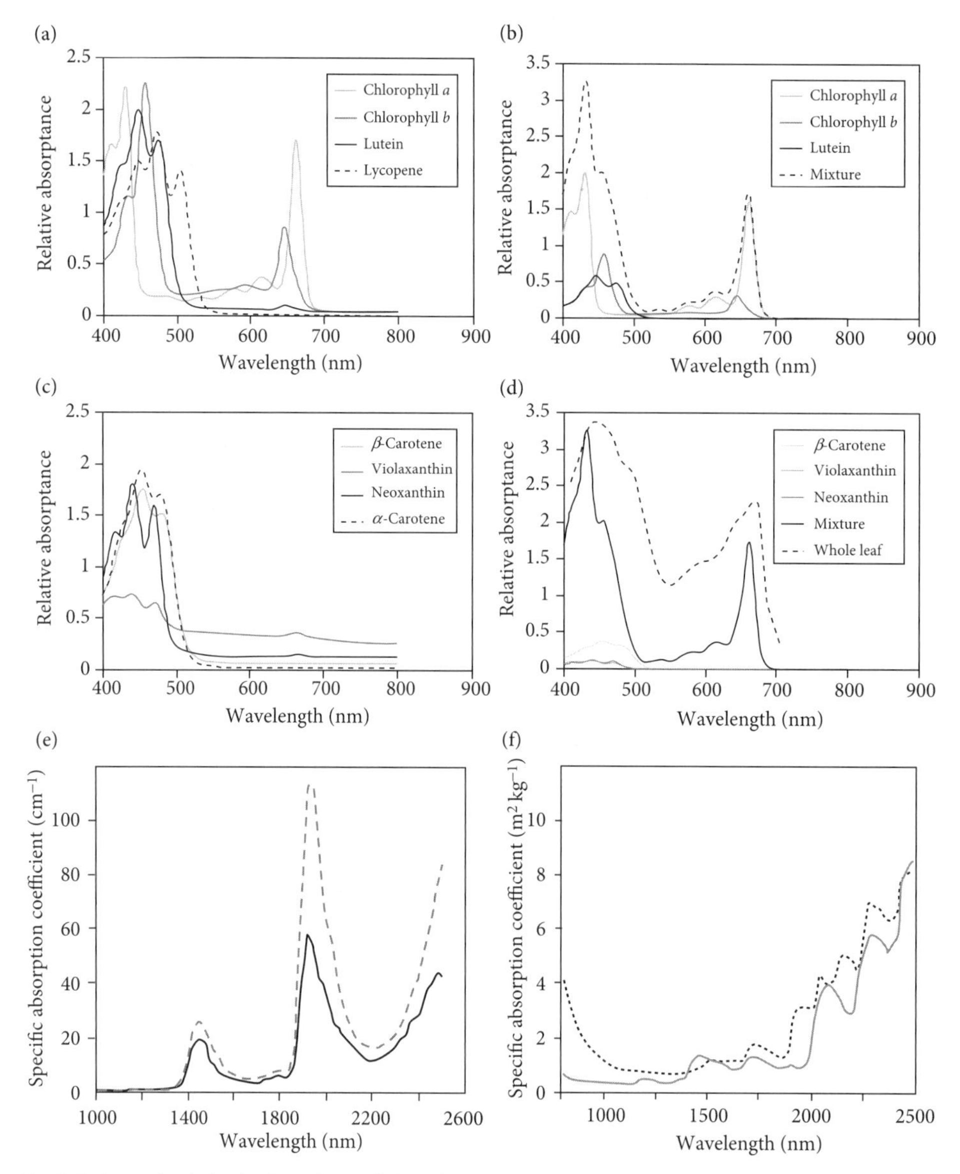

Fig. 3.5. Spectral variation in absorption coefficients for a number of key pigments and other components of leaves: (a, c) absorption by the most important leaf pigments (chlorophyll-*a*, chlorophyll-*b*, lutein, carotene, xanthophylls, and lycopene) measured in acetone extracts and scaled so that total area under the curves are equal (data courtesy of C. Buschmann); (b, d) relative absorption of the different pigments scaled according to their typical concentrations in a leaf, together with the sum of the absorption by the various extracted pigments and the spectral absorption of the leaf from which extracted (courtesy C. Buschmann); (e) specific absorption coefficients for leaf water (data from Jacquemoud and Baret, 1990) and for pure water (- - - -) (data from Curcio and Petty, 1951); (f) specific absorption coefficients for leaf protein (. . . . . . .) and cellulose + lignin (——) (redrawn from Jacquemoud *et al.*, 1996).

(giving predominantly red colours) become more apparent, for example as the developing autumn colour of deciduous trees. Differences in pigment composition can be used as a powerful tool for species discrimination on the basis of leaf colour.

In contrast to the situation in the visible region, water is the dominant chemical contributor to the observed radiative properties in the infrared, where most pigments do not absorb significantly. Water only absorbs strongly at wavelengths longer than about 1100 nm

(Fig. 3.5(e)) with characteristic absorption bands in the mid-infrared at c.1450 nm, 1950 nm, and 2500 nm. The low, but finite, absorptance in the near infrared is interesting as neither chlorophyll nor water contributes to this and it is assumed to be more dependent on structure and on some other component in the cell walls. As one would expect, increasing the water content per unit area, as a result of increased leaf thickness or, water content per unit area, leads to increasing absorption in the water bands with a corresponding reduction in mid-infrared reflectance (Fig. 3.3(a); see also p. 183). Proteins, cellulose, and lignin all contribute to radiation absorption in the infrared (Fig. 3.5(f)).

The most important absorption bands in spectra of plant leaves and their chemical origins are summarized in Table 3.1. When referring to absorption by biochemical components we can quantify this as a *specific absorption coefficient* (a_i). This is defined as the absorption of energy flux per unit incident flux of radiation (at a given wavelength) per unit thickness of an infinitely thin layer of absorbing medium. When the concentration of the material of interest is expressed in kg m^{-3}, the units would be m^2 kg^{-1}, though for water it is usual to use units of cm^{-1}. The absorption bands at shorter wavelengths (<800 nm) are primarily determined by electronic transitions; for example the process of photosynthesis depends on the excitation by absorbed radiation of electrons in pigments such as chlorophyll from their ground energy state to higher-energy orbitals. The utilization of this trapped energy in subsequent electron transfer is what drives photosynthesis. At longer wavelengths, where the energy content per quantum is lower, absorption tends to be explained by rotation and stretching of chemical bonds between light atoms (C, O, H, N) resulting in the characteristic broad absorption features due to water, cellulose, protein, and lignins (e.g. Baret, 1995; Curran, 1989). Note that when the absorption spectra are measured at low temperature (e.g. in liquid N_2 at 77 K) there is a marked sharpening and a detectable increase in the structure of the bands. The pattern of spectral reflectance for leaves, with its sharp transition at c. 700 nm (the *red-edge*), contrasts with the relatively smooth change of soil reflectance over the same spectral region (see Section 3.1.2). The characteristic absorption and hence reflectance features shown by different constituents of leaves can provide a very powerful means for obtaining information on leaf biochemistry; statistical regression approaches have been shown to identify critical wavelengths from reflectance and transmission spectra that can then be used to estimate contents of at least chlorophyll, protein, cellulose, and lignin. The results are generally found to be most robust using dry material and optically thick samples (stacked leaves) (e.g. Jacquemoud *et al.*, 1995; Kokaly and Clark, 1999), though clearly this is not relevant to the data obtained in many remote-sensing situations.

In the laboratory at least, the information on chemical composition of plant leaves can be enhanced by extending measurements into the thermal infrared

Table 3.1 Absorption features in visible and near IR related to leaf components (data from Curran, 1989). Less significant bands are shown in parentheses and the main electronic transitions or vibrations are indicated. Note that many of the absorption bands in the mid-infrared are subject to substantial atmospheric absorption and so have rather limited use for remote sensing, being of greater value in laboratory or close-field situations.

Wavelength (μm)	Chemical	Electronic transition or bond vibration
0.43, 0.46, 0.64, 0.66	Chlorophyll	Electronic transitions
0.97, 1.20, 1.40, 1.94	Water	O–H bond stretching
1.51, 2.18 (0.91, 1.02, 1.69, 1.94, 1.98, 2.06, 2.13, 2.24, 2.30, 2.35)	Protein, nitrogen	N–H stretching and bending, C–H stretching
2.31 (0.93, 1.02)	Oil	C–H stretching and bending
1.69 (1.12, 1.42, 1.94)	Lignin	C–H stretching
1.78	Cellulose and sugar	

range using ATR (*attenuated total reflectance*) spectroscopy that makes use of complex and specific absorption features related to organic constituents (see Ribeiro da Luz, 2006) with spectra that are very characteristic of particular species. This technique involves placing the sample in contact with a crystal having a high index of refraction and using a thermal infrared beam incident at an angle greater than the critical angle, so it is totally reflected; the attenuation of this totally reflected radiation by the sample is characteristic of the crystal.

Dependence on leaf structure

Reflection from leaves (or absorption) is a more complex process than one might at first sight imagine; this is because a simplistic single layer of homogeneous absorber only poorly approximates the behaviour of real leaves. Although one intuitively imagines reflection as occurring at the surface of the leaves, in reality only a proportion of incident radiation is reflected at the upper cuticle surface with the remainder entering the leaf and being either transmitted, absorbed by chlorophyll or other pigments, or scattered, especially at air/water interfaces at the surface of cells (Fig. 3.6) or at other sites where there are substantial discontinuities in refractive index. This scattering process at the interfaces within leaves is known as *volume scattering* and it depends on the differing values of refractive index for components within leaves, varying from 1 for air at 1 μm to about 1.33 for water and about 1.4 for hydrated cell walls (Gausman *et al.*, 1974). The arrangement of cells and of the air/water interfaces is crucial in determining their influence on leaf reflection, particularly in the infra-red, and in determining the angular dependence of reflection properties (see Chapter 8 for more detail on the implications for canopy reflectance). Nevertheless, for present purposes we will initially assume that leaves behave as perfect scatterers (that is their reflectance is diffuse or Lambertian).

The optical boundaries created by abrupt changes in refractive index not only reflect light but can change the direction of light by refraction. Such boundaries include the surface of the leaf and the interfaces between cells and intercellular spaces. A substantial fraction of reflection, especially in the infrared, occurs at the upper cuticular surface. In leaves with a hairy upper surface, the additional air/water interfaces can also lead to an increased reflectance in the photosynthetically active region (400–700 nm); increasing the value from around 14% to as

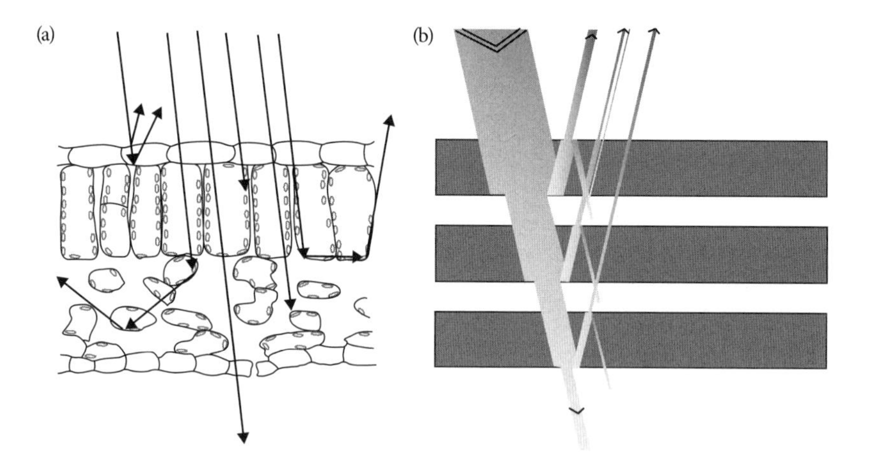

Fig. 3.6. (a) Schematic cross-section of a typical dicotyledonous leaf (cotton) illustrating diagrammatically the diversity of paths for radiation interaction with the leaf. A substantial proportion of incident visible radiation is absorbed by chloroplasts but a large fraction of the infrared is reflected or scattered either at the surface or especially at the interfaces between cells and intercellular spaces. The upper layer of densely packed photosynthetic cells represents the palisade layer while the lower porous layer is known as the spongy-mesophyll. (b) Illustration of the 'plate' model for representing radiation interaction in a leaf. In this approximation the leaf comprises a number (n) of individual homogeneous chlorophyll-containing absorber plates separated by an airspace, with reflection/diffraction occurring at the air/cell interfaces (first- and second-order interactions only are shown). Increasing numbers of discrete layers can be used to simulate an increasing proportion of spongy-mesophyll.

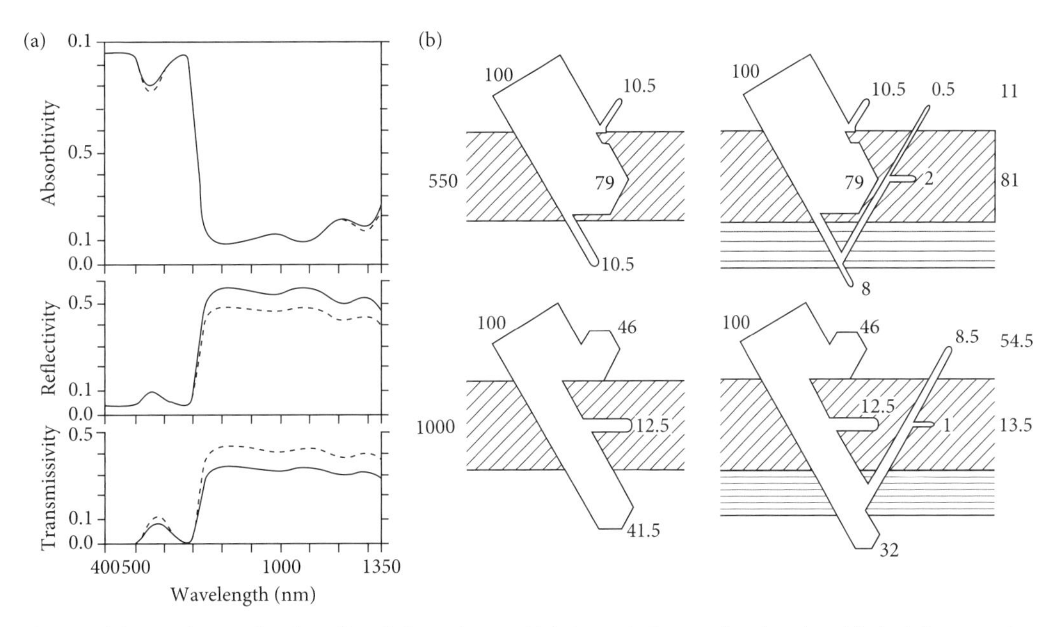

Fig. 3.7. (a) Spectral properties of *Tussilago farfara* L. leaves with hairs on the lower surface (——) or with the hairs removed (- - - - -); (b) diagrammatic illustration of the radiation fluxes at 550 nm (top) and 1000 nm (bottom) without or with the lower hairs (Eller, 1977). This illustrates the much greater effect of hairs in the infrared than in the visible.

much as 70% (Ehleringer *et al.*, 1976). Even hairs on the lower surface can substantially enhance reflectance, as shown in Fig. 3.7, by returning some initially transmitted rays back into the leaf.

Modelling leaf optical properties – PROSPECT and LIBERTY

As we pointed out above, the majority of reflection from leaves occurs at the interfaces where there are discontinuities in refractive index and it is therefore very dependent on leaf structure. This 'structural' component of the overall reflectance complicates determination of the chemical constituents of leaves from observed reflectance or transmission spectra as the shape of the spectra can be substantially modified from that occurring in pure solutions. As can be appreciated from Fig. 3.6(a), radiative transfer within leaves is extremely complex with multiple sites of absorption and scattering. Therefore detailed descriptions of the leaf structure that would allow a complete description of all the radiation paths in a leaf, for example by using ray-tracing methods, would require an unfeasibly large number of parameters in their definition so are of limited use where one is hoping to derive a general method for the extraction of key canopy biophysical parameters from remotely sensed data. Conveniently, however, it has proved possible to develop simple physically based models that approximate real leaves without losing the power to describe their essential properties and that allow one to separate chemical and structural contributions to the overall spectral properties (Jacquemoud and Baret, 1990).

A useful approximation was introduced by Allen *et al.* (1969 and associated papers) who showed that the leaf can be represented by a number of layers of homogeneous absorbing tissue separated by airspaces at which the majority of reflection or diffraction occurs (Fig. 3.6(b)). Increasing the number of layers (N) simulates an increasing proportion of airspaces; in general, monocotyledonous leaves such as grasses tend to have rather compact mesophyll tissue with few airspaces (and hence less scattering), while dicotyledonous leaves (broad-leaved plants) have a dense photosynthetic palisade layer on the upper surface, where most of the chloroplasts are situated, overlying a rather porous spongy-mesophyll tissue (Fig. 3.6(a)). The value of N should be looked on as a parameter rather than as a direct measure of the number of layers and has been generalized so that it can be regarded as a continuous variable in the model PROSPECT (Jacquemoud and Baret, 1990). This model describes the reflectance spectrum of a leaf in terms of only three parameters: a structural parameter, N, a pigment concentration, C_{chl}, and a water content, C_w. These, together with the necessary spectra for refractive index of leaf tissue and for the absorption coefficients of water and leaf pigments, are used to simulate reflectance spectra of different leaves. The approach has been extended to estimate other biochemical components such as carotenoids, anthocyanins, proteins, and lignin, though generally with lesser accuracy (Feret *et al.*, 2008; Jacquemoud *et al.*, 1996), while a comprehensive reworking and updating of the model with improved and extended parameterization has been reported (PROSPECT-5: Feret *et al.*, 2008). The structure of conifer needles and canopies is rather different from that of the flat leaves modelled using PROSPECT, so an alternative model (LIBERTY) has been developed for conifer needles based on radiative transfer between spherical 'cells' (Dawson and Curran, 1998).

The average transmissivity or reflectance of the plate model is relatively unaffected by angle of incidence between 0° (normal) and about 50° (Jacquemoud and Baret, 1990) and decreases as the effective refractive index increases. It is also worth noting that though the reflectance increases as you increase the number of layers (or stack leaves on top of each other) it rapidly reaches an asymptote when N exceeds about 8, though transmission and absorption saturate rather more slowly. This behaviour is illustrated in Fig. 3.8 and simulates well the behaviour of real leaves of increasing thickness, showing that remote sensing of reflectance may not be particularly powerful at discriminating between thick leaves (or canopies) where the signal is saturated.

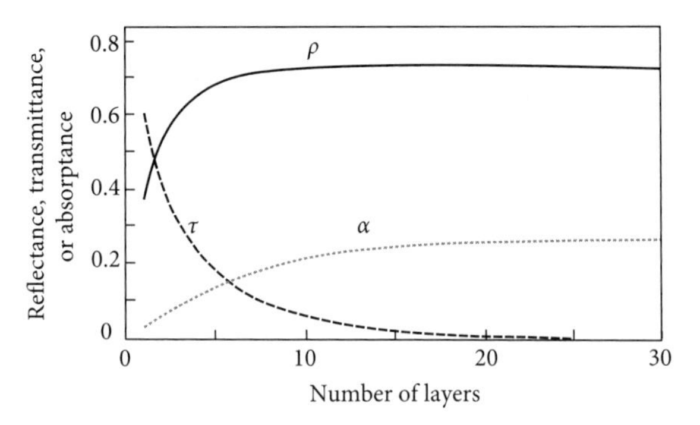

Fig. 3.8. Illustration of the effect of increasing leaf thickness (number of layers) on transmission (τ), absorption (α), and reflection (ρ) from a leaf. It is clear from this that reflection reaches an asymptote much earlier than do transmission or absorptance (redrawn from Jacquemoud and Baret, 1990).

Inversion of models such as PROSPECT and LIBERTY can be used to estimate biochemical composition of leaves. For application at the canopy scale it is necessary to account for radiative transfer in the canopy; this can be achieved by combining a biochemistry model such as PROSPECT with a canopy radiative-transfer model such as SAIL (see Section 8.3.1).

Integrated reflectance and absorptance

Some typical values of shortwave reflectance (ρ_S) and absorptance (α_S) integrated over the shortwave spectrum and of the corresponding reflectance (ρ_{PAR}) for the photosynthetically active or visible wavebands, are summarized in Table 3.2 for single leaves. Note that transmittance (τ)

Table 3.2 Typical broadband reflection (ρ) coefficients for single leaves in the shortwave (S) and photosynthetically active (PAR) regions with typical absorption coefficients (α) in the shortwave (e.g. from Jones, 1992). More detailed information for a range of specific surfaces may be found in Gates (1980); Stanhill (1981); and in a range of websites (e.g. the ASTER Spectral Library, the LOPEX93 and OTTER databases; and the USGS Digital Spectral Library splib06a – see appropriate websites).

	ρ_S	ρ_{PAR}	α_S
Typical crop leaves	0.29–0.33	0.10	0.4–0.6
Conifer needles	0.12	0.08	0.88
Deciduous broad leaves	0.23–0.29	0.10	
White pubescent leaves	0.39	0.20	0.55

can be obtained as $\tau = 1 - (\rho + \alpha)$. The corresponding values for canopies and other land surfaces will be discussed below (Section 3.1.3). It is worth noting that a greater fraction of radiation is absorbed (ρ is smaller) if one considers only radiation within the photosynthetically active wavebands. Remember also that the effective albedo depends on the spectral distribution of the incoming radiation, but is usually reported for sunlight. It is interesting to note that differences between species may often be smaller than differences between samples of the same species (e.g. as a function of leaf age and nutrient status).

Fluorescence

A small proportion of the radiation emitted from leaves and other surfaces is not strictly reflection but represents *fluorescence* or *luminescence*, though fluorescence usually represents no more than 2–5% of the emitted energy even within the narrow fluorescence emission bands (Meroni *et al.*, 2009). Fluorescence is the rapid re-emission of absorbed radiative energy, usually at characteristic wavelengths corresponding to specific energy transitions in the pigments. This re-emission usually occurs within about 10^{-9} s, though some excited molecules decay by a process of *delayed fluorescence* (or *luminescence*) that has a half-life of the order of seconds. Fluorescence always occurs at longer wavelengths (lower energy) than the exciting wavelengths with the difference in energy being typically lost as heat. In photosynthetic organisms, as we shall see in Chapter 4, there are three competing processes by which excited chlorophyll molecules can lose their excitation and return to the ground state: these are reradiation as fluorescence, dissipation as heat, and by the coupling of energy dissipation to chemical reactions in *photochemistry*, through stimulation of electron transport. Fluorescence from chlorophyll primarily arises from photosystem II in the chloroplasts with emission peaks in intact leaves around 690 nm and 735 nm (Fig. 3.9). This figure also shows how the chlorophyll emission spectrum shifts to longer wavelengths as the concentration of chlorophyll increases. This shift is more apparent than real and results from the increased reabsorption of the shorter-wavelength emissions by the chlorophyll as its concentration increases. Many other leaf pigments can also give rise to characteristic fluorescence emission spectra, and, as we shall see in Chapter 11, this information can be used to provide useful diagnostic information on plant stresses.

3.1.2 Radiative properties of soil and water

For remote-sensing purposes, all radiation incident on soil is either absorbed or reflected. Although the absolute reflectance may be variable between soils, they all tend to show a rather smooth increase in reflectance from the visible to the near infrared (Fig. 3.10); a pattern that contrasts markedly with the very sharp increase around 700 nm generally observed for vegetation. This difference provides the basis for much spectral remote sensing as we shall see in Chapter 7. As is illustrated in Figs 3.11 and 3.12, the detailed spectral reflectance of soils can be modified substantially by a range of chemical and physical characteristics, with the following being the most important

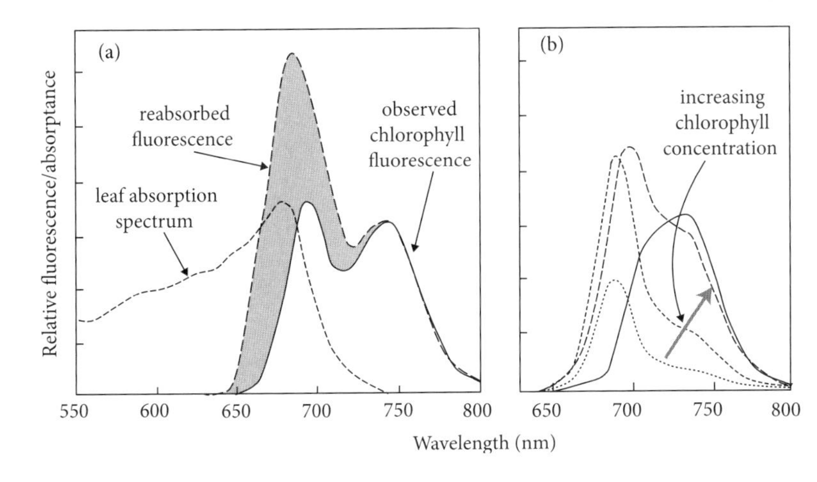

Fig. 3.9. (a) Illustration of chlorophyll fluorescence reabsorption (shaded area) by leaf pigments that absorb the shorter fluorescence wavelengths (based on data for an *Ulmus* leaf from Lichtenthaler and Rinderle, 1988), (b) fluorescence emission spectra for suspensions of isolated chloroplasts with the chlorophyll content ranging from 3, through 13, 51, and 153 μg chlorophyll/ml showing the initial increase in fluorescence followed by the successive quenching by reabsorption of the shorter fluorescence wavelengths (data from Lichtenthaler and Rinderle, 1988).

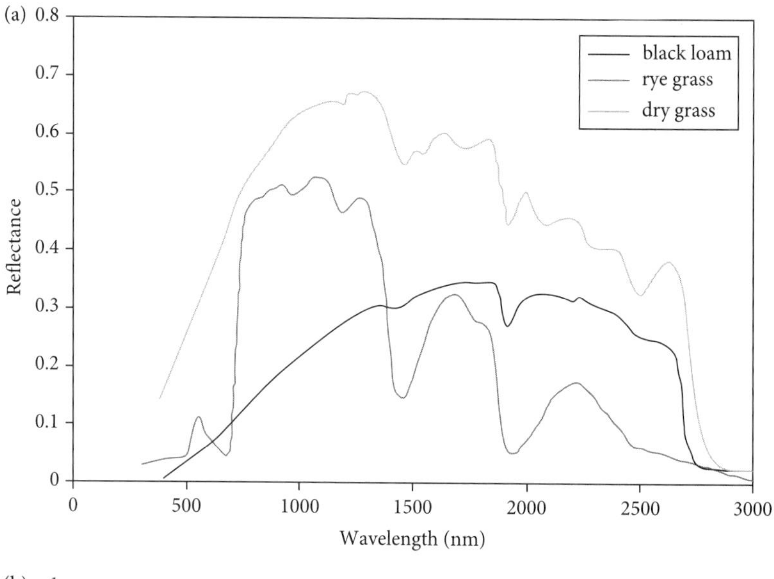

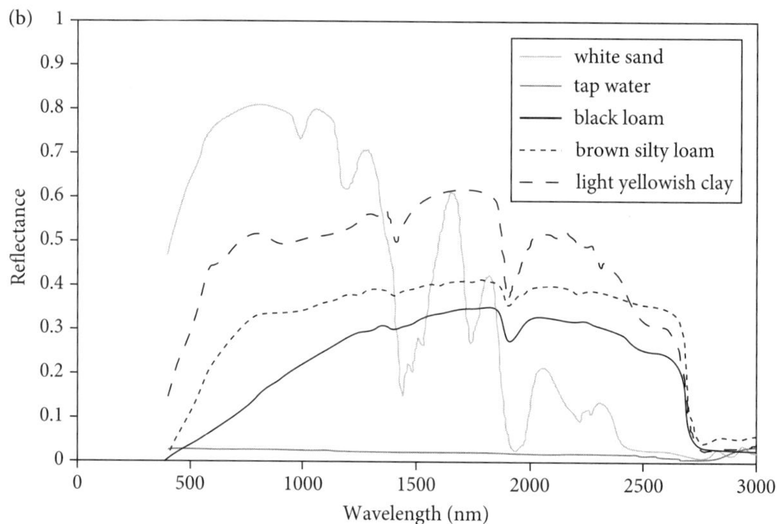

Fig. 3.10. (a) Illustration of typical spectral reflectance characteristics of soil as compared with vegetation, showing the reflectance spectrum for a rye-grass sward, with the sharp increase at c.700 nm, and a black-loam soil, together with the reflectance spectrum for a desiccated grass sward illustrating the loss of the sharp transition at 700 nm; (b) spectral reflectances of a range of different soils and water over the 400–3000 nm waveband (data from ASTER spectral library; Baldridge *et al.*, 2009).

(in approximate order of decreasing importance): (i) increasing soil-moisture content decreases the reflectance in the water absorption bands due to water absorption and in the remaining wavebands due to internal reflections within the water film covering the soil particles; indeed we readily recognize wet soils from the fact that they appear darker (less reflective) than dry soils, (ii) increasing organic content also gives darker (less reflective) soils, (iii) changing soil texture from clay to sand increases reflectance, (iv) decreases in surface roughness, as occur for example with the development of a soil crust (Eshel *et al.*, 2004), slightly increase reflection, (v) increasing iron oxide content gives many soils their characteristic brick red colour, which implies an increased reflection in the red and a decrease in the green (Ben-Dor, 2002).

A feature of many soils is that there is commonly a close relationship between the red and near infrared reflectance over a wide range of mineral soils and soil-moisture contents: this is known as the *soil line.* Peaty soils, however, tend to have greater near infrared reflectance at any value of red reflectance (Fig. 3.11). When soils are dry it is possible to distinguish specific soils and minerals by using characteristic spectral features arising from specific mineral and organic components (Adams and Gillespie, 2006).

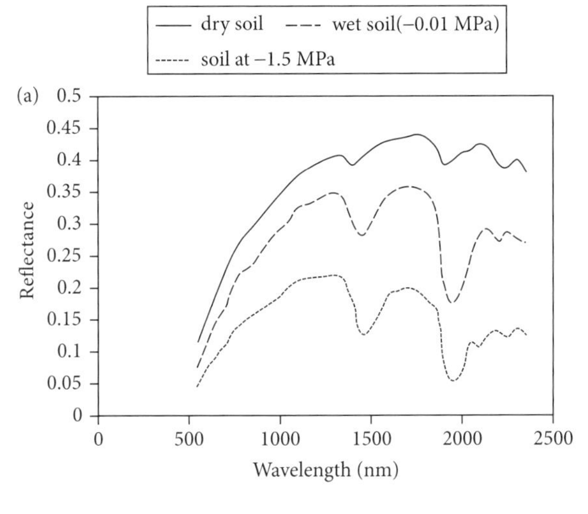

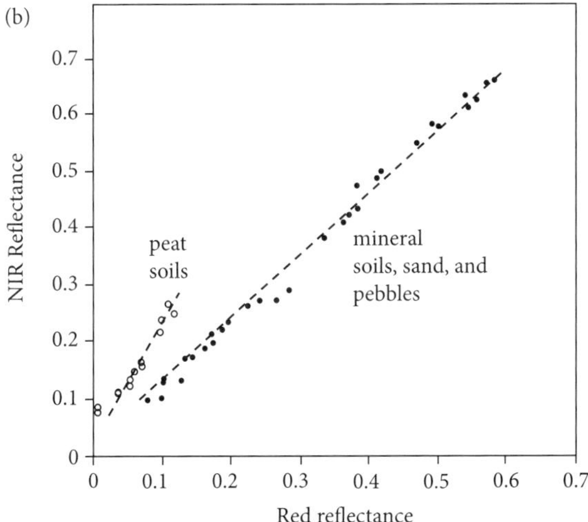

Fig. 3.11. (a) Dependence of soil spectral reflectance on soil-moisture content for a Typic Hapludalf soil, showing the increasing absorption bands in the near to mid-infrared as water content increases from wet soil (soil moisture tension = −0.01 MPa) through soil at 'wilting point' (soil moisture tension = −1.5 MPa) (redrawn from Baumgardner *et al.*, 1985), (b) relationships between R and NIR reflectances for different soils as water content changes, showing that the slope of the line differs between organic (peaty) soils and mineral soils (soil reflectance data from Baret *et al.*, 1993).

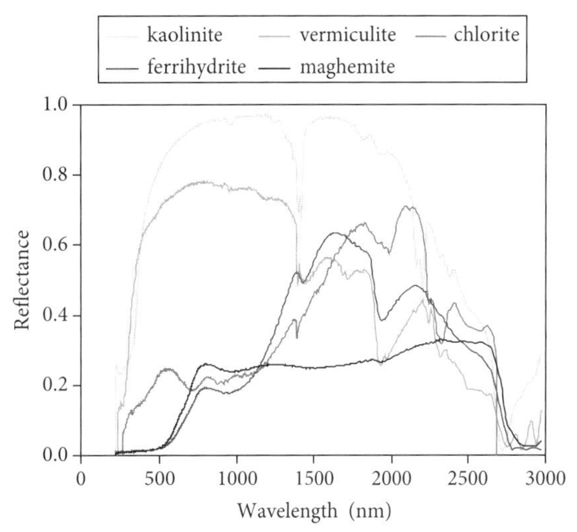

Fig. 3.12. Spectral reflectance of a sample of some contrasting soil minerals (data from Clark *et al.*, 2007; http://speclab.cr.usgs.gov/spectral-lib.html).

The reflectance of most soils is strongly anisotropic and varies as a function of the angle of view in relation to the angle of illumination, as shown in Fig. 3.13, with the greatest reflection usually occurring back at the angle of illumination. The pattern of reflection also varies with surface roughness and wetness, with smoother soils having an increased proportion of specular reflection and hence a greater influence also on both the polarization of the reflected light and the directional reflectance (Fig. 3.13). The directional properties of reflectance will be developed further in Chapter 8.

In contrast to soils and vegetation, pure water tends to reflect a much smaller proportion of incident radiation (Fig. 3.10), and is almost completely transparent to shorter wavelengths of visible radiation, with significant absorption only at wavelengths >1100 nm. Generally, reflectance remains less than about 3% over the whole visible, near and mid-infrared region. The situation is complicated, however, by factors such as the water turbidity, its chlorophyll content, the surface roughness (which has a major impact on the nature of the reflection process – specular or diffuse), and the water depth together with the nature of the substrate below the water body. Increasing turbidity markedly affects absorption and reflection with suspended sediments increasing reflectance (and reducing transmission) in the visible (especially the red, which is why water with suspended silt appears brownish). High concentrations of chlorophyll, as in phytoplankton, increase reflectance in the green (but decrease it in the blue).

3.1.3 Radiative properties of canopies

Reflection of radiation from vegetated systems depends both on the radiative properties of the individual components of the vegetation (leaves, stems,

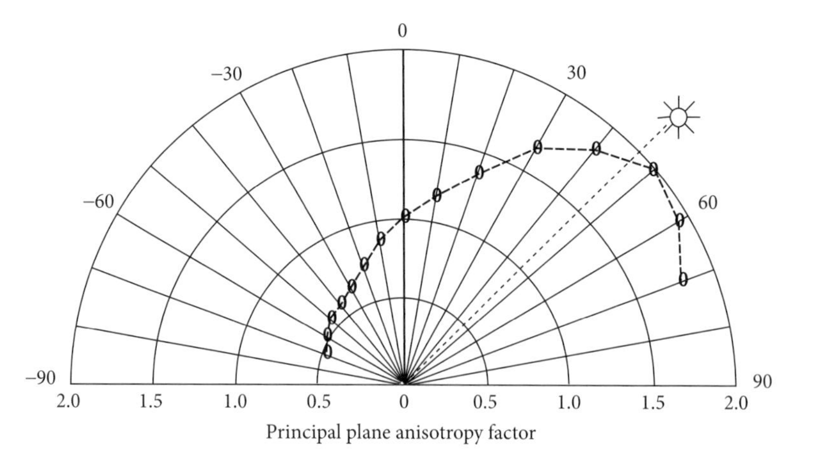

Fig. 3.13. A polar plot showing the anisotropy of directional reflectance in the near infrared along the solar principal plane above a fine loamy soil of medium roughness, illustrating the non-Lambertian character of bare soil, with highest reflectance in the direction of the sun (indicated by the dashed line and symbol) (data from Irons *et al.*, 1992).

soils, water, etc.) and also importantly on the detailed canopy architecture or spatial organization in relation to the angular distribution of the incident radiation and the orientation of the sensor. We have already touched on some of the consequences of these effects; here we will concentrate on presenting the basic principles underlying radiation propagation in vegetation, noting that for simplicity it is common to ignore the contributions made by stems, even though they may be quite substantial contributors to canopy radiative properties. We will leave a detailed consideration of the orientation-dependent behaviour for Chapter 8.

The most important plant characteristic determining radiation absorption and transmission by canopies is the *leaf-area index* (*LAI*, abbreviated to *L* when used in equations). This is usually defined as the 'one-sided' leaf area per unit ground area, though as pointed out by Campbell and Norman (1998), it is more generally useful to use the alternative definition of a *hemi-surface area index* (*HSAI*, defined as half the actual surface area of leaves per unit ground area). Although *HSAI* and *LAI* are equivalent for flat leaves, they differ substantially for conifer needles and some other plant parts. Campbell and Norman (1998) have argued persuasively that *HSAI* is a more generally useful measure of leaf area than *LAI* in the description and modelling of the properties of canopies with complex leaf shapes. Nevertheless, the two terms can be used interchangeably in much of what follows.

For many purposes, especially in studies of photosynthesis and canopy energy balance, one is primarily interested in determining how much radiation is absorbed by a canopy. This quantity can be derived from either the reflected or the transmitted radiation. The traditional approach used in plant physiological and agronomic studies has been to derive this information from measurements of radiation transmission through the canopy to the ground (using instruments such as the LiCor 2000, Sunscan, Sunfleck Ceptometer, subcanopy hemispherical photography, etc., as described in Chapter 8). These measurements of radiation transmission are inverted on the basis of an appropriate model for radiation transmission through plant canopies to derive the canopy variables such as leaf-area index and leaf-angle distribution (*LAD*). The remote-sensing approach aims to derive the same information from what is reflected by the canopy rather than what is transmitted, but is dependent on the same underlying physical principles.

Some of the many possible trajectories of radiation scattering and penetration in a plant canopy are illustrated in Fig. 3.14; these processes have substantial similarities to the volume scattering processes occurring within leaves (Fig. 3.6), but differ in scale. The incident sunlight may either be directly reflected back to the sky from a leaf (A), with a small fraction transmitted through the leaf, or else it may be involved in secondary (B), or even tertiary (C) reflections before finally being reflected back to the sky. Similarly, some of the reflections may involve the soil (D). Since some energy is absorbed at each scattering event, the intensity of radiation decreases with increasing numbers of interactions. Because much of the light reflected from a canopy has undergone more than one reflection, it follows that the overall reflectance of a dense canopy is usually substantially less than the reflectance measured for a single leaf (compare Tables 3.2 and 3.3). As leaf-area index decreases, however, the

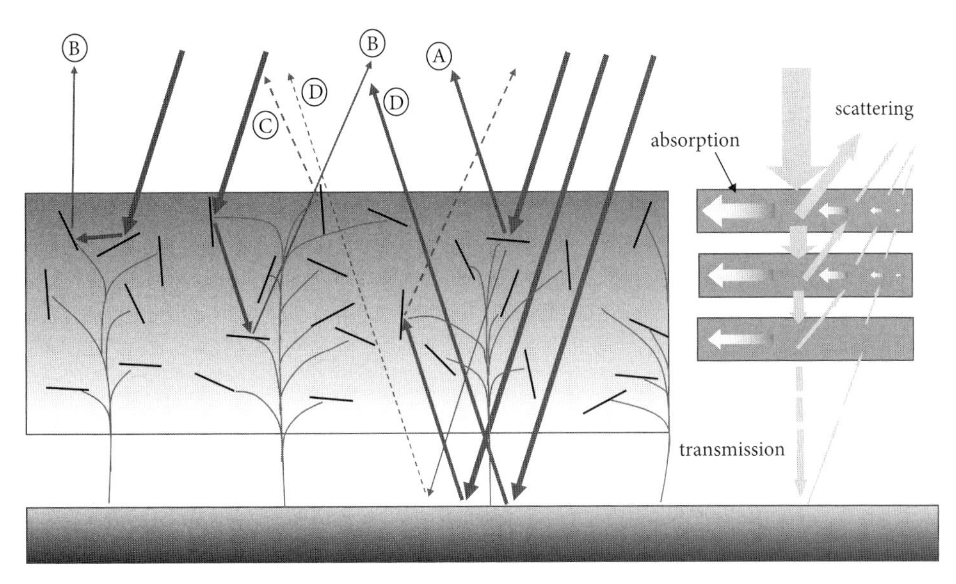

Fig. 3.14. (a) Illustration of some of the possible interactions of radiation with a plant canopy with randomly oriented leaves, showing the multiple scattering events. The incident sunlight may either be directly reflected back to the sky from a leaf (A), with a small fraction transmitted through the leaf, or else it may be involved in a secondary (B), or even tertiary (C) reflection before finally being reflected back to the sky, Similarly some of the reflections may involve the soil (D). (b) Illustrates a notional simplification of the real canopy, where the canopy is treated as a set of thin layers, each treated as infinitely thin, where the downward radiation is attenuated by absorption and scattering at each layer, while the upward radiation flux is the sum of all the upwardly scattered radiation.

Table 3.3 Approximate values for solar reflectivity or albedo for different land surfaces. These values can only be assumed to be approximate as they vary substantially with stage of the vegetation, with illumination conditions, view angle, and with canopy structure.

	ρ_S	Comments	Ref.
Crops		Generally lower than corresponding values for single leaves	
Wheat	0.20 (±0.08)		1, 2
Cotton	0.20 (±0.08)		1
Maize	0.19 (±0.04)		1, 2
Orange	0.16 (±0.05)		1
Grassland	0.17-0.26		2, 3
Sugar beet	0.18		2
	0.16–0.26		

(Continued)

Table 3.3 (Continued)

	ρ_S	Comments	Ref.
Trees and forest			
Deciduous forest	0.10–0.20		2
Aspen (BOREAS)	0.156 (0.116 bare)	0.214 (with snow)	4
Grass (BOREAS)	0.197	0.75 (with snow)	4
Coniferous forest	0.05–0.15	Jack pine ρ doubles with snow	2, 4
Spruce/poplar (BOREAS)	0.081	0.108 (with snow)	4
	0.05–0.20		
Other natural vegetation			
Tundra	0.15–0.20	In winter can be 0.80 with fresh snow	2
Desert	(0.1) –0.35–(0.7)	Dependent on moisture content and mineral content	2, 5
Rainforest	0.10–0.13		2, 7
Other surfaces			
Water	0.02–0.05 (sun overhead)	increases rapidly to c. 0.90 with specular reflection at low angles; increases with suspended matter	6
Dry soil	0.13–0.18		2
Wet soil	0.08–0.10		2
Sand (dry, white)	0.35		2
Snow (fresh)	0.75–0.95		2
Snow (old)	0.40–0.70		2

References: 1. Stanhill (1981); 2. Campbell and Norman (1998); 3. Ripley and Redman (1976); 4. Betts and Ball (1997); 5. Bowker (1985); 6. Campbell (2007); 7. Monteith and Unsworth (2008)

canopy reflectance tends to that of the underlying soil (Fig. 3.15). The broadband reflectance of shortwave radiation by a canopy or other surface is commonly referred to as its albedo (ρ_S), though more rigorous definitions will be presented in Chapter 7.

Because of the differing reflectance of leaves in the visible and near infrared wavelengths, it also follows that the relative attenuation in these wavebands will depend on the number of reflections at leaf surfaces. Radiation becomes enriched in the infrared both as the number of reflections at leaf surfaces increases and with increasing depth in the canopy.

It should be noted that, because the canopy albedo depends critically on canopy structure, it also follows that albedo changes as a function of time of day as the solar angle changes and as a function of cloudiness as the proportion of diffuse and direct radiation changes. The values presented in Table 3.3 emphasize

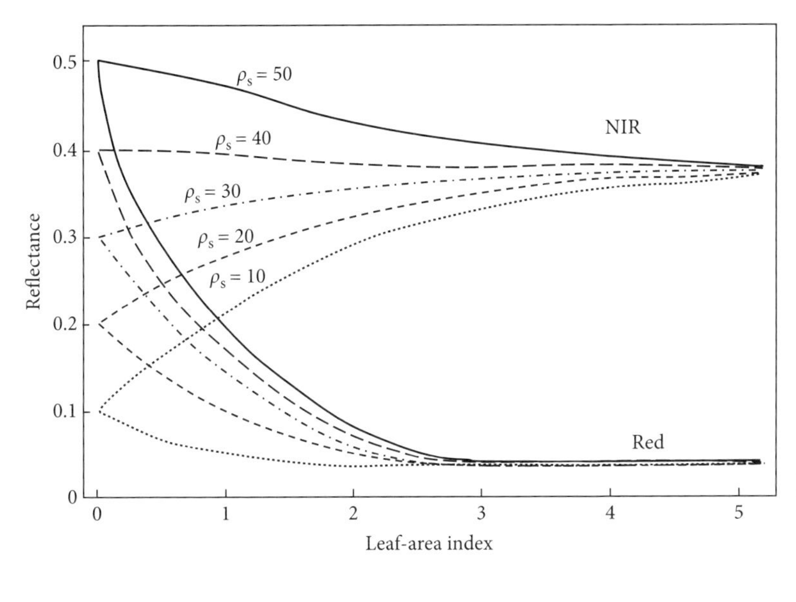

Fig. 3.15. Simulated influence of soil brightness on the relation between canopy reflectance in the red and NIR wavebands and the canopy leaf-area index (modified from Leblon, 1990). The simulations were done using the SAIL (Verhoef, 1984) radiative-transfer model.

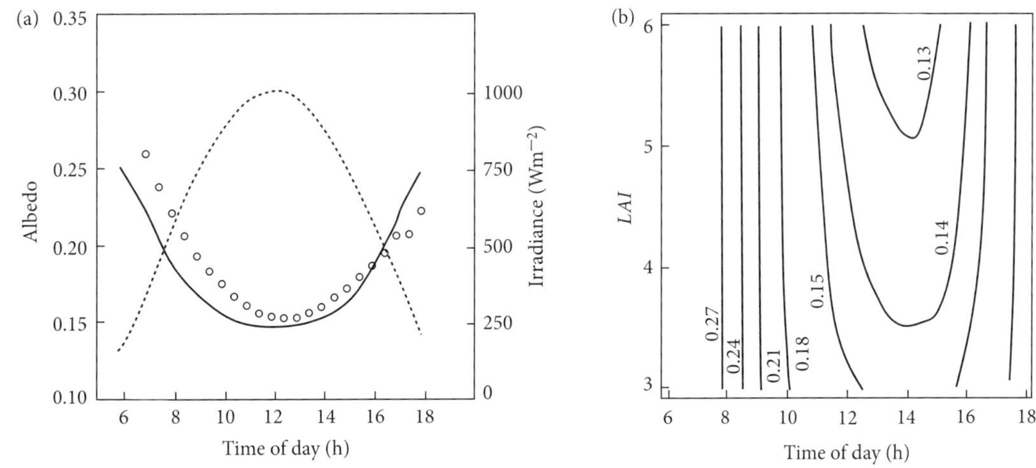

Fig. 3.16. (a) Diurnal trend in actual (o) and simulated (——) albedo for a wheat crop with LAI = 3.0. Solar irradiance (. . . . .). (b) Simulated diurnal trends in albedo for wheat canopies over a range of LAI subject to wind from the east forcing leaves to recline at the standard angle of 30° to the west (redrawn from Song, 1998). The lines represent constant albedos as indicated by the numbers.

the typical diurnal average albedo observed for different surface types. An example of diurnal variation is illustrated for a cereal canopy in Fig. 3.16, where we see that the albedo may vary from near 0.15 around midday to 0.30 or more near sunrise and sunset. A similar diurnal range has been reported for other species such as oak forests (Rauner, 1976). Figure 3.16 also shows that wind direction, which modifies the leaf orientation, can affect the albedo, leading to diurnal asymmetry if the wind is blowing consistently from an appropriate direction. There can also be a slight effect of albedo increasing as the ratio of direct to diffuse radiation increases (e.g. Betts and Ball, 1997). Typically the visible albedo for Canadian grassland during summer may be around 0.076 (range with varying climatic conditions being 0.055–0.097) with a corresponding NIR albedo of around 0.225 (0.21–0.25) (Wang and Davidson, 2007). The strong decrease in soil reflectance as it gets wet (Fig. 3.11) can have a substantial effect on overall canopy albedo for sparse canopies. Typical ranges of albedo for some natural surfaces are summarized in Fig. 3.17.

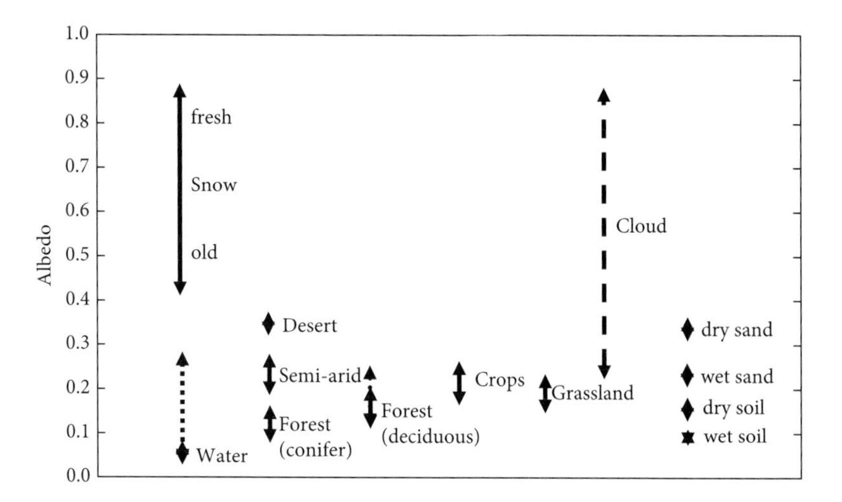

Fig. 3.17. Typical ranges of average daily albedo for a range of natural surfaces.

Analytical treatment of radiation transmission by canopies

In what follows we develop the basic theory underlying the study of radiation scattering and transfer in plant canopies. Although the radiative properties of real plant canopies are rather complex, there are a number of convenient approximations that allow us to make progress with an analytical treatment of radiation transfer in canopies (Campbell and Norman, 1998; Jones, 1992; Monteith and Unsworth, 2008; Ross, 1981).

(i) Homogeneous canopy

The simplest approximation is to assume that the canopy is homogeneous, that is that the absorbing components are evenly distributed and small in comparison with the size of the canopy. Where the absorbers are black, so that the radiation is absorbed and not scattered, radiation is attenuated with depth reducing exponentially according to Beer's Law as was introduced in eqn (2.11):

$$\boldsymbol{I}_z = \boldsymbol{I}_0\, \mathrm{e}^{-k\,z}, \tag{3.1}$$

where $\boldsymbol{I}_z$ is the radiant flux density at any level, z, in the canopy, $\boldsymbol{I}_0$ is the radiant flux density at the canopy surface, k is an *extinction coefficient* (or *attenuation coefficient*) that determines the rate of attenuation and z is the distance travelled in the medium (e.g. depth in the canopy).

One approach to the derivation of this relationship is to consider the canopy as being composed of a large number of infinitely thin layers as illustrated in Fig. 3.14(b), but to assume that the components are non-reflecting perfect absorbers (black). For any one of these layers of thickness dz, the fractional change in flux density across the layer ($\mathrm{d}\boldsymbol{I}/\boldsymbol{I}$) is given by

$$\mathrm{d}\boldsymbol{I}/\boldsymbol{I} = -k\,\mathrm{d}z, \tag{3.2}$$

where the minus sign implies that flux density decreases with increasing distance, z. Integration of eqn (3.2) over depths from 0 to z, gives eqn (3.1). This exponential relationship works well in the atmosphere and in water where the effective particle size of the attenuators is small. In plant canopies, however, the relatively large size of the individual leaves leads to it being only approximate.

In such a system with perfect absorbers, clearly only single scattering/absorption events can occur, and no radiation would be reflected upwards. In this case the extinction coefficient is equivalent to an absorption coefficient (γ_a = the fraction of incident radiation absorbed per unit thickness, but calculated on the basis of an infinitely thin layer). In real canopies, however, a proportion of the incident radiation is reflected or scattered upwards, and some is scattered forwards or transmitted. This leads to the possibility of multiple scattering, which requires a rather more sophisticated analysis. The basic theory for this was developed by Kubelka and Munk (1931). Again for an infinitely thin layer, the proportion scattered in the reverse direction is defined as the *backscattering coefficient* (γ_s = the fraction of incident radiation scattered in the return direction per unit thickness). As shown in Fig. 3.18, therefore, the sum of the absorption and backscattering coefficients is equal to the extinction coefficient, k, while the fraction transmitted in the z-direction

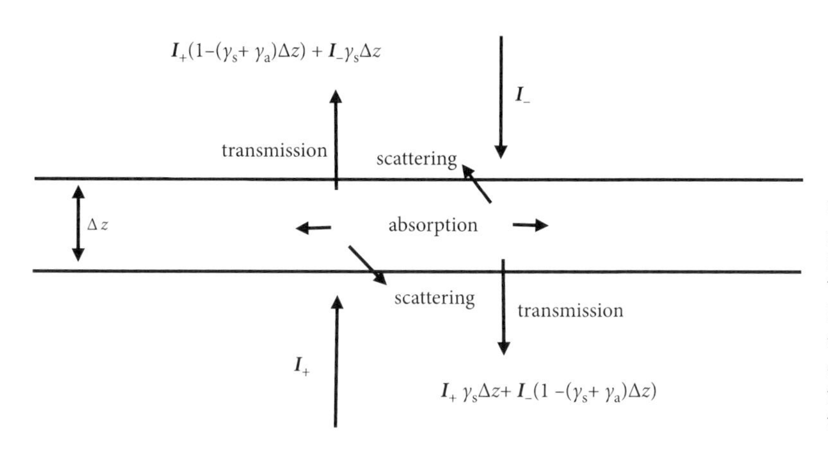

Fig. 3.18. Simple diagram showing scattering and absorption by an infinitesimal layer of thickness Δz. The radiation passing through the thin layer in the +ve direction is the sum of the transmitted radiation in the +ve direction ($= I_+(1-(\gamma_s+\gamma_a)\Delta z)$) and the backscattered radiation originally travelling in the −ve direction ($= I_-\gamma_s\Delta z$).

(forward scattering and transmission) is therefore $1 - (\gamma_a + \gamma_s)$. For a thin layer situated within the canopy it follows (e.g. Rees, 2001) that the net attenuation of radiation propagating in the $+z$ direction (I_+) across the layer is given by

$$dI_+/dz = -(\gamma_a + \gamma_s)\, I_+ + \gamma_s\, I_- \quad (3.3)$$

where the first term on the right represents losses due to absorption and scattering, while the second term on the right represents a gain from the forward scattering of the backscattered radiation stream. This latter term reduces the value of the extinction coefficient when considering the overall radiant flux in one direction.

Solution of this equation allows one to estimate the canopy reflectance. For the extreme case of an infinitely thick canopy where no radiation penetrates to the ground, this equation can be solved to give, for the forward flux in eqn (3.1), an appropriate extinction coefficient

$$k = \sqrt{\gamma_a^2 + 2\gamma_a\gamma_s}. \quad (3.4)$$

The reflection coefficient (ρ) is given by

$$\rho = (\gamma_a + \gamma_s - \sqrt{\gamma_a^2 + 2\gamma_a\gamma_s})/\gamma_s. \quad (3.5)$$

For 'black' canopy components these reduce, respectively, to $k = \gamma_a$, and $\rho = 0$.

(ii) Horizontal leaves

For the simple case where the canopy is assumed to consist of randomly arranged horizontal leaves one can again notionally divide the canopy into a number of horizontal layers each containing equal small areas of non-overlapping leaves (dL). This gives us an alternative way of envisaging how radiation interception and radiation attenuation in canopies are related.

By analogy with the Beer's law derivation given above, if one assumes that the area of these small and opaque leaves in any layer (dL) is small in relation to ground area, the probability that any ray of light will be intercepted by that layer is ($k'\,dL$), where k' is a *shape factor* relating the shadow area on a horizontal surface to the (one-sided) leaf area. For a horizontal canopy the shadow area equals the leaf area, whatever the orientation of the incident radiation. In this case, therefore, $k' = 1$ and $k'\,dL = dL$. The probability that it will be transmitted is $(1 - k'\,dL)$. The probability that this transmitted light will then be transmitted by a second layer is now $(1 - k'dL)^2$, and the probability of transmission after N layers is given by the binomial $(1 - k'\,dL)^N$. For limitingly small leaves the fraction transmitted below N layers can be approximated as

$$I/I_0 = (1 - k'\,dL)^N \approx e^{(-N k\, dL)} = e^{-k'L}, \quad (3.6)$$

which for a horizontal leaved canopy reduces to

$$I/I_0 = e^{-L}. \quad (3.6a)$$

Although superficially similar to Beer's Law, in this case the irradiance at any level within or below the canopy is actually the average of unattenuated areas (sunflecks) and completely shaded areas as illustrated in Fig. 3.19. It follows therefore that for completely opaque horizontal leaves e^{-L} is the fraction of the area on a horizontal plane below leaf area L that is sunlit; conversely the fraction of the plane below the canopy that is shaded is equal to $(1 - e^{-L})$. For many remote-sensing applications we are interested to determine the fraction of the incident radiation that is intercepted by the canopy; this is also given by $(1 - e^{-L})$;

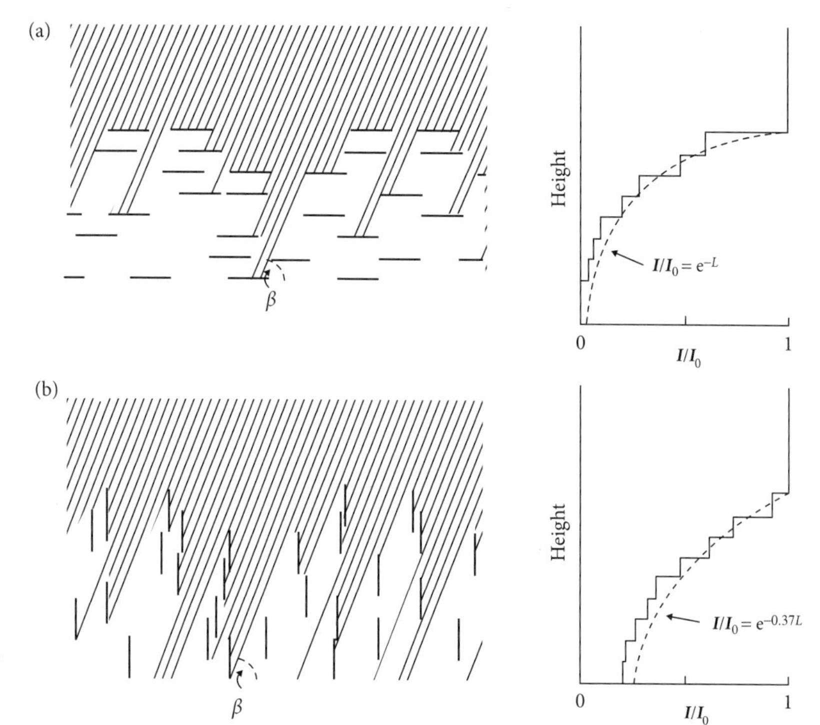

Fig. 3.19. Illustration of attenuation of radiation from near the zenith (zenith angle, $\theta = 30°$; $\beta = 60°$) in (a) a horizontal-leaved canopy (extinction coefficient = 1) and (b) a vertical-leaved canopy (extinction coefficient = 0.37 for this solar angle), with equivalent leaf-area indices, together with the corresponding profiles of radiation attenuation through the canopy (from Jones, 1992).

for a horizontal leaved canopy this term also equals the total sunlit leaf-area index of the canopy (L_{sun}).

It is straightforward to show (e.g. Monteith and Unsworth, 2008) that the shape factor k' is equivalent to the extinction coefficient k, so we will use the latter symbol from now on. Although the value for the extinction coefficient, k, for black horizontal leaves is unity, this factor allows one to correct for the more typical real situation where some scattering is occurring, or for other leaf angles where the shape factor (the ratio of the shadow area to the actual leaf area) is not unity. Goudriaan (1977) showed that the extinction coefficient should be multiplied by $\sqrt{\alpha}$ where α is the absorptance) to take approximate account of the effect of multiple scattering. He also showed that the reflectance from a deep horizontal leaved canopy (ρ_h) can be approximated by

$$\rho_h = (1 - \sqrt{\alpha})/(1 + \sqrt{\alpha}), \tag{3.7}$$

which approximates the values given by eqn (3.5).

(iii) Other leaf-angle distributions

In practice, real canopies do not have leaves all at one angle and they vary from having predominantly horizontal leaves (planophile) to having predominantly vertical leaves (erectophile); clearly radiation penetration and hence reflectance properties of the two types of canopy are rather different. For canopies where the leaves are not all horizontal, the Beer's law model can be easily extended by adjusting the value of the extinction coefficient. The general principle is to make use of the shape factor introduced above; that is to project the shadow of the leaves onto a horizontal plane and to define k as the ratio of projected shadow area to actual (hemispheric) leaf area. The situation for vertical leaves is illustrated in Fig. 3.19 and illustrates the much reduced rate of attenuation of a nearly vertical beam of light compared with that which one finds for horizontal leaves. A detailed discussion of the effect of different leaf-angle distributions on the attenuation of incident radiation may be found in environmental biophysics texts such as Ross (1981), Monteith and Unsworth (2008), and Campbell and Norman (1998).

Spherical (random) leaf-angle distribution

A common type of canopy is where the leaves are approximately randomly oriented in space; that is they have an equal probability of any orientation (both in terms of azimuth and elevation). In such a case we can notionally rearrange the leaves in space so that they would evenly cover the surface of a sphere; in this case the shape factor can be obtained in the usual way by projecting the sphere onto a horizontal plane. An interesting, though somewhat counter-intuitive, property of such an arrangement is that although there is an equal chance of leaves having any azimuth (that is direction such as N, E, S, etc.), erect leaves are more common (occurring all around the equator) than horizontal leaves (occurring only at the poles). It follows from our definition of k that for a spherical distribution k is given by the ratio of the shadow area of a sphere when projected onto a horizontal plane ($= \pi r^2/\sin\beta$) to its hemisurface area ($2\pi r^2$). This reduces to $1/2 \sin\beta$ ($= 0.5 \operatorname{cosec}\beta$), where β is the elevation of the illuminating beam. Note that we use the hemisurface area of the sphere as the other half of the sphere area represents the lower/unilluminated sides of the leaves.

The use of the shape factor is a very valuable parameter in canopy radiation-transfer models, and as was pointed out by Campbell and Norman (1998), can be generalized to leaf-angle distributions that are more predominantly vertical or horizontal by assuming that the leaves can be rearranged not onto a sphere but onto an ellipsoid that is either elongated upwards (prolate) for a vertical-tending distribution or squashed (oblate) for a predominantly horizontal distribution (see Fig. 3.20(a)). In this approach we use a single parameter, $x, = b/a$ (the ratio of the horizontal (rotational) semi-axis to the vertical semi-axis of the spheroid), to describe differences in leaf-angle distribution. Oblate spheroids ($x > 1$) that are flattened at the poles have a greater proportion of horizontal leaves, while prolate or elongated spheroids ($x < 1$) would have a greater proportion of vertical leaves giving more erectophile canopies (Fig. 3.20(b)).

By varying Campbell's leaf distribution parameter x one can simulate quite well the radiation-transfer properties and hence the angular dependence of both the extinction coefficient and of the reflectance of a wide range of real plant canopies; typical values for x for different canopies range from values as low as 0.8 for soybean and some grass or cereal canopies to values greater than three for more horizontal leaved canopies such as strawberry (Campbell and van Evert, 1994).

There are a number of alternative ways in which leaf-angle distributions have been modelled for radiation-transfer studies. A common approach, particularly in older remote-sensing studies, has been to derive a mean inclination or *average leaf angle* (ALA; the angle between the vertical and the normal to the upper side of the leaf). For example, the mean inclination for a spherical distribution is 57.4°, and it is 71.1° and 38.5°, respectively, for ellipsoidal distributions with $x = 2$ and 0.5 (Wang and Jarvis, 1988). The use of average leaf angles gives rise to conical distributions that can then be substituted into the calculations for radiation transfer (Monteith and Unsworth, 2008). Among other functions that have been used for approximating

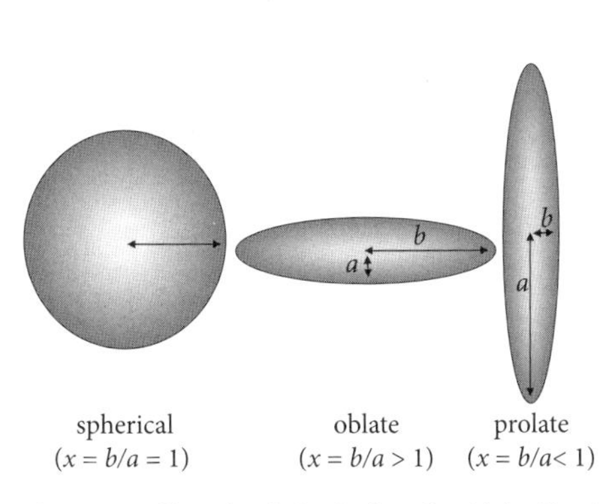

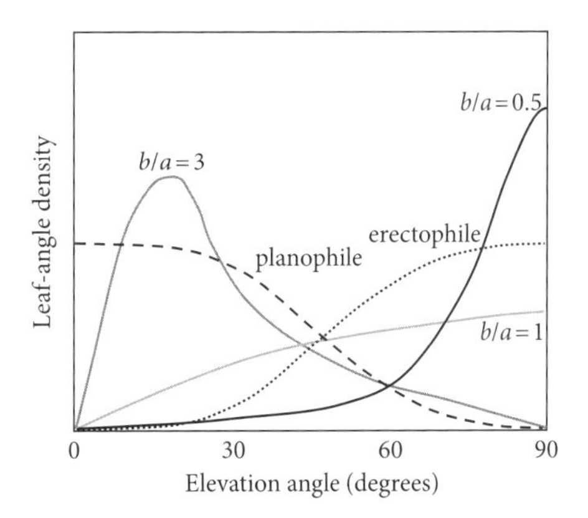

Fig. 3.20. Effect of variation in the spheroidal/ellipsoidal leaf-angle distribution shape on leaf-angle distribution, showing the effect of variation in Campbell's x parameter (= the ratio of the horizontal semi-axis, (b), to the vertical semi-axis, (a)) from $x = 3$ representing a predominantly horizontal leaved canopy to $x = 0.5$ representing a predominantly vertical leaf-angle distribution. For comparison the corresponding planophile (leaves tending to horizontal) and erectophile (leaves tending to vertical) distributions derived using the beta function are also shown.

real leaf-angle distributions are two-parameter beta distributions (Goel and Strebel, 1984), and triangular distributions. Although the ellipsoidal distribution has been found to be adequate for many canopies, there are cases (such as for some of the canopies shown in Fig. 3.21) and for plagiotropic canopies (leaves distributed around an angle of 45°) where the beta distribution is superior.

It is commonly assumed that variation in azimuth and elevation of leaves behave independently, and that azimuthal orientation is random. This latter assumption is generally considered not to introduce much error,

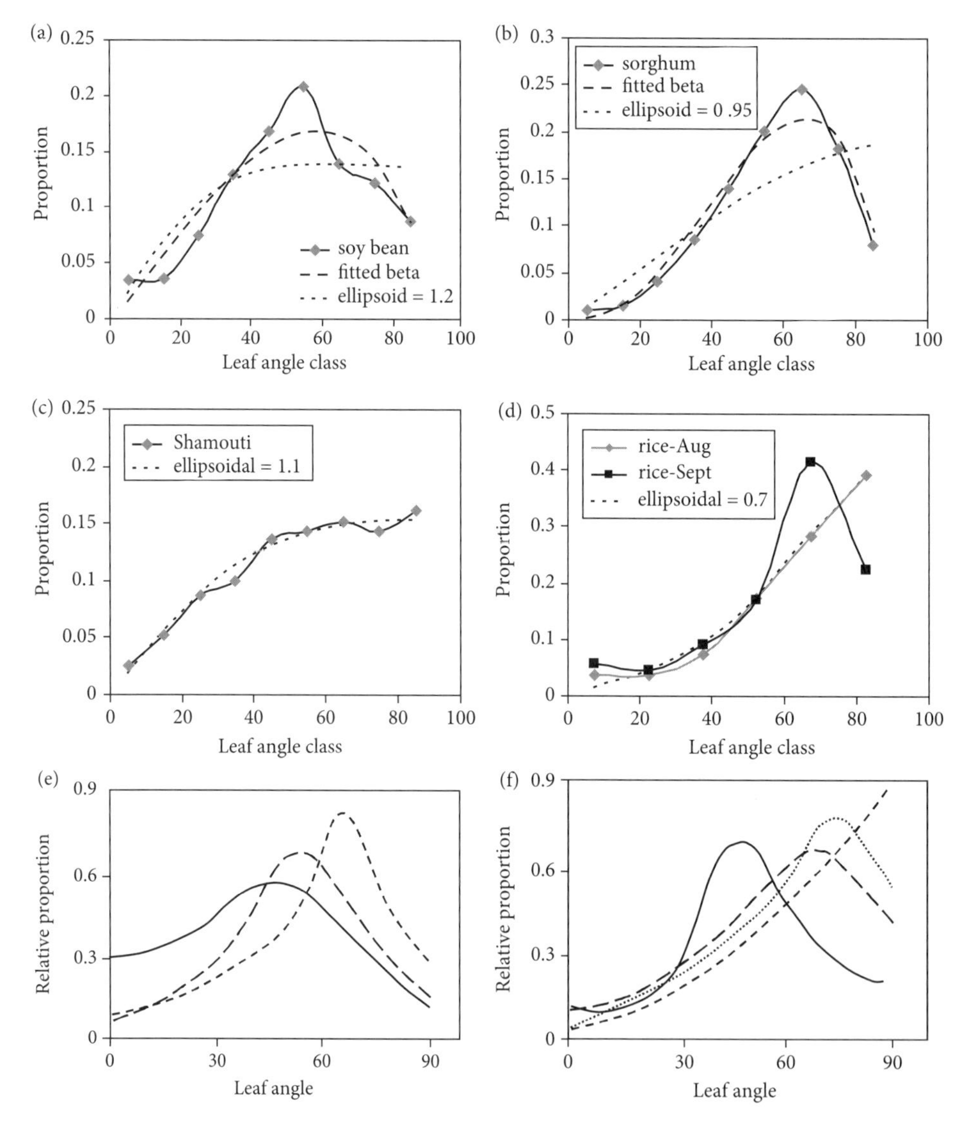

Fig. 3.21. Observed leaf-angle distributions and their variation for a range of different canopies together with fitted distributions using the Campbell ellipsoid function or the two-parameter beta distribution: (a) soybean and (b) sorghum data from Goel and Strebel (1984), (c) Shamouti orange from Cohen and Fuchs (1987), (d) rice data from Uchjima (1976). Campbell leaf distribution function calculated using the eqation in Campbell (1990); (e) data on leaf-angle distributions for different maize canopies at the 9th leaf stage from Holland (—), Tajikistan (– – –) and Estonia (- - - -) (data from Ross, 1981); (f) variation of leaf inclination distribution at different levels in a dense maize canopy (at 0–20 cm ——, at 40–60 cm – –, at 80–100 cm - - - -, at 120–140 cm ····· (data from Ross, 1981).

but it can be important for those species where the leaf-angle distribution is sensitive to environmental stresses, especially those leading to water deficits and leaf wilting. There are many species, including important agricultural crops such as sunflower and beans, where the leaves move in relation to the sun; that is they show heliotropic movements that can be modified in response to limited soil moisture (Ehleringer and Forseth, 1980; Oosterhuis *et al.*, 1985). When well watered, the leaves tend to follow the sun with the leaf blade perpendicular to the solar beam (diaheliotropic movements), but as water deficits increase the leaf blades tend to orient parallel to the beam (paraheliotropic movements) minimizing radiant energy absorbed. For heliotropic leaves, therefore, complete canopy radiation models require information not only on the *LAD* but also on the azimuthal distribution or orientation of the leaves.

Appropriate values for the extinction coefficient (k) and its variation with beam angle for different leaf-angle distributions are summarized in Table 3.4 and illustrated graphically in Fig. 3.22.

Formulae that allow calculation of quantities such as the fractional irradiance below a given cumulative leaf area, the fraction of radiation intercepted, and the sunlit leaf-area index are given in Box 3.1.

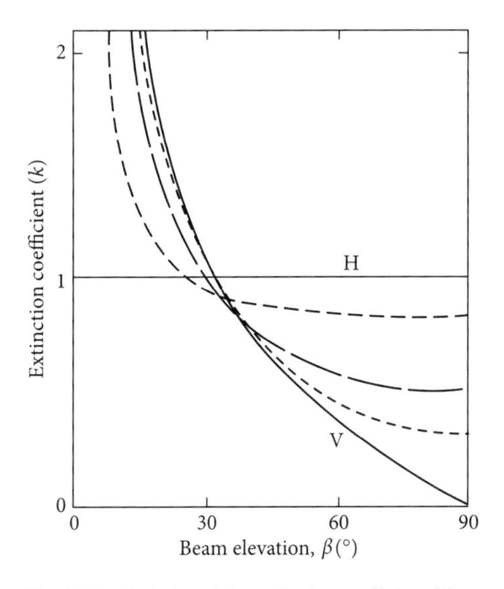

Fig. 3.22. Variation of the extinction coefficient (k) with beam elevation (β) for different values of the Campbell ellipsoidal parameter (x) from $x = 0.5$ - - - - - -, through $x = 1$ (spherical) —— to $x = 3.0$ – –, and including horizontal (H) and vertical (V) leaf-angle distributions (from Jones, 1992).

BOX 3.1 Radiation penetration in canopies in relation to leaf-area index

Useful formulae for the calculation of radiative properties of canopies from; values of the extinction coefficient k are functions of the beam zenith angle (θ) according to the different leaf-angle distributions in Table 3.4

fractional irradiance below leaf area L	$= e^{-kL}$
fraction of sunlit leaves at depth L	$= e^{-kL}$
fraction of radiation intercepted	$= (1 - e^{-kL})$
sunlit leaf-area index	$= (1 - e^{-kL})/k$

Table 3.4 Dependence of the extinction coefficient (k) on beam elevation (β) or zenith angle (θ), for a range of commonly used leaf-angle distributions (from Jones 1992). These relationships can all be derived on the basis of the shape factor (the ratio between the shadow area and the hemi-surface area as outlined for the spherical distribution in the text).

Leaf-angle distribution	Extinction coefficient (k) as a function of β or θ
Horizontal	$k = 1$
Vertical	$k = (2 \cot \beta)/\pi = 2/(\pi \tan \beta) = (2 \tan \theta)/\pi$
Spherical (= random)	$k = 1/(2 \sin \beta) = 1/(2 \cos \theta)$
Ellipsoidal#	$k = (x^2 + \cot^2\beta)^{1/2}/(Ax) = (x^2 + \tan^2\theta)^{1/2}/(Ax)$
Diaheliotropic	$k = 1/ \sin \beta = 1/\cos \theta$

where x is the ratio of the horizontal to vertical semi-axis of the ellipsoid and $A \approx (x + 1.774x^2 + 1.182)^{-0.733})/x$ (N.B. values of x greater than 1 equate to a more horizontal leaf-angle distribution)

3.1.4 Measurement of leaf-angle distributions

As might be expected, the various theoretical leaf-angle distribution functions are only approximations to real leaf-angle distributions. Although leaf-angle distribution is an important parameter for incorporation in radiation-transfer models of canopies it is rather tedious to measure accurately. Because of this it is common to estimate leaf-angle distributions through the inversion of radiation-transfer models as described in Chapter 8, but even then it is essential to be able to validate the estimates by use of real canopy data. The most straightforward direct approach is to align an inclinometer with each leaf in the canopy, and to record the angles, but this is particularly slow, and inaccurate for leaves that have a pronounced curvature such as maize. A range of instruments have been developed to improve the precision of the measurements and to ease their recording. An early example of a suitable instrument was proposed by Lang (Lang, 1973) where the tip of a 4-segment jointed arm was used to point to positions on each leaf. The absolute position in space of the tip was derived from the angles of each joint (measured electronically using potenti ometers) by simple geometry. The orientation of each leaf or leaf-segment could be obtained by defining the vertices of triangles representing each leaf. Alternative sensors involve the use of sound propagation or current induction in magnetic fields to locate the sensor and its orientation. The latter method (Sinoquet *et al.*, 1998) is particularly robust and allows one both to locate any leaf in space and to record its orientation. Leaf-angle distributions can also be estimated from stereographic imagery.

Strictly the leaf-angle distribution is defined as the probability density of the distribution of leaf normals with respect to the upper hemisphere. Horizontal leaves, therefore, have the zenith angles of their normals equal to zero. We consider normals in relation to the upper hemisphere irrespective of the actual anatomical orientation of the leaves (many leaves such as grasses often have their anatomically lower surface oriented upwards).

Some representative examples of measured leaf-angle distributions for different canopies are shown in Fig. 3.21, together with some fitted theoretical distributions. Although the shape of the distribution is characteristic for different species, it is worth noting that the detailed shape changes both with growth stage and with position in the canopy, as shown for the maize and rice examples in Fig. 3.21.

3.2 Thermal region

The thermal radiation actually *emitted* by a surface depends on the surface temperature and its emissivity according to the Stefan–Boltzmann equation (2.4), but the total radiation *leaving* the surface also includes a proportion of thermal radiation emitted by the surroundings and then *reflected* by the surface. Furthermore, as we have seen the radiance at the sensor is further modified by absorption, emission, and scattering by the intervening atmosphere. Although modifications due to the atmosphere can generally be ignored for ground-based and laboratory studies, they cannot be ignored for airborne or satellite-based sensors. These three radiation streams are illustrated in Fig. 3.23; summing their individual contributions allows us to write for the radiation received at the sensor

$$\boldsymbol{L} = \tau\,[\varepsilon\sigma \times (T_s)^4 + (1-\varepsilon)\,\boldsymbol{L}_{\text{background}}] + \boldsymbol{L}_{\text{atm}}, \quad (3.8)$$

where the transmissivity of the atmosphere (τ) modifies both streams from the surface of interest. For close-range sensing, where we can ignore the contribution of the atmosphere, eqn (3.8) reduces to

$$\boldsymbol{L} = \varepsilon\sigma \times (T_s)^4 + (1-\varepsilon)\,\boldsymbol{L}_{\text{background}}. \quad (3.9)$$

Note that thermal radiation is normally emitted from the surface of the object, though for water bodies the volume emitting may be as deep as 100 μm (Robinson, 2003). In this context it is useful to note that the water surface is frequently at least a few tenths of a degree cooler than the water body (due to evaporative cooling and thermal IR losses) so that radiometric temperatures may underestimate bulk water temperatures. This *skin effect* may be disrupted by waves though the temperature difference may re-establish within a few seconds (Robinson, 2003).

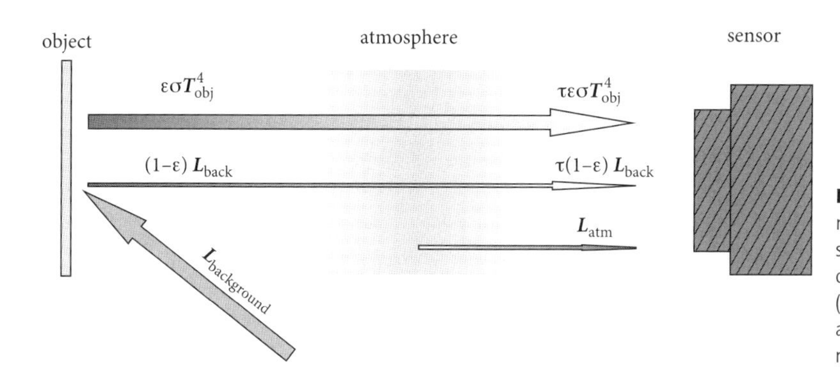

Fig. 3.23. The total longwave radiance arriving at the sensor is the sum of the radiation emitted by the object and that reflected by the object (both modified by transmission through any intervening atmosphere), plus any radiation emitted by the atmosphere.

The error in derived surface temperature introduced by errors in the emissivity of the surface depends on the background temperature, as well as on the emissivity. Substituting into eqn (2.4) shows that with no atmospheric absorption a 1% error in emissivity gives rise to an error of approximately 0.75 K in an estimated temperature at 300 K. As indicated in Fig. 3.23, however, this simple calculation ignores the contribution of the reflected background radiation. For clear-sky conditions the incoming sky radiance is often approximately equivalent to about 20 K lower than the actual temperature (Section 2.7.2), reducing the error to about 0.2 K/%.

The emitted thermal radiation tends to be highly anisotropic with the apparent temperature varying with view azimuth by amounts that may range from about 4 K for many agricultural crops and coniferous forest through 9.3 K for sunflower canopies to as much as 13 K (Kimes *et al.*, 1980; Lagouarde *et al.*, 1995; McGuire *et al.*, 1989; Paw and Meyers, 1989) largely as a result of varying proportions of (hotter) soil in the field of view of the sensor. The apparent temperature recorded at each angle therefore weights differently the temperatures of the components of the surface. Remotely sensed surface temperatures in the TIR only indicate the mean temperature of the surfaces that can actually be seen by the sensor, but the energy exchanges between the surface and the atmosphere also depend on temperatures of surfaces that may not be visible (e.g. the lower canopy leaves) (see Section 9.2.2 for further discussion of this effect). It also follows that the total thermal radiation emitted from the canopy over the complete hemisphere is difficult to estimate from the directional measures usual in remote sensing, though Otterman *et al.* (1995) have suggested that measurements made at an angle of 50° from the zenith should estimate the hemispherical emission within a few per cent.

3.2.1 Emissivities of canopy components

An accurate estimate of the emissivity is crucial for accurate determination of surface temperature from measurements of thermal radiation received at a detector. A range of approaches to measurement are possible. For laboratory measurement of spectral emissivities, it is common to measure the spectral reflectance of flat surfaces in an integrating sphere (see Fig. 3.4(b)) using a TIR spectrometer; the reflectance is then converted to a directional-hemispherical emissivity by the use of Kirchhoff's law (which states that $\varepsilon_\lambda = (1 - \rho_\lambda)$). The most straightforward field method for estimation of broadband emissivity for any surface with a thermal sensor, however, is to measure accurately the temperature of the surface (e.g. using thermocouples) together with the background radiation and to substitute these values into eqn (3.9) and solve for ε. Because it is difficult to measure accurately the background radiation it is common to isolate the test material in a highly reflective box (ε close to zero). For a detailed explanation of the method see Rubio *et al.* (2003). Many infrared thermometers or thermal cameras do not detect all the thermal radiation emitted by any source as they tend to be more sensitive to specific wavelengths (often to avoid wavelengths where radiation is absorbed by the atmosphere).[12] As a result they generally need calibration, frequently by means of an internal reference surface whose temperature is known.

[12] Many thermometers are based on bolometric sensors that are non-selective; wavelength selection in scanners can be achieved by using a reflection grating and placing the sensor at an appropriate point in the spectrum.

For most surfaces, emissivity is a weak function of the wavelength (Sutherland, 1986), while effects of surface structure on the emissivity have been reviewed by Norman *et al.* (1995). Some typical values reported for broadband emissivities of different surfaces relevant to the study of plant canopies are summarized in Table 3.5 (3–5 μm, and 8–14 μm) and in Fig. 3.24. Differences between species probably reflect both the chemical composition of the surface and the differences in microtexture, with the relatively smooth upper surface of *Acer rubrum* leaves having an emissivity of at least 0.02 greater than the abaxial surface with its rough wax coating (Ribeiro da Luz and Crowley, 2007). Some examples of spectral emissivities of different surfaces are presented in Fig. 3.25. This figure shows that, at least at leaf level, there are clear differences in the spectral variation in ε between species; these can be related to constituents such as xylans and celluloses.

The emissivity of bare soil varies widely with typical values for dolomite of 0.958 and those for granite and sand, respectively, being reported as 0.815 and 0.6 (Bramson, 1968; Sabins, 1997). The barren emissivity value in the NASA emissivities map, where the emissivity is a function of wavelength, are between 0.84 ($8 < \lambda < 9$ μm) and 0.92 ($\lambda > 16$ μm) (Gupta *et al.*, 1999). The emissivities of soil and leaf surfaces are also functions of water content, among other factors (Bramson, 1968; Lagouarde *et al.*, 1995; Salisbury and Milton, 1988; Sutherland, 1986).

Table 3.5 Broadband directional-hemispheric emissivities for a range of different materials relevant in remote-sensing studies. All values for 8–14 μm unless otherwise indicated; see also Figs 3.24 and 3.25.

Material	ε (8–14 μm)	ε (3–5 μm)	Reference and notes
34 leaf samples	0.975 (0.949–0.995)		1
3 crop canopies	0.966–0.974		2
5 leaf samples	0.964 (0.956–0.981)	0.959 (0.925–0.984)	3
Various plant samples	0.984–0.991	0.985–0.989	4
26 different plant samples	0.979 (0.971–0.994)		5 Aim to mimic canopy arrangement in space
Sands	0.888–0.914	0.823–0.937	1, 3, 6
Various soils	0.947–0.966	0.924	1, 2, 3, 6
12 different soils	0.959 (0.929–0.979)		5
Water	0.984	0.973	4
Fine snow	0.994	0.989	4
	ε^a (10.8–11.3 μm)	ε^b (11.8–12.3 μm)	
Evergreen needle forest	0.989 (0.975–0.992)	0.991 (0.978–0.994)	7 Averaged values
Green broadleaf forest	0.987 (0.975–0.995)	0.990 (0.978–0.995)	for different land
Green grass savannah	0.987 (0.974–0.994)	0.991 (0.977–0.996)	communities
Senescent sparse shrubs	0.970 (0.924–0.987)	0.975 (0.932–0.993)	
Bare arid soil	0.966 (0.925–0.983)	0.972 (0.934–0.990)	

[a]MODIS band 31; [b] MODIS band 32

References: 1. Idso *et al.* (1969); 2. Sobrino *et al.* (2004); 3. MODIS UCSB emissivity library (http://www.icess.ucsb.edu/modis/EMIS/html/em.html); 4. ASTER spectral library (http://speclib.jpl.nasa.gov/); 5. Rubio *et al.* (2003); 6. Sutherland (1986); 7. Snyder *et al.* (1998)

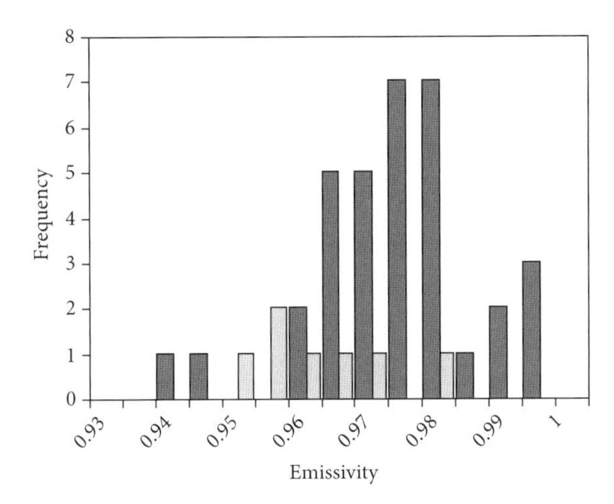

Fig. 3.24. Frequency of different observed values for broadband thermal infrared emissivity (8–14 μm) for individual leaves: dark bars for 34 species (data from Idso *et al.*, 1969); pale bars for seven species (data from Ribeiro da Luz and Crowley, 2007).

3.2.2 Emissivities of canopies

In general, the emissivity of a crop canopy or, indeed any rough surface, tends to be greater than that for a smooth sample of the material. The reason for this is readily understood if we remember that radiation emitted from the rough/complex surface will include a substantial component of radiation that has undergone secondary reflections from surfaces in the canopy. As the canopy gets deeper the proportion of scattered radiation that originates from within the canopy increases, so that the total radiation emitted by the surface ($\boldsymbol{L}$) tends to

$$\boldsymbol{L} = \varepsilon\sigma\,(T_{leaf})^4 + (1-\varepsilon)\,\sigma\,(T_{leaf})^4 = \sigma\,(T_{leaf})^4, \qquad (3.10)$$

that is the value of ε tends to one. As a consequence, the effective emissivity of most dense plant canopies is usually between 0.98 and 0.99, with coniferous

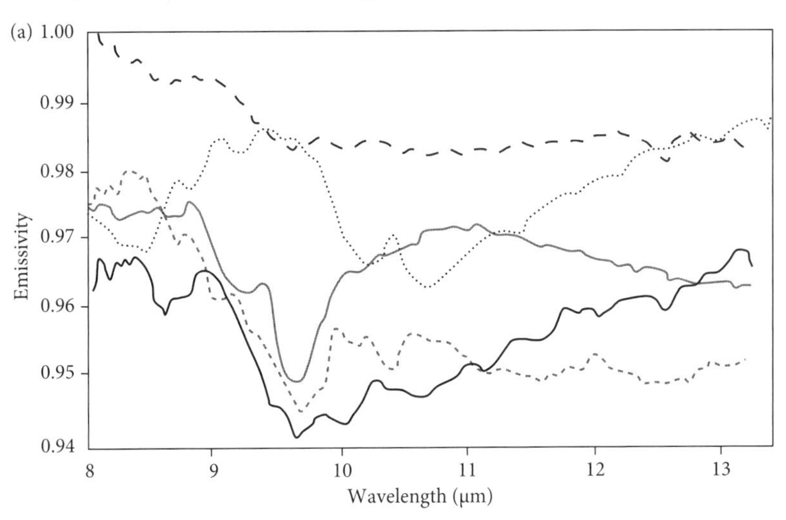

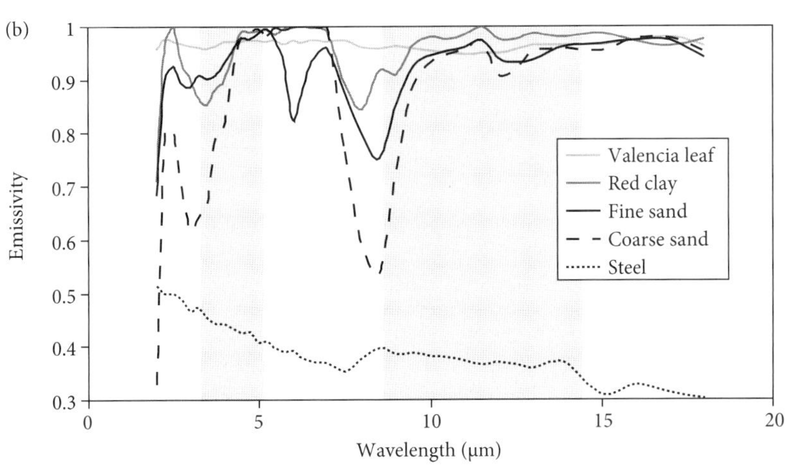

Fig. 3.25. (a) Some detailed spectral emissivity data for a range of leaves: *Fagus grandiflora* (black solid line – laboratory value, long-dashed line – field data obtained at a distance of 5.5 m), light solid – *Zea mays*, short-dashed line – *Prunus serotina* (data from Ribeiro da Luz and Crowley, 2007), together with data (dotted line) for yellow poplar from Salisbury and Milton (1988).
(b) Spectral emissivity for some typical leaf and soil surfaces (data from Sutherland, 1986). The shaded bands are the typical wavelengths used for 'shortwave' and 'longwave' thermal imagers. Because of interference by reflected sunlight, daytime measurements outdoors are normally restricted to the longer wavelengths.

canopies being particularly high, even though the emissivities of the component single leaves may be as low as 0.93. Nevertheless, some canopies can have emissivities as low as 0.94, especially where vegetation cover is thin partially exposing soil beneath. The effect is nicely illustrated by Fig. 3.26, which shows the effect of increasing viewing distance when measuring emissivity of trees, where the close range values of around 0.95 primarily represent values for isolated leaves, but the values rapidly increase as the area and complexity of the canopy viewed increases. The degree of sensitivity observed depends on canopy structure.

Where the nature of the surface is known it is possible to use laboratory-derived estimates of emissivity for the specific surface (such as those in Table 3.5 and in the ASTER and MODIS emissivity libraries). In practice, however, there are difficulties with applying this information. These include:

(i) Scale effects and the difference in behaviour of canopies and single leaves. Not only does the structural difference between single leaves and canopies affect the observed emissivity, as indicated above, but there is also a substantial potential scale effect related to the fact that most calibration data in the field are only obtained for small areas of at most a few square metres.

(ii) Related to this is the error that arises when the surface in the field of view comprises an ensemble of surfaces at different temperatures and with different emissivities (e.g. soil and leaves). Even if each has an emissivity close to unity the wavelength distribution of the emitted thermal radiation from the ensemble will not correspond exactly to that of a black body. Temperature differences greater than 10 K between soil and canopy can consequently lead to errors in radiometric estimates of surface temperature of the order of 1 K (Norman *et al.*, 1995a).

(iii) It is difficult to identify surface composition (and hence the appropriate ε) from remote images.

(iv) In addition to the angular variation in radiation field that results from changing proportions of soil and vegetation in the field of view as view angle changes, there is also evidence that emissivity of homogeneous surfaces can also vary with angle.

Approaches to the estimation of ε from satellite data, including the use of 'split-window' algorithms, or approaches based on the relationship between ε and spectral indices or derived biophysical parameters will be discussed in Section 6.2.3, where we consider further how emissivity varies with surface type and with view angle.

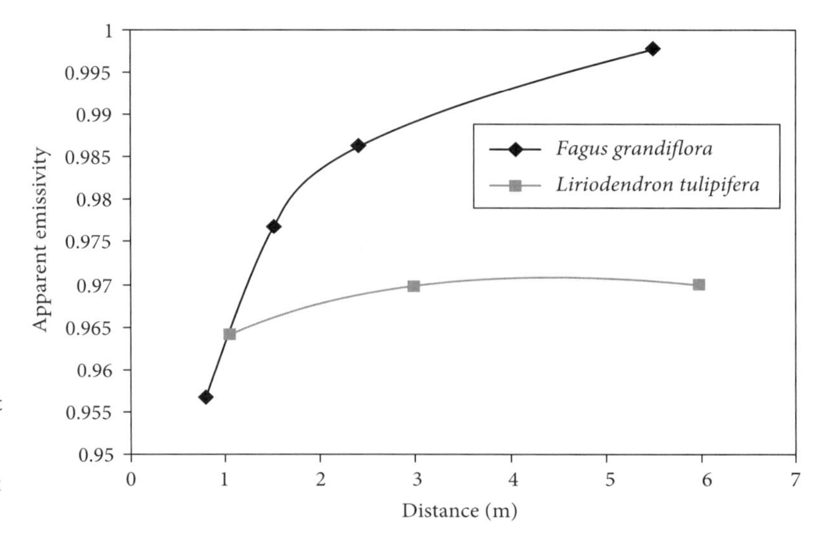

Fig. 3.26. Illustration of the apparent change in broadband (8–14 μm) emissivity as one views canopies from further away (data from Ribeiro da Luz and Crowley, 2007).

3.3 Microwave region

In Chapter 2 we looked at the physical properties of microwave radiation and its transmission through the atmosphere, while in Chapter 5 we will look at the technology and use of microwaves to derive information about the land surface properties. Here, we will consider in more detail the ways in which microwave radiation interacts with natural surfaces, in particular with soils and vegetation, and how it may be used to study them. The microwaves *emitted* from a surface depend mainly on the temperature of that surface but also on the emissivity, which is a function of both structure and of moisture content of the surface (see Section 2.5). Passive microwave sensing, for example, effectively measures the temperature of the vegetation canopy but at wavelengths at which the canopy is more transparent than in the thermal region. In active sensing, however, an artificially generated microwave signal is scattered back to the detector. This scattering is influenced mainly by the texture of the surface as well as by its dielectric constant and by the polarization of the waves. In this case, the structure of the canopy plays a very significant role. As in optical sensing, the interactions are complex as there are many different scales involved, but it is important to remember that microwave scattering is not sensitive to the chemical composition or pigmentation of leaves, but mainly to their physical structure and moisture content.

3.3.1 Microwave emissivity

The principles that determine emissivity at microwave wavelengths are the same as for emissivity at other wavelengths. For example, emissivity increases with surface roughness. Because emissivities can be strongly wavelength dependent it is particular important to take account of this fact in microwave sensing where observations cover a very wide range of wavelengths varying by an order of magnitude. As we shall see in Chapter 5, microwave observations are frequently made at angles far away from the normal so we have to take account of the angular variation in emissivity, which itself is dependent on the polarization of the waves.

The dielectric constant (ϵ) is a particularly important determinant of the emissivity of a material, and as water has a particularly high dielectric constant (around 80 at terrestrial temperatures and low frequencies but falling to c.6 at 300 GHz) compared with soil minerals (around 3–4), the dielectric constant of a surface is strongly dependent on water content. The strong interaction between dielectric constant of the material, incidence angle, frequency, and polarization means that it is difficult to make generalizations. The reflection coefficient, and hence by Kirchhoff's Law, the emissivity, may be derived from the Fresnel coefficients as incorporated in eqns. (3.11) and (3.12).

$$\varepsilon_V = 1 - \left| \frac{\epsilon \cos\theta - \sqrt{\epsilon - \sin^2\theta}}{\epsilon \cos\theta + \sqrt{\epsilon - \sin^2\theta}} \right|^2 \qquad (3.11)$$

$$\varepsilon_H = 1 - \left| \frac{\cos\theta - \sqrt{\epsilon - \sin^2\theta}}{\cos\theta + \sqrt{\epsilon - \sin^2\theta}} \right|^2 \qquad (3.12)$$

where ε_V and ε_H are the emissivities for vertical and horizontal polarization, θ is the incident zenith angle, and ϵ is the relative permittivity. Examples of the dependence of emissivity on zenith angle and polarization are presented in Fig. 3.27. This shows the very different behaviour of horizontally and vertically polarized radiation. Note the very strong peak of emissivity (corresponding to high absorption) of vertically polarized microwaves at an incidence angle that depends on the dielectric constant (70° for $\epsilon = 8$).

The emissivity of most natural surfaces does not vary very much with wavelength over the range used in remote sensing, but, when combined with the effects

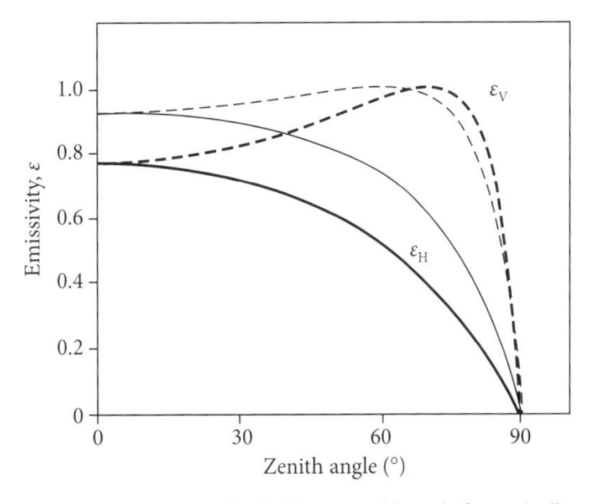

Fig. 3.27. Dependence of emissivity on zenith angle for vertically (V) and horizontally polarized (H) radiation for surfaces with dielectric constants of 3 (within the range for mineral soils – fine lines) and 8 (within the range for plant material – thick lines) as calculated from the Fresnel equations.

of roughness and moisture, there may be considerable differences in the emitted or scattered signal at different wavelengths. Some typical examples of the dependency of emissivity on wavelength of the microwaves are presented in Fig. 3.28.

Measurement of microwave emissivity

The only practical way of measuring emissivity is by using a contact thermometer in conjunction with a radiometer. This may be done in controlled experiments for individual leaves, plants, and small communities either in the laboratory or in-field. Even so, the scaling up of such measurements to real-world scenes is fraught with difficulty. For large areas as used in passive microwave sensing it is usual either to use a radiative-transfer model (e.g. Weng *et al.*, 2001) or to estimate emissivity on the basis of alternative surface temperature estimates (e.g. using MODIS thermal bands).

3.3.2 Microwave backscatter

In active microwave remote sensing one is usually interested in measuring the backscattering (reflection) of microwaves in the direction of incidence. The parameter that determines the characteristics of the microwave signal that is returned from a target is the *backscattering cross-section*, σ.[13] This is influenced

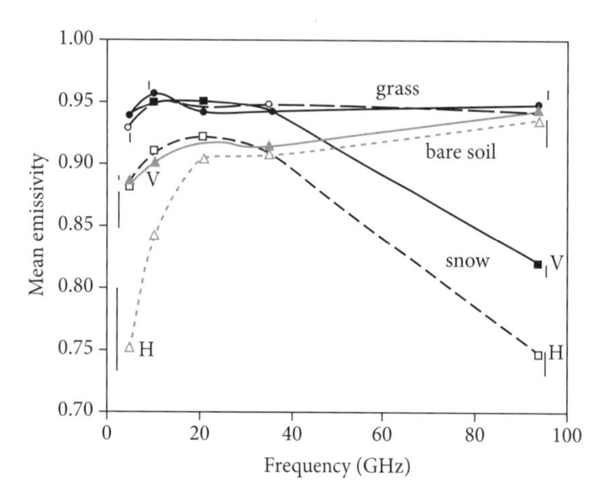

Fig. 3.28. Dependence of observed emissivity at an incidence angle of 50° on frequency for bare soil, grass, and powder snow for vertically (V – continuous lines) and horizontally (H – dashed lines) polarized radiation. Also shown are the standard deviation of different measurements for representative points. Data of Mätzler (1994) from Weng *et al.* (2001).

by the viewing geometry, the dielectric constant, and the surface roughness as well as by the wavelength used. The backscattered signal combines components due both to *surface scattering*, and for materials such as vegetation where some microwaves penetrate the canopy and are absorbed and scattered as they pass through the different layers, a component due to *volume scattering*. Especially when viewing vegetation from near nadir or at longer wavelengths some microwaves may penetrate as far as the underlying soil and interact with that before being scattered back. All the various components of the radiation will suffer more interactions on their way back through the canopy. Thus, the signal detected will contain much information about the moisture content and the structure of the canopy.

For surface scattering, we would expect from what we have learned so far that the backscattered signal from a smooth surface (where specular forward scattering dominates) will be much less than from a rough surface. An extreme example of a surface where backscattering is strong is one consisting of surfaces perpendicular to one another; for such a surface (e.g. as one might find in urban areas) radiation is strongly reflected back in its direction of incidence, irrespective of what the incidence direction is; this is known as a 'corner reflector'.

Some illustrative examples of the dependence of the backscatter coefficient on angle of incidence for a range of surfaces are shown for X-band radar in Fig. 3.29. Note that, because of the rather large amount of attenuation between the incident signal and the detected signal, often over several orders of magnitude in many microwave sensing applications, it is common to express the attenuation in terms of *decibel* (dB) where a reduction by a factor x (= power of source/power detected) equates to $10 \log_{10}(x)$ dB. This figure illustrates the relatively effective backscattering from urban areas with their high population of corner scatterers as compared with 'smooth' concrete surfaces. The backscatter from vegetation that includes a large volume scattering component is intermediate. Where volume scattering is important the backscatter is less dependent on incidence angle than where specular scattering dominates.

[13] Strictly speaking, in a practical application such as vegetation monitoring, where the scatterer is not an individual object but is an extended target, it is the *normalized backscatter coefficient*, $\sigma°$, also called the *normalized radar cross-section, NRCS*, which should be used. This is a unitless quantity that is a property of the target, not of the instrument used nor its viewing geometry.

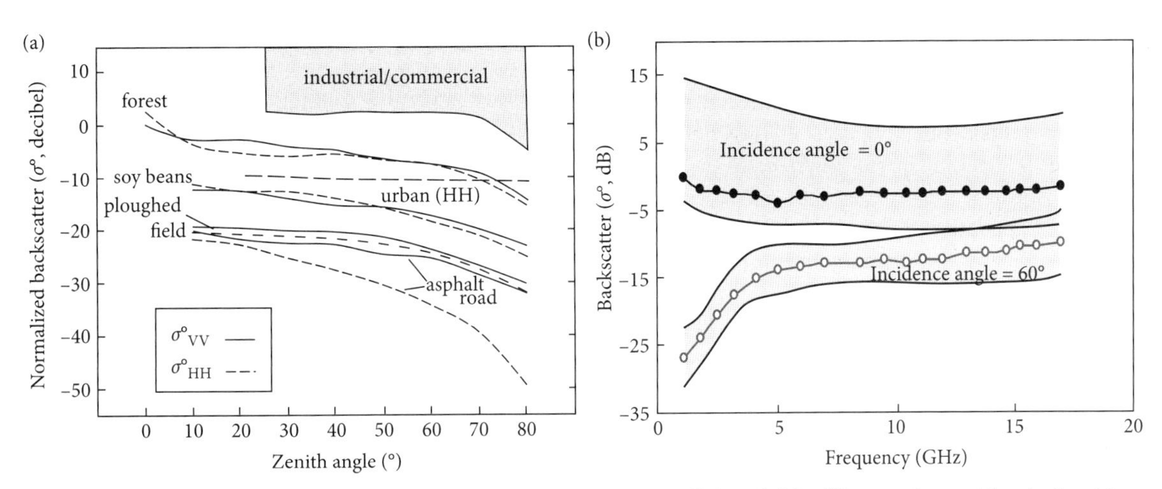

Fig. 3.29. (a) Sample variation of normalized (mean or median) backscattering coefficients (σ^o) for different surfaces at X-band using either VV polarization (where both the incident and detected radiation are V-polarized) or HH polarization (where incident and detected radiation are H-polarized) (data from Long, 1983; Ulaby *et al.*, 1982); (b) Frequency dependence of normalized backscattering coefficients (σ^o) for agricultural crops at either 0° or 60° incidence angle. The solid line shows median data and the shaded areas indicate the range of 90% of the observations (data from Ulaby *et al.*, 1982).

Soil

Soil consists of loose grains of solid material, water, and air. In general, the amount of scattering increases when the soil gets wetter. Also, as the grain size is usually much smaller than the wavelength used, it is the behaviour of the bulk dielectric properties, not the microstructure, which is relevant. A further complication is that some of the water molecules can bind to the surface of the grains, which inhibits their freedom to rotate, whereas the remainder can have their rotational states excited and thus contribute to the attenuation of radiation. At low concentrations of water, most of the molecules bind to the grains, enabling the wave to penetrate into the soil. As the concentration increases, more and more molecules remain free, thus the *penetration depth* decreases. Soil therefore acts as a volume scatterer. The scattering is also affected by the water content due to its effect on the texture of the surface. Longer wavelengths penetrate further into the soil and the shorter wavelengths are therefore more sensitive to the surface layers, but it is the moisture profile which is more significant than the average wetness.

Because the soil dielectric constant varies strongly with soil-moisture content, the backscatter is dependent on soil moisture and therefore potentially of use for studying soil moisture variation. Unfortunately, retrieval of soil moisture is complicated by the effect of surface roughness on the backscatter coefficient. The surface roughness parameters most commonly used in soil moisture retrieval models are the standard deviation of surface heights (s), the correlation model used, and the surface correlation length (l). In agricultural systems the roughness (especially the parameter s) is related to tillage of the surface while the value of l is related also to both the incidence angle and the polarization. The difficulty of estimation of appropriate soil roughness parameters means that it is difficult to estimate soil moisture directly using microwave data; this is an area of active research. Useful calibration of empirical algorithms (see Wigneron *et al.*, 2003) for soil moisture retrieval that correct, for example, for vegetation cover can be obtained using ground-based microwave radiometers (Laymon *et al.*, 1999). Passive microwave sensing of soil moisture from satellites is limited by the low spatial resolution available (e.g. 10s of km with the Advance Microwave Scanning Radiometer on the Aqua satellite).

Vegetation

A vegetation canopy forms a complex, heterogeneous volume consisting of components of different sizes, shapes, and orientations. The individual scattering components consist of leaves, stems, branches, and trunks, together with scattering by the underlying soil of the radiation that penetrates the canopy. Forest canopies usually tend to be so dense and so complex that they can often be modelled as a volume composed of random idealized identical objects so that an average behaviour can be assumed (Woodhouse, 2006). Volume scattering will be particularly strong when the wavelength of the radiation is of the same order of magnitude as that of the plant components. Volume scattering by canopies and leaves is larger for shorter wavelengths, $\lambda \approx 2$–6 cm, and this is usually greater than the scattering from the soil surface. A continuous medium will only absorb and emit radiation, while a discontinuous medium, consisting of discrete elements, will also scatter the radiation. When, however, the number of infinitely small scatterers gets very large, as in a dense canopy, the target begins to behave like a continuous medium and the internal scattering becomes minimal compared to the absorption so that the scattered signal can be considered to be from the canopy crown. Longer wavelengths, $\lambda \approx 10$–30 cm, which penetrate the canopy quite well, are best for detecting larger branches and tree trunks (see Chapter 11). The elevation angle will also affect the signal as there will be greater penetration at smaller angles to the vertical (pathlength is greater for larger angles). There may be a strong return from tree trunks due to the corner reflection, as where the trunks meet the ground at right angles, a wave may suffer a double reflection, once from the trunk and once from the ground (or *vice versa*). The resulting ray will be reflected back almost along the incident direction giving a strong response. These long-wavelength microwaves are very sensitive to the volume of the tree structure, showing a positive relation between backscattering and biomass density. The density of the vegetation also affects the depolarizing effect. The H/V contrast reduces as the amount of vegetation increases, while bare soil has a high H/V contrast.

In agriculture, where crops may be growing in rows, the orientation of the rows will also significantly affect the return, with stronger scattering when the look direction is at the face of the rows rather than along the rows. At L-band, there is an increase in σ° as biomass (plant water content and leaf-area index) of broad-leaf crops (such as sugarbeet and sunflower) increases, but remains fairly constant for narrow-leaf crops (such as wheat and alfalfa). Differences in leaf shape and orientation also affect the polarization contrast. Potatoes and similar plants that have large flat leaves produce lower H/V contrast than do the long vertical stalks of wheat. Colour composite displays (see Section 6.3.3) using two or three different polarizations will produce distinctive colours for different crops enabling some sort of classification to be performed. Similarly, colour composites formed from radar images using different wavelengths may also enable different vegetation types to be recognized.

The water-cloud model

Because a vegetation canopy can be considered to consist of a collection of randomly oriented scatterers, the *water-cloud model* (originally developed by meteorologists for the scattering due to water droplets in the atmosphere) has been adapted to model the microwave response of vegetation canopies. The water-cloud model represents the power backscattered by the whole canopy as the incoherent sum of the contributions of the vegetation and the underlying soil. If we consider a sparse volume-scattering layer (the canopy) with a low value of transmissivity τ_v above a surface layer, there are three components in the backscattered signal to consider. The signal scattered by the surface, σ°_s, is modified by a factor τ_v^2, squared because the radiation makes two transits through the canopy. The second component is that scattered by the canopy itself, σ°_v. The third component corresponds to that part of the wave that is scattered in the volume, travels downwards, and is scattered by the surface. This 'forward-scattering' component is usually ignored in the model. All the components depend on the incidence angle of the wave, θ, an important consideration in radar observations. Hence, the *NRCS* is given by the expression:

$$\begin{aligned}\sigma^{\circ}_{canopy}(\theta) &= \text{(attenuated surface emission)} + \text{(direct volume emission)}\\ &= \sigma^{\circ}_s(\theta)\,\tau_v^2(\theta) + \sigma^{\circ}_v\\ &= \sigma^{\circ}_s(\theta)\,\tau_v^2 + AW\cos\theta\,(1-\tau_v^2(\theta)),\end{aligned} \quad (3.13)$$

where $\tau^2 = \exp(-2BW/\cos\theta)$, W is the volume water content and A and B are constants.

The water-cloud model does not explicitly include the effects of polarization, although a separate model can be generated for each polarization if needed.

3.3.3 Advantages of microwaves for remote sensing of vegetation

It is apparent from the above that the range of information about the surface properties that is available using microwave sensing at different frequencies and at different polarizations complements extremely well the information that can be obtained by conventional remote sensing in the optical and thermal regions. Critical features of microwaves for remote sensing include: (i) their ability to penetrate canopies and soils to some degree, thus providing volumetric information about the system that is not available in the visible, (ii) their strong sensitivity to water content, and hence their use as indicators of water status, (iii) the fact that microwave sensing can continue even with cloud cover, or at night, while optical sensing is not possible under such conditions. In addition multiangular sensing with microwaves provides a powerful tool for the study of soils and vegetation, with shallow look angles providing a useful different perspective than normal photography, and the simultaneous operation at several frequencies may have advantages.

3.4 Other types of radiation

It is also worth noting in passing that other types of radiation, including both highly energetic β- (Jones, 1973) and γ-radiation as well as terahertz radiation can be used for close-range non-destructive estimation of water content of leaves from changes in permittivity and of attenuation when transmitted through leaves or stems of plants. Terahertz radiation is particularly promising because it is strongly absorbed by water, but not by non-polar organic leaf material, the shorter wavelengths than for microwaves gives a higher spatial resolution, and it is not subject to radiation protection provisions required for radioactive sources (Jördens *et al.*, 2009).

Further reading

The classic text on optical properties of plant leaves and vegetation, packed with examples, is that of Gates (Gates, 1980). Very clear descriptions of radiative properties of leaves and canopies are available in Monteith and Unsworth (2008) and Campbell and Norman (1998), while large databases of spectral properties of different materials may be found in the websites listed below. A particularly readable text on microwave sensing is that by Woodhouse (2006) while Rees (2001) has much useful information.

Websites

Several detailed and excellent spectral reflectance datasets are available. These include

(1) ASTER spectral library version 2 (reflectance data for a range of natural surfaces) (Baldridge *et al.*, 2009): **http://speclib.jpl.nasa.gov/**

(2) USGS Digital Spectral Library splib06a (reflectances for a range of soils vegetation, rocks, minerals) (Clark *et al.*, 2007): **http://speclab.cr.usgs.gov/spectral.lib06/**

(3) LOPEX93 (Leaf Optical Properties Experiment 93): **http://ies.jrc.ec.europa.eu/index.php?page=data-portals**

(4) DAAC ORNL (gives access to a number of useful spectral datasets including OTTER and ACCP – the Accelerated Canopy Chemistry Program): **http://daac.ornl.gov/holdings.html**

(5) MODIS UCSB Emissivity library: **http://www.icess.ucsb.edu/modis/EMIS/html/em.html**

Sample problems

3.1 Assuming the incoming longwave radiation received (from the sky, neighbouring leaves, and the soil) by a leaf equates to an average environmental temperature of 10 °C, and the true leaf temperature is 20 °C (with an emissivity of 0.95), calculate the effect on estimated leaf temperature (when using an infrared thermometer) of (a) ignoring the reflected radiation, and (b) of a 1% error in the assumed value of emissivity on the leaf temperature.

3.2 Assuming an emissivity for fresh snow of 0.8 and of boreal forest of 0.98, estimate the percentage change in emitted radiance when forest at 0 °C is covered by a fall of snow at −5 °C. Why might this calculation overestimate the change in emitted radiance? If the true ε of the snowy forest is 0.95, estimate the error in calculated surface temperature from assuming ε is 0.98.

3.3 For a solar zenith angle of 20°, calculate the fraction of incident radiation intercepted by (a) a horizontal leaved canopy with a leaf-area index of 1.2, (b) by a vertical leaved canopy with an *LAI* of 2.6 and (c) by a canopy with randomly oriented leaves and an *LAI* of 2.6. What is the sunlit leaf-area index in each case?

3.4 Calculate the fractional radiation transmission for a plant canopy with an ellipsoidal leaf-angle distribution (x = 0.5) and a leaf-area index of 2.5 for solar zenith angles of 0°, 30°, and 60°.

4 Plant and canopy function

4.1 Introduction

This chapter introduces the key aspects of plant function that we might hope to gain some understanding of by using remote sensing. The emphasis in this chapter is on the main non-radiative ways in which plants and vegetation interact with the aerial environment; these include the transfer of matter (such as the uptake of atmospheric carbon dioxide (CO_2) in photosynthesis or the loss of water in evaporation), and of heat and momentum. Unfortunately, these processes are generally not directly detectable by remote sensing and can only be inferred; therefore they are rarely treated in any depth in remote sensing texts. Nevertheless, these exchanges are critical to understanding the functioning of plants and vegetation canopies and knowledge of their regulation is necessary for interpreting and using information obtained by remote-sensing techniques for the diagnosis and management of agricultural crops and natural vegetation and for studying the role of vegetation in climate change.

Following an introduction to the fundamental physiological processes involved in CO_2 exchange and water loss from vegetation, we outline the basic mechanisms involved in heat and mass transfer between the vegetation, or other surfaces, and the atmosphere, including exchanges of CO_2, water vapour, and heat. For further details of the physiological processes involved the reader is referred to more specialized texts: these would include general plant physiology texts (e.g. Raven *et al.*, 2005; Salisbury and Ross, 1995; Taiz and Zeiger, 2006) or environmental plant physiology texts (e.g. Fitter and Hay, 2001; Jones, 1992; Nobel, 2009). The ways in which information on these mass and energy fluxes can be obtained by remote sensing will be developed in Chapter 9.

4.2 Plant anatomy and function

Although there is enormous diversity in the world of plants, there is much commonality in their structures, especially amongst the terrestrial plants which are the main topic of this book. The structure and functioning of plants are largely determined by their adaptations to the specific physical environments in which they grow, coupled with their requirement for water, nutrients, and light to survive and grow. For terrestrial plants, the air provides access to the gases such as carbon dioxide (CO_2) and oxygen (O_2) that they require, while the soil or rooting medium usually provides the necessary nutrients and water. A problem with the terrestrial habit, however, and a key problem that was resolved in the evolutionary migration of organisms from aquatic to terrestrial habitats is the tendency to desiccation. The free gas exchange required to allow uptake of CO_2 in photosynthesis comes with a corresponding tendency to lose water by evaporation with consequent desiccation and loss of essential tissue water. Terrestrial plants have adapted to life on land by developing a range of adaptations that allow them to acquire and retain water while photosynthesizing.

As shown in Fig. 4.1 (see also Fig. 3.2), typical plants therefore consist of a more or less extensive and sophisticated root system that absorbs water (and also nutrients) from the soil and a vascular system to conduct it to the above-ground leaves. There is also a supporting structure of stems and branches that spreads out the photosynthetic tissue in an optimal way to absorb energy from the sun used to drive photosynthesis. In 'lower' plants such as mosses and bryophytes that tend to inhabit more humid environments the water acquisition structures (simple unicellular rhizoids) and mechanisms for the control of water loss are less well developed than in 'higher' plants. As well as the development of roots to aid water uptake, the evolution of impermeable cuticles combined with stomatal pores that have the power to open and close and regulate gas exchange, and the development of intercellular gas spaces were also critical. The evolution of desiccation tolerance in the vegetative state (especially in the lower plants) also contributed. In addition to these vegetative structures, we are often concerned, especially in agricultural contexts, with the reproductive structures such as flowers and fruits. These are often visibly distinct from the vegetative structures so can often be detected remotely (e.g. flowers can have quite distinct spectral properties from leaves).

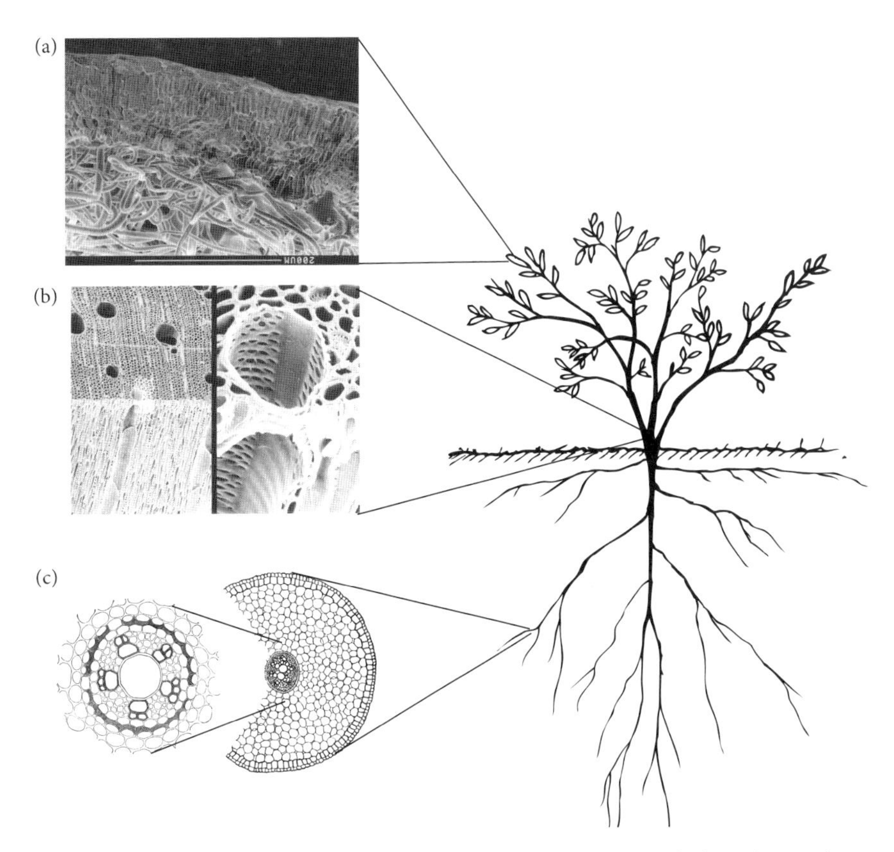

Fig. 4.1. Typical form of a higher plant, showing the extensive root system, a conductive and supportive stem and branch structure, and a canopy of photosynthetic leaves that may include reproductive structures (flowers). The enlarged images show (a) a scanning electron micrograph of a *Ziziphus mauritiana* leaf, showing the extensive layer of leaf hairs on the underside, (b) scanning electron micrographs of cross-section of the woody stem of a *Prunus* stem showing the xylem vessels, and (c) a cross-section of a root showing the thick outer cortex and an expanded view of the conducting stele at the centre.

4.2.1 Photosynthesis and respiration

The key process underlying life on earth and the functioning of most ecosystems is the fixation of atmospheric carbon dioxide by plants in the process of *photosynthesis*. This produces sugars that act both as building blocks for all plant, and ultimately animal, material, and as a source of energy for the synthesis of more complex organic molecules. Because almost all living organisms are ultimately dependent on the energy from the sun that is trapped during the process of photosynthesis, this is referred to as *primary production* and plants are the *primary producers*. In addition to fixing CO_2, photosynthesis releases O_2; this has led over billions of years to the generation of the aerobic environment that we currently enjoy. Photosynthesis, and its reverse, the release of carbon dioxide in the process of *respiration*, are major determinants of the global carbon balance, though combustion of fossil fuels (themselves based on historic photosynthesis) is a significant component of the increasing CO_2 concentration in the atmosphere. Over geological time volcanic emissions and soda-springs can also add CO_2 to the atmosphere, while marine sedimentation and subduction of organic C removes CO_2.

Most photosynthesis in plants occurs in the mesophyll of leaves (Fig. 4.2; see also Fig. 3.2 for some leaf structures found in different types of plant). Incoming radiation from the sun is absorbed by the light-absorbing green pigments, the chlorophylls, together with some additional absorbers known as accessory pigments, located on the thylakoid membranes of the chloroplasts; this absorbed *photosynthetically active radiation* (PAR, between about 400 and 700 nm) excites electrons in the chlorophyll molecules in the chloroplast membranes. This excitation is collected in the antenna pigments and transferred to the photosynthetic reaction centres where the energy is used to drive electron transport in the chloroplast membranes leading to the splitting of water and the generation of reducing power (reduced nicotinamide adenine dinucleotide phosphate or NADPH) and chemical energy (adenosine tri-phosphate or ATP). These high-energy molecules are then used to drive the series of biochemical reactions that reduce atmospheric CO_2 to form a 3-C compound (a triose phosphate known as phospho-glyceric acid or PGA) and subsequently sugars (Fig. 4.3). The conventional summary equation for photosynthesis is a little misleading as the process involves two rather weakly linked reactions (Fig. 4.3): the absorption of photosynthetically active radiation

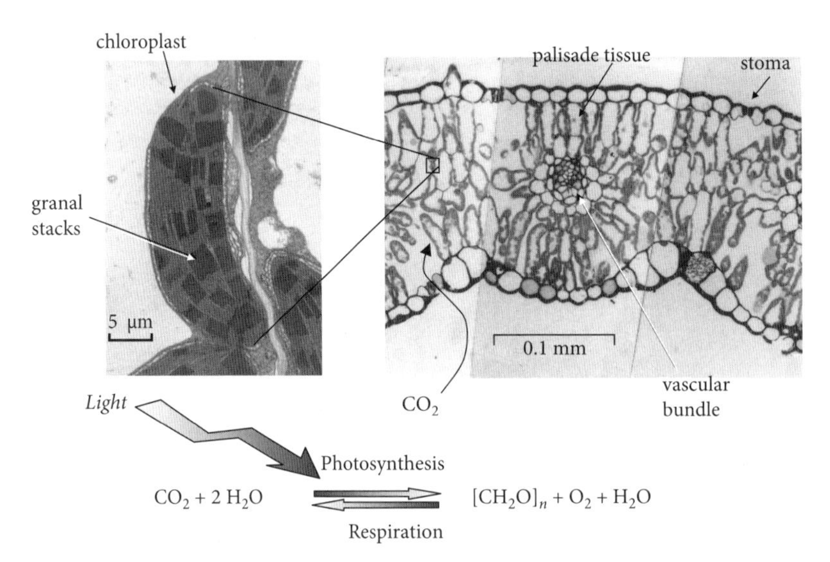

Fig. 4.2. A cross-section of a leaf, showing as an inset an electron micrograph of a chloroplast with its complex of tightly stacked lamella membranes containing chlorophyll. The overall equation for photosynthesis is also shown and indicates the fixation of atmospheric CO_2 into carbohydrate (written as $[CH_2O]_n$). The reverse process that releases CO_2 and generates energy for metabolism is respiration.

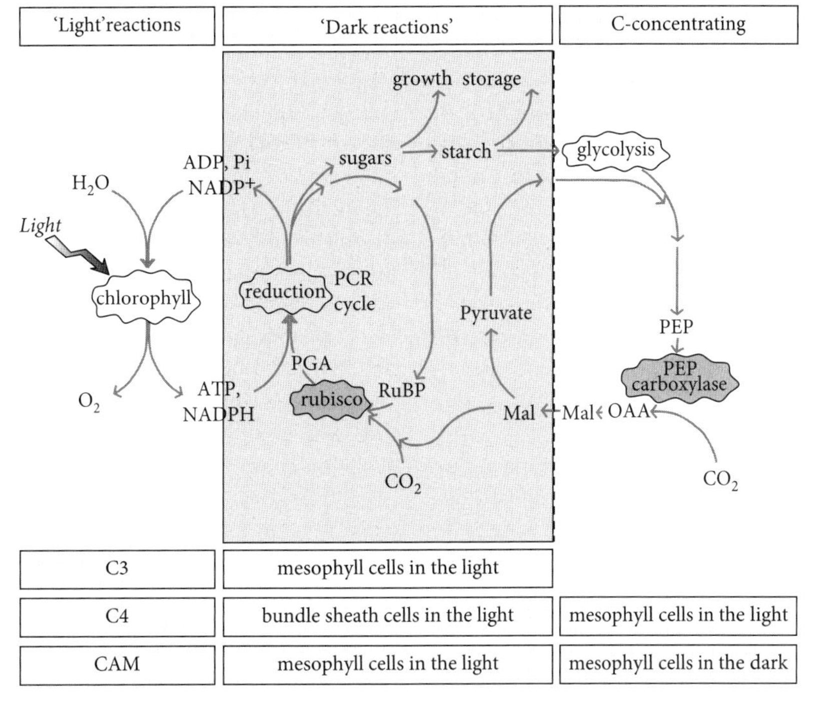

Fig. 4.3. An illustration of the photosynthetic process in C3 plants where sunlight is absorbed by chlorophylls and accessory pigments in the chloroplast lamella membranes leading to the generation of reducing power and ATP ('light reactions'). These high-energy compounds are then linked to the 'dark reactions' where these energetic molecules are used by rubisco for the fixation of atmospheric CO_2 and its conversion to sugars. Carbon-concentrating mechanisms in C4 and CAM plants are illustrated in the right-hand panel.

and the subsequent electron transfer processes are conventionally referred to as the *photosynthetic light reactions*, while the biochemical reactions utilizing the resultant ATP and reducing power to fix carbon dioxide and synthesise sugars are termed the *photosynthetic dark reactions*. The light reactions involve the absorption of energy from sunlight by two linked photosystems (photosystems I and II, abbreviated as *PSI* and *PSII*); this process drives the electron transport required for energy transduction and synthesis of energy-containing molecules. The key chloroplast enzyme involved in the dark fixation of CO_2 in the thylakoid is ribulose bis-phosphate carboxylase-oxygenase or *rubisco*, which is probably the most abundant protein in the world as it comprises around a quarter of the protein in a typical plant leaf. The rate of CO_2 fixation also depends on the rate of its diffusion from the atmosphere, through the leaf boundary layer, then through the stomata and the intercellular spaces and finally in the liquid phase to the chloroplasts; at the same time the O_2 produced has to diffuse in the opposite direction. When the stomata are closed this supply pathway can act as a significant limitation to the rate of photosynthesis.

Photosynthetic pathways

The majority of plants from temperate regions, including most trees and many of the major crop plants such as wheat, rice, potato, and all legumes, fix CO_2 using the so-called *C3-pathway* of photosynthesis where the initial fixation product is the 3-carbon PGA as indicated in Fig. 4.3. In the hotter regions of the world many plants such as maize and sorghum have an alternative fixation pathway, the so-called C4-pathway where CO_2 is initially fixed using a different enzyme (phosphoenolpyruvate carboxylase, PEP carboxylase) with the initial product being a 4-carbon compound. Though there are several variants of the *C4-pathway* they all have in common a specialized anatomy ('Krantz' anatomy) where the photosynthetic tissue is distributed between the general mesophyll cells where the primary fixation process involving PEP carboxylase occurs, and the bundle sheath cells that contain the usual C3 (rubisco) pathway. This characteristic structure acts as a CO_2-concentrating mechanism where the C4 products are transferred to the bundle sheath cells and decarboxylated to release CO_2 that is trapped within this tissue and builds up to high concentrations, thus enhancing the secondary CO_2 fixation by the usual rubisco fixation mechanism. This two-stage process increases the overall affinity of the leaf for CO_2 and thus acts to increase the ratio of photosynthesis to water loss (the water-use efficiency, WUE). It is therefore a valuable adaptation in high temperature and arid environments where water conservation may be a priority.

A further improvement in WUE is achieved in desert plants such as cacti and many other succulents that have the *crassulacean acid metabolism* (CAM) pathway of photosynthesis. In this case there is no obvious structural separation of the C3 and C4 steps, but separation occurs temporally, with the initial fixation of CO_2 again using PEP carboxylase, but in this case at night when the stomata are open. During the day time the stomata are closed (thus minimizing water loss) and the normal C3-pathway operates to fix the internally released CO_2. Although the maximal photosynthetic rate of CAM plants tends to be rather low, the water-use efficiency can be very high indeed as stomata are closed during the time of greatest evaporative demand. In this case the initial fixation is completely decoupled from the light reactions.

Net primary production

The sugars and other carbohydrates produced in photosynthesis are used both as building blocks for more complex organic molecules used in the synthesis of whole organisms and as substrates for respiration. This latter process releases CO_2 but retains some of the energy in active molecules such as ATP that are then used to drive the synthetic biochemistry and fuel the various maintenance reactions that are needed to retain the integrity of living cells. We can define the term *gross photosynthesis* (P_g) as the rate of fixation of CO_2 by the carboxylation enzyme, rubisco. Gross photosynthesis is generally proportional to linear electron transport so chlorophyll fluorescence (see below) can provide a powerful remote probe of photosynthetic functioning, even though there are a number of possible electron acceptors other than CO_2. In practice, while photosynthetic carbon fixation occurs in the light, some respiratory CO_2 loss from *mitochondria* in the leaf cells continues simultaneously to provide energy for maintenance of leaf function and growth. In many plants (C3 plants) there is further CO_2 loss in the light through a process known as photorespiration. *Net photosynthesis* (P_n) is the instantaneous balance between gross photosynthesis and respiratory losses from the plant leaves. Over longer time periods the net photosynthesis during the daylight periods is further diminished by respiratory losses at night, giving the net daily fixation of carbon by the plants; this is commonly referred to as the *net primary production* (*NPP*). When scaling it up further to an ecosystem level, the respiratory fluxes to the atmosphere also include respiration from non-photosynthetic parts of plants, from soil organisms, from decay processes, and from animals, so the *net ecosystem exchange* (*NEE*) is smaller than the *NPP*.

Chlorophyll fluorescence

Only a fraction of the light energy absorbed by chlorophyll is used to drive electron transport and photosynthetic carbon assimilation; a significant fraction is lost as heat and a rather small fraction (about 1–2%) is reradiated at a longer wavelength as chlorophyll fluorescence (see Section 3.1.1). This phenomenon is of particular relevance for remote sensing of photosynthetic function in plants. The quantum yield of fluorescence (Φ_F), or the fraction of incident quanta that are dissipated by fluorescence, is given by

$$\Phi_F = k_F/(k_F + k_D + k_P), \tag{4.1}$$

where k_F, k_D, and k_P, respectively, are the rate constants for de-excitation through fluorescence, thermal dissipation as heat, and energy transfer to photosynthesis. As photosynthetic electron transport (k_P) increases, fewer electrons are available to give rise to fluorescence; therefore the amount of fluorescence can be used as an (inverse) indicator of electron transport and hence of photosynthesis, being approximately inversely related to the electron flow in *PSII* that is used in splitting water.

Where excess radiant energy is absorbed by the photosynthetic systems beyond that which it is possible for the plant to use in photosynthetic electron transport (for instance under drought conditions when CO_2 supply is limited by stomatal closure) it is necessary for the plant to have mechanisms

BOX 4.1 Some useful terminology/definitions

Gross photosynthesis **(P_g).** The instantaneous rate of CO_2 fixation by rubisco (mol m^{-2} s^{-1}). Can alternatively be expressed in kg m^{-2} s^{-1} or in energy equivalents.

Net photosynthesis **(P_n), *or net carbon exchange* (*NCE*).** The instantaneous net flux of CO_2 into a leaf (mol m^{-2} s^{-1}) which is the instantaneous value of gross photosynthesis minus simultaneous respiratory losses from the leaf (both by photorespiration and any 'dark' respiration occurring in the light). [Maximum instantaneous rates of P_n are as much as 30–40 μmol m^{-2} s^{-1}.]

Net primary productivity **(*NPP*).** The rate at which energy (usually expressed in terms of biomass) is accumulated by vegetation and integrates photosynthesis and respiration by the plant per unit area of ground over at least a daily cycle. *NPP* is less than P_n because it includes respiratory losses by non-photosynthetic tissues and at night. Note that although net above-ground accumulation of biomass is readily measured, *NPP* should strictly also include any root growth. [Maximum *NPP* may be as much as 7500 g m^{-2} $year^{-1}$ for a well-fertilized sugar cane crop, with typical values for tropical rain forest being around 2000 g m^{-2} $year^{-1}$; environmental and biological stresses lead to values for other crops and ecosystems being less than these values.]

Net ecosystem exchange **(*NEE*).** The net rate of carbon accumulation by an ecosystem. This equates to net primary productivity minus all the respiratory losses from the system including from other organisms in the soil.

for dissipating the excess excitation energy. This is because the excess energy can lead to 'oxidative stress' and the generation of a range of highly reactive oxygen species (ROS) that are potentially very damaging to the cells; these include the superoxide radical ($O_2^{\cdot -}$), hydrogen peroxide (H_2O_2), singlet oxygen ($^1O_2^*$) and, especially, the hydroxyl radical ($HO^{\cdot}$). The most important mechanism for enhanced dissipation of this excess energy and photoprotection of the photosynthetic system is as heat that results in 'quenching' of the fluorescence signal below its unattenuated value. The xanthophyll pigments in the chloroplast play an important role in this 'non-photochemical quenching' or NPQ, with the de-epoxidated form (zeaxanthin) quenching fluorescence much more strongly than does the epoxidated form (violaxanthin), so that de-epoxidation leads to substantial heat dissipation and NPQ. It is possible to sense the epoxidation state of the xanthophylls by means of their characteristically different absorption/reflection spectra around 531 nm; this provides the basis for the *photochemical reflectance index* (*PRI*) discussed in Chapter 7. The degree of quenching of fluorescence below the maximum value obtained when all the photosynthetic reaction centres in the chloroplast are in the 'open' state provides a powerful diagnostic tool for photosynthetic function, and one that is potentially very suitable for study by remote-sensing techniques (see Section 9.7.3).

4.2.2 Water relations, evaporation, and water loss

Plant water status

Another critical mass exchange between vegetation and the atmosphere is the water lost by evaporation. The water status of a plant at any time depends on the balance between water uptake from the soil and the rate of water loss from the leaves (and other above-ground tissues). Water is an essential plant component being required as a solvent for most of the biochemical reactions that are critical for life and that depend on an aqueous environment; it often comprises as much as 90% of a plant's fresh weight. The tissues of all plants and animals are composed of functional units called cells: the living central material or protoplast is bounded by a semi-permeable lipid bilayer membrane. Outside this plasma membrane in plants is the cell wall, a structure largely composed of cellulose fibrils and that, when combined with the internal pressure in the cell generated by means of *osmosis*, provides rigidity to soft structures such as leaves. Osmosis is the process whereby water moves between two compartments separated by a semi-permeable membrane (one that lets water through but not larger solute molecules). When two compartments are separated by such a semi-permeable membrane, water tends to move from the compartment with a higher water content to one with a lower effective water content (for example from the cell wall water into a cell that generally has a higher concentration of solutes); this generates a pressure in the cell known as *turgor pressure*. In addition to the chloroplasts that we have discussed already, there are a number of other sub-cellular compartments or organelles, also bounded by membranes.

Although it is tempting to describe the water status of a plant tissue such as a leaf in terms of the absolute water content (whether expressed per unit area or per unit mass), such a measure can be readily seen to be of little direct relationship to cell function. For example, the water content per unit area of a thin cereal leaf, even when the plant is well watered, is substantially less than the corresponding value for a succulent leaf, such as a cactus. This does not imply that the cereal is necessarily subject to any greater water-deficit stress. Physiologically a much more useful measure of leaf water content (see Jones, 1992) is the *relative water content* (*RWC*) defined as

$$RWC = (\text{fresh mass} - \text{dry mass})/(\text{turgid mass} - \text{dry mass}), \quad (4.2)$$

where the turgid mass is the mass obtained after allowing the leaf cells (but not the intercellular spaces) to saturate by floating the leaf tissue on water for about 24 h at low temperature and under low light to inhibit respiratory losses. The fresh mass is the value obtained on weighing a freshly cut leaf and dry mass is obtained after drying at about 80 °C until no further change in mass is observed. The specific advantage of *RWC* is that it normalizes water content for leaf thickness and for differences in tissue elasticity.

In much of the plant physiological literature, plant water status is not expressed in terms of water content (or *RWC*) but in terms of the free energy of water in the plant tissue, expressed as a *leaf water potential* (ψ_l; expressed in pressure units as MPa). Gradients in leaf water potential determine the direction of water movement, i.e. from the soil through the stem to the leaves and out into the atmosphere in a transpiring plant, while the value of ψ provides a measure of the degree of water stress to which a plant is subject at any time. A *RWC* of 1.0 (sometimes expressed as 100%) approximately corresponds to a *leaf water potential* of 0 MPa, which is the value expected for an unstressed plant when water is freely available; as a plant becomes increasingly deficient in water, and hence more 'water stressed', the water potential falls further below zero until at the time of wilting it is often of the order of −1.5 MPa (see Jones (1992) and Chapter 11 for further discussion of water stress and its consequences in plants).

In simple terms it is possible to partition the total water potential for any tissue into components including those relating to pressure (ψ_p), to the solute concentration (ψ_π), and to gravity (ψ_g) according to

$$\psi = \psi_p + \psi_\pi + \psi_g, \quad (4.3)$$

where the osmotic component (ψ_π) is always negative as it arises from the dissolved solutes (and matric binding to surfaces) lowering the free energy of the water, while the pressure component is positive in living plant cells where it is known as turgor pressure, but can be substantially below atmospheric (i.e. negative) in the transpiration stream in the conducting dead xylem vessels. The ψ_g results from differences in potential energy due to a difference in elevation above a reference plane, and in plants only becomes important when considering water flow to the top of tall trees.

Although the total leaf water potential (ψ_l) may not be as useful a measure of plant water status as was once thought, it is still of great value, both in the study of water flow, and more importantly because of its generally close correlation with turgor pressure, which is the quantity that primarily determines many physiological responses to drying (Jones, 2007).

Water potential of water in solution, air, or in plant tissue may be measured by means of a psychrometer that measures the wet bulb depression of air in equilibrium with the tissue or solution (i.e. is a measure of the air humidity) (e.g. Jones, 1992; Nobel, 2009). Unfortunately psychrometers tend to be laboratory instruments and are not convenient for routine use in the field, so an alternative approach based on the *pressure chamber* introduced by Dixon and Scholander is more widely used. Its use involves cutting a leaf from a transpiring plant and sealing it in a pressure vessel with the cut end of the petiole emergent from the seal. Pressure is increased slowly until a meniscus of water appears at the cut surface; the pressure at this point is assumed to equal the magnitude of the tension in the conducting xylem vessels before the leaf was excised. The greater the tension, the more stressed is the plant. Although this method strictly estimates the xylem pressure potential it is a good estimate of plant water potential, and when the measurement is made predawn it gives a good indication of the water availability in the soil (Jones, 2007).

Evaporation and water loss

The rate of evaporative water loss is a key determinant of plant water status; at least it is somewhat more amenable to study using remote sensing than is CO_2 exchange. Evaporation of water requires a substantial input of energy to satisfy the latent heat of vaporization (λ), which for water is 2.454 MJ kg^{-1} at 20 °C. This is over three orders of magnitude greater than the amount of heat required to raise the temperature of the same amount of water by 1 K, as given by the specific heat capacity (c_p = 1.01 kJ kg^{-1} K^{-1}). The energy exchange associated with evaporation therefore provides, through its effect on surface temperature, a powerful tool for remote sensing of evaporative fluxes. Indeed as we shall see below, the energy fluxes associated with evaporation can be of the same order of magnitude or even bigger than convective or radiative fluxes.

Water lost by evaporation from within the leaf intercellular spaces and then out through the stomata is termed *transpiration*, while the total water lost from a canopy (including both transpiration and any evaporation from the surface of wet leaves or evaporation from the soil) is termed *evapotranspiration* (we will use the symbol $\boldsymbol{E}$ to refer to evaporation, whatever its source, though many authors use *ET* to distinguish evapotranspiration). The importance of water to life is perhaps not surprising given that life originally evolved in an aqueous environment. It follows that in the evolutionary step when plants advanced from aquatic to terrestrial environments, adaptations were needed to allow plants to maximize the uptake of water from the environment (e.g. roots) and to minimize the losses of water by evapotranspiration (e.g. the development of stomata and an impermeable cuticle).

Unfortunately, the process of photosynthesis requires the stomatal pores in the leaf to be open to permit inward diffusion of CO_2 and an inevitable consequence of this is the potentially serious loss of water vapour out of the leaf, diffusing from the saturated surfaces inside the leaf to the generally drier air outside (Fig. 4.4). Although one might have expected evolution to develop a membrane or other adaptation that is preferentially permeable to CO_2 as compared with water, no such material appears to have evolved, so optimization of the ratio between photosynthetic carbon uptake and water loss has been a key factor in evolution and is a key target of plant breeders selecting for performance in dry environments. As outlined below, the fluxes of CO_2 and H_2O depend on the driving forces for the fluxes (the concentration differences between inside and outside the leaves), and on the resistances to flow in the pathway.

4.2.3 Other exchange processes (e.g. momentum and pollutants)

In addition to mass transfers between the atmosphere and the vegetation the forces exerted on plants by the

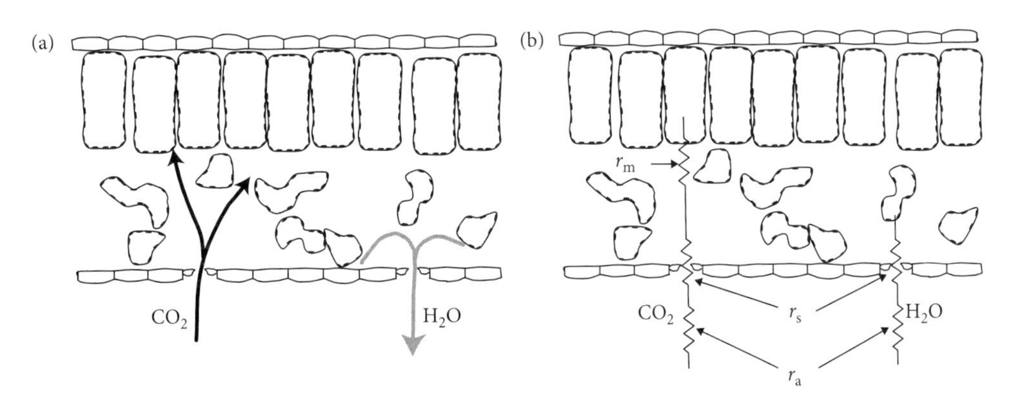

Fig. 4.4. (a) Illustration of the comparable pathways for water vapour and CO_2 exchange by plant leaves. Both pathways go through the stomata, but while water vapour largely originates at the surfaces of the cells near to the stomata, CO_2 also has a component of transfer through the cell wall to the chloroplasts in the liquid phase.
(b) Corresponding resistance networks showing resistances to transfer in the mesophyll (r_m – CO_2 only), in the stomata (r_s) and in the boundary layer surrounding the leaf (r_a).

wind are a manifestation of *momentum transfer*. The wind applies a shearing force on the surface, and the efficiency of transfer of this shearing force and the corresponding air velocity gradient are functions of the air viscosity. An understanding of velocity gradients and momentum transfer is particularly useful for mass-transfer studies and especially for derivation of the atmospheric transfer coefficients (see Section 4.3.1) because of the close similarity between transfer coefficients for different atmospheric properties (heat, mass, and momentum). Momentum transfer is difficult to estimate remotely and we generally have to rely on indirect estimates, though wind shear can be estimated from radar scattering over oceans, through effects on wave height.

A further mass-transport process that is relevant to many vegetation situations is the transfer of atmospheric pollutants to plant canopies; many of these are naturally occurring but have been substantially increased by human activity. The deposition of potentially toxic dry gaseous pollutants such as ammonia (NH_3), sulphur dioxide (SO_2), nitrogen oxides (NO_x), and ozone (O_3) occurs by a similar atmospheric transfer process to that for the transfer of CO_2. The key differences relate to the actual absorption process at the leaf surface, which, depending on the pollutant and whether the surface is wet (as occurs after rain or dew), or dry, may be through the stomata or involve direct absorption through the cuticle. In some cases pollutants such as NH_3 and sulphates can dissolve in rainwater and also be deposited as wet pollution. Another important gaseous exchange that is critical to models of climate change is the loss of methane (CH_4) from many aquatic and tundra ecosystems. However, as for momentum, direct monitoring of these exchange processes is not amenable to current remote-sensing technologies.

4.3 Principles of mass and energy transfer between vegetation and atmosphere

4.3.1 General transport equation

The spontaneous transfer of mass, or of other entities such as heat, momentum, or even electric current, occurs from a region of high '*concentration*' to one of lower 'concentration'. The rate of transfer in any situation is a function both of the concentration difference, which acts as the driving force for the transfer, and of a proportionality constant, sometimes referred to as a *transfer coefficient*, which measures the effectiveness of the transfer process. This relationship is commonly known as the *general transport equation* and can be written in the form

$$\text{Flux density} = \text{proportionality constant} \times \text{driving force}. \tag{4.4}$$

The magnitude of the proportionality constant, and hence of the rate of transfer, depends on the mechanism involved: at the molecular scale transport is by *diffusion* and transfer is a rather slow process, while at the canopy scale transfer is largely by *convection*, which can be much faster.

Units. Although fluxes have historically been expressed in mass flux terms (e.g. kg m^{-2} s^{-1}), it is often more convenient and is becoming more widely accepted, especially in the plant physiological literature, to express them in molar terms (using a caret (ˆ) to distinguish molar fluxes and concentrations). Both molar and the more conventional mass units will be used in this text with usage depending on context. Division by molar mass (M) converts mass to moles while other conversions are outlined in Appendix 1.

4.3.2 Diffusion

At the molecular scale transfer is primarily by *diffusion*, which depends on the rapid thermal motions of the individual molecules leading to a random rearrangement of their position, which, when the fluid is non-homogeneous, will lead to a net transfer from areas of high concentration to lower concentration. In one dimension the rate of this transfer by diffusion is described by *Fick's first law of diffusion* as being proportional to the concentration gradient

$$\hat{J}_i = -\mathbf{D}_i\,(\partial \hat{c}_i / \partial z). \tag{4.5}$$

In this equation $\hat{J}_i$ is the molar flux density (mol m^{-2} s^{-1}) of the material being considered, $\partial \hat{c}_i / \partial z$ is the gradient of concentration (mol m^{-3} m^{-1}) and represents the driving force and $\boldsymbol{D}_i$ (m^2 s^{-1}), is a transfer coefficient, which in this case is termed a *diffusion coefficient* and determines how rapid the transfer is. Equation (4.5) can alternatively be expressed in terms of the corresponding mass units (J and c_i). The diffusion coefficient increases proportionally to increasing (Kelvin) temperature as the molecular motions increase in amplitude and increases for smaller molecules according to *Graham's law* (which states that $\boldsymbol{D}$ is proportional to $M^{-1/2}$; see Appendix 1).

In many practical situations it is more convenient to measure concentrations at two positions in the system rather than to attempt to derive the concentration gradient at a point, so that Fick's Law is often applied in its integrated form (equivalent to eqn (4.4)) as

$$\hat{J}_i = (\boldsymbol{D}_i/\ell)\,(\hat{c}_{i1} - \hat{c}_{i2}) = \hat{g}\Delta\hat{c}_i, \tag{4.6}$$

where $\hat{c}_{i1} - \hat{c}_{i2}$ ($=\Delta\hat{c}_i$) is the driving force, ℓ is the distance between the points, and the proportionality constant $\boldsymbol{D}_i/\ell$ is commonly called a *conductance* ($\hat{g}$, m s^{-1}).

4.3.3 Convection and turbulent transfer

As we have seen above, mass or heat transfer in a still fluid (such as in the air within the intercellular spaces of a leaf) involves the individual molecules moving by diffusion down a concentration gradient solely as a result of their random thermal motions at the molecular scale. In still air fluxes for different entities are then proportional to their diffusivities. In laminar conditions relative fluxes are approximately proportional to $(\boldsymbol{D}_1/\boldsymbol{D}_2)^{0.67}$ (see Jones, 1992). For plant surfaces outdoors, however, wind and the consequent air movement over the leaf or other surfaces can greatly enhance the rate of transfer of heat and mass. This enhanced flux depends both on the fact that the moving air stream entrains the molecules removing them from the region of the surface and consequently preventing a build up of concentration (and thus maintaining the driving force) and on the more rapid transfer in the eddies that build up. Conveniently, the form of the equations defining turbulent transfer is the same as for diffusive transfer, just with a larger transfer coefficient.

In practice, the air in the lower atmosphere is never really still with both a horizontal component of wind and many random movements of small eddies (Fig. 4.5). Because these eddies tend to be at least similar in scale to the size of surface irregularities it follows that they are several orders of magnitude larger than the molecular movements giving rise to diffusion and are correspondingly greater contributors to mass and heat transfer. The magnitude and importance of these eddies also increases as the wind speed increases, so that conversely, as the wind is attenuated as it approaches the ground surface, the size and contribution of the eddies to mass and heat transfer decreases. Very close to the plant surfaces, wind speeds get so low that the turbulent eddies may disappear altogether and any flow becomes parallel to the surface forming the so-called *laminar boundary layer*. Within such a laminar layer, transfer between layers is essentially diffusive.

At a larger scale, the pattern of air movement also depends on the convection regime that exists: where the air is quite still it can be heated by contact with hot surfaces and expands. This decreased density causes the air to rise, transporting both heat

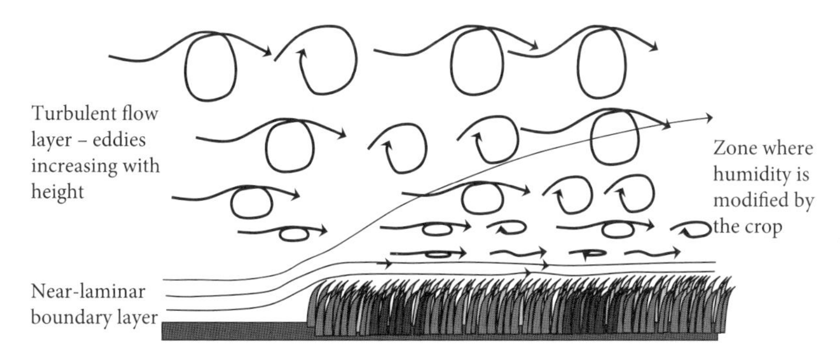

Fig. 4.5. A diagrammatic illustration of the turbulent structure of air flowing over a field showing a near-laminar boundary layer near the surface and an increasing amount of turbulence and eddy size with increasing height above the surface. The further downwind one goes, the greater the depth of the zone where the air's properties (such as humidity) are modified by transpiration from the crop.

and mass in a process known as *free convection*. On the other hand, air movements may be generated by external pressure gradients, leading to what is termed *forced convection*. The eddy size then depends on the *roughness* of the surface, with forests, for example, generating relatively greater turbulence than comparatively smooth surfaces such as grassland. In windy conditions forced convection dominates. When the surfaces are warmer than the overlying air the atmosphere is *unstable* with the surfaces heating the air, which then rises by free convection, thus transfer in such conditions may be dominated by free convection, while when the surface is cooler than the overlying air (as during a temperature inversion such as commonly occurs at night especially during a radiative frost event) the atmosphere is *stable* and any transfer is dominated by forced convection processes. The difference between free and forced convection can be illustrated with reference to conventional space heating in houses: a fan heater forces the air over the heating element giving heating by forced convection, while a conventional 'radiator' transfers the heat in air movements generated by the heating of the air near the radiator (*free convection*). The actual transfer of heat from the heater surface to the air, however, is primarily by *conduction* from the solid to the adjacent air molecules.

4.3.4 Resistances and conductances

In plant studies the proportionality constant in eqn (4.6) (= $\mathbf{D}_i/\ell$ for mass diffusion) or the corresponding transfer coefficient for turbulent flow is conventionally called a conductance and given the symbol g or $\hat{g}$. In many situations (especially where dealing with conductances acting in series such as with the diffusion through stomata and the boundary layer adjacent to a leaf) it is algebraically more convenient to use the reciprocal of the conductance, termed a *resistance* with the symbol r (Figs 4.4 and 4.6). It is therefore useful to become familiar with both terms as they are used interchangeably throughout this book. Note that both conductances and resistances are *system properties* (and hence use the -ance ending; see Box 2.1). This is because they combine both the property of the medium (the diffusivity or diffusion coefficient) and of the dimensions of the system (ℓ).

4.3.5 Units and analogies between different transfer processes

The transfer of quantities as different as heat, water vapour, CO_2, electric current and momentum can all

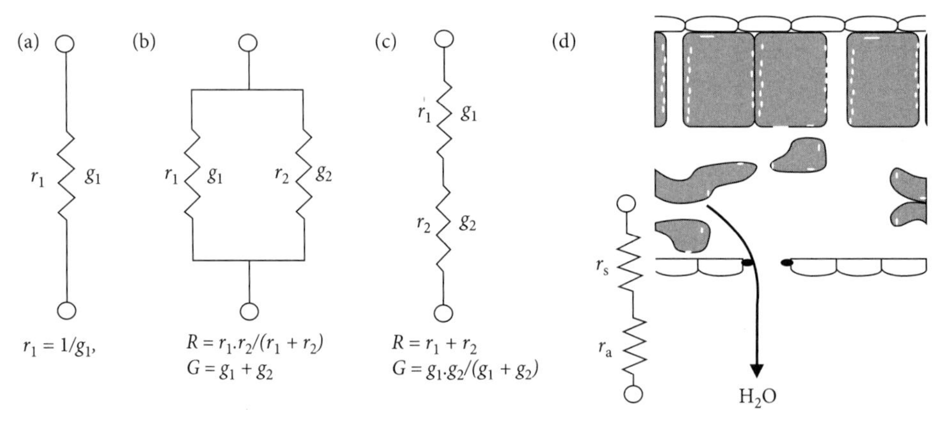

Fig. 4.6. The use of electrical circuit analogies to show the arithmetic relationships between resistances (r) and conductances (g), for parallel and series arrangements of resistors, where R and G refer to the total resistances or conductances across the relevant circuit. (a) shows a simple component resistor, (b) shows flow through parallel components where the driving force across each component is the same, and the flux through each is proportional to the ease with which it can move (conductance), (c) illustrates the situation for components in series where the same flux flows through each and the driving force across each component depends on its resistance, (d) illustrates for a leaf the series arrangement of resistances to water vapour loss which flows sequentially through the stomatal resistance (r_s) and then a boundary layer resistance (r_a). Where a leaf has stomata on both upper and lower surfaces the resistances of these two surfaces would be in parallel.

be described in a similar way with only the driving force and the constant of proportionality changing. The analogies between these processes are summarized in Box 4.2 showing the appropriate units to be used for each. For example, for mass transfer of CO_2 (as in photosynthesis or respiration) the driving force is the CO_2 concentration difference between two points (e.g. within the leaf and in the bulk air outside the leaf), while for evaporation, the driving force will be the difference in air humidity between the inside and outside of the leaf. These 'concentration' differences can be expressed in a number of different ways ranging from the conventional concentration (mass per unit volume, or density) through molar concentration to measures such as mole fraction (x_i) (Box 4.3). As shown in Box 4.2 for mass transfer, each formulation of concentration leads to differing proportionality coefficients.[14] The important conversion between molar and mass units for conductance is given by $\hat{g} = (P/\mathfrak{R}T)\, g$, so that $\hat{g}$ (mmol m^{-2} s^{-1}) is approximately equal (at 25 °C) to $40 \times g$ (mm s^{-1}).

4.4 Canopy–atmosphere exchanges

4.4.1 Energy balance – steady state

Central to all canopy transfer processes is the canopy *energy balance*; this is the balance between those processes, such as insolation that tend to heat the canopy, and those such as evaporation that remove heat (see Fig. 2.14). Since the law of energy conservation says that energy cannot normally be created or destroyed, the sum of all the energy fluxes at the surface must be zero, so we can write

$$\mathbf{R}_n - \mathbf{C} - \lambda\mathbf{E} - \mathbf{M} - \mathbf{S} = 0, \qquad (4.7)$$

BOX 4.2 Analogies between different integrated forms of the transport equation for different transport processes (see Jones (1992) and Campbell and Norman (1998) for full explanation and derivations). Some conversions between mass units and molar units are given in Appendix 1.

Transfer process	Flux density	= Driving force	× Conductance
Electric charge (*Ohm's law*)	I (A)	$= V$ (W A^{-1})	$\times 1/R$ (A^2 W^{-1})
Mass transfer (*Fick's law*)			
(mass units)	J_i (kg m^{-2} s^{-1})	$= \Delta c_i$ (kg m^{-3})	$\times \mathbf{D}_i/\ell\ (= g)$ (m s^{-1})
(molar units – 1)	$\hat{J}_i$ (mol m^{-2} s^{-1})	$= \Delta\hat{c}_i$ (mol m^{-3})	$\times \mathbf{D}_i/\ell$ (m s^{-1})
(molar units – 2)	$\hat{J}_i$ (mol m^{-2} s^{-1})	$= \Delta x_i$ (dimensionless)	$\times P\mathbf{D}_i/\ell\mathfrak{R}T\ (= \hat{g})$ (mol m^{-2} s^{-1})
Heat transfer (*Fourier's law*)	C (J m^{-2} s^{-1})	$= \Delta T$ (K)	$\times K/\ell$ (W m^{-2} K^{-1})
(Campbell and Norman units)	C (J m^{-2} s^{-1})	$= \hat{c}_p \Delta T$ (J mol^{-1})	$\times \mathbf{D}_H/\ell$ (W m^{-2} K^{-1})
Momentum transfer (*Newton's law of viscosity*)	τ (kg m^{-1} s^{-2})	$= \Delta u$ (m s^{-1})	$\times \eta/\ell$ (kg m^{-2} s^{-1})

where c_i is the (mass) concentration (kg m^{-3}); $\hat{c}_i$ is molar concentration (mol m^{-3}); $\mathbf{D}$ is a diffusion coefficient (m^2 s^{-1}), K is thermal conductivity (W m^{-1} K^{-1}); η = is dynamic viscosity (kg m^{-1} s^{-1}); τ is shearing stress (kg m^{-1} m^{-2}); u is wind velocity (m s^{-1}); x_i is mol fraction (mol mol^{-1}); I is current (A); V is voltage (V); R is resistance (ohm); $\mathfrak{R}$ is the gas constant. Note that for turbulent transfer, $\mathbf{D}$ is replaced by a larger *transfer coefficient* that depends on the effectiveness of the turbulence.

[14] A good argument in favour of molar units for conductance or resistance is that $\hat{g}$ is relatively independent of the properties of the air, being independent of P and approximately proportional to T, while g is inversely proportional to P and proportional to T^2 (Jones, 1992).

BOX 4.3 Measures of gas concentration

In order to understand driving forces for transfer processes between vegetation and the atmosphere we need to consider what is meant by 'concentration'. Although concentration is usually thought of in terms of a volume fraction (e.g. %) or as mass per unit volume (= density, ρ; kg m^{-3}), it is often more convenient to express concentrations in terms of molar concentration ($\hat{\rho}_i$, moles of the quantity per unit volume) or even better as mole fraction (x_i). A mole of substance i contains Avogadro's number of molecules (6.02252×10^{23}), with the molecular mass (often incorrectly called 'weight') being the mass of one mole (M_i). Conveniently, the volume of one mole of gas is the same for all gases and at the standard temperature of 0 °C and pressure of 101.3 kPa equals 0.0224 m^3 mol^{-1} (or 22.4 litres), with the corresponding molar density, ($= n/V = 44.6$ mol m^{-3}). Since the volume of a gas changes in proportion to the (Kelvin) temperature, T, and inversely in proportion to pressure, P (the *Universal gas law*), we can write for the partial pressure, p_i: $p_i\ V = n_i\ \Re T$, where $\Re$ is the *universal gas constant* (8.3143 J mol^{-1} K^{-1}). Therefore the molar density at any temperature and pressure can be obtained from

$$\hat{\rho} = n_i/V = p_i/\Re T = 44.6\,(p/101.3)\,(273.15/T). \quad \text{(B4.3.1)}$$

Mass concentration or density

$$c_i = \rho_i = m_i/V = n_i\,M_i/V = p_i\,M_i/\Re T. \quad \text{(B4.3.2)}$$

Molar concentration

$$\hat{c}_i = n_i/V = c_i/M_i = p_i/\Re T = \hat{\rho}_i. \quad \text{(B4.3.3)}$$

Mole fraction

$$x_i = n_i/\Sigma n = p_i/P = n_i\,\Re T/PV = \hat{\rho}_i\,\Re T/V = n_i/n_a = c_i\,M_a/c_a\,M_i.$$

[*Mixing ratio*

$$w_i = m_i/(\text{total mass} - m_i)]$$

Conversions for water vapour When describing water vapour in air, it is common to express its concentration in terms of its partial vapour pressure (e) or even in terms of the relative humidity (e/e_{sat}) where e_{sat} is the saturation vapour pressure at air temperature. Because the partial pressure of a gas is proportional to the molar density of that gas in the mixture, it follows from the fact that the molar volume is fixed that the partial pressure of that gas divided by the total pressure equals the mole fraction x_i of that gas component, i.e. $x_i = n_w/n = e/P$. A range of other units for water vapour concentration in air such as dew point temperature, wet bulb temperature, or water potential are widely used for specific purposes (Jones, 1992).

[Information on the temperature dependence of saturation vapour pressure is provided in Appendix 3.]

where $\mathbf{R_n}$ is the net radiation flux absorbed by the canopy and includes both the shortwave radiation absorbed ($= \alpha \mathbf{R_S}$) and the net longwave absorbed ($= \mathbf{R_L}$down $- \mathbf{R_L}$up), $\mathbf{C}$ is the loss of heat from the surface as a 'sensible' heat flux through conduction and convection, $\lambda \mathbf{E}$ is the loss of heat in evaporation as a latent heat flux, $\mathbf{M}$ is a metabolic term indicating the net heat stored as chemical energy in biochemical reactions such as in photosynthesis (when heat is generated as a result of exothermic reactions such as respiration, $\mathbf{M}$ is negative), and $\mathbf{S}$ is the heat flux into physical storage that acts to raise temperatures (for heat flux into soil this is usually given the symbol $\mathbf{G}$).[15] It is convenient to express all these fluxes per unit area of ground (or of the projected area of leaves for leaf studies) to give all energy exchanges in units of flux density (W m^{-2}).

For most environmental studies the metabolic term is usually ignored as it is commonly about two orders of magnitude smaller than the other fluxes. The canopy temperature reaches equilibrium when the net energy flux is zero (gains equal losses) with no flux into or out of storage. The value of the temperature at any time is an important variable that can be sensed remotely using thermal sensors; it therefore provides a key tool in the monitoring of energy fluxes to and from canopies.

The control of canopy temperature is quite complex: not only do the magnitudes of the various fluxes

15. It is worth noting that the standard meteorological convention with all downward (absorbed) fluxes being negative and upward fluxes (losses) being positive is somewhat different from the surface-based terminology used here. The meteorological convention arises because they are concerned with gains or losses by the atmosphere rather than the surface.

determine canopy temperature, but of course canopy temperature itself determines the magnitudes of many of the fluxes. Indeed temperature is a key linking variable in this feedback. Canopy temperature affects both the sensible heat flux term and the latent heat term (through its effect on the vapour pressure of water in the leaf spaces). The radiative terms in this equation have been discussed in detail in Chapters 2 and 3, so in what follows we address the sensible and latent heat terms and the overall solution of this equation for leaf temperature and **E**.

4.4.2 Sensible heat flux

Using the conventional form of Fourier's law for heat conduction (Box 4.2) we can write the sensible heat flux (**C**, though note that many texts use the symbol **H**) as a function of the temperature difference between the canopy surface (T_s) and the bulk atmosphere (T_a) as

$$\mathbf{C} = K/\ell\,(T_s - T_a) = g_H\,\rho c_p\,(T_s - T_a), \tag{4.8a}$$

where the symbols are explained in Box 4.2. This can be written alternatively in molar units (Campbell and Norman, 1998) as

$$\mathbf{C} = \hat{c}_p\,\hat{g}_H\,(T_s - T_a) = \hat{c}_p\,(T_s - T_a)/\hat{r}_H, \tag{4.8b}$$

where $\hat{c}_p$ is the molar specific heat of air (= 29.3 J mol^{-1} K^{-1}). The boundary-layer conductance, or its inverse the resistance, is that which is appropriate for heat transfer from the surface to the reference level at which air temperature is measured.

4.4.3 Evaporation

The rate of evaporation from plant canopies is determined by (i) *environmental conditions* such as the air temperature, the availability of energy, the vapour-pressure difference between the canopy and the atmosphere, and by the transfer resistance in the atmospheric boundary layer, and (ii) by *plant factors* such as leaf area, canopy structure, and especially by the stomatal aperture. As pointed out in Section 4.2.2, **E** impacts substantially on the energy balance and is therefore critical to the control of canopy temperature. According to Box 4.2, for evaporation from a leaf (transpiration, **E**; mol m^{-2} s^{-1}) we can write

$$\mathbf{E} = \hat{g}_W\,(x_s - x_a) = (x_s - x_a)/\hat{r}_W \tag{4.9}$$

where $\hat{g}_W$ is the total conductance (mol m^{-2} s^{-1}) to water vapour that includes both the pathway through the stomata and in the atmospheric boundary layer around the leaf and canopy, x_s is the humidity (expressed as a mol fraction, which equals the vapour pressure, e, divided by air pressure, P) in the intercellular spaces within the leaf, and x_a is the corresponding water vapour pressure in the air. The value of x_s is usually assumed with very little error to be equal to the saturation mole fraction of water at leaf temperature, so is an exponential function of leaf temperature (see Appendix 1). This equation can be expanded to partition the overall resistance to water loss, $\hat{r}_W$, into a diffusive component representing water loss through the stomatal pores in the plant leaves ($\hat{r}_s$), and a primarily convective component ($\hat{r}_a$) relating to transfer through the leaf boundary layer (e.g. Fig. 4.4) and also, where relevant, the boundary layer above the canopy:

$$\mathbf{E} = (x_s - x_a)/\hat{r}_W = (x_s - x_a)/(\hat{r}_s + \hat{r}_a) = (x_s - x_{epidermis})/\hat{r}_s = (x_{epidermis} - x_a)/\hat{r}_a, \tag{4.10}$$

where $x_{epidermis}$ is the humidity (mol fraction) at the epidermal surface of the leaf and the principle of continuity means that in a steady state, the flux through each of a series of resistances in series must be the same so that the last two equalities in this equation hold. The corresponding rate of heat loss directly associated with the water loss is given by $\lambda\mathbf{E}$ (W m^{-2}).

Units for evaporation[16]

For energy-balance studies evaporation is usually expressed in terms of equivalent energy transfer (= $\lambda\mathbf{E}$; W m^{-2}), but for most agricultural or ecological purposes the favoured units for evaporation are mm of water per h, while for compatibility with physiological studies it is often useful to use alternative units expressed in terms of mol m^{-2} s^{-1}. In an arid environment, **E** may reach 1 mm h^{-1}, which equals 1 kg m^{-2} h^{-1} or 0.278 g m^{-2} s^{-1} or 682 W m^{-2} (at 20 °C) or 15.4 mmol m^{-2} s^{-1}.

16. The appropriate conversions between different units for evaporation at 20 °C are: 1 mm = 1000 cm^3 m^{-2} = 0.01 ML ha^{-1} = 1 kg m^{-2} = 55.55 mol m^{-2} ≅ 2.454 MJ m^{-2}.

4.4.4 Penman–Monteith combination equation

Solution of the steady-state energy balance of a leaf or plant canopy (eqn (4.7)) to obtain an estimate of $\boldsymbol{E}$ requires not only information on the environmental conditions such as radiation, air temperature, humidity, and transfer resistances, but it also requires a knowledge of canopy surface temperature. Unfortunately the surface temperature both *determines* the latent heat ($\lambda \boldsymbol{E}$) and the sensible heat ($\boldsymbol{C}$) exchanges (eqns (4.10) and (4.8)) and is itself *determined* by these exchanges. The energy balance can be solved to estimate both leaf temperature and evaporation rate by making use of a linear approximation originally proposed by Penman (1948). The approach is to express the leaf–air humidity difference that drives evaporation ($x_s - x_a$) in terms of the leaf–air temperature difference and the humidity deficit of the air ($x_{sat(Ta)} - x_a$) = D_x by using the following approximation (illustrated in Fig. 4.7):

$$(x_s - x_a) = x_{sat(Ts)} - x_a \cong (x_{sat(Ta)} - x_a) - \hat{s}\,(T_a - T_s) = D_x + \hat{s}\,(T_s - T_a), \qquad (4.11)$$

where humidities are expressed as mole fractions ($x = e/P$) and $\hat{s}$ is the slope of a curve relating saturation water vapour mole fraction and temperature (at $(T_a - T_s)/2$).

Most frequently this substitution is used to derive an equation, the Penman–Monteith equation, for estimation of evaporation. The usual derivation substitutes eqn (4.11) into eqn (4.9), and then eliminates ($T_s - T_a$) from this equation and eqn (4.8) to give

$$\boldsymbol{E} = \hat{g}_W\,(D_x + \hat{s}\,\boldsymbol{C}/\,\hat{c}_p\,\hat{g}_H). \qquad (4.12)$$

Elimination of $\boldsymbol{C}$ from this and the energy balance (eqn (4.7) – assuming for the steady state that flux into storage is zero) and rearranging gives

$$\lambda \boldsymbol{E} = \frac{\hat{c}_p\,\hat{g}_H\,D_x + \hat{s}\,\boldsymbol{R}_n}{(\hat{c}_p\,\hat{g}_H/\lambda\,\hat{g}_W) + \hat{s}} = \frac{\hat{\gamma}^*\lambda\hat{g}_W\,D_x + \hat{s}\,\boldsymbol{R}_n}{\hat{\gamma}^* + \hat{s}}, \qquad (4.13)$$

where $\hat{\gamma}^*$ (= $\hat{c}_p\,\hat{g}_H/\lambda\,\hat{g}_W$) is known as the modified *psychrometer constant* and corrects for differences between the conductances of water vapour and heat. A particular feature of this equation is that it eliminates the requirement for knowledge of surface temperature, but it does have a term representing the surface conductance to water vapour ($\hat{g}_W$) that includes control exercised by the stomata and the boundary layer.

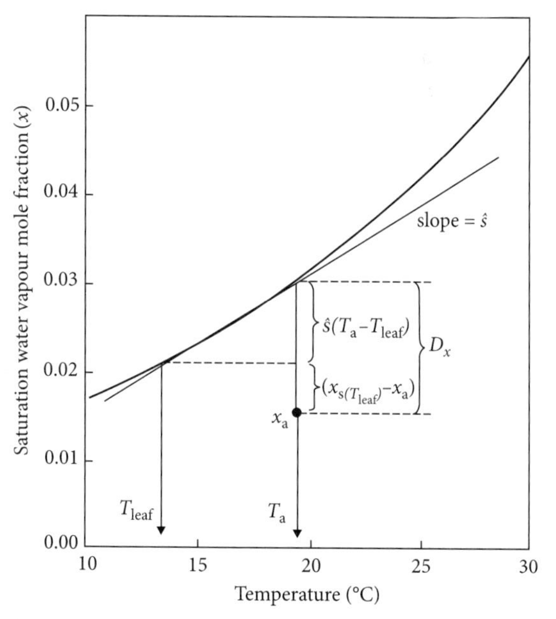

Fig. 4.7. An illustration of the Penman transformation where the solid curve represents the relationship between saturation vapour mole fraction and temperature, and the straight line of slope, $\hat{s}$, represents the average slope between the leaf surface temperature (T_{leaf}) and T_a. Water vapour mole fraction is calculated as e/P where P is an atmospheric pressure of 10^5 Pa. The surface to air vapour mole fraction difference ($x_{s(T_{leaf})} - x_a$) is given by D_x of the ambient air minus the difference between the saturation vapour mole fractions at T_a and T_s (following Jones, 1992).

An alternative arrangement of this equation gives an expression for canopy temperature. Elimination of $\boldsymbol{E}$ from eqns. (4.7)–(4.9) and again using the Penman approximation (eqn (4.11)) gives after some rearrangement (see, e.g. Jones, 1992) an expression for the leaf (or canopy) temperature as

$$T_{leaf} = T_a + \frac{\boldsymbol{R}_n - \hat{g}_W\,\lambda\,D_a}{(\hat{c}_p\,\hat{g}_H + \lambda\,\hat{s}\,\hat{g}_W)} = T_a + \frac{\boldsymbol{R}_n\,\hat{r}_W - \lambda\,D_a}{(\hat{c}_p\,\hat{r}_W\,/\,\hat{r}_{aH} + \lambda\,\hat{s})}. \qquad (4.14)$$

Comparable equations to (4.13) and (4.14) in conventional units are given in Appendix 1. There are two important assumptions in the derivation of these equations. The first is that the rate of change of saturation mole fraction with temperature is constant over the temperature range involved; this introduces negligible error for most temperature differences. The second is that net radiation is unaffected by leaf temperature, but it actually depends on leaf temperature through its effect on longwave emission; for accurate work especially in modelling this can be corrected by using the concept of *isothermal net radiation* (R_{ni}) (Jones, 1992).

Isothermal net radiation

Isothermal net radiation is defined as the net radiation that would be received by an identical surface in an identical environment if it were at air temperature. The difference from the net radiation results from the difference in the longwave emitted radiation, which is a function of surface temperature, so it can be shown (Jones, 1992) that

$$R_{ni} = R_n + \hat{g}_R \hat{c}_p (T_s - T_a), \quad (4.15)$$

where $\hat{g}_R$ is known as a radiative-transfer conductance ($= 4\varepsilon\sigma T_a^3/\hat{c}_p$; see Jones, 1992) and occurs in parallel with the normal boundary layer pathway to heat transfer. The total heat transfer conductance ($\hat{g}_{HR}$) is therefore given by the parallel sum of the two components

$$\hat{g}_{HR} = \hat{g}_H \hat{g}_R / (\hat{g}_H + \hat{g}_R). \quad (4.16)$$

For convenience we often use the reciprocals (resistances, where $r_{HR} = r_H + r_R$). Substituting $\hat{g}_{HR}$ for $\hat{g}_R$ and R_{ni} for R_n into eqns. (4.14) and (4.15) gives improved versions for modelling purposes.

4.4.5 Canopy models

Thus far we have used simple leaf analogies where the resistances to water-vapour transfer are treated as a simple series involving stomata and boundary layer and have implicitly applied these to plant canopies. In a real canopy the heat and water-vapour fluxes differ for the individual canopy components with, for example, lower leaves not being exposed to the sky, but having greater radiative and mass exchanges with the underlying soil. Nevertheless, for most agronomic or remote-sensing applications we can follow the approach used thus far and simplify the complex vertical (and often also the spatial), heterogeneity. The simplest approach is the 'big-leaf' approximation (Fig. 4.8(b)), where the whole canopy is treated in a similar way to a leaf, with both heat and mass fluxes exchanging with a single 'average' surface. For remote sensing this will often be the assumption of one average 'big-leaf' for each pixel. It is clear, however, that especially in more typical heterogeneous canopies the total canopy evaporation is the sum of evaporation from the sunlit (and probably hotter) leaves at the top of the canopy (receiving high solar radiation), leaves at the bottom of the canopy (relatively shaded), and from the soil. Much of this complexity can be adequately approximated using the 'two-source' model (Fig. 4.8(c)) where soil and canopy are treated separately as a simple network of sources of water/energy, with the within-canopy and above-canopy transfer resistances being treated separately.

For remote sensing we are usually concerned with the average flux per pixel, though especially at the smaller scales, there can be substantial interactions between fluxes in neighbouring pixels. Indeed, a full treatment of canopy transfer and energy balance, especially in

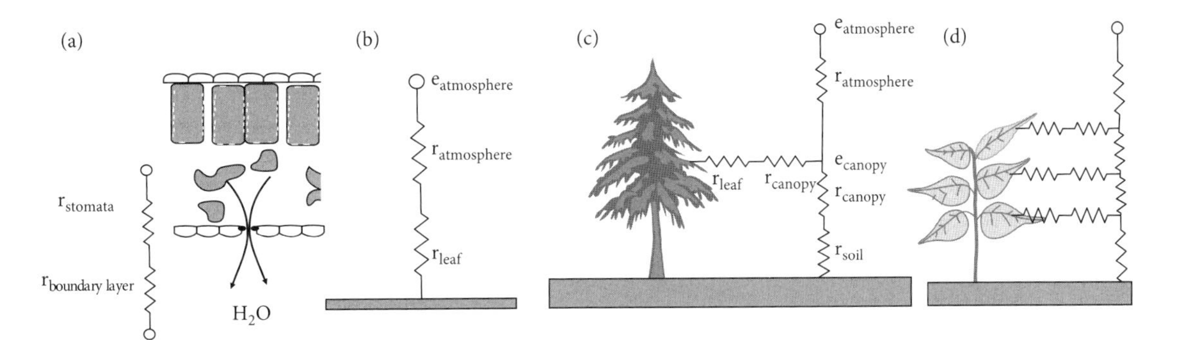

Fig. 4.8. (a) Simple leaf model for water-vapour exchange showing the stomatal and boundary layer resistances. (b) The simplest ('big-leaf') model of a canopy, where the average behaviour of the leaf and soil mixture is approximated by an average surface and the resistance to water-vapour transfer consists of a 'leaf' resistance and a boundary layer resistance in series, while the resistance to heat transfer consists only of the boundary layer resistance. (c) A rather more realistic 'two-source' model for partitioning evaporation and energy fluxes from a canopy that separates vegetation and soil, but all the vegetation is treated the same. (d) A multilayer model that allows for different energy balance and different transfer resistances for different layers within the canopy.

semi-arid or very heterogeneous systems requires one to allow for this advection of energy between pixels. A common effect is that evaporation of an isolated crop in an arid environment can be substantially greater than the amount of radiant energy received by the crop as a result of heat and dry air brought in from neighbouring areas by advection.

4.4.6 Non-steady state

A particularly powerful remote sensing tool that gives useful information about canopies, soils, and soil moisture is provided by the study of thermal dynamics. So far we have made the simplification that equilibrium has been reached and that temperatures are constant. In the real world, however, temperatures are continually changing as the inward and outward energy fluxes are rarely in balance and there is a resulting energy flux into or out of storage. For canopies, the energy storage in the plant material is generally rather small, and reaches effective equilibrium rapidly, with the major storage flux being into the soil (often given the symbol, $\boldsymbol{G}$).

Thin leaves

For thin materials such as leaves (especially when highly thermally conductive) the rate of temperature change ($\mathrm{d}T/\mathrm{d}t$) is given by

$$\mathrm{d}T/\mathrm{d}t = \boldsymbol{S}/c_{\mathrm{area}} \tag{4.17}$$

where c_{area} (J m^{-2} K^{-1}) is the specific heat capacity per unit area of the leaf ($= \rho c_{\mathrm{p}} \ell^*$; where ρ is the density, c_{p} is the specific heat capacity of the leaf, and ℓ^* is the thickness) and indicates how much heat is required to change the temperature of the body by one degree. For such a thin object, combining eqn (4.17) with the full energy balance equation (4.7) gives an expression for the rate of change of leaf temperature in terms of the difference between leaf temperature (T_ℓ) and the equilibrium temperature (T_{e}) for the environmental conditions at that time (see, e.g., Jones, 1992) as

$$\mathrm{d}T/\mathrm{d}t = (T_{\mathrm{e}} - T_\ell)(\rho c_{\mathrm{p}}/c_{\mathrm{area}}) \cdot f(r), \tag{4.18}$$

where $f(r)$ is a function of the boundary layer and surface resistances ($f(r) = 1/r_{\mathrm{HR}} + (s/\gamma(r_{\mathrm{aW}} + r_{\ell\mathrm{W}}))$, γ is the psychrometric constant ($\rho c_{\mathrm{p}}/0.622\ \lambda$) and s is the slope of the saturation vapour pressure curve. This allows T_ℓ to be expressed at any time after a change in environmental conditions from environmental conditions that give an equilibrium temperature (T_{e1}) to a set of conditions giving T_{e2} as

$$T_{\mathrm{l}} = T_{\mathrm{e2}} - (T_{\mathrm{e2}} - T_{\mathrm{e1}}) \exp(-t/\tau), \tag{4.19}$$

where τ ($= c_{\mathrm{area}}/(\rho c_{\mathrm{p}}[(1/r_{\mathrm{HR}}) + [s/\gamma(r_{\mathrm{aW}}+r_{\ell\mathrm{W}})]])$) is a time constant indicating the time for a fraction equal to $(1 - (1/e))$ of the total change (i.e. c.63%) to occur. The time constant indicates how rapidly the surface tracks changes in environmental conditions, with average values between about 30 s and 50 s for leaves and increases with c_{area} and also with both increasing boundary layer and stomatal conductances.

Thick leaves, soils, etc.

The calculation of time constants in the previous section assumes rapid heat transfer within the body whose temperature is changing, and hence a rapid approach to temperature equilibrium throughout the body; this is approximately true for thin leaves, but for thicker surfaces such as soils, or even objects such as cacti, one also has to take into account the rate at which heat is conducted away from the surface to heat the body of the object as we pointed out in Section 2.4 when we introduced the concept of thermal inertia. The slow rate of heat transfer results in a developing temperature gradient in the soil with the diurnal or seasonal temperature fluctuations being greatest at the surface and being increasingly damped as depth increases. In parallel with the increased damping is an increasing lag behind the surface temperature changes at depth as shown in Fig. 4.9.

The actual temperature dynamics at any point can be derived by solution of Fourier's equation for heat flow, which can be written in one dimension as

$$\partial T/\partial t = \boldsymbol{D}_{\mathrm{H}}\ \partial^2 T/\partial z^2, \tag{4.20}$$

where $\boldsymbol{D}_{\mathrm{H}}$ is the soil thermal diffusivity ($= K/(\rho_{\mathrm{s}} c_{\mathrm{s}})$; (often given the symbol κ by soil scientists), in m^2 s^{-1}, where K is the soil thermal conductivity (W m^{-1} C^{-1}), and ρ_{s} and c_{s} are the density and specific heat capacity of the soil. Readers are referred to appropriate texts such as Campbell and Norman (1998) for further detail and derivations. Given a sinusoidal fluctuation in the energy input to the soil, and hence a sinusoidal fluctuation in the surface temperature, a situation that is approximated over both a daily cycle and an annual

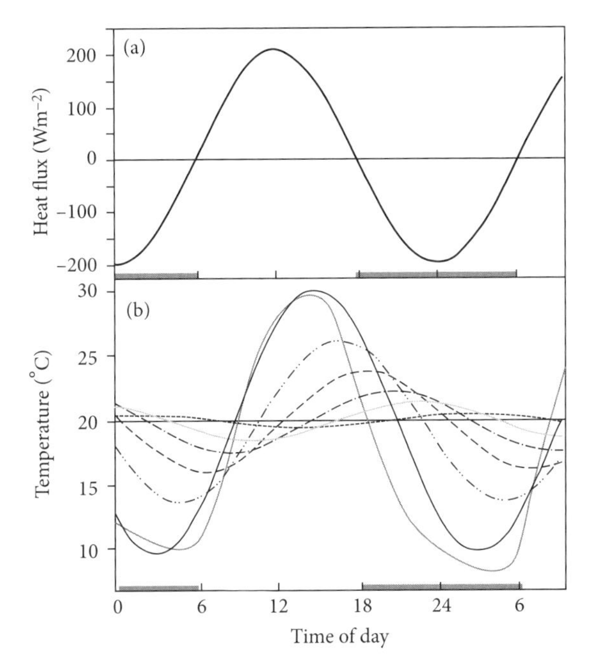

Fig. 4.9. Diurnal trends in (a) the imposed soil heat flux at the surface resulting from solar energy input on a clear day (with positive values equalling a positive flux, $\boldsymbol{G}$, into the soil) and (b) the corresponding fluctuations in soil temperature at various depths calculated using eqn (4.21) (with D = 0.102); surface (——), 5 cm depth (–··–), 10 cm (– – –), 15 cm (–·–), 20 cm (—), and 30 cm (- - - -). A typical actual surface temperature trend is also shown (pale solid line).

cycle, eqn (4.20) can be solved to give the temperature ($T_{z,t}$) at any depth (z) and time (t) as

$$T_{z,t} = T_{ave} + (\Delta T/2)\exp(-z/Z)\sin(\omega t - z/Z), \quad (4.21)$$

where T_{ave} is the average temperature over the cycle and $\Delta T/2$ is the half-maximum amplitude of the cycle, and ω is the angular frequency (= 2π/period) of the driving energy input that for diurnal cycles equals 7.3×10^{-5} s^{-1}, and for annual cycles equals 2×10^{-7} s^{-1}. The term Z is known as the *damping depth* and is a measure both of how energy is absorbed by the surface and of how deep the effect goes. It therefore determines the rate at which temperature fluctuations are attenuated with depth and how much the temperature at any depth lags behind the change at the surface. The damping depth is given by

$$Z = \sqrt{2D_H/\omega}. \quad (4.22)$$

The behaviour of this equation is illustrated in Fig. 4.9. It can be shown that the amplitude falls to only 5% of that at the surface at three times the damping depth. It is clear from the above discussion that as well as depending on the energy input and its variation over time, the magnitude of temperature fluctuations at the surface depends on a combination of the specific heat capacity and on the thermal conductivity of the material. The effect of these two quantities on the fluctuation in surface temperature for a homogeneous material can be expressed as the *thermal admittance* (μ), or, as we indicated in Section 2.6, the equivalent *thermal inertia* (P). The latter term is commonest in the remote sensing literature, where eqn (2.19) can be written:

$$\mu = P = \sqrt{K\rho_s c_s} = \rho_s c_s \sqrt{\boldsymbol{D}_H}. \quad (4.23)$$

This term provides a useful measure of the tendency of a material to resist changes in surface temperature. Rather counter-intuitively, surfaces with highly conductive materials show smaller/slower changes in surface temperature as, for example, radiation input changes. This effect results from the fact that with increasing thermal conductance, the absorbed radiant energy can penetrate further into the surface and therefore needs to heat up a larger volume of material; this leads to smaller and slower temperature changes at the surface. It follows that insulators such as wood have a low thermal admittance (low inertia) and show rapid surface temperature changes as compared with highly conductive materials such as metals. Conversely, it is worth noting that materials with a high thermal inertia show greater temperature changes at depth in the material. Over a diurnal cycle, surface temperatures often approximate the sinusoidal curve quite closely (Fig. 4.9). Some typical diurnal curves for different surfaces are shown in Fig. 4.10. It is apparent from this figure that a much larger diurnal temperature range is found for sand (or for other dry soils) than for vegetation, with water surfaces having even more stable surface temperatures (and hence higher thermal inertia).

Over a full diurnal cycle, the amount of heat stored in a surface during the sunlight hours is approximately equal to the amount lost at night and is proportional to the thermal inertia or thermal admittance of that surface. It can be seen from Fig. 4.9 that the surface temperature lags behind the energy input by 1/8th of a cycle. The amplitude (the peak-peak difference between maximum and minimum) of the surface temperature change (ΔT) is given by $2\boldsymbol{G}_{max}/(P\sqrt{\omega})$, where

G_{max} is the maximum diurnal soil heat flux. For bare soil surfaces, the value of G_{max} is often around 40% of the daily maximum net radiation ($R_{n\text{-}max}$), with the rest of the heat being lost as sensible and/or latent heat. For vegetated surfaces, G_{max} is much smaller (see Chapter 9).

For a dry soil where latent heat can be ignored, the partitioning of absorbed radiant energy between soil heat flux (G) and sensible heat flux (C) depends on the relative thermal admittance of the soil and the atmospheric transfer processes, i.e. $G/C = \mu_{soil}/\mu_{atm}$ (or P_{soil}/P_{atm}). The former depends on soil type and on its insulating properties, while the latter depends on wind speed and turbulence. For dry soil G/R_n ranges from 0.98 for still air, through 0.33 for calm conditions to <0.1 for very windy conditions, while corresponding values for a soil covered in insulating mulch might range from 0.95 to less than 0.04 (Table 8.1 in Campbell and Norman, 1998).

Some typical thermal properties of soils and soil components are summarized in Table 4.1, together with values for the thermal inertia (P; J s$^{-1/2}$ m^{-2} K^{-1}) of typical vegetated surfaces. Values range from about 500 for dry soils through 1500 for water to 2400 for some soil minerals to as much as 20 000 for some metals.

The remote sensing of thermal dynamics can therefore provide much useful information, though in many remote-sensing applications we do not have the full information on the energy fluxes or the surface parameters (K, ρ_s, or c_s) necessary for calculation of P. Therefore, the temperature difference (ΔT) between two images (at different times such as midday and midnight) is often instead used to derive an *apparent thermal inertia* (*ATI*). Figure 4.10 shows the temperatures obtained from two satellite images in the early morning and early afternoon, approximately at the time of extreme surface temperatures, from which *ATI* can be calculated according to

$$ATI = (1 - A)/\Delta T, \tag{4.24}$$

where A is the solar albedo; this equation can include a modification to allow for varying incident radiation. The albedo is used to account for the effects that

Table 4.1 Soil thermal properties, where T is in °C (data from Campbell and Norman, 1998; Rees, 1999); for further information see also (Hillel, 1998; Ochsner *et al.*, 2001) and Appendix 1.

Material	ρ_s (kg m^{-3})	c_s (J kg^{-1} K^{-1})	K (W m^{-1} K^{-1})	c_v ($=c_s\rho_s$) (J m^{-3} K^{-1})	P (J s$^{-1/2}$ m^{-2} K^{-1})
Soil minerals	2.65×10^3	870	2.5	2.31×10^6	2400
Granite	2.64×10^3	820	3.0	2.16×10^6	2550
Organic matter	1.3×10^3	1920	0.25	2.50×10^6	790
Water	1.0×10^3	4180	$0.56 + 0.0018T$	4.18×10^6	1580
Air (at 101 kPa)	$1.290 - 0.0041T$	1010	$0.024 + 0.00007T$	$1300 - 4.1T$	5.57
Sand (dry)	1.436×10^3		0.25 – 0.4	1.38×10^6	590 – 740
Sand 30% water			1.87	2.63×10^6	2220
Loam (dry)	1.330×10^3		0.23 – 0.4	1.39×10^6	560 – 745
Loam 30% water			0.98	2.42×10^6	1540
Organic soil dry	0.26×10^3		0.05	0.32×10^6	126
Organic soil (30% water)			0.22	1.58×10^6	590
Organic soil (70% water)			0.45	3.21×10^6	1200

different surface absorptivities have on the surface temperature – dark materials absorb more sunlight than do light materials, hence increasing their temperatures more during daytime and thus exaggerating their ΔT. Information on the thermal inertia of different surfaces can be very informative about the surface properties. Surfaces with a low thermal inertia such as bare sandy soils have a large diurnal temperature range, while surfaces with a high thermal inertia such as water bodies and forests have a much smaller temperature range. The situation is further complicated by the fact that for vegetated surfaces latent heat fluxes are also a major component of the surface energy balance that vary diurnally under stomatal control leading to substantial damping of the diurnal temperature fluctuations with an increased *ATI* (Fig. 4.10).

The Heat Capacity Mapping Mission (HCMM), which flew 1978–1980, was designed specifically to record day and night thermal images with a pixel size of 600 m. The equatorial crossing times occurred shortly after midnight and in the early afternoon, corresponding to periods of minimum and maximum surface temperatures. Analysis of suitably registered images allowed maps of *ATI* to be produced that provided information on the different landcovers and geological formations. There are several more sophisticated algorithms available, some of which make use of three or more satellite thermal images at different times of day (Sobrino *et al.*, 1998; Xue and Cracknell, 1999).

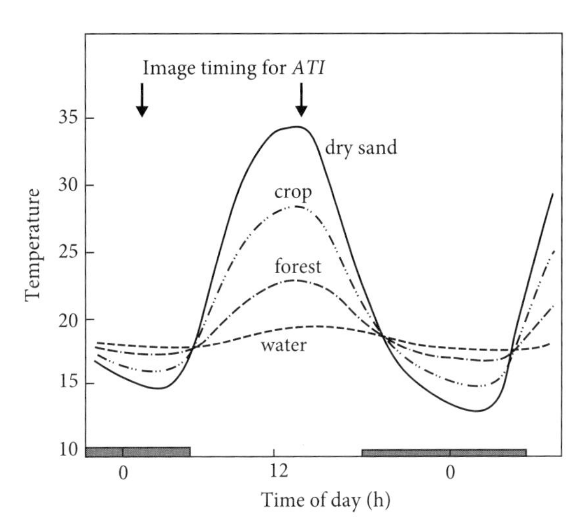

Fig. 4.10. Schematic diagram showing the differences in diurnal temperature cycles for different surfaces as they might be derived from two satellite images at 02.00 and 14.00 h. The differences in *ATI* relate both to soil heat fluxes and to differing contributions of latent heat exchange to the energy balance.

4.5 Measurement of fluxes

Remote-sensing approaches for the estimation of fluxes will be discussed in detail in Chapter 9, but as they are almost always rather indirect they need to be calibrated against more conventional approaches. A useful description of the main micrometeorological approaches to the estimation of fluxes may be found in Monteith and Unsworth (2008). Here, we very briefly outline three main ways in which mass and energy exchange between vegetation and the atmosphere may be estimated as a basis for calibration or validation of remote-sensing data: (i) use of individual organs enclosed in small chambers or 'cuvettes' and subsequent 'scaling-up' through the use of a vegetation–environment model coupled with a knowledge of the number and spatial distribution of the organs measured; (ii) measurement of the gas exchange from larger enclosed areas of canopy, either using transparent chambers, or in the case of evaporation, by using weighing lysimeters; and (iii) using micrometeorological methods such as eddy covariance or calculation from meteorological data.

4.5.1 Cuvette and chamber approaches

Although it is possible to measure mass exchanges, for example of CO_2 or water vapour, at a leaf or plant scale using gas exchange in enclosed chambers, it is extremely difficult to scale these up effectively to a canopy scale, partly because of the difficulty of choosing representative leaves or trees, and partly because it is difficult to mimic environmental conditions (especially of wind speed) inside leaf or canopy chambers. In particular, it is not possible to mimic the within- and above-canopy transfer resistances in such systems. For short dense canopies such as grass, or for isolated trees, it is possible to use canopy chambers or weighing lysimeters where small areas of crop are grown in what is effectively a very large weighable pot. The scaling-up process is illustrated in Box 4.4. For larger areas of vegetation, the most reliable measurements of fluxes are obtained using micrometeorological approaches.

4.5.2 Micrometeorological and meteorological methods

Flux-gradient approaches

Historically, flux-gradient approaches were used (Fig. 4.11). These are based on the fact that heat or mass flow to or from the surface is effectively one-dimensional, so that for substantial height differences above canopies, transport can be described by the conventional transport equation with the vertical gradient of mole fraction being proportional to the flux density. The constant of proportionality or transfer coefficient is frequently assumed to be similar for different gases and for heat when transfer is occurring by turbulent transfer. For this the transfer coefficient is conveniently calculated from the gradient of wind speed (momentum). An alternative approach to the use of data from more than one measurement height is the *Bowen-ratio* method. This approach, rather than using the profile measurements of wind speed to estimate the transfer coefficients, assumes that transfer coefficients

BOX 4.4 Scaling up of cuvette data

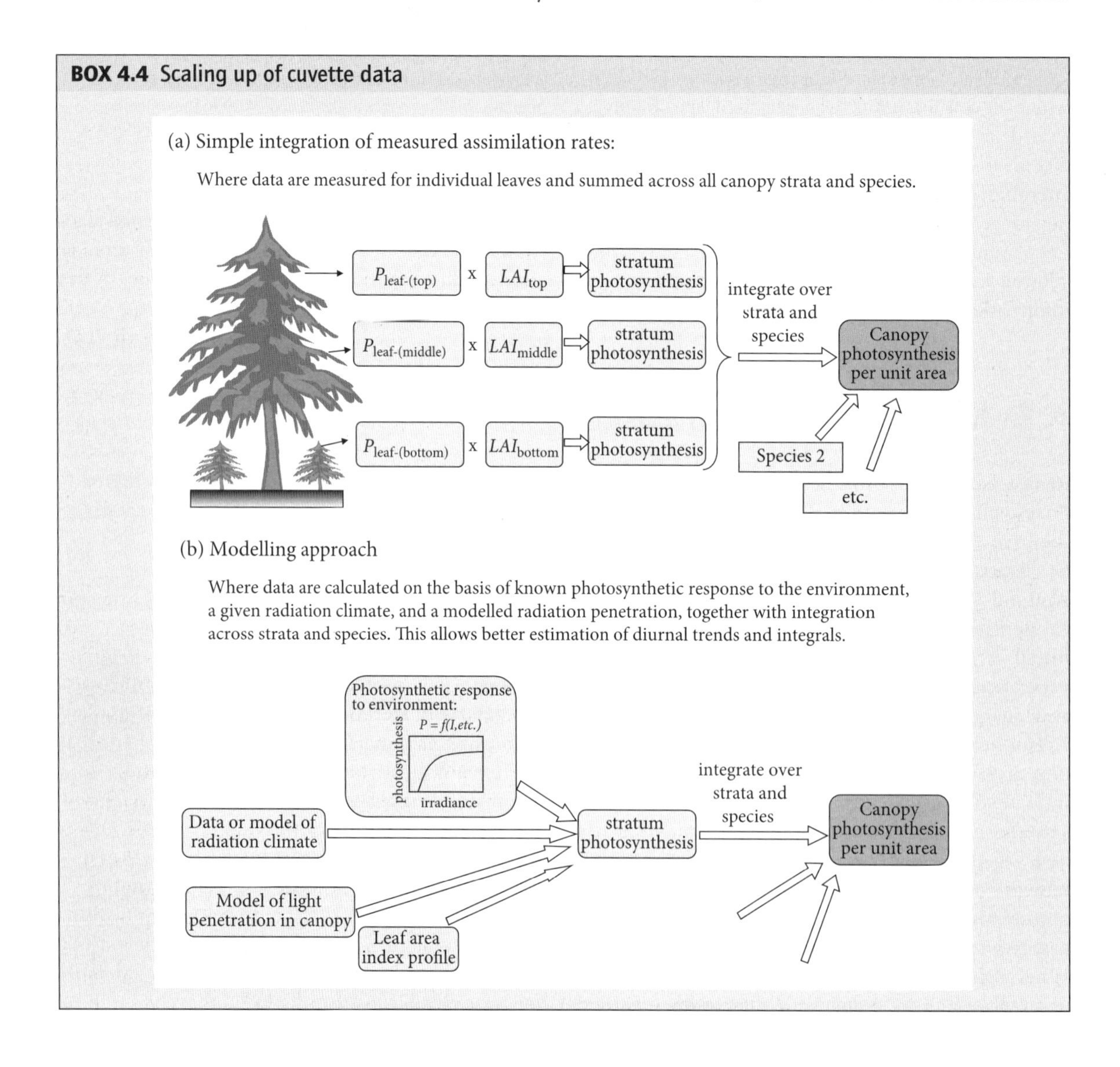

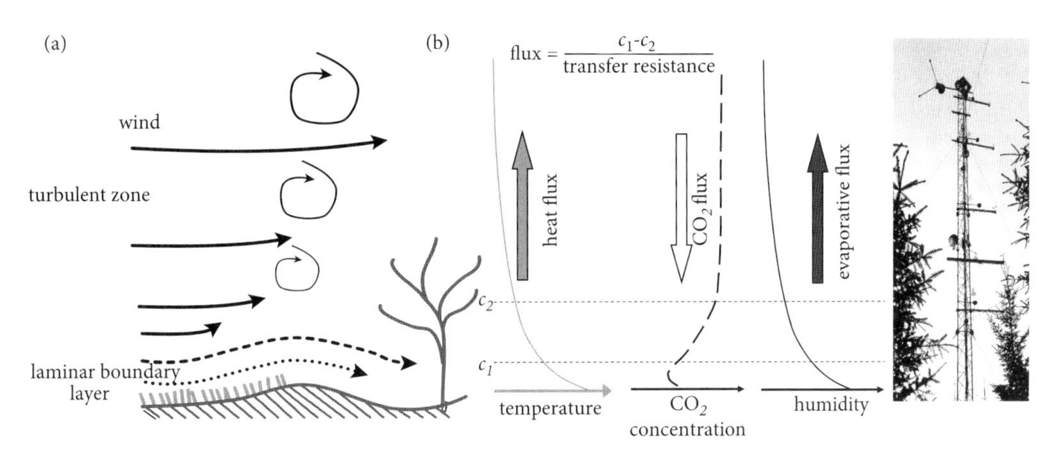

Fig. 4.11. (a) Illustration of the wind-speed profile over vegetation, showing the transition from a laminar flow regime near the surface to a turbulent regime. Also shown are the gradients of air temperature, CO_2 concentration, and atmospheric humidity showing the direction the corresponding fluxes of heat, CO_2, and water vapour. The flux of each entity can be derived from the concentration difference between any two heights and the transfer resistance over that zone. (b) A meteorological tower, showing environmental sensors at different levels.

for heat and water vapour are similar. Rearrangement of the energy balance equation then gives

$$\begin{aligned}\lambda \boldsymbol{E} &= (\boldsymbol{R}_n - \boldsymbol{G})/(1+\beta) = (\boldsymbol{R}_n - \boldsymbol{G})/(1+\mathbf{C}/\lambda\mathbf{E}) \\ &= (\boldsymbol{R}_n - \boldsymbol{G})/(1+\gamma\partial T/\partial e), \qquad (4.25)\end{aligned}$$

where β is the Bowen ratio ($=\boldsymbol{C}/\lambda\boldsymbol{E}$), γ is the psychrometric constant and $\partial T/\partial e$ is the ratio of the temperature and humidity gradients in the boundary layer above the crop. More detail of the approach may be found in Monteith and Unsworth (2008) and Thom (1975) where approaches for taking account of differing atmospheric stability conditions are outlined. The atmospheric stability depends on the profile of temperature variation with height where strong temperature decreases with height lead to strong convection processes and atmospheric instability, while temperature inversions lead to stable conditions.

Eddy covariance

The more recent development of high-frequency sonic anemometers and fast-responding sensors has allowed the development of a more direct approach termed *eddy covariance* where vertical flux density is obtained from the time-averaged product of vertical wind speed (u') and concentration (x'). The basic principles are outlined in Fig. 4.12, though it should be noted that accurate estimates depend on careful operation with, for example, a requirement for an adequate fetch (that is the area of homogeneous canopy upwind of the sensor) and flow distortions in the vicinity of the mast, especially on uneven terrain, can lead to errors. Footprint analysis enables one to estimate the canopy area that gives rise to the observed eddy covariance signal. Accurate operation depends on averaging over periods of tens of minutes or longer and depends on constancy of atmospheric conditions over the period.

Scintillometry

Large-aperture scintillometers provide a tool for derivation of area averages of heat fluxes over larger scales that better correspond to satellite observations than do eddy flux approaches (e.g. Hemakumara *et al.*, 2003). Scintillometers measure the fluctuations in infrared radiation attenuation (e.g. at 0.94 μm) over a pathlength of up to several km in the surface layer; these fluctuations are a measure of the turbulent intensity of the refractive index of air over the viewed path, which itself is related to the turbulent intensities for temperature and humidity. The similarity of transfer processes for different entities means that the turbulent intensity of refractive index can be used to estimate sensible heat flux given appropriate meteorological data.

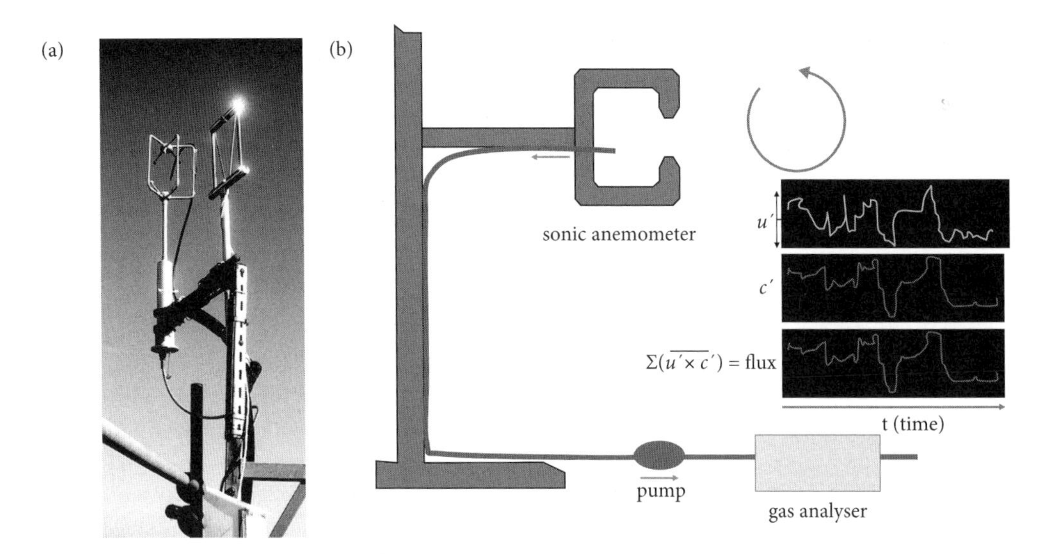

Fig. 4.12. An illustration of (a) sensors for an eddy covariance system (showing on the left a 3D sonic anemometer on the right a krypton hygrometer) mounted on a meteorological mast and (b) the basic principles of operation of an eddy covariance system. The sonic anemometer measures the rapid changes in direction and velocity of air passing the sensor, infrared sensors measure the corresponding changes in concentrations of water vapour and CO_2 and a thermistor measures the corresponding rapid changes in air temperature. Together, these measurements allow calculation of the fluxes of heat, water vapour, and CO_2 according to $\boldsymbol{F} = \int (u \times c)\, dt$, where $(u \times c)$ is the instantaneous product of the vertical velocity (u) and the concentration (c) of the entity being transferred.

Surface renewal

In addition, a number of simpler micrometeorological approaches have been proposed for estimating $\boldsymbol{E}$ that use cheaper sensing systems. One of the most promising is *surface renewal*, which is based on the premise that heat builds up steadily in the surface layer of vegetation and is only removed in sharp bursts, so that the rate of heat (and hence mass) transfer can be inferred from the rate of temperature increase in the near-surface layer during the steady ramps and the frequency of the bursts (Snyder *et al.*, 1996). An increasing number of studies have shown good correlations between the results obtained with surface renewal techniques over a range of both homogeneous and clumped vegetative surfaces and data obtained by scintillometry and eddy covariance (Anandakumar, 2009).

Meteorological and other approaches to estimation of E

A wide range of methods is available for *estimation* of evaporation rates from canopies using meteorological data; these are largely based on the concept that evaporation is determined by the environment, and in particular by the amount of energy available to evaporate the water. There is also some influence of wind speed, temperature, and atmospheric humidity. The approaches vary from totally empirical relationships with either temperature (and often daylength; Thornthwaite, 1948) or radiation (Makkink, 1957) through to those having a more rigorous theoretical basis such as those proposed by Priestley and Taylor (1972). Details of these and other methods are summarized by Brutsaert (1982). The standard approach in agrometeorological work, however, is now the Penman–Monteith method (P–M; based on eqn (4.13)). Details of its practical application may be found in Allen *et al.* (1999). Not surprisingly the latter approach, which is the most reliable, also has the greatest requirement for data.

In general, estimates of $\boldsymbol{E}$ from meteorological data, as with estimates from measurement of water loss from free water surfaces such as in evaporation pans, actually estimate a *potential evaporation*; the actual evaporation rates from plant canopies ($\boldsymbol{E}_c$) may be substantially less than this potential as a result of (a) limited ground cover or (b) stomatal closure.

This potential evaporation is a measure of the evaporating power of the atmosphere and is expressed as the *reference evapotranspiration* ($\boldsymbol{E}_0$), that is the evaporation that would occur from a short grass surface not short of water (see Allen *et al.*, 1999). The conversion of reference, $\boldsymbol{E}_0$, to the actual $\boldsymbol{E}$ for any surface makes use of a largely empirical multiplier or 'crop coefficient', K_c that takes account of differences in the aerodynamic properties of the surface vegetation (e.g. height), and especially of its effective ground cover or leaf-area index, and also physiological differences such as stomatal closure under drought conditions (values of K_c for a wide range of crops may be found in Allen *et al.* (1999). Where less complete meteorological data are available, the simpler approaches may provide useful approximations to $\boldsymbol{E}_0$, or alternatively one may directly estimate a reference evaporation using an evaporation pan, which is a standard-sized free water surface. Different types of evaporation pan have slightly different conversion factors to convert the results to $\boldsymbol{E}_0$.

Further reading

More information on the basic structure and functioning of plants may be found in any introductory plant science text, though the following are particularly useful (Raven *et al.*, 2005; Salisbury and Ross, 1995; Taiz and Zeiger, 2006). The basics of plant interactions with the environment are discussed in more detail in environmental plant physiology texts such as those by Fitter and Hay (2001), Jones (1992), and Nobel (2009), with the latter being particularly recommended as a good text covering the basic biophysics of aspects such as cell–water relations and photosynthesis. Excellent introductions to the more physical aspects of plant–environment interactions can be found in the books by Monteith and Unsworth (2008), and Campbell and Norman (1998), as well as in Jones (1992).

Sample problems

4.1 Given that the osmotic potential inside a leaf cell is −1.2 MPa, and that the total leaf water potential is −0.3 MPa, what is the turgor pressure in leaf cells? If one assumes an inelastic cell wall, what would be the total water potential at which turgor pressure reaches zero and the leaf wilts? How would the wilting point change if a more realistic elastic cell wall was assumed? Explain your answer.

4.2 Given a leaf with stomatal conductances to water vapour for the upper and lower surfaces of 50 mmol m^{-2} s^{-1} and 200 mmol m^{-2} s^{-1}, respectively, and boundary-layer conductances to water vapour for each surface of 500 mmol m^{-2} s^{-1}, calculate (a) the overall leaf conductance to water vapour, (b) the overall leaf resistance to water vapour. (c) Assuming a temperature of 20 °C, what would the total leaf conductance to water vapour be in units of m s^{-1}? (d) Estimate the leaf boundary-layer conductance to heat transfer.

4.3 (a) Using Fig. 4.7, estimate the water vapour mol fraction difference between the intercellular spaces of a leaf at 18 °C and the air where air temperature is 24 °C and the relative humidity is 50%. (b) At what leaf temperature would the vapour-pressure difference reach zero? (c) Using the Magnus formula in Appendix 1, calculate the water vapour pressure of air at 60% relative humidity at 40 °C.

4.4 (a) Calculate the net isothermal radiation for a leaf receiving net radiation of 400 W m^{-2} at 24 °C where the air temperature is 20 °C. (b) Assuming that there is no evaporative loss, calculate the boundary-layer conductance to heat transfer.

4.5 (a) Estimate the thermal time constant for a leaf 100 μm thick with a boundary-layer conductance of 0.1 m s^{-1} and a stomatal conductance of 0.02 m s^{-1}. (b) What would be the limits for a non-transpiring leaf and for a leaf with stomata fully open?

5 Earth observation systems

We saw in Chapter 2 that remote sensing requires the detection of electromagnetic radiation. In most cases, this also requires the recording of the detected signal so as to produce a permanent record of the scene observed for later analysis. A wide range of instruments can be used, from the early aerial camera to the modern hyperspectral scanner. The particular type of instrument that is best for any particular case depends on a number of factors, such as the use to which the data will be put, the most suitable wavelengths of electromagnetic radiation for that particular application, the frequency at which observation is required, and the scale and variability in the scene observed.

The sun acts as a natural source of visible radiation, which is the reason why so-called *optical remote sensing*[17] was the original form of remote sensing and why it is still the most common. The earth emits mainly heat energy which can be detected by airborne and space-borne thermal detectors and can be used to make temperature maps of the surface. Both these types of remote sensing rely on naturally produced radiation and this is therefore called *passive* remote sensing. We saw in Section 2.7 that the earth also emits a small amount of radiation in the short microwave part of the electromagnetic spectrum that can also be passively sensed. But most remote sensing in the microwave region uses radiation that is artificially generated, usually on board the aircraft or satellite that carries an instrument that senses the radiation reflected or scattered from the surface below. Because human participation is required, this is known as *active* remote sensing. Remote sensing can be carried out actively in the optical region by using lasers as the source of radiation (Section 5.6), but such use is limited by energetic or health and safety considerations.

Observations can be carried out from a whole range of observing platforms, from hand-held or tractor-mounted instruments through towers, to aircraft and satellites. It is therefore essential to consider the platform on which the instrument is mounted as this may affect the spatial and temporal resolution of the images obtained and the ways in which they can be used. In principle, any instrument can be carried on any platform (prototypes of those carried on satellites have usually been tested out on aircraft before being launched into space so that the required analysis techniques can be perfected), but in practice one usually associates a certain instrument with a particular satellite, such as the *Advanced Very High Resolution Radiometer* (AVHRR) on the NOAA series and the Thematic Mapper (TM) on Landsat. Such a package of different components acting together, such as platform plus detector, is referred to as a sensing *system*. In general terms, the instrument determines the character of the image, and the platform determines its location.

Since the first environmental satellites were launched in the 1960s, hundreds have been flown, often for specific applications such as meteorology or altimetry. In many cases alternative uses have subsequently been found for data that were never envisaged by the designers, such as vegetation monitoring using AVHRR. In this chapter we will consider some of the

[17] The term 'optical' usually encompasses a range of wavelengths that is broader than the strictly 'visible' region, stretching from the ultraviolet well into the near 'reflected' infrared. In this range, lenses and mirrors can be used in the detectors and also photographic emulsions are sensitive to these radiations.

most common systems that are currently providing data of relevance to vegetation monitoring. Many of the available systems are well documented in textbooks such as Campbell (2007), and Lillesand *et al.* (2007), and in the *Survey of Missions and Sensors* by Kramer (2002), but it should be borne in mind that, in this fast-developing science, such texts rapidly become out of date.

5.1 Principles of system design

5.1.1 Pictures and pixels

Most remotely sensed images nowadays exist in digital form. As electronic detectors become cheaper, even domestic cameras are now usually digital instruments, and the use of photographic film is becoming rarer and rarer. There are a number of reasons for this change in technology. Originally, aerial photographs were recorded on film. This posed few problems then as the film could be unloaded for processing when the aeroplane returned to earth. Early space-borne remote sensing in the 1960s was undertaken using cameras on modified sounding rockets, and the first pictures of the earth's surface from the Mercury manned spacecraft were colour photographs – all physically returned to earth. But the advent of unmanned satellites, which did not return to earth on a regular basis, required a different recording technique using electronic detectors, the output from which could be transmitted back to earth as a radio signal. But even then, the signal was usually in analogue form, and was used to modulate a cathode ray tube display or a laser film writer for visual analysis. Advances in technology have led to almost complete adoption of digital transmission and processing as such data are readily subject to mathematical manipulations by a computer. Initially, these data were stored on magnetic tape, which was bulky and was slow to access, but cheap storage devices have now been developed that can store the huge volumes of data that constitute modern images.

Analogue images, such as photographs, are continuous, both in their spatial extent (they can be enlarged almost without limit) and radiometrically (there is a continuous range of shades of grey). The word 'picture' is usually used for such an image. On the other hand, a *digital image*[18] is spatially and radiometrically discrete. The grey levels increment in a stepwise fashion, and the scene is made up from an array of individual elements ('picture elements', abbreviated to *pixels*), each of which is represented by one of the discrete grey levels. Such a pixel cannot be subdivided, and enlargement merely produces larger pixels, which contain no more information than the original ones. We are familiar with this effect on our television or computer screen – the picture we see consists of an array of dots of light, the density of which determines the screen resolution.

5.1.2 Resolution

The concept of resolution has been around for a long time in association with photography. It implies the ease with which small detail can be discriminated in the photograph.[19] But for a digital image, we can broaden this concept and consider several different types of resolution. *Spatial resolution* now tends to refer to the size of the ground element represented by an individual pixel. This is usually determined by the optical system of the sensor used to capture the data (i.e. its field of view) and on the height of the sensor (see Section 5.4 for further discussion of spatial resolution in relation to different sensors). The *ground sampling distance* (GSD) is the distance between pixel centres as determined by the scanner's sampling rate. *Radiometric resolution* refers to the number of distinct levels into which the intensity of the signal is divided – the so-called *grey levels* (i.e. 256 for 8-bit resolution or 4096 for 12-bit). The useful radiometric resolution, however, is determined by the signal-to-noise ratio of the sensor and the analogue-to-digital conversion electronics. For temperature sensors the radiometric resolution is often converted to the equivalent thermal resolution.

18. Strictly speaking, pictures and images are not the same thing, although one frequently sees in the press/news media reference to a 'picture from space'.

19. Resolution actually refers to the ability to distinguish between two closely spaced objects in an image, not the size of the smallest object seen. The ability of a system to distinguish features can be quantified in terms of the concept *resolving power*, but even so is still fairly subjective. The ability to distinguish a small object depends critically on the contrast between it and its surroundings.

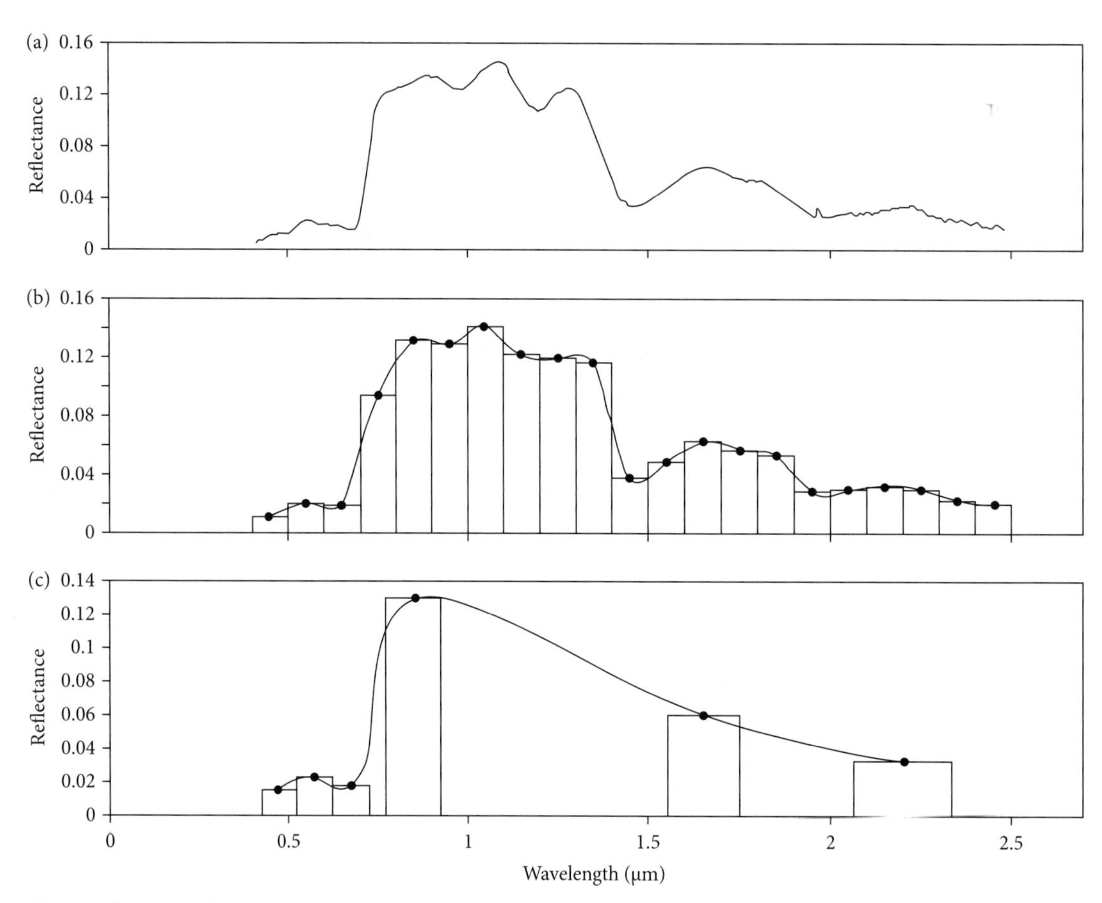

Fig. 5.1. Illustration of spectral resolution showing the differing ability of sensors with different spectral resolutions to detect the true spectral signature of an object. (a) Spectral reflectance of a Douglas fir canopy (USGS library at c. 10 nm resolution), (b) curve sampled with sensor channels covering 100-nm bands, and (c) curve sampled with sensors having the spectral sensitivity of the Landsat TM instrument.

Spectral resolution refers to the ability of a sensor to define fine wavelength intervals and hence to discriminate between different component wavelengths in the scene, and is determined by the number and width of the individual spectral bands that are recorded (Fig. 5.1). Finally, *temporal resolution* is the time between successive images, as set, for example, by the subsequent overflights of an aeroplane or by the orbital repeat of a satellite.

5.1.3 Principles of measurement

The basic initial analysis of even digital images is usually carried out by visual examination. The interpreter may look at a photographic print or display an image on a screen to check for location, visual quality, presence of features, etc., before subjecting it to more scientific scrutiny. The particular analysis technique then employed depends on the type of image and the type of information that is required. *Photogrammetry* requires the *spatial* analysis of the image, where the geometric fidelity is of paramount importance and it is the location of a feature that is the parameter measured. This is a well-established art, and mapping was one of the earlier uses of air photographs. *Photointerpretation*, on the other hand, involves the visual examination of the picture for the extraction of attributes and to identify features and to judge their significance. Such visual interpretation is a common experience, as we do it subconsciously whenever we look at a picture, a painting, or even a holiday snap. The process has been refined over the years and professional photointerpreters use a number of key indicators for features, such as size, shape, tone (or colour), texture, context, etc., in order to improve consistency of interpretation (e.g. Campbell, 2007; Jensen, 2005). If the images are in digital form, both of these forms of interpretation can be assisted by the use of computers. Digital image processing plays a major part in the extraction of information from remotely sensed images, and we will return to it later in Chapter 6.

5.2 Sensing platforms

The purpose of the platform is to position the detector over the area of interest. The type of platform is therefore determined by the requirements of the measurements to be made. Simple hand-held instruments, or instruments mounted on fixed platforms (towers, masts, etc.) may be sufficient for small-scale, infrequent monitoring or for calibration purposes. Aircraft provide flexible platforms, in that the height and flight direction can be varied as can the timing and frequency of flights. Orbiting satellites provide very stable platforms that enable regular, periodic measurements to be made from positions that are precisely known. Other platforms may include cherry-pickers, balloons, helicopters, microlites, unmanned aerial vehicles, or even model aircraft. Each platform may have its own advantages, such as cost, availability, high spatial resolution, flexibility of use, etc., but also disadvantages, such as difficulties in mosaicing images due to changing geometry and radiation. The choice of which system to use is fundamental to a research project being undertaken. In this section, however, we will mainly concentrate on the characteristics of airborne and satellite platforms.

5.2.1 Static and 'within-field' platforms

Important data can be collected in the field using hand-held or vehicle-mounted instruments. Such data may be collected for calibration or as part of a field campaign and are of interest in their own right for applications such as *precision agriculture* (Section 11.3). The growth characteristics of individual plants or species can be monitored, perhaps under different conditions or from different viewpoints, either as a study in itself or to be used to help identification in an aircraft or satellite image. Larger areas can be monitored from fixed towers or mobile platforms. One important advantage of close-range imaging is that it opens up opportunities for analyses based on pixel-to-pixel variation at small scales, as it eliminates the averaging involved in large pixels. Applications include the inversion of radiative-transfer models to extract canopy structure information from the fractions of shaded and unshaded leaves (Section 8.6.3). Another advantage of such close-range (*proximal*) remote sensing is that little if any atmospheric correction is needed, though it is important to also record the downwelling radiation at the same time, as illumination alters continually as cloud cover or sun angle changes. When measuring reflectance it is common for detectors to have dual sensors, one pointing downwards and the other upwards (in order to measure incoming and reflected radiation), though in other situations absolute reflectances may be estimated by using a reflectance standard as target.

5.2.2 Aircraft

A great deal of remote sensing has been done, and still is being done, using aircraft of various forms as the platform. One very great advantage of aircraft is their versatility. They can be flown at short notice where and when required (weather permitting). Their height can be altered to adjust the scale of the photo/image or to fly under cloud cover. Also, their flight lines can be arranged for specific purposes such as to cover a specific area, to observe that area from a particular angle or to produce overlapping images for stereoscopy. Imagery, at least from small areas, can be quite inexpensive. They are not without disadvantages however. They are not universally available, and have to be deployed to the test site that may be some distance from their base. The area that can be covered in a realistic time may not be large, particularly if high resolution is required, and the weather may be, and often seems to be, inclement. Aircraft do not provide a very stable platform, and images may suffer from various distortions due to drift, yaw, roll, and pitch and the positioning of the aeroplane may be slightly uncertain and is not reproducible. Satellite navigation can improve the positional accuracy to some extent, and gyroscopic mounts can help keep the recording instrument stable.

Aircraft remote sensing comes into its own when short-term variations are present in the parameters to be measured, such as coastal and vegetation monitoring when tidal or diurnal variations may require frequent observations during the day, or when very high spatial resolution is required, or perhaps to study a single event.

All types of aircraft can be used, from UAVs (unmanned aerial vehicles) and balloons, to light planes and the high-flying reconnaissance planes operated by NASA. Helicopters can be used for low-level sensing, for example Lidar profiling, and useful work has been done in the fields of agronomy and archaeology using microlites, model aircraft, or balloons for close study of crop marks and other features. Drones and unmanned vehicles have been used to access inhospitable places and are now being developed for more routine applications. Much can be achieved with automated flight-path control and automatic mosaicing of images as achieved with instruments such as the SmartOne mini-UAV described by Rydberg *et al.* (2007).

For an aircraft campaign, careful flight planning is necessary in order that the whole test site may be covered at the required height, at the required time, and from the required angle. Great skill is needed by the pilot to keep the plane stable in a continuous flightline; in fact different skills are required for taking aerial photographs from those required for using a line scanner. Since a camera records each image almost instantaneously, the flightline can be continually corrected as the photographs can be matched up in the laboratory. But a scanner builds up a continuous image line by line, and so it is important that there are no sudden changes in direction between lines.

5.2.3 Satellites and orbits

Johannes Kepler, a German astronomer (1571–1630), analysed the planetary observations made by his early mentor, Tycho Brahe, which enabled him to quantify the laws governing planetary motion in 1620 and that are still accepted today. The first two laws recognized the elliptical nature of the orbits, while the third one related the distances of the planets from the sun to their orbital periods. The same equations pertain to the trajectory of artificial satellites around the earth, and therefore make it possible to predict precisely their path and height.

The *universal law of gravitation* states that any two bodies are attracted to one another by a force that depends on the masses of the bodies and on their distance apart. A moving body, according to *Newton's laws*, will tend to move in a straight line unless acted on by a force. In order for a body to travel in a circular path, there must therefore be a force constantly pushing the body sideways. This force is always directed towards the centre of that circular path, and is called a *centripetal force*. For a satellite to remain in a circular orbit around the earth, it is the gravitational force between the satellite and the earth that provides this force. A simple mathematical analysis (see Box 5.1) shows that the orbital period T (the time taken for the satellite to complete one orbit and return to its original position) is related to the radius of the orbit r, by the expression:

$$T^2 = [4\pi^2/(GM_e)]r^3. \quad (5.1)$$

Substituting values for the gravitational constant, G, and the mass of the earth, M_e, gives the period (in seconds), when r is expressed in km, as

$$T = (9.952 \times 10^{-3})r^{3/2}. \quad (5.2)$$

Note that eqn (5.1) does not contain the mass of the satellite, and therefore holds for any, from the lightest to the heaviest. Substitution into this equation shows that for a value of $r = 384\ 000$ km, the distance of the moon from the earth, T would be about 28 days. For $r \approx 42\ 000$ km, the orbital radius of a geostationary satellite at a height of about 36 000 km, T would be about 24 h (equal to the earth's rotational period) and for $r = 7000$ km, the orbital radius for a satellite about 700 km above the surface of the earth, T is about 100 min.

In reality, the path of a satellite is not circular but slightly elliptical, but the theory can easily be extended to account for this. What is more important is that the orbits are perturbed due to a number of factors. The main factor is that the earth's gravitational field is distorted by the earth's oblate shape (flattening at the poles and bulging at the equator), and by the geological non-uniformity of the earth. Other perturbations due to the gravitational attraction of the moon and sun, electromagnetic interaction with the earth's magnetic field and the effects of solar winds and ionizing particles are small – but are still worthy of consideration if the precise orbit of the satellite is to be known. At lower altitudes there is also the effect of air resistance

BOX 5.1 Derivation of Kepler's laws of planetary motion

The centripetal force F_c that is required to keep a mass, m_s, travelling at a speed v, moving in a circle of radius r is

$$F_c = m_s v^2 / r.$$

The gravitational attraction F_a between two masses (M_e and m_s) is

$$F_a = G M_e m_s / r^2,$$

where G is the gravitational constant.

For circular motion, the centripetal force is provided by the gravitational attraction, so

$$m_s v^2 / r = G M_e m_s / r^2.$$

Hence, $v^2 = G M_e / r$,

but $T = 2\pi r / v$,

where T is the time taken for m_s to complete one orbit. Hence $v^2 = 4\pi^2 r^2 / T^2 = G M_e / r$, and therefore

$$T^2 \propto r^3.$$

Although neither G nor M_e have been measured to very high accuracy, their product has been, and its value is 398 603 $km^3 s^{-2}$. Substitution shows that for a value of r = 384 000 km, the distance of the moon from the earth, T would be about 28 days. For r = 42 170 km, the orbital radius of a geostationary satellite at a height of 35 786 km, T would be 23 h 56 min (approximately equal to the earth's rotational period) and for r = 7000 km, the orbital radius

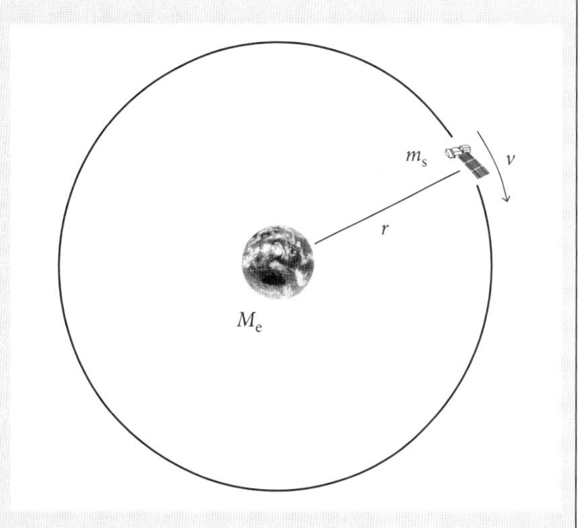

for a satellite about 700 km above the surface of the earth, T is about 100 min.

Actually, planetary orbits are not circular but elliptical, and Kepler's laws may be stated as:

1) Each planet moves in an elliptical orbit, with the sun at one focus of the ellipse.
2) A line from the sun to each planet sweeps out equal areas in equal times.
3) The periods of the planets are proportional to the 3/2 power of the major axis lengths of their orbits.

Law 2) simply follows from the conservation of angular momentum.

reducing the speed of the satellite. If this were not constantly adjusted by the firing of small rockets on board, the satellite would soon spiral into the lower atmosphere and burn up.

Satellites are placed in orbits that are designed for the specific purposes of the mission, and to suit the particular characteristics of the instrument(s) carried. There are two main classes of orbits that are generally used for earth observing satellites, namely the *geostationary* orbit and the *low earth* or *near-polar* orbit.

Geostationary orbits – At a height of about 36 000 km, the orbital period of a satellite is nearly 24 h. If a satellite is placed in an orbit exactly over the equator at this height, and is travelling in the same direction as the earth is rotating, then, as the satellite progresses, the earth spins underneath it with exactly the same rotational velocity, and so the satellite appears to remain in position over one point on the equator. A slight change in the speed of the satellite causes it to lose synchronism with the earth,[20] and it will appear to migrate eastwards or westwards across the sky. Although this orbit was proposed by the science fiction writer Arthur C Clarke in 1945, the first geostationary satellite was not launched until 1963 because of the technical complexities of the operation (Verger

20. In this orbit, the satellite is travelling at about 3 km s^{-1}. A change of only 0.1 per cent in this speed will change the orbital period by over 4 min.

BOX 5.2 Orbit precession

If a satellite were to make an integral number of orbits in an integral number of days, it would arrive at exactly the same longitude at which it started, and must repeat exactly the same ground track each time.

If it were to make $(N+k/m)$ orbits per day (where k and m are integers), the orbit track repeats every m days after making $(mN + k)$ orbits.

If k and m have no common factors, all m of the ground tracks will be traversed.

e.g. for Landsats 1, 2, and 3 there were $13^{17}/_{18}$ orbits per day.

If the period were 103.27 min, then the distance between adjacent tracks at the equator would be 2874 km. Since the swath width for the MSS was only 185 km, then only a small portion of the equator would have been covered each day. But, because of the inclination of the orbit to the earth's axis, precession caused a daily drift of −1.43° or 160 km west, which meant that swaths would overlap on successive days since their width was greater than the drift. Hence, in 18 days the satellite had observed every point on the equator.

For SPOT, there are $14^{5}/_{26}$ orbits per day, leading to a 26-day repeat.

The repeat cycle is very sensitive to the semi-major axis of the orbit. If the height of Landsat were decreased by only 19 km, it would have made exactly 14 orbits per day, there would have been no westward progression and so not every part of the earth's surface would have been observed.

et al., 2003). Such an orbit, sometimes referred to as the *Clarke orbit*, is favoured for communications satellites as they remain in view of small fixed antennae and give continuous reception.[21] There are also five or six meteorological satellites (e.g. Meteosat, GOES, GOMS, etc.) distributed around the equator, each viewing nearly 40% of the earth's surface and providing almost continuous coverage of the global weather patterns. Data from these meteorological satellites are often used for earth observation for large-scale monitoring, but because of their large field of view these satellites generally have low spatial resolution (e.g. 1 to 5 km pixels at nadir for GOES and Meteosat). The fact that these satellites are about 50 times the distance from the earth as are low-earth orbit satellites means that they receive much weaker signals.

Low-earth orbits – Most earth observing satellites are in near-polar orbits that range from about 600 to 2000 km above the ground. Remembering that the earth's radius is about 6300 km, eqn (5.2) tells us that such satellites will take about 100 min to complete one orbit. If such an orbit passed over the earth's poles, the subsatellite track would always pass through the same points on the earth's surface, and only that track would be observed by the instrument on board. If, however, the orbit were slightly inclined away from the poles, then the orbit would *precess* with respect to the earth and subsequent tracks would be displaced by an amount that depends on the angle made by the plane of the satellite orbit and the earth's rotational axis (see Box 5.2). The rate of precession will determine the number of orbits required before the satellite track repeats itself, and the time taken to do this will be the satellite revisit time, i.e. the time between successive observations of a particular point on the earth's surface. If the satellite is to provide complete earth coverage, the distance between adjacent orbits at the equator must be equal to or less than the swath width of the instrument on board (see Section 5.3).

Such orbits are called *polar orbits*, or more correctly *near-polar orbits*. A special case of a polar orbit is one where the orbital precession is exactly equal to the earth's solar precession, so that the satellite crosses the equator at exactly the same local solar time each orbit. Such an orbit is referred to as a *sun-synchronous orbit*.[22] The advantage of this type of orbit is that the solar angle is approximately the same each time a point is imaged, and hence variability of illumination and

[21] In 2007 there were about 320 active geostationary satellites in orbit, mostly for communications, which means that they were spaced on average only about 700 km apart, and there are many more decommissioned ones still in orbit. About 40 or so are launched each year, some of which possibly replace earlier satellites.

[22] A sun-synchronous orbit is one lying in a plane at a fixed angle with the earth–sun direction. This means that the orbital plane has a precession equal to 360° per year, or 0.986° per day. Such a precession corresponds to the mean displacement of the earth around the sun, hence giving a constant local time for passage through a given location, thereby guaranteeing almost constant illumination, varying only with the seasons.

shadow angles will be minimized. This makes them particularly suitable for vegetation monitoring.

Other orbits – Whereas most earth observation satellites are in sun-synchronous orbits, other orbits may be used for special missions. The US manned missions (Spacelab, Skylab, Shuttle, etc.) have usually been placed in lower orbits, between 200 and 450 km high, and at a much higher angle to the earth's axis. Such an orbit will give coverage only between about 50° north and south. The orbits of satellites dedicated to radar altimetry of the oceans (e.g. TOPEX/Poseidon) are usually at a greater height (about 1300 km) and cover a much greater portion of the earth's surface, as sun-synchronism is not required. Cryosat, a European satellite dedicated to measuring polar ice stability, needs to go almost over the poles but does not need to give good earth coverage, and therefore precesses very little.

In order to overcome the problems of lack of reception at high latitudes by geostationary satellites, the USSR developed the use of highly elliptical orbits (*Molniya orbits*), having an apogee of 400 km and a perigee of 40 000 km. In such an orbit the satellite appears to hover for much of its orbit at its perigee, for over 8 h, requiring only three satellites to give continuous coverage of that area. These were originally designed for communications purposes, but Russia now has plans to launch up to five satellites in such orbits, carrying optical and radar systems, to monitor the Arctic.

5.2.4 The ground segment

The ground segment is an integral part of the remote-sensing system. A satellite is not just put into orbit and left alone. Even at the height of most low-earth orbits there will be some atmospheric drag on a satellite, and, left to itself, the satellite's orbit would decay and it would eventually burn up in the denser atmosphere. Also, any slight perturbation may affect the satellite's stability and it would eventually start spinning. Ground control will therefore constantly monitor the conditions and take steps to rectify such problems by firing on-board rockets to boost the orbit and to make sure the detector is pointing in the direction required. Ground control will also control the post-launch deployment of solar panel and antennae and commission the on-board instruments. It will constantly monitor the performance of the instruments and take such action as is necessary to remedy any problems, often by reprogramming or switching circuits (there is often a certain amount of built-in redundancy). The positions of geostationary satellites may be changed by suitable adjustment to their speeds, allowing them to migrate around the equator before being placed back in geostationary orbit in a new location – Meteosat 5 is now positioned over the Pacific Ocean for Indian Ocean Data Coverage (IODC). Occasionally the orbits of satellites have been lowered in order that they could be serviced by Shuttle, as happened to Hubble to allow camera replacement.

5.3 Sensing instruments

Robinson (2003) specifies that there are only four primary quantities that can be observed from space, namely 'colour' (considered in its wider sense to include near to mid-infrared reflected radiation), temperature, roughness, and height, and these, in most cases, relate only to a fairly thin surface layer of the earth. Any subsurface properties, and many physiological processes (see Chapters 4 and 9), have to be inferred indirectly from the various surface features. Remote sensing detectors have been developed to exploit specific properties of electromagnetic radiation, such as wavelength, polarization, interaction with different surfaces, and speed of propagation in order to enable the interpreter to extract such quantities from the data produced.[23] In addition to the usual optical and microwave sensors, some in-field sensors are based on other technologies such as ultrasonic sensing (Reusch, 2009).

Apart from the human eye, of course, the first remote sensing instrument to be developed was the *camera* nearly 150 years ago. The very early ones were flown on balloons (Lillesand *et al.*, 2007) but, as soon as aircraft became available, the art of aerial photography

23. The detector produces just a digital number related to the particular parameter being measured at that instant, e.g. intensity of red light. It is the task of the analyst to interpret that number in terms of the value of a particular physical quantity related to that specific location.

was born (see Appendix 4 on history). The photographs obtained were initially used to aid mapmaking, but, during the First World War, air photographs were found to be useful for reconnaissance purposes. By the time of the Second World War, the techniques and equipment had become more sophisticated – near infrared photography was used for camouflage detection, for example, and other non-visible wavebands such as thermal infrared and microwaves started to be exploited. The availability of aerial colour film and colour infrared film led to the use of air photographs for studying vegetation and its diseases, pioneered in the 1950s by Colwell (1956) and this in turn led to the use of these techniques for environmental studies, from whence the art or science of remote sensing evolved. The name remote sensing (RS) was coined in 1960; the term earth observation (EO) is now commonly used for environmental remote sensing, especially from satellites.

There are many ways to try to categorize different detectors. They can be classified according to the wavelengths in which they operate, whether they are active or passive, in terms of their intended application, or whether they are imaging or non-imaging instruments.

Most detectors used in earth observation are *imaging instruments*, that is to say they produce a reasonably recognizable spatial representation of the scene below. *Non-imaging instruments*, on the other hand, tend to make sequential measurements of a particular parameter below the platform as it progresses. Examples of this kind would be a microwave or laser altimeter (which determines the height of the instrument above a point on the ground vertically below), or a microwave radiometer (which measures the microwave radiation emitted by the earth or the atmosphere).

5.3.1 Cameras

Cameras, and video cameras, are *central perspective* imagers. This means that the whole of a scene is imaged at one instant of time from the viewpoint of the camera lens. Objects at the centre of the scene will be viewed normally, while those towards the edge will be viewed obliquely. The angle of view increases towards the periphery and tall objects will appear to lean away from the centre. On the other hand, many detectors used in remote sensing operate on the line-scanning principle (see below) whereby narrow strips of ground immediately beneath the instrument are imaged successively as the platform progresses. The result is that all features are viewed normally, at least in the direction of flight. This has important advantages, particularly for vegetation monitoring, since this minimizes the range of view angles. This is important, as we shall see in Chapter 8, because the reflected radiation varies anisotropically, varying with the angle of view.

Cameras have played a major role in remote sensing, and still serve a useful purpose, particularly now that digital cameras are available with the necessary resolution. Apart from utilizing more sophisticated technology, cameras used for aerial photography are similar in principle to ordinary amateur cameras. A lens focuses the light onto a detector, either film or electronic, and a shutter mechanism provides the necessary exposure. High geometric and radiometric accuracy are required, and a high-quality lens is used. The film normally used has a larger format, usually 23 cm square, especially for photogrammetry, although useful photographs can be obtained using 35 mm film and modern digital cameras. The mounting in the aircraft holds the camera firmly in a vertical position, and is often stabilized against vibration or sudden movement. For very accurate work, a mounting that tilts during film exposure may be used so as to keep the camera pointing at a fixed point in order to reduce motion blurring. Haze filters or coloured filters can be mounted on the lens as required. Multispectral images can be obtained by mounting several cameras coaxially, each with its appropriate filter, but care is required to ensure that all the cameras are exactly aligned.

5.3.2 Radiometers

A radiometer is the basic element of all electro-optical and microwave sensors. Simply, it is a device for measuring the intensity of the electromagnetic radiation falling on its detector within a defined spectral range. The technical details depend upon the particular part of the spectrum in which it is used, but all radiometers comprise three elements (Rees, 1999): an optical system to focus the radiation and to select the wavelength, detectors that produce an electrical signal, and a signal processor to provide an output.

The simplest radiometers are *non-imaging detectors* that integrate the radiation that arrives from within a defined field of view and within a specific waveband. Hand-held and platform-mounted spectroradiometers with high spectral resolution are often used in the field to measure the reflectance spectrum of particular targets or vegetation species, often as part of the calibration process, whereas radiometers mounted on aeroplanes or satellites will measure the radiation from individual large areas of ground along the flightline or subsatellite track. An increasing range of *imaging detectors* is coming available. In the optical region, standard lenses and mirrors are used to focus the radiation onto a light-sensitive detector that is often a photodiode or more usually a charge-coupled device (CCD) array with wavelength selection by means of diffraction gratings or filters. In the thermal wavelengths, heat detectors such as bolometers or photon detector materials such as indium antimonide are used for the sensors. As these wavelengths are less energetic than the optical wavelengths it used to be common to cool the sensor to minimize thermal noise, but a range of new imagers, including microbolometers with sensitivities of the order of 0.05 K when used at room temperature are becoming available for use in thermal imagers. Because standard glass is opaque to thermal radiation, where lenses are used they need to be made of materials such as germanium, which transmit radiation up to 14 μm, or else dispersion gratings can be used. Thermal imagers are commonly designed to use atmospheric windows in the 3–5 μm range (shortwave infrared imagers) or 8–14 μm (longwave imagers). The former cannot be used in the daytime as there is substantial reflected shortwave solar radiation in these wavelengths, so daytime measurements require longwave imagers where there is negligible interference by sunlight (cf. Figs. 2.4 and 2.5). Owing to the low energy of the radiation, frequent calibration is usually required, for example by the use of a chopper (a rotating mirror) to direct the beam onto a blackened plate at a known temperature in order to calibrate the signal. The output signal then alternates between that which is to be measured and the calibration target.

In a microwave system, the signal is received by an antenna and amplified using standard electronics before being detected by a diode that converts the microwave energy into an electrical signal. A passive radiometer, which effectively measures the temperature of the surface, will usually have a large instantaneous field of view, many tens of kilometres across because the amount of microwave energy emitted by the earth is very small (Section 2.5).

5.3.3 Line scanners

Most remote-sensing imagers carried on aircraft and satellites, rather than taking instantaneous images, work on a *scanning* principle. A two-dimensional spatial representation (an image) of the strip of ground below the platform is built up line by line as the platform progresses. This can be done in two different ways, namely using *across-track* (*whiskbroom*) scanning or using *along-track* (*pushbroom*) scanning as shown in Fig. 5.2.

Whiskbroom scanning – In this method, an oscillating or rotating mirror is used to scan the ground below in a direction at right angles to the direction in which the platform is travelling.[24] Successive lines are scanned as the platform moves forward. The energy directed towards the scanner from the *instantaneous field of view* (IFOV) falls on a single electronic detector and is recorded. The output from the detector is sampled as the IFOV is swept along the scan line thus producing a series of discrete values for successive parts of the scan line. This is repeated for successive contiguous lines, thus building up a two-dimensional array of data values, each relating to the energy received from the corresponding ground element. These can be displayed as pixels in the image representing the terrain below. The speed of the mirror must be carefully chosen so that the mirror completes one sweep in the time that the platform takes to travel a distance equivalent to the size of the ground element in order that the lines of data will be contiguous. If the mirror is too slow, parts of the ground will be missed, if too fast, parts of the ground elements will be imaged twice in successive lines. The length of the scan line defines the *swath width*.

Pushbroom scanning – The pushbroom scanner does not use a mechanical scanning system but uses a line of detectors to detect the energy from a complete scan line. Individual detectors measure the energy coming from adjacent discrete parts of the scan line. The forward motion of the platform again allows a two-dimensional image of pixels to be built up. Usually charge-coupled devices (CCDs) are used as detectors. These are very sensitive and can be made very small. A typical array, as used on SPOT, contains several thousand individual CCDs yet is only a couple of centimetres long. This type of scanner has a number of

[24] A number of instruments have used conical scanning, notably the ATSR/AATSR (Section 5.8.1), the Soviet multispectral MSU-SK, some helicopter-mounted Lidar scanners and some airborne scatterometers.

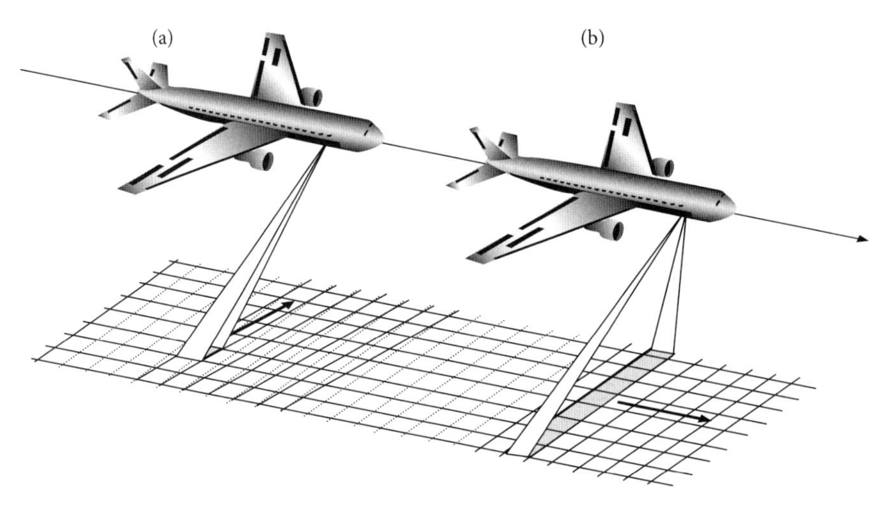

Fig. 5.2. Illustration of (a) whiskbroom and (b) pushbroom scanning principles. The whiskbroom scanner has only a single detector that views pixels across the track in turn, while the pushbroom scanner has a line of detectors that view the ground as the sensor moves over the ground. For multispectral and hyperspectral sensors a full spectrum is obtained for each pixel, illustrated by the stacked pixels in Fig. 5.3.

advantages. It is much lighter than an opto-mechanical scanner, and uses less power. It has no moving parts to go wrong, and can be very sensitive. Against this, however, is the fact that the several thousand individual CCDs need to be intercalibrated, rather than the single one in the whiskbroom scanner, and CCDs can only detect a restricted range of wavelengths in the visible and near infrared.

Multispectral and hyperspectral scanning – In a multispectral scanner, a prism or diffraction grating is used to split the detected beam into its constituent spectral components. Separate detectors are used to record discrete ranges of wavelength, each detector chosen to be most sensitive to those wavebands detected. The size of the detector and its position determines the amount of spectrum intercepted and therefore the width of the waveband recorded. Early instruments, such as the MultiSpectral Scanner (MSS) on Landsat, used only four detectors recording four contiguous broad wavebands. Later instruments have used more (e.g. 12 on the airborne thematic mapper, ATM), smaller, detectors to sample selected narrow wavebands. Multispectral pushbroom scanners use separate lines of detectors, each recording the chosen waveband. The early SPOT HRV systems collected three broad wavebands, while the more recent MERIS and MODIS collect 15 and 36 wavebands respectively. In instruments such as MSS and TM, a number of lines are scanned simultaneously using six or more sets of detectors in order to increase the dwell time.

Multispectral scanners collect data in a few fairly broad wavebands. Hyperspectral scanners, on the other hand, may collect data in over 200 very narrow contiguous bands, and are often referred to as *imaging spectrometers*. Such a detector may be envisaged as an extension of the pushbroom concept, only now using a two-dimensional array of detectors rather than just a few strip detectors. The signal is dispersed across the rows of detectors, each row detecting the waveband component that falls on it, thus producing as many bands of data as there are rows of detectors. This obviously produces huge quantities of data, which poses severe problems in transmission from satellites. Such instruments are usually flown on aircraft where the data can be recorded for later processing in the laboratory, but even so, many of these are programmable, enabling the operator to select either a restricted number of wavebands suited to their requirements, or more wavebands at reduced spatial resolution. Even though hyperspectral instruments *can* detect hundreds of wavebands, rarely do they do so in practice. CASI (the Compact Airborne Spectrographic Imager) provides 288 wavebands but only over a very restricted swath, whereas it can be programmed to give full-resolution images in just 19 pre-selected bands. PROBA, which was the UK's first microsatellite (weighing 94 kg), launched in 2001, carried CHRIS (Compact High Resolution Imaging Spectrometer). This was one of the highest resolution imagers of its type operating in space at that time, providing 19 spectral bands in the

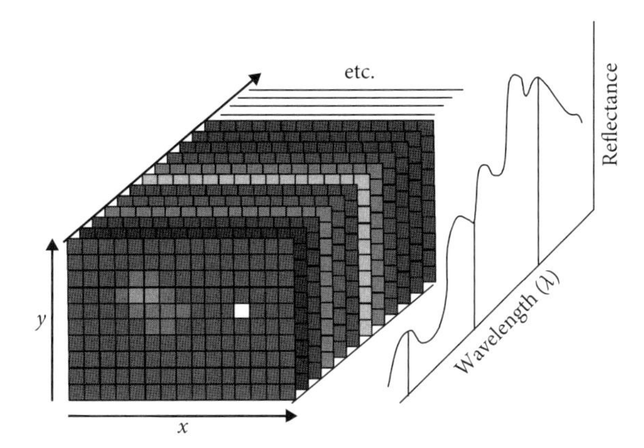

Fig. 5.3. Illustration of a hyperspectral image cube, where there may be a series of several hundred images at different wavelengths stacked to form a three-dimensional cube with two spatial dimensions (x, y) and a spectral dimension (λ). The spectrum illustrates the change in reflectance with wavelength for one pixel position. (See Plate 5.1)

visible and near infrared range at 25 m resolution and a swath width of 19 km, or it can be reconfigured to give 62 spectral bands at 50 m resolution.

These multiple images (one for each waveband) can be stacked to obtain what is sometimes referred to as an 'image cube' (Fig. 5.3, Plate 5.1). The advantage of hyperspectral over multispectral images is that they contain more information and can be used to detect quite subtle differences in the pattern of spectral reflectance between different surfaces and hence can be used to discriminate rather sensitively between them. The enormous amount of data from imaging spectrometers opens up analytical approaches not possible with multispectral data, though specialist software is required (see Chapter 7).

There are other ways to collect hyperspectral data. VIFIS (Variable Interference Filter Imaging Spectrometer), designed and built in the University of Dundee, is based on a standard shuttered video camera, but incorporates a variable interference grating in its optics (Sun and Anderson, 1993). In this, the grating spreads out the spectral components in each frame in the flight direction. Each line of pixels in a frame therefore contains one spectral component, from say blue at one end to red at the other. A 'blue' image, for example, may be built up using software that takes the first line of pixels from successive frames. If fewer, broader, wavebands are required, several lines of pixels may be grouped together. It should be noted that, unlike the conventional hyperspectral scanner that operates in a set of pre-determined wavebands, in VIFIS the full spectral information is captured and stored and can be processed at leisure to suit the differing requirements of users as and when needed. A similar principle is adopted in more sophisticated cameras such as the Hyspex (Norsk Elektro Optikk AS, Lørenskog, Norway).

Thermal scanning – Thermal scanners are simply multispectral scanners operating in the thermal part of the electromagnetic spectrum. Occasionally a thermal channel is incorporated into a multispectral instrument. The thermal signal is separated from the visible signal using a reflection grating and directed towards a separate detection system. Detectors in the thermal region are on the whole less efficient than those in the optical region, and also the energy in a quantum of thermal radiation is less than that for optical wavelengths (see Section 2.2.3), so that the size of the pixels in such a system are usually bigger than those in the visible bands in order that sufficient sensitivity is obtained. In the Landsat Thematic Mapper, the six visible bands have 30-m pixels but the thermal channel has 120-m pixels (60 m in later versions). Usually, thermal data are used in a quantitative fashion to measure the temperature of ground features. This will require calibration of the system, since, even though pre-launch calibration has been carried out, the detection efficiency of the detectors may deteriorate over time. The AVHRR, for example, uses an on-board black body maintained at a temperature that is measured using thermocouples. This black body is scanned as the mirror rotates, thus giving a reading of a known temperature once every scan line. Calibration is not so important if only relative temperatures are required. Landsat thermal data can be used in a qualitative way for studying variations in thermal emission from geological features and the Heat Capacity Mapping Mission (HCMM) was designed to differentiate between rock types by detecting their different thermal inertias.

Other types of scanning – All the line scanners described above rely on the motion of the platform in order to provide the second spatial component of the image. This type of scanning would therefore not be possible on geostationary platforms. Meteosat, for example uses a spin-scan mechanism in its Visible and Infrared Spin-Scan Radiometer (VISSR). The whole satellite (which is shaped like a barrel) spins on its axis. The detector points out from its side and therefore its field of view is continually sweeping out a complete circle that is designed to encompass the earth. The detector therefore 'looks' at a strip of the surface of the earth during each rotation. A tilting mechanism is used to step the angle of the detector by a certain amount between each rotation so that each scan line is displaced from the previous one by just the dimension of the pixel. VISSR takes 25 min to scan from the top to the bottom of the earth, creating about 2500 lines of data. During the next 5 min, the satellite is stabilized and the mirror reset to scan the top of the earth. Images are therefore collected every 30 min.

5.4 What is in a pixel?

The above brief description presents only a rather simplistic explanation of the scanning principle. In practice, the field of view is not rectangular and the pixels produced are not the simple representations of the square area of ground depicted in the final image, but instead are artificially created from the data received at the scanner. There are a number of complicating factors introduced by the scanning mechanism itself that need to be considered if the image content is to be fully understood. The essential terminology is summarized in Box 5.3 and discussed further by Cracknell (2008). Much of the terminology (especially the use of the term instantaneous field of view, IFOV) relating to an individual pixel derives from the operation of scanners; with CCD-array imagers there is a closer relationship between the pixel on the ground and the pixel in the image.

BOX 5.3 Some useful concepts relating to pixels and imaging (see also Fig. 5.4)

Ground sampling distance **(GSD)** – Distance between pixels, and hence the pixel width, set by the instrument's sampling rate. This may not be the same as the GIFOV.

Footprint – equivalent to the GIFOV of a sensor, usually used in preference in the case of a non-imaging sensor.

Ground resolution cell **(GRD)** (equivalent to the GSD) – Pixel size as set by the scanner's sampling rate or the area of elements on a CCD detector.

Field of view **(FOV)** – The total acceptance angle of the detector covering the whole image width. For a line scanner the projection of the angle covered by the whole of the scan line onto the ground is equivalent to the swath width.

Instantaneous field of view **(IFOV or 'angular IFOV')** – The angle over which an individual sensor is sensitive to radiation.

Ground instantaneous field of view **(GIFOV)** – The projection of the IFOV onto the ground. Note that there is commonly pixel overlap, so the GIFOV is often larger than the GSD. Also note that the GIFOV for a fixed angular IFOV increases from the nadir to more oblique angles of view.

Point spread function **(PSF)** – In the context of remote sensing we are often concerned to know the spread function relating the relative contribution of radiation from different parts of the GIFOV to the value recorded at the detector. This is often referred to as the PSF, though more usually in optics the term refers to the distribution of energy, which comes from a single point, across a sensor's image plane.

Swath – The strip of ground beneath the airborne or satellite sensor from which data are collected. The swath width is determined by the length of the scan line (i.e. by the FOV) of the detector.

Nadir – Looking vertically downwards.

Zenith – Vertically upwards.

Solar azimuth – Angle that the line of sight to the sun makes with the north in the horizontal plane.

Solar elevation – Vertical angle that the line of sight to the sun makes with the horizontal.

Dwell time – Time required for a detector IFOV to sweep across a GRC.

Ground track – The track of the satellite projected on the ground below.

Revisit time – Time interval between two successive overpasses of a given area.

The area on the ground whose properties are sensed and allocated to one pixel is known as the *ground instantaneous field of view* (GIFOV); this is essentially the projection of the IFOV on to the ground. As shown in Fig. 5.4 this does not exactly equate to the ground resolution cell. It seems to be implied that the scanning mechanism somehow pauses while the information from a ground resolution cell is recorded before moving on to the next one. This is of course impossible, and data are recorded continuously across the sweep, being divided into time segments by the gating mechanism of the detector (similar to the opening of the shutter of a camera) and allocated to a sequence of pixels by the computer. Given that the GIFOV of the detector is circular (at nadir) and moves across the scan direction over the short period of time that the detector gate is open, the information that is collected during this interval comes from an oval-shaped area of ground often larger than the size of the pixel to which it is eventually allocated. Adjacent pixels in the image will therefore contain information about the same points on the ground. Similarly, the satellite does not pause while a line of data is collected, before moving on to collect the next one. Instead, the satellite continues to move forward as the line is scanned so that the line of pixels is not perpendicular to the flight line of the satellite. Again, because of the oval GIFOV there is overlap between scan lines. The point spread function illustrated in Fig. 5.4 describes the relative weighting of contributions from different areas on the ground to the pixel value.

At the nadir point, the IFOV of a scanner is circular, but as the line is scanned, the IFOV intersects the earth's surface at an increasing angle, becoming more elliptical towards the end of the scan line (see Fig. 5.5). If all data were allocated to similar-sized square pixels, the image would become more and more distorted (compressed) towards the end of the scan line (see also Fig. 6.1). Instead they must be allocated to longer and longer rectangular pixels in order to preserve geometric fidelity. Similarly, because the distance from the platform increases with angle, the size of the GIFOV increases, which means that there is greater overlap between adjacent scan lines. These effects are aggravated by the curvature of the surface of the earth, particularly for detectors with large swath widths such as the AVHRR. The nadir pixels of the AVHRR are about 1.1 km square, whereas those at the extremity of the scan line are about 6 km long.

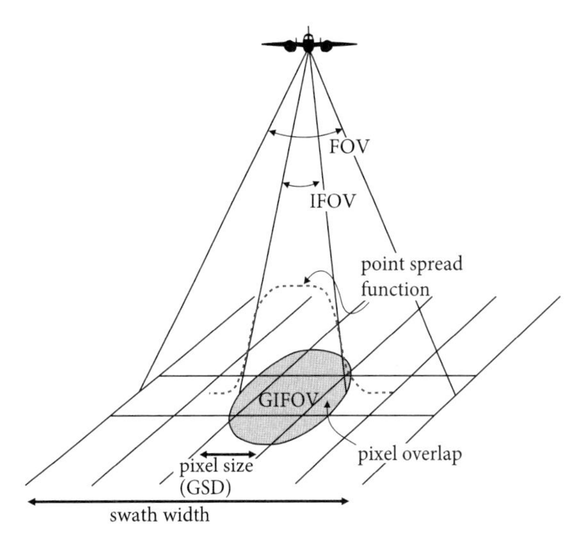

Fig. 5.4. Illustration of key terms used in imaging, showing the relationship between FOV and swath width and between IFOV and GIFOV and pixel size. Because the GIFOV can overlap pixels, there is usually some contamination of the signal in one pixel by signal from adjoining pixels. The point spread function illustrates the fact that not only does the GIFOV not necessarily match the nominal pixel size, but that the contribution of different areas on the ground to the pixel value varies with position.

Further discussion of the correction of these and a number of other geometric and radiometric problems may be found in Section 6.2 and in Cracknell (2008). Objects on the ground are rarely exactly the same size or shape as the hypothetical rectangular pixel. They are

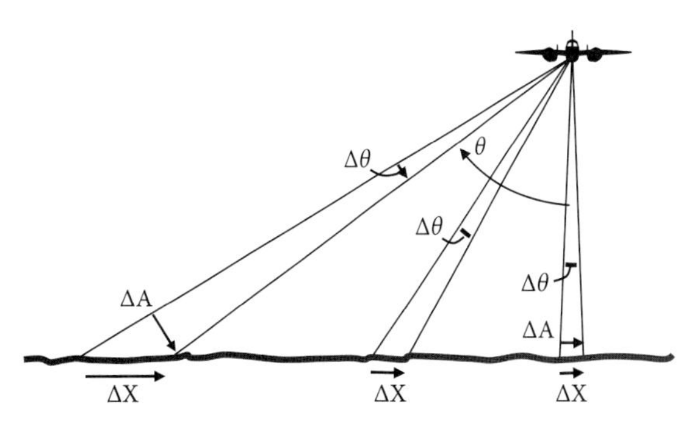

Fig. 5.5. View-angle effects, showing that for constant IFOV ($\Delta\theta$) at the sensor, not only does the area viewed on the ground (the GIFOV represented by ΔX) increase as the angle, θ, increases (according to $\Delta X(\theta) = H(\tan(\theta) - \tan(\theta - \Delta\theta))$, but the area of the normal cross-section from which the signal is received also increases (according to $H \tan(\theta)/\cos(\theta)$), as does the pathlength of the radiation through the atmosphere.

neither aligned with their edges parallel or perpendicular to the scan direction nor do their centres coincide with the centre of the pixels. If there is more than one feature in the field of view, then the detector response will be an average of the responses from the individual features weighted according to the proportion of the field of view they occupy (*mixed pixels* – see Section 7.4.3). Conversely, if an object spans more than one pixel, its signature will affect all those pixels. Because of diffraction at the lens of the instrument, a point object does not produce a point image, so the total intensity at any point in the image is the sum of the contributions from many adjacent points. The shape, location with respect to the optical system, and the electronic response of the detectors also all affect the point spread function of the instrument and hence the radiometric fidelity of the image.

Yet another complication arises from *resampling* the pixels (see Chapter 6). The size and orientation of pixels can be artificially changed for example when rectifying an image or registering it to a map. The radiometric values of new pixels are generated from suitably weighted proportions of a number of adjacent pixels and allocated to the new rectangular grid. The new pixels are thus admixtures of the old ones and hence the information content is blurred out and the real spatial resolution, as opposed to the nominal spatial resolution, is reduced.

All the above factors will affect the spatial resolution of the final image. One might naively think that the resolution is just the size of the pixel. But what exactly does that pixel represent? New, smaller, pixels can be generated by subdividing the original, but that does not mean to say that we are actually getting finer spatial detail. Even the original pixel contains information from a larger area than that to which it has been allocated in the image. In fact, an object generally has to be several times the size of a pixel, and to sharply contrast with its surroundings, in order to show up on an image.

Consideration also needs to be given to the spectral sensitivity of the detectors. This is not constant across the specific wavelength interval defined by the filters and diffraction gratings used. These do not define strict limits, and the sensitivity is a maximum near the centre of the band and decreases towards the extreme limits (see Fig. 5.6). Consequently, it is usual to define the spectral sensitivity in terms of the *full width half-maximum* (FWHM) value. This therefore

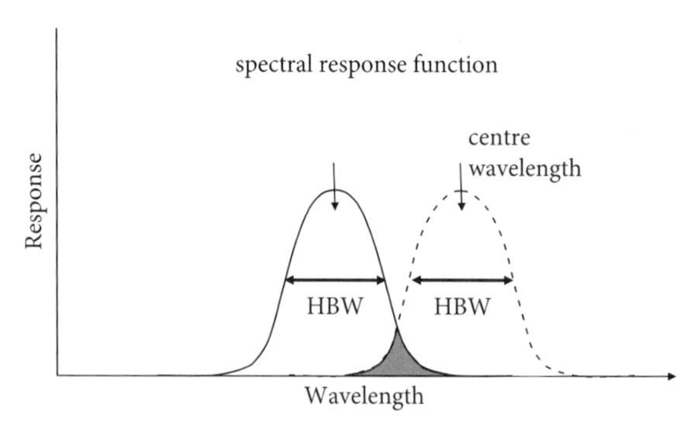

Fig. 5.6. Illustration of the spectral response function showing how the sensitivities (responses) of two detectors vary with wavelength of the received radiation. The response of each detector is defined in terms of the central wavelength and the full width half-maximum bandwidth (often abbreviated to HBW or half-bandwidth). The spectral resolution is set by the half-bandwidth and not by the absolute number of channels. For example, an instrument with channels centred at 1-nm intervals but with a half-bandwidth of 10 nm will have a lower spectral resolution than a sensor with one fifth the number of channels at 5-nm intervals, but with a half-bandwidth of 5 nm.

defines the *spectral resolution* of the instrument, the narrowest spectral interval that can be resolved, even though it is sensitive to wavelengths beyond these values.

A final complication arises in the radiometric values of the pixels. Detectors are calibrated prior to launch, but their sensitivity may deteriorate during their lifetime. Some instruments such as the AVHRR have on-board calibration facilities for the thermal band (see Section 5.3.2). Others, such as SPOT, are periodically recalibrated by using their response over known areas of uniform illumination, such as a desert. Often measurements for calibration are made by the user during field work in order to correct the particular images being used in that project. Intercalibration between the many individual detectors on an array in a pushbroom scanner is vital so that their individual responses can be suitably adjusted to prevent the image appearing stripy. If data from different instruments in a series (e.g. from archived AVHRR images), or from different instruments (e.g. Landsat TM and SPOT HRV) are to be meaningfully compared then intercalibration is essential. Calibration is also essential when comparing images taken at different times or whenever the data are used quantitatively in physical and biophysical models.

5.5 Using microwaves

The heritage of microwave remote sensing lies in the early studies of the generation and transmission of radio waves in the late nineteenth century. Early in the twentieth century, radio technology and basic radar[25] concepts were being developed. After World War II, scientists started using the newly developed microwave techniques and instruments to study radiation of extraterrestrial origin, and many significant discoveries in radioastronomy have subsequently been made. The development of airborne radar really stems from the development of the magnetron, a radio valve for generating shortwave radio signals. Before that, the long aerials required at longer radio wavelengths were not suited for mounting on aircraft, but now small antennae could be used that generated sufficient power for enemy detection, etc. Meteorologists further developed the techniques for ground-based observations of the atmosphere, thunderstorms, and rainfall. The realization that microwaves could be used for reconnaissance as well as for navigation led to the development of imaging radars. Side-looking airborne radar (SLAR) became available for military purposes in the 1950s and the technique of Doppler beam sharpening soon followed. The declassification of these military systems led to their first civilian use in the late 1960s and to the first deployment of synthetic aperture radar (SAR) in space in 1978 on SeaSat (see Section 5.5.2). Since that time, a number of satellites have deployed both microwave radiometers and imaging radars, and, in spite of the complexity of the data, the parallel developments in computers have enabled images from these to be used for many environmental studies.

Many of the concepts of microwave remote sensing are probably more difficult to grasp than those of other forms of remote sensing, and a certain amount of mathematics and physics is required in order to understand many of the topics. The recognition of features in an optical image is fairly intuitive, but we have no everyday experience of how microwaves interact with matter to aid our understanding of the information content in a radar image. The geometry of a radar image is unnatural – objects are mapped as a function of their slant range (i.e. their physical distance from the instrument) rather than their look angle as in an optical system, and hence they give no indication of the direction to the target. A georeferenced image has to be constructed using the delay time between sending and receiving the signal, and any change in frequency which may occur (Doppler shift). The interpretation of radar images is also not straightforward. The physical parameters that influence the strength of the signal detected are things like water content and roughness, so different areas of vegetation, for example, would be distinguished by the lushness of leaves, the presence of dew, the texture and the structure of the plants (see Section 3.3).

The use of microwaves in remote sensing has escalated over the last few years. As was pointed out in Chapter 2, the major advantage of microwaves is that they are not affected by cloud and light rain, and so provide an 'all-weather capability'. Furthermore, at least the longer wavelengths can also penetrate vegetation canopies and so provide information about the underlying layers and the soil. Since they do not occur naturally in any quantity, they often have to be generated artificially and so their use neither depends on, nor is affected by, natural sources such as the sun – hence they can be used at night as well as by day. Microwaves interact with surfaces in a different way from visible radiation. Their wavelengths are of the order of the size of physical objects, and so the main mechanism of interaction is by scattering. Microwave interactions with different surface features depend on their wavelength. The scattering may also depend on the direction of polarization of the microwaves and on the moisture content (and even the salinity) of the surface soil. It is frequently possible to combine images obtained using different wavelengths and polarization, or obtained at different times, in a similar way to that in which multispectral visible data are used to form composite images to help analysis. But the concept of

25. *RADAR* is an acronym for 'RAdio Detection And Ranging'. Pulses of microwaves are sent out from a transmitter and the component that is scattered back is detected by a receiver. The time delay between sending and receiving gives a measure of the distance travelled and thus of the distance to the scattering object. By scanning the microwave beam, it was possible to build up a spatial distribution of the scattering points, which could be aircraft, on a cathode ray tube, thus enabling the location of the aircraft in space to be determined. Active remote sensing using microwaves is often referred to as *radar remote sensing*, particularly when used in imaging mode as, for example, in *synthetic aperture radar*.

'colour' no longer has a meaning, and we talk about 'multispecular' images rather than 'multispectral' ones. Indeed the amount of energy of any wavelength scattered from a green leaf depends on its size, shape, and water content, rather than the amount of chlorophyll or its 'greenness' (Woodhouse, 2006). Since microwaves are absorbed very strongly by water, the presence of even dew on vegetation will markedly alter the scattering parameters and complicate analysis. For all these reasons, this form of remote sensing is a difficult subject. It requires very bulky and expensive equipment, a lot of computing, and much experience to interpret, but it can produce very high resolution and accurate quantitative information.

5.5.1 Passive microwave sensing

The intensity, or brightness, of the microwave radiation naturally emitted by the earth's surface is used in *passive remote sensing* to provide information related to the surface temperature and its dielectric properties. As was pointed out in Chapter 2, the microwave emissions from the earth can be detected from space using a microwave radiometer; these emissions can be used to map surface brightness temperatures (T_B) over large areas (although the atmospheric emissions contribute a significant amount of noise to the received signal). The intensity of the radiation is very small, and the sensitivity of detectors at these wavelengths is limited, which reduces the spatial resolution of the measurements, and so the temperature maps are at much lower resolution (typically 50–70 km) than those obtained using thermal infrared wavelengths. But, nevertheless, microwaves do have an advantage when mapping sea ice, for example. In the thermal region, both sea ice and seawater at the same temperature have similar emissivities, but in the microwave region they are different, 0.8 and ~1, respectively. Thus, microwave radiometry can distinguish between ice and water even if they have the same kinetic temperatures, and so is used extensively for mapping in Arctic and Antarctic regions. Platform-mounted radiometers can also be used in the field for studying particular plant species or for making measurements over small areas. The main applications of passive microwave sensing have been in ocean studies, to measure sea-surface temperatures, salinity, surface winds and the distribution of sea ice. Across the oceans, the emissivity is fairly constant so that variations in brightness temperature correspond closely to changes in the physical temperature of the water. Over land, however, the emissivity varies strongly, and so it is not so easy to map land temperatures as it is water temperatures. The emitted microwaves, however, do contain information about soil and snow properties, in particular their water content. But because the footprints of passive radar systems are of the order of tens of kilometres, their use tends to be limited to regional rather than to global scale applications.

A range of approaches has been developed to retrieve information on near-surface soil moisture from passive microwave data. Low-frequency (between about 1 and 6 GHz) or long-wavelength measurements tend to be most useful for soil-moisture measurement because they are less affected by the vegetation layer (and hence well approximated by simple radiative-transfer models) and they are also less susceptible to atmospheric interference than are higher frequencies. Methods need to take account of the effects of the many factors that affect microwave emission including temperature, vegetation cover, topography, and soil roughness (Chapter 3, and see, e.g., Wigneron *et al.*, 2003). There is a range of approaches available to retrieve soil moisture from microwave measurements: these include (a) the use of ancillary information such as landcover maps or optically derived information on vegetation density or optical depth, together with thermal infrared temperature data, and (b) studies where the essential additional information is obtained from multiconfiguration microwave measurements involving multiple frequencies, polarimetric, and/or multiangular instruments (e.g. ADEOS-II). However this additional information is derived, its use to improve estimates of soil moisture can be based either on empirical statistical relationships, or better on the inversion of physically based radiation-transfer models. Multiconfiguration microwave sensors allow the simultaneous derivation of soil moisture and vegetation depth and possibly also the effective surface temperature.

Atmospheric sounding – Rather than aiming to utilize atmospheric windows where absorption and scattering of radiation are minimal, it is also possible to make use of those wavelengths where radiation is more or less strongly absorbed so that the radiation emitted from the atmosphere gives information about the atmospheric profile with depth. For example, the measurement of microwave emissions from atmospheric components at different wavelengths provides information about atmospheric composition and atmospheric chemistry. This is known as *microwave sounding* and many meteorological satellites have carried microwave radiometers for this purpose. Apart from their meteorological application, these data can be useful for the atmospheric correction of remotely sensed images (see Section 6.2).

5.5.2 Active microwave sensing

For many reasons mentioned above, microwaves are more often used in an active fashion for studying the earth. Repetitive pulses of microwave energy are generated on board an aircraft or satellite and a beam is directed downwards (or more usually obliquely downwards) towards the earth. The component that is reflected or scattered back towards the transmitter is detected, often using the same antenna. Examples include *radar altimetry*, which is used to profile forest stands and estimate canopy heights, *scatterometry*, which gives information, albeit on a coarse scale, about surface texture and roughness for agriculture and forestry, and imaging radar (SAR) which provides high resolution information about size, shape, texture, and water content, rather than, say, the amount of chlorophyll present (i.e. its 'greenness').

The design of the active microwave system involves, among other things, decisions about the appropriate system geometry. In particular, diffraction of the microwaves as they are emitted from the antenna causes the beam to spread out in a cone (Section 2.6), the semi-angle of which, θ, depends upon the wavelength and the diameter, d, of the antenna according to $\theta \approx \lambda/d$. Hence, the diameter of the area illuminated (the *footprint*) is, for small values of θ of the order of $H\theta$, or $H(\lambda/d)$, where H is the height of the platform above the surface. Substituting typical values for X-band microwaves ($\lambda = 3 \times 10^{-2}$ m), an antenna diameter of 1 m and a height of 5000 m would produce a footprint of diameter about 150 m, whereas from a height of 800 km, the orbital height of many environmental satellites, the diameter would be about 24 km. Since it would not be practical to mount antennae of much larger diameter than this on an aircraft or satellite, the spatial resolution of microwave instruments would be severely compromised, unless special steps were taken to artificially improve it. In altimeters, electronic means are used to pick off only a small portion of the footprint by frequency modulating the emitted pulse and by using only a small part of the returned signal (*pulse limiting*). The altimeter on ERS-1 used an antenna of diameter 1.2 m yet was able to produce a footprint of only a few kilometres. In imaging radars, at low frequencies, an ordinary linear aerial can be mounted along the length of an aircraft, but usually the resolution is improved by artificially synthesizing a much larger antenna using the Doppler effect (*synthetic aperture radar*, SAR) and employing a large number of small microwave transmitters (*a phased array*) instead of a dish.

The proportion of the transmitted energy that is returned from the target to the detector determines the sensitivity of the system to small signals, and is itself determined by the *radar equation*. The parameter related to the surface and that is eventually measured by the radar system is the *backscattering cross-section*, σ, which can be related to the physical and environmental parameters being investigated (see Section 3.3.2). The radar equation takes into consideration the following factors: the power leaving the antenna, P_t, the distance to the target (the range) R, since the intensity of the signal reduces with increasing distance, the directional sensitivity of the antenna (the gain) G, and σ, which determines how much of the incident power is retransmitted back to the detector. The power reaching the target, P_s, is given by the power transmitted in the direction of the target, P_tG, reduced by the factor $\sigma/4\pi R^2$ (the ratio of the effective area of the target to the surface of a sphere of radius R)

$$P_s = P_t G / 4\pi R^2. \quad (5.3)$$

This power is scattered in all directions, and so the fraction intercepted by the receiving antenna is this power reduced by another factor $4\pi R^2$. The effective area of the antenna is given by $G\lambda^2/4\pi$, so that the signal received by the detector, P_r, is finally given by the expression, the radar equation

$$P_r = P_t G^2 \lambda^2 \sigma/(4\pi)^3 R^4. \quad (5.4)$$

The most important thing to note from this equation is that the detected signal drops off as the fourth power of the distance to the target. By tripling the height of the platform, for example, the signal decreases by a factor of 81! It is usual to define the sensitivity of the system in terms of the power received from the target as a proportion of the noise signal received, N_o, giving a *signal-to-noise ratio* (*SNR*), where

$$SNR = P_r/N_o = P_t G^2 \lambda^2 \sigma/(4\pi)^3 R^4 N_o. \quad (5.5)$$

It can be seen that doubling the transmitted power doubles the received signal, whereas doubling the size of the aerial or doubling the wavelength will each increase the signal by a factor of four. For any particular system, the signal received is therefore directly proportional to the scattering cross-section, where the proportionality constant depends purely on the design parameters of the system.

Altimeters

Altimeters are used to measure height, or rather the distance below the platform to the reflecting (backscattering) object. This means that it is essential to know exactly where the platform is when it is making a measurement. This is not as trivial as it sounds, which is why it is necessary to know satellite orbits as accurately as possible, or for airborne platforms to be equipped with accurate GPS. Pulses of radiation are emitted vertically downwards below the platform and the time taken for the pulse to return is measured. In principle, this enables the distance travelled to be calculated, but, as in all walks of life, this is not a simple procedure. A microwave beam fans out when it is emitted from an antenna, so that from the height of a satellite, its footprint on the ground below will be several tens of kilometres across, so only an average value over this area will be obtained. The spatial resolution can be improved for example by reducing the pulse length or modulating the frequency of the pulse, but problems arise when the height of the surface varies across the footprint. The microwaves reflected from different parts of the rough surface will travel different distances and hence arrive back at the detector at different times, distorting the shape of the returned signal and making it difficult to know at which point on the pulse to measure. Various corrections also need to be made to account both for the effects of atmospheric temperature, humidity, and ionization on the speed of the pulse and for the shape of the earth's geoid from which height is measured. Even so, over the oceans, relative heights can be measured to a few centimetres. Over land, the roughness usually precludes meaningful measurements except over extensive deserts or ice sheets. The altimeter on Cryosat-2 (designed to determine variations in thickness of the earth's continental ice sheets and sea ice) uses an interferometric technique, combining the phase and amplitude information from two antennae one metre apart, to achieve centimetre height resolution at about 250 m × 250 m spatial resolution. The use of Lidar (see Section 5.6) provides an alternative and powerful approach for investigating the 3D structure of plant, especially forest, canopies.

Scatterometers

Scatterometers also send out a series of microwave pulses, but usually at an angle to the platform so that they illuminate the ground obliquely. If the surface were a perfect reflector of microwaves, these pulses would be reflected away from the platform. If it is a rough surface, however, some signal will be scattered back and be detected by a receiver on the platform. The intensity of the signal received will give a measure of the roughness of the surface. The beam can be scanned (often a conical scan is used from aircraft) so that a spatial representation can be built up of the area studied. The scatterometer was originally designed to be used over water and has been used extensively from space to map sea-surface roughness, wave height and speed, and direction. Because water waves are originally generated by wind friction on the surface, wind speed and direction can be inferred, so these measurements are invaluable for mapping wind fields over large areas.

Over land, the coarseness of the spatial resolution (tens of kilometres) limits the use of spaceborne scatterometers to regional studies, but their high radiometric accuracy and stability, together with the varying incidence angle, have enabled soil-moisture dynamics, ground thaw in Siberia, and ice-sheet properties to be studied. Much more useful information has been obtained from scatterometers on aircraft and helicopters. The scattered signal depends not only on surface roughness but also on soil moisture. Microwaves can penetrate tree and vegetation canopies, and have been used to map vegetation cover since the texture parameter depends not only on the gross structure of the vegetation – size and shape of trees and shrubs, for example, but also on the internal structure such as leaf shape and orientation (Section 3.3.2). Much useful work has been done using ground-based scatterometers to verify scattering models and to generate databases of scattering properties.

Imaging radars

We saw above that the footprint generated by a beam of microwaves is large. The radar beam suffers diffraction when it is emitted from the antenna according to eqn (2.20), and the area illuminated also depends on the distance between the antenna and the ground. Over the oceans, this poor spatial resolution is not a problem because the wind and wave fields observed usually have small variability over large areas, but the spatial variability of features on land is usually very much greater. So, in order to produce high-resolution maps of microwave backscatter, it is necessary to consider ways in which the spatial resolution of microwave systems may be improved as well as how they can be used as imaging devices.

The original type of imaging radar, *side-looking airborne radar* SLAR (sometimes referred to as *real aperture radar*, RAR; Fig. 5.7) observed a swath of terrain obliquely to the side of the aircraft. The distance from the aircraft to different points in the illuminated strip varies with look angle, so the signal that returns from the different parts of the strip arrives at different times. If the signal is sampled electronically, each sample contains information about a different part of the strip, or *ground resolution cell*, and so a spatial image may be built up from successive swaths. But the strip of ground illuminated by the conical beam of the microwave pulse gets wider the further it is from the nadir, and so the size of the ground resolution cell increases with distance from the sensor. This has to be compensated for when the image is constructed if a reasonable geometric representation is to be obtained.

The resolution along the ground in the direction of the beam, the range resolution, R_r is given by

$$R_r = \tau.c/2\cos\beta, \qquad (5.6)$$

where τ is the pulse length (e.g. 1 μs), c is the speed of light, and β is the depression angle to the ground at a particular distance. As can be seen from Fig. 5.7(b), a given pulse length will correspond to a longer distance on the ground in the near range, and hence a poorer range resolution in this area than at distance. Similarly, the resolution in the direction of the line of flight is determined by the width of the beam, as this determines the distance between points that can be resolved. It is clear from Fig. 5.7(c) that this resolution is better in the near range than further from the sensor.

It is possible to reduce the size of the resolution cell in the across-track direction by using shorter pulses and smaller sampling times to pick off the returns, but the spreading of the beam is a physical consequence of diffraction and can only be reduced by using either a shorter wavelength or a larger antenna. Microwaves have only a small range of wavelengths and so the scope to improve the situation in this way is very limited. There is also a limit to the physical size of an antenna that can be mounted on an aeroplane or a satellite, so that in practice they are limited to less than about 10 m. It is, however, possible to produce electronically the same effect as that obtained by increasing the size of the antenna – a 'synthetic' antenna. This requires the forward movement of the system and is done by making use of the *Doppler shift*, or the change in frequency that occurs when either the source of a wave or the detector are in relative motion and its magnitude depends on the relative speed between the source and detector. The frequency of the return received from an object in the field of view will vary depending on whether the platform is approaching or receding from it and on their relative speed. This relative speed is greatest at the time that the object enters the field of view, is zero when the object is to the side of the platform, and has a maximum negative value when the object leaves the rear of the field of view. Thus, by tracing the frequency history of the point while it is in the field of view it can be located very accurately. In fact, the resolution obtained is that which would be obtained by an antenna that is as long as the distance travelled by the platform while this point remains in the field of view, which may be tens of kilometres. We have therefore 'synthesized' an antenna that is far larger than its physical size; hence this is referred to as *synthetic aperture radar*, or SAR. The spatial resolution of SAR is a considerable improvement over that of RAR, and, what is more, is constant across the range direction. The ASAR image from Envisat has a resolution cell of about 25 m.

ASAR is an example of a polarimetric radar. Microwaves can be launched from the instrument in one of two modes, vertically (V) or horizontally (H) polarized, and detected in either of these two modes. Different agricultural crops have different polarimetric

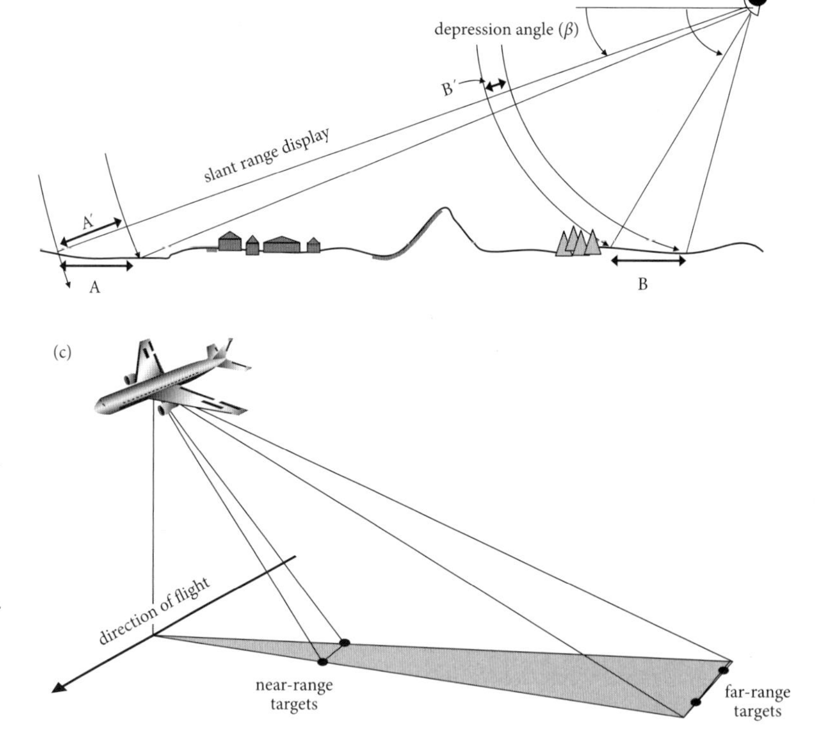

Fig. 5.7. (a) Schematic diagram of SLAR or SAR imagery, showing how the timing of the return signal can be used to allocate a particular intensity signal to a distance on the ground. Each pulse gives rise to a scan line and as the platform progresses an image of the surface can be derived. The brightest signals are from the built-up area that comprises many corner reflectors and the face of the mountain that is nearly perpendicular to the beam, while the microwaves are largely specularly reflected away from the sensor by the water surface so it appears dark. There are no returns at the time expected for the land shaded by the mountain. (b) The apparent foreshortening of near-range objects; both the regions A and B are the same dimensions on the ground, but because of the differences in time of return depend on the angle of view, the nearer area is foreshortened. All modern radar systems reproject the image to correct this effect, but note that this effect also leads to the range resolution being larger (less good) for near-range objects. (c) The increasing beam spread with distance from the antenna leads to a reducing ability to resolve objects at a given distance apart.

responses (i.e. the scattering depends on the polarization of the incident radiation), whereas full forest canopies tend to be completely depolarizing. Hence different surfaces can often be distinguished by using different combinations of polarization, such as HV, where the incident signal is horizontally polarized and it is the vertical component in the scattered signal that is detected, VH, HH, or VV. When two or three such images are displayed on the different guns of a computer monitor, a colour composite is formed in which different vegetation types will be depicted in different colours, thereby helping in their classification.

Because of their extremely stable nature, microwaves can be used to perform interferometry. Their phase remains constant over extended periods of time, and so the returns from two different pulses are coherent and can be made to interfere. If the pulses have travelled different distances, they will add or subtract, depending on their phase differences. This can be used to provide high-resolution *digital elevation models* (DEMs) at centimetre vertical precision from space (see Section 6.6). After ERS-2 was launched in 1995, for several months it was run in tandem with ERS-1, approximately 35 min behind, with the ground track of ERS2 coinciding exactly with that of ERS-1 24 h earlier. This enabled pairs of SAR images of an area to be obtained both from different viewpoints and at different times. In all, about 110 000 pairs of images were obtained during the 9-month tandem campaign. In fact, the microwave signal from ASAR on Envisat is so stable that interferograms can be made from images obtained several months apart. The Shuttle Radar Topography Mission in 2000 used two receiving antennae, one located on the end of a 60-m mast, to provide the necessary separation and hence phase differences and provided data to produce high-resolution topographic maps of much of the world.

The use of interferometric SAR (InSAR) for obtaining digital elevation models (DEMs) is a key application of the technology. It should be noted that the measured elevations represent the ground-plus-canopy surface and not the ground surface alone. The scattering phase centre that determines the phase difference geometry, and thus the height value obtained for the DEM, is some distance above the actual ground surface, by an amount that depends on the density of the vegetation and the wavelength of the microwaves used. The use of interferometric SAR for the study of trees and tree heights will be explored further in Chapter 11.

The potential importance of interferometry has been highlighted by Woodhouse (2006) who stated that "the impetus for future radar satellites is the high-precision topographic data available from interferometric imaging radar and the statistical relationship between forest biomass and L- and P-band radar backscatter which offers the potential to help map and monitor carbon stocks and terrestrial carbon dynamics".

Rain radar

As was pointed out in Chapter 2, though water vapour and water droplets in cloud and haze scatter microwaves very little, when the droplets coalesce to form rain their size approaches that of shorter wavelength microwaves, and scattering starts to become important. The scattering coefficient also tends to increase with rain rate, and so the microwave backscatter from rain cells can be used as an indicator of rainfall distribution and intensity. High-frequency microwaves are therefore employed in rain radars to detect and study precipitating rain. A beam of microwaves is scanned in a circle just above the horizon and the scattered signal is detected in much the same way as that in aircraft control. Rain cells up to 200 km away can be detected, depending on the flatness of the terrain, and so only ten or twelve radars would be needed to cover the entire British Isles. These are the source of the 'radar images' shown on television weather forecasts. Optical remote sensing from satellites can only detect the presence of clouds (from above) but cannot distinguish those that are precipitating from the ones that are not.

5.6 Laser scanning and Lidar

The use of lasers in remote sensing has increased over the recent years. Originally they were used for profiling and bathymetry but, the advent of accurate (differential) *global positioning systems* (GPS) and *inertial stabilizing systems* (ISS) in the 1990s made it possible to measure absolute heights very precisely, and the use of very short pulses and rapid scanning, using a mirror to deflect the beam from side to side, opened up the possibility of producing three-dimensional images of the terrain and vegetation. In addition, portable ground-based laser systems are increasingly being used for studies of canopy structure from the ground.

Lidar (light detection and ranging) systems employ similar principles to radar systems. The laser sends out a series of very short pulses of a very narrow beam of coherent light, in a precise waveband; the time delay of the reflected pulse can then be used to determine the distance between the sensor and the reflecting surface.

The first sensors recorded only one echo per pulse, but with vegetation, even with small footprints of <0.5 m, there may be echoes from leaves at different levels, branches, and the underlying soil (Fig. 5.8). Multi-echo sensors may detect several returns for one pulse including the first and last returns, representing typically the top of the canopy and the underlying ground surface, respectively. With a nadir view, therefore, the difference between the first and last returns estimates the canopy height. The accuracy of measurement can be of the order of a few centimetres (if the height of the aircraft is known that precisely), but the main use is in profiling the surface below, where a knowledge of the absolute height is not so important. Such systems can be used in vegetation monitoring, to profile forest stands and provide information on the heights of canopies and their component vegetation layers.

The intensity of the returned echoes, which is related to the reflectance of the surface components, can also be used to characterize the surfaces. In many airborne systems the laser beam diverges by only about 0.1–2 mrad and therefore illuminates footprints of 20 cm to 5 m, depending on the height of the aircraft. Since the detectors can sample the reflected waveform with a time interval of about 1 ns, this gives a height accuracy of about 15 cm, while the horizontal positional accuracy can be 50 cm or better. Satellite systems such as the Geoscience Laser Altimeter System (GLAS) on ICESat, however, with its beam divergence of c. 0.12 mrad has a 'large' footprint of 70 m in diameter, spaced at 170 m intervals along the orbit (see Fig. 5.10).

As a result of the small footprints of *airborne laser scanning* (ALS) systems, they can generate *3D point clouds* (Fig. 5.8) describing the positions of canopy components in space. The density of these point clouds may reach 25–100 points m^{-2} in some systems. The overall point density depends on flying height, aircraft speed, and on the pulse repetition frequency. Even with laser footprint sizes of <0.5 m, laser pulses are likely to be reflected by several discrete objects (leaves, branches, soil) potentially yielding complex backscatter signals known as waveforms. As it is possible to time the return of the pulses to within 1 ns, some discrete-return systems can detect up to five returns from each pulse. In afforested areas the first reflections come from the top of the canopy, but since some of the laser beam can pass through the branches, reflections also occur from lower in the canopy, from the understorey, shrubs, etc., and finally from the ground surface. This technique allows the generation of 2D and 3D images of the vegetation canopies, providing information on tree heights, biomass, timber volume, and so on. By filtering out the above-ground reflections, the ground reflections will provide a measure of the terrain surface and thus enable the creation of a DTM or relief model. This method is superior to using photogrammetric methods, which are not very good for vegetated areas.

In *full-waveform* mode, instead of just sampling the returned pulse at a few points, recent developments have enabled the whole of the waveform to be analysed using a form of Gaussian decomposition to extract the component echoes as shown in Fig. 5.8. From the intensity measurements, files of *x*, *y*, and *z* coordinates of the reflecting objects can be recorded. In addition to providing 3D coordinates of scatterers in the canopy, analysis of the full-waveform data can provide useful biophysical information on the nature of the scattering surface (Morsdorf *et al.*, 2006). The heights can be colour coded and presented in a 3D perspective as a point cloud (Fig. 5.8). Small-footprint full-waveform ALS systems can map vegetation in three dimensions with a spatial sampling of about 0.5–2 m in all directions enabling tree features to be computed and tree species to be identified by unsupervised classification methods (see Sections 7.4 and 11.5). Estimation of *LAI* from discrete return Lidar requires some linking model (Lefsky *et al.*, 1999; Richardson *et al.*, 2009). Small-footprint, multiple-return systems are becoming more widely available, and have been shown to be capable of estimating *LAI* in single-species-dominated stands and/or in stands with a small range of *LAI* values (Morsdorf *et al.*, 2006; Riano *et al.*, 2004). Analysis of full-waveform pulses from space-based laser altimeter systems, such as GLAS on ICESat, are expected to improve our ability to retrieve canopy heights on a global scale and to monitor temporal changes. Another use of lasers in terrestrial laser scanning provides fast and reliable three-dimensional point cloud data acquisition for forest inventory from the ground.

Ultrasonic sensors (Reusch, 2009) work in a similar fashion to Lidar, using the timing of echoes after a signal pulse, and can be used to derive information on the height(s) of the major horizontal canopy components when used with a nadir view, and other aspects of canopy structure and density. Such sensors are, however, restricted to rather short-range in-field applications as a resultof the rapid attenuation of sound in the atmosphere. Similarly, the new generation of

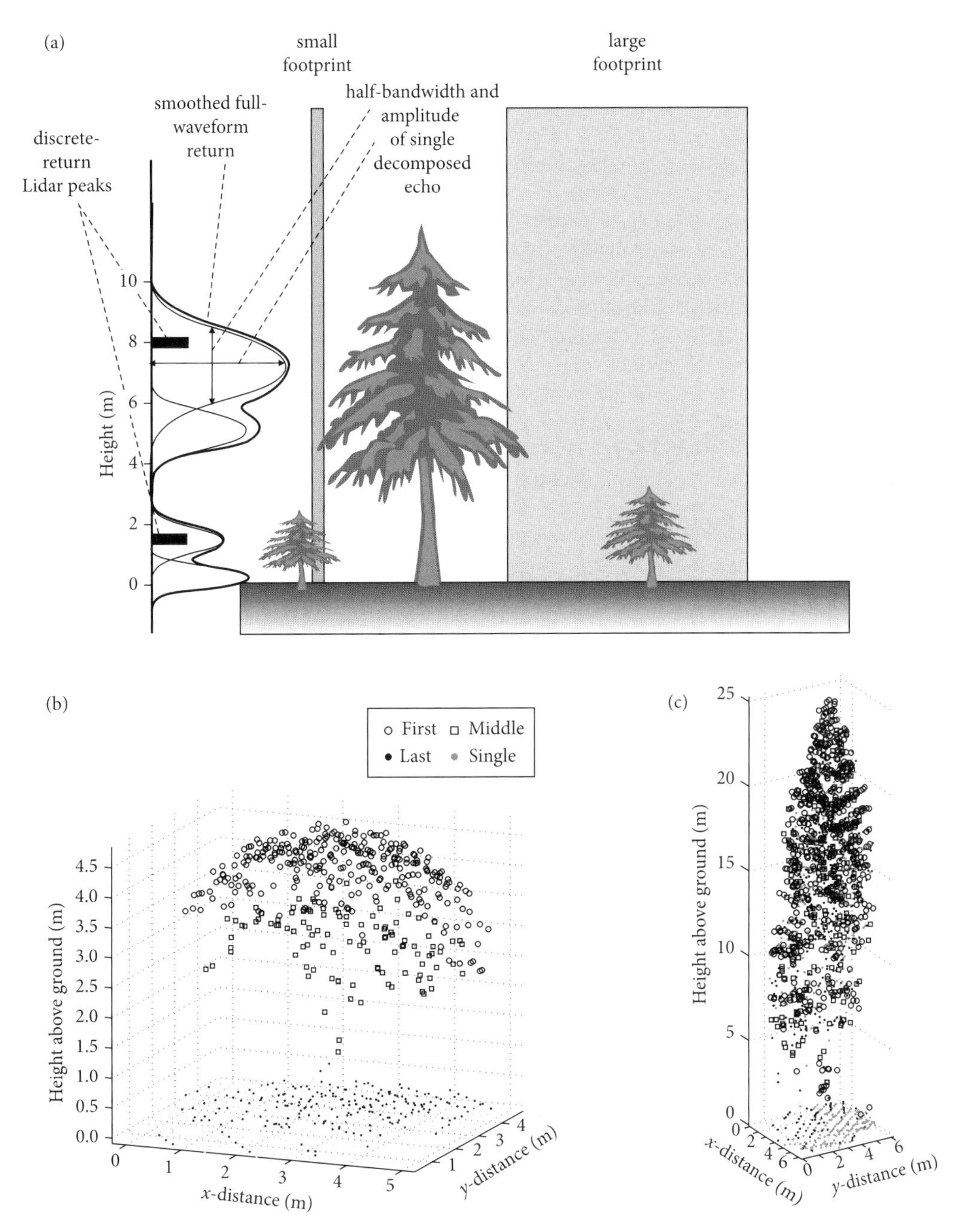

Fig. 5.8. Illustration of the use of small-footprint and large-footprint laser scanners. (a) The pattern of echoes for both discrete-return and full-waveform scanners are illustrated, together with the Gaussian decomposition giving information on the height, amplitude, and half-bandwidth of echoes from specific canopy and ground components. (b) and (c) show small-footprint full-waveform Lidar point clouds for (b) a beech tree and (c) a spruce tree in the Bavarian Forest National Park. Points were grouped into three categories: 'first' and last represent all points derived from the first and last detected peak, while 'middle' refers to all other points. Also shown are single returns (plotted using data from Reitberger *et al.*, 2008).

'time-of-flight' optical cameras can directly give 3D imagery from the phase of the return signal. For such cameras, as with Lidar, it is necessary to have a rapidly modulated illumination pulse with the time for the signal to be received at each pixel depending on the distance to the object.

Laser fluorosensing: Fluorescence occurs when a molecule absorbs a photon and subsequently emits another photon with a longer wavelength. Vegetation and many minerals fluoresce when illuminated with UV light, emitting light at visible wavelengths. The source of the stimulating radiation can be the sun, but may also be an active laser. Vegetation illuminated with UV may emit radiation at characteristic wavelengths of 440 nm (blue), 525 nm (green), 685 nm (red), and 740 nm (far red). The ratios of intensities of pairs of fluorescence peaks can be used as an indication of chlorophyll content. A system that both stimulates and analyses the emission is called a *laser fluorosensor*. It is possible to measure the decay times of the fluorescent signal using very high speed detectors and this can be used to measure stress in vegetation (see Chapter 11).

5.7 Observing system principles

5.7.1 Sensing logistics

The strength of the signal generated by a detector is a function of both the sensitivity of the detectors (the sensing elements and the electronic circuitry) and the amount of electromagnetic energy falling on it. The sensitivity of a given instrument therefore depends upon a number of factors:

Instantaneous field of view – Both the physical size of the sensitive element of the detector and the effective focal length of the scanner optics determine the IFOV. A small IFOV is required for high spatial resolution but also restricts the amount of energy received by the detector.

Energy flux – The amount of energy reflected or radiated from the surface of the earth is the *energy flux*. For visible detectors, the reflected flux depends on the illumination of the surface and is lower on a dark day than on a sunny day. In the thermal infrared, the individual quanta have only about one twentieth of the energy of quanta of visible light ($E = hf$). This, coupled with the lower efficiency of detectors at these wavelengths, means that usually thermal scanners have a larger IFOV than do visible scanners.

Altitude – For a given ground resolution cell, the amount of energy reaching the detector varies inversely with the square of the distance it has travelled. At greater altitudes the signal strength is therefore correspondingly weaker.

Spectral bandwidth – The signal is stronger for detectors that respond to a broader bandwidth of energy. For example, a detector that is sensitive to the entire visible range will receive more energy than a detector that is sensitive to only one narrow band, such as visible red.

Dwell time – The time required for the detector IFOV to sweep across a ground resolution cell is the *dwell time*. A longer dwell time allows more energy to fall on the detector, which creates a stronger signal.

5.7.2 User requirements

Until recently, most satellite systems have been planned, developed, built, and operated by national agencies. Although user communities may have been consulted in the planning stage, almost universally the system developed was designed to be used by a wide range of users, yet was not optimum for any particular application. Data were made available, and the users had to do what they could with it. Airborne instruments, on the other hand, were often designed, and sometimes built, in individual research laboratories for use in specific research projects. This 'bottom-up' approach, as opposed to the agencies' 'top-down' approach, has obvious advantages. In the last ten years, though, the European Space Agency has tried to adopt the latter approach in its Earth Explorer programme. It has periodically issued calls for proposals for 'Core' missions (large missions costing about 400 MEuro, and 'Opportunity' missions (smaller missions costing less than 100 MEuro). These are assessed by committees of experts, who select several for further feasibility and technical investigation. From these, one or two are chosen to go forward to be built and operated by the agency in collaboration with the Principal Investigator (Cryosat, mentioned above, was the first Opportunity Mission chosen in this way).

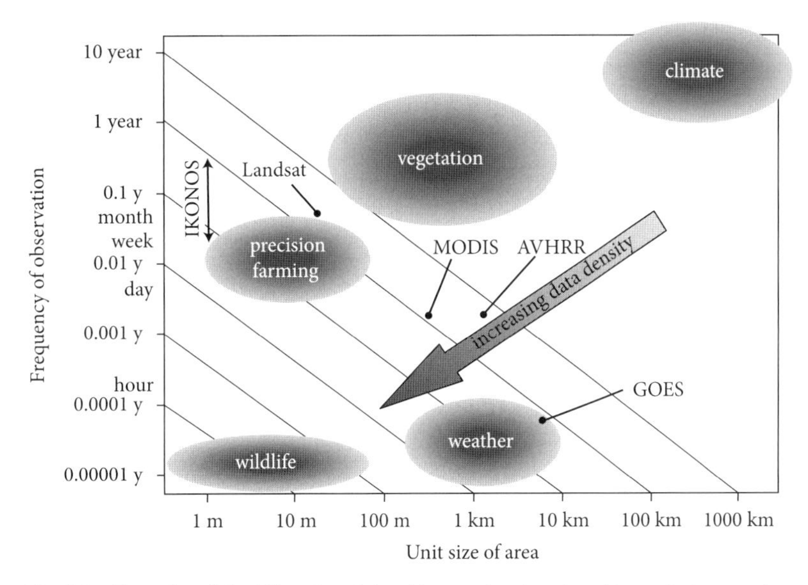

Fig. 5.9. Illustration of the different spatial and temporal scales relevant to particular aspects of remote sensing. The pixel size and frequency of return coverage are marked for some typical satellite/sensor systems (GOES, NOAH-AVHRR, MODIS, Landsat-TM, and IKONOS; note that a range of frequencies is shown for IKONOS owing to the pointing capability of the sensor). The diagonal lines are lines of equal data volume and illustrate the potential trade-offs between high frequency and high spatial resolution; the total data volume increases towards the bottom left-hand corner. High spatial/radiometric resolution comes at the cost of image size (area covered) and/or frequency of image collection.

In general, detectors provide data, but the user requires information. A particular landowner may need to know the crop yield he might expect for forward-planning purposes. He does not care whether the data comes from Landsat or SPOT, or whether supervised or unsupervised classification was used in the analysis. He wants to know the number of tonnes that will be produced – together with an indication of the accuracy of that number. It is the role of the analyst (the 'remote sensor') to provide this value-added product. The user also requires timely data supplied at a price he can afford in a usable format.

5.7.3 Data limitations

Although it may be counter-intuitive, high (= fine) spatial resolution may not always be desirable; for global and regional studies one is often interested in average values over large areas. Although the 30-m pixel resolution in a Landsat TM image may be very useful for many purposes, such as crop monitoring, it is not so useful in a large desert, partly because there may be no ground control points to allow accurate location of the 180 × 180 km image. In this situation, the 1-km pixels in an AVHRR image that covers a swath of about 2800 km may be much more useful, especially since such images can be obtained every few hours rather than every few weeks. Indeed there is always a trade-off between the different resolutions (radiometric, temporal, and spatial; see Fig. 5.9).

Fine spatial resolution is usually a characteristic of a small image because of the amount of digital data that a large number of pixels would generate. This implies infrequent image acquisition because of the number of orbits required to give full earth coverage while avoiding view angles that are too oblique. Similarly, a large number of narrow wavebands both increases the total amount of data that would need to be transmitted to earth in real time[26] and also reduces the radiometric sensitivity of the detector, unless the pixels are made

[26]. The MSS on the early Landsat satellites produced 80 m pixels in four wavebands. The Thematic Mapper that superseded it produced 30 m pixels in seven wavebands. By decreasing the size of pixels by a factor of nearly three, and increasing the number of wavebands from four to seven (and also increasing the radiometric resolution from 6 bit to 8 bit), the volume of data increased by a factor of 50 (60 Mbps to 3000 Mbps).

correspondingly bigger. In spaceborne hyperspectral instruments the number of wavebands used in high-resolution mode is usually restricted, in contrast with airborne instruments that do not need to transmit the data but can store it on board. An extreme example of the trade-off between spatial and temporal resolution occurs in satellite altimetry where heights are sampled as the satellite proceeds in its orbit. With the GLAS Lidar system on ICESat, for example, the satellite has near global coverage and has a repeat cycle of 91 days. The individual footprints are approximately 70 m diameter and 170 m apart along track. The 183 day ground track repeat cycle provides 15 km track spacing at the equator and 2.5 km at 80° latitude, so it follows that only a very small fraction of the earth's surface is covered (Fig. 5.10)

Full earth coverage at 1-m spatial resolution, in 100 wavebands and 16-bit data, every day, may sound desirable, but would produce terabits of data that would be beyond our present capacity to store – and who is going to use it anyway? Only a fraction of all the data collected today is actually analysed, but even so much of it is archived for possible future use. The analysis of archived data often provides valuable information about trends and changes (Dundee University Satellite Data Receiving Station has an archive of European AVHRR data going back to 1978, which is still providing a valuable resource to researchers – www.sat.dundee.ac.uk/). One of the major problems nowadays is deciding which datasets to archive and which to discard.

A number of limitations and problems have been already mentioned, such as the trade-off between the different resolutions. One way of improving the temporal resolution yet retaining high spatial resolution is that incorporated into the HRV detector on SPOT which has the ability to point away from nadir by up to 27° on either side. Not only does this enable areas of

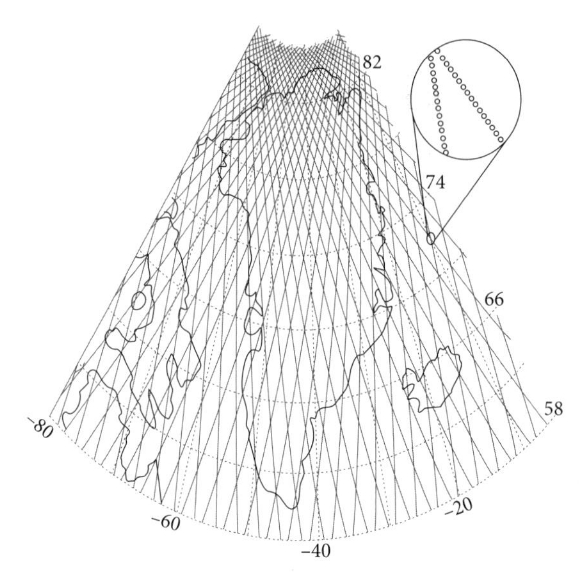

Fig. 5.10. An illustration of ICESat tracks over Greenland during its validation phase with an enlargement of one small area showing the 70-m diameter individual footprints which are 170 m apart. This indicates the very low fraction of the earth's surface that is actually sampled by this very narrow beam sensor.

ground to be imaged more frequently than the normal 26 days, as little as once every 5 days at the equator and even more frequently at higher latitudes, but it also provides stereo possibility. Another use that could be made of this oblique viewing mode is to study bidirectional reflectance.

One serious problem with operating instruments on spacecraft is generating sufficient power for them. There is a limit to the size of solar panels, particularly on small satellites. Even though the large solar panel on Envisat generates 6500 W, the microwave ASAR instrument on board consumes so much power that it is operated for only about 30 minutes in each orbit, mainly over land or areas of interest. KOMPSAT, a Korean radar minisatellite, acquires data for only 20 min per orbit.

5.8 Current systems

There are far too many sources of remotely sensed data to be able to do full justice here, and in any case the information will rapidly become outdated, so we will concentrate here on some typical systems that have been used extensively for vegetation monitoring, emphasizing the principles underlying the choice of system for specific vegetation monitoring activities. Extensive information about hundreds of satellites, detectors, airborne systems, and campaigns may be found in the book by Kramer (2002), which is the 'bible' of remote-sensing systems.

Perhaps the most useful way to categorize systems is by resolution – spatial and spectral. We can conveniently consider satellite systems as falling into three categories: low-resolution systems, such as those designed primarily for meteorology or oceanography (Meteosat, NOAA/AVHRR, ATSR, SeaWiFS, etc.); medium-resolution systems (Landsat, SPOT, MERIS, MODIS, etc.); and high-resolution systems (IKONOS, Quickbird, etc.). From a spectral perspective, we might differentiate these multispectral systems from hyperspectral systems such as CHRIS and Hyperion. Microwave systems will be considered separately. These and some others will be briefly discussed below, and the reader is referred to Appendix 3 for more information about their specifications.

5.8.1 Low-resolution systems

Atmospheric (and oceanographic) phenomena exist over far larger spatial scales and vary much more rapidly than do land-based phenomena. Consequently, spatial resolutions of 1–5 km are quite adequate, and high temporal resolution is essential – fortunately both criteria are compatible. High spectral resolution is not necessary for meteorology, and these detectors usually have a few broad wavebands in the visible (for cloud brightness), mid-infrared (for water vapour) and thermal infrared (for cloud temperatures). Although these meteorological satellites have poor spatial and spectral resolutions, they have much better temporal resolutions than higher-resolution systems, and may be useful for frequent monitoring of large areas and for large-scale and generic mapping of land surfaces. The two major current meteorological systems are the geostationary satellites (Meteosat for Europe and GOES-E and GOES-W for USA), and the polar orbiting NOAA series, carrying the AVHRR. Meteorological satellites are perhaps the only truly 'operational' systems, namely their continuity is so important that nations have collaborated over the years such that there are always replacements available should one fail. By contrast, 'scientific' systems (most of the rest) have been either 'one-off' systems or at best have only a few in the series (such as Landsat, although that has been quite a long time series).

The spin-scan radiometer on Meteosat, for example, has 2.5 m spatial resolution in the visible band and 5 m resolution in the water vapour and thermal channels. But because of the vantage point of the satellite (over the Greenwich meridian at the equator), only the central portion of the image is viewed normally, pixels becoming increasingly elongated as they approach the rim because of the curvature of the earth's surface. In fact, pixels too far away from the centre are not much use for land-surface monitoring because of the obliquity, and use is limited to about 60° N and S. But the 30-min repeat time makes the data very useful as it often enables more frequent observations in cloud-covered areas. The new Meteosat Second Generation (MSG), launched in 2005, has improved specifications of 15-min repeat at 1-m resolution in 12 channels, which makes it much more suitable for land observations. The GOES imager, the Multispectral Imaging Radiometer, uses a two-axis scanning system and provides five channels, one in the visible and four thermal bands.

The NOAA series of satellites, now numbering 18, has been supplying images at 1 km resolution for many years. They are in sun-synchronous polar orbits and evolved over many years after the launch of NOAA's first satellite, TIROS-1, in 1960. The current instrument is the Advanced Very High Resolution Radiometer. This is a whiskbroom instrument operating in six channels, in the red, near infrared, mid-infrared and two in the thermal band (to permit atmospheric correction of the measured temperatures), although the two mid-infrared channels 3a and 3b do not operate simultaneously. The field of view of the AVHRR is about 110°, giving a swath width of about 2400 km, thus providing twice daily coverage, one overpass during the day and one at night. There are always two satellites in orbit at any one time, so the nominal repeat period is about 6 h at the equator. In the northern latitudes, of course, where the orbits are converging, many more of the 14 orbits per day can be seen giving almost hourly coverage of north Europe. Because of the wide swath in comparison to the satellite's height (~850 km), and the curvature of the earth, ground resolution cell size changes markedly along the scan line, in fact the pixels at the end of the scan line are about 6 km long; whereas at nadir they are 1.1 km. Corrections have to be made for this distortion in the final image. Another consequence of this is that the angles and distances through which the atmosphere and land surfaces are viewed vary greatly across the swath.

Metop, launched by ESA in 2006, also carries the AVHRR, and will provide the morning overpass when the NOAA Polar Orbiting Environmental Satellite (POES) series of satellites is phased out. The US NPOESS programme, designed to replace the POES programme, was due to start in 2006 but because of slippage and considerable budget overrun will not commence until beyond 2013. This potentially puts its whole future at risk and with it the continuity of weather and climate data. With a 40% chance of a launch failure, there could well be a gap in data lasting several years, making intercalibration of its instruments against earlier ones impossible.

Although designed for meteorological use, AVHRR data have found considerable use in both oceanographic and land applications. The *normalized difference vegetation index* (*NDVI*; see Chapter 7) that uses the red and near infrared bands has proved particularly valuable for vegetation monitoring using this instrument. The frequent data acquisition has enabled large scale monitoring and mapping to take place, particularly by using maximum value composites. The level 1B data product has particularly good quality control and accurate geolocation.

The Moderate Resolution Imaging Spectroradiometer (MODIS) is flown on both Terra (morning) and Aqua (afternoon) EOS platforms. Again the orbits are sun-synchronous at an altitude of 705 km. The EOS project is designed to observe and monitor the surface of the earth for 15 years. The MODIS instruments are scanning imaging radiometers with a swath of 2330 km giving full-earth coverage every one to two days in 36 wavebands, 10 in the visible range, 6 in the NIR, 14 in the MIR and shortwave TIR, and 6 in the longwave thermal range. Twenty nine of the wavebands are at 1000 m resolution, 5 at 500 m (to map land/cloud/aerosol properties) and two at 250 m resolution (to map land/cloud aerosol boundaries).

One of the datasets provided by EOS is the MODIS Land Surface Temperature Product. This is produced using a split-window atmospheric correction algorithm (see Section 6.2.2), corrected for view angle. The temperatures are obtained from the 11- and 12-μm channels and emissivities are inferred using a classification-based emissivity method that simulates the scene emissivity from the proportions, surface structure, and spectral emissivities of the components in the scene. The advantage of MODIS data over AVHRR data is that it has more finely defined visible and near infrared bands and has one of the most accurate calibration subsystems ever flown on a remote-sensing instrument. The calibration allows the raw brightness values to be converted to true percentage reflection or radiance measurements (Wan, 2008). It also has a higher radiometric resolution than most other sensors (12-bit quantization in all bands) rather than AVHRR's 10 bit.

The Along Track Scanning Radiometer (ATSR) on ERS and Envisat has specifications similar to those of AVHRR, but uses a conical scanning system providing two look angles for atmospheric corrections. Other low-resolution instruments designed primarily for oceanographic work but that have found use in land monitoring have been the Coastal Zone Color Scanner and SeaWiFS. These have spatial resolutions of about 1 km, but much better spectral resolution (8 channels on SeaWiFS). The Vegetation instrument on SPOT (see below) has a spatial resolution of 1 km.

5.8.2 Medium-resolution systems

There is a large jump in scale between the low- and the medium-resolution data. Recently, attempts have been made to bridge this divide with MERIS (Medium Resolution Imaging Spectrometer) that provides 15 waveband continuous 1200 m 'reduced' spatial resolution, as well as 300 m full spatial resolution data for up to 20 min per orbit mainly over the land and coastal zones.

Most of the earlier medium-resolution systems were designed with vegetation monitoring in mind. The first such system was the MSS on Landsat that had 80 m resolution over a swath of 185 km and 18-day repeat. The green, red, and infrared bands were used to develop ways of detecting, classifying, and monitoring a range of vegetation phenomena and to study the seasonal changes taking place as well as to detect stress and disease. Although launched in 1972, the system was conceived and designed some years earlier and so contained 1960s technology. With the experience gained from using this data, by 1982 technology and understanding had improved to such an extent that the Thematic Mapper on Landsat 4 had a resolution of 30 m in six optical bands which were tuned to specific spectral features (plus 120 m in a thermal band). The Enhanced Thematic Mapper, ETM+, on Landsat 7, launched in 1999, has similar seven bands (with the thermal band resolution improved to 60 m) plus a panchromatic band at 15 m resolution. At the time of writing, Landsat 7 is still providing data in spite of problems with stabilization. In about 2011, it is planned to fly the Operational Land Imager (OLI) on the Landsat Data Continuity Mission to provide Landsat-type imagery. If this mission lasts for the planned ten years, then there will have been a continuity of compatible data extending over a period of about 50 years!

The MSS and TM instruments both used conventional whiskbroom scanning. The High Resolution Video system (Haute Resolution Video, HRV) on the French SPOT-1 satellite, launched in 1986, made use of pushbroom scanning using an array of sensors in order to increase the dwell time sufficiently to enable pixels of 20 m (10 m panchromatic) to be obtained. There were two independent HRV scanners on board, each of which had a swath width of 60 km, and each could be programmed so as to be directed up to 27° off nadir. This could improve the temporal resolution from the normal 26 days to a nominal 5 days at the equator. It also permitted users to take advantage of gaps in the cloud and provided stereo-type coverage. The early HRV detector had only three wavebands, but later models increased this first of all to four plus a pan band, and then to five plus two pan bands. On SPOT 5, launched in 2002, the three optical bands had 10 m resolution, the shortwave infrared band 20 m resolution and the two pan bands each had 5 m resolution. These two pan bands could be combined in the so-called *supermode* to give 2.5 m resolution. From SPOT-4 onwards, the satellites also carried the *Vegetation* instrument. This is a four-band instrument (three of the bands the same as the HRV with a blue band in place of the green band), with a large field of view giving a very wide swath of 2250 km at 1 km resolution and daily coverage. This is therefore similar to the AVHRR, but the advantage is that it is flown in combination with HRV, thus giving coincident coarse and fine resolution data in compatible wavelengths.

Other medium-resolution systems that are relevant to the study of vegetation are the Indian Remote Sensing series of satellites (IRS). These carried a Wide Field Sensor (188 m spatial resolution in two or three vegetation bands) and the Linear Imaging Self Scanning (LISS) instrument. The current one, LISS III, has a spatial resolution of 23.5 m in three optical bands and 70 m in the shortwave infrared.

Some recent line scanning instruments have been specifically designed to point in a non-nadir direction. One of these is the Multi-Angle Imaging Spectro-Radiometer (MISR) flown on the Terra satellite. This collects simultaneous multidirectional data at four different angles fore and aft in four spectral bands in the visible and near infrared regions that can aid in bidirectional reflectance studies (Chapter 8).

The first civilian spaceborne hyperspectral sensor was Hyperion, flown on the NASA EO-1 platform that had 220 spectral bands at 30 m spatial resolution. (It also carried the Advanced Land Imager, ALI, which is a Landsat-type instrument to provide some continuity of data.)

5.8.3 High-resolution systems

Remote sensing entered a new era in the twenty-first century with the introduction of high-resolution (hyperspatial) and hyperspectral sensors on board satellites. Recently, there has been a move to provide high-resolution data for very specific groups of users over small areas using smaller, special purpose satellites. The first commercial hyperspatial satellite was IKONOS, producing 1 m panchromatic and 4 m multispectral images using linear array technology in 13 km swaths. The repeat time (at the equator) is 11 days, but the system is extremely manoeuvrable and the instrument can be pointed at up to 45° to the vertical in both along-track and cross-track directions. This gives both stereo capability and can decrease the revisit time. The satellite itself can be programmed either to follow meandering features or to collect data to order from a particular area.

This was soon followed in 2001 by Quickbird, with a similar four-band multispectral sensor at 2.4 m and panchromatic at 0.6 m. It has a 1–3.5-day revisit time, depending on latitude and a swath width of 16.5 m. Worldview-1, launched in 2007, provides half-metre panchromatic imagery with a revisit time of 1.7 days, and Worldview-2 will provide eight bands of multispectral imagery in the visible and near infrared.

5.8.4 Smallsats

The very first satellites (Sputnik, Vanguard-1) were small in weight (83 and 1.6 kg, respectively) and size (58 and 16 cm diameter) and carried no more than a radio transmitter. After that, the tendency was for them to become larger, more complex and more expensive, culminating in the biggest of all, Envisat, weighing 8000 kg and carrying 10 instruments. Each spacecraft was specially designed from scratch (in shape, size, stabilization methods, power provision, instrument mounting techniques, thermal control, onboard data handling, data communications, etc.) (Kramer and

Table 5.1 Classification of smallsats.

Satellite class	Mass	Cost US$*	Comments
Large satellites	>1000 kg	>20 million	Observatories, etc.
Minisatellites	100–1000 kg	5–20 million	Small satellites (smallsats)
Microsatellites	10–100 kg	2–5 million	
Nanosatellites	1–10 kg	<1 million	
Picosatellites	0.1–1 kg		
Femtosatellites	1–100 g		Satellite-on-a-chip

* at 1999 prices

Cracknell, 2008), leading to long lead-times (up to 10 years). This required lengthy manufacture and testing procedures, leading to delays and cost overruns. The fact that many of them were multisensor platforms meant that a failure of the whole system meant the death of many different projects.

More recently, the tendency has been to focus on available and existing technologies using a general-purpose bus and off-the-shelf components (COTS – commercial off-the-shelf technology) and instruments to achieve cost reductions. Also, developments of lightweight materials and miniaturization of electronics[27] has enabled smaller, lighter, and more reliable satellites to be built. There is now a growing trend towards using small, dedicated satellites rather than the larger, general-purpose ones of the past. This, together with the availability of new, small low-cost launchers, has made it affordable for many more countries, as well as universities, to develop their own remote-sensing systems, and constellations of satellites are now feasible.

Small, low-cost satellites have actually been around since the 1960s, as the preserve of radio amateurs, being regarded as 'toys' by the 'professional' community. But, thanks to the pioneering work of Surrey Satellite Technology Ltd (SSTL), a spin-off from Surrey University, hundreds of small satellites have been developed and are now being used extensively around the world for communications and for environmental monitoring. The reader is referred to the very comprehensive, up-to-date review of the small satellite scene by Kramer and Cracknell (2008) and the article by Xue *et al.* (2008), which includes an overview of smallsat literature as well as applications.

Satellites can be classified by function, type of orbit, cost, size, etc., but a good indicator of size is that given in Table 5.1, based on standard scientific prefixes (mini-, micro-, nano-, etc.). Another classification could be in terms of cost, this was adopted at the United Nations Conference UNISPACE III in 1999.

The high-resolution satellites such as IKONOS and Quickbird as well as OrbView, EO-1, and IRS, are all examples of smallsats. Most were designed for a single application, such as the Bispectral InfraRed Detector (BIRD) for active fire detection (see Section 11.6) and PROBA (Project for On-Board Autonomy) (Section 7.2.1) for the collection of multiangular data. Another of relevance here is IRSUTE (InfraRed miniSatellite Unit for Terrestrial Environment), designed to improve water-budget estimates. It is intended to provide thermal imagery for the determination and analysis of soil/vegetation/atmosphere processes at field scale, which will provide data for the scaling-up of these processes from local to regional scales (Becker *et al.*, 1996) by providing 50-m imagery in five visible and NIR channels and three thermal channels. Fluxes should be retrieved to an accuracy of about 50 W m^{-2} with a repeat time of one to three days.

Because of their cheapness, it is feasible to fly multiples of the same instruments on smallsats that can be flown in formation and constellations. This can decrease the revisit time to about one day. The first such operational constellation was the Disaster Monitoring Constellation (DMC), coordinated by SSTL, to provide daily global imaging capability at medium

[27]. The first spaceborne microprocessor was flown on SeaSat in 1978.

resolution (32 m) in three or four spectral bands using pushbroom technology. So far, five satellites are in orbit, operated by UK, Algeria, China, Nigeria, and Turkey, and there are plans to increase the number, and the resolution to 2 m. Other examples are RapidEye, a constellation of 5 minisatellites, to provide high-resolution multispectral imagery in five wavebands at 6.5 m resolution on a daily basis for evaporation monitoring and yield predictions and updating of field maps for agricultural producers, Pleiades (CNES), two high-resolution satellites with 2.8 m MS and 0.7 m pan resolution and a swath width of 20 km, and SAR-Lupe (Germany), five minisatellites in three orbital planes providing high-resolution SAR data at X-band.

Nano- and pico-sats, at present some way off for commercial applications, have very limited capability, but their low cost and speed of design and build make their use very attractive, particularly for using in clusters. With the advent of nanotechnologies, femto-sats are becoming possible although still experimental, and promise intriguing uses.

5.8.5 Microwave systems

A lot of early work was done on developing the use of microwaves, particularly for studying vegetation, using ground-based and airborne microwave systems. The first satellite to carry an imaging radar was SeaSat in 1978. The synthetic aperture radar system used 23-cm microwaves (L-band) and had a ground resolution cell of 25 m. It suffered an electrical systems failure after only 98 days, but even so a lot of valuable data were collected, although much of it was never analysed. It also carried three microwave radiometers as well as a radiometer that operated in both the visible and the infrared.

Between 1981 and 1994 there were three Space Shuttle campaigns that carried SAR. The first two (SIR-A and SIR-B) carried L-band instruments with resolutions of about 40 m, and the third, in 1994, carried both the Shuttle Imaging Radar (SIR-C), operating at 6 cm and 23 cm and a 3-cm German X-SAR system with resolutions from 10–200 m. There was also a Shuttle Radar Topography Mission in 2000 that used C- and X-band interferometric instruments that acquired topographic data over nearly 80% of the earth's land mass between 60°N and 56°S with up to 16 m absolute vertical accuracy. The second instrument was mounted on a 60-m arm to provide the 'stereoscopic' capability.

After the demise of SeaSat, it was over twelve years before the next radar system was flown on a satellite. ERS-1 (European Resource Satellite) was launched by the European Space Agency in 1991, followed by ERS-2 in 1995. These carried 6-cm C-band radars with ground resolution of 30 m. Because of the extremely high data rate of these SARs, the data could not be stored on board and could only be acquired within range of a receiving station. Both satellites also carried scatterometers, altimeters, the Along-Track Scanning Radiometer (ATSR), and a microwave and an infrared radiometer. For a year after ERS-2 was launched, both satellites were flown in tandem collecting pairs of SAR images for interferometric use. The satellites were 35 min apart and their orbits were adjusted so that the ERS-2 track coincided exactly with that of ERS-1 24 h earlier.

The Canadian Radarsat was developed mainly for monitoring ice conditions in their northern shipping lanes because the poor illumination throughout most of the year and bad weather conditions made other forms of remote sensing impossible for much of the year. It was said that it would pay for itself in improving the economy in only a few years! It operated at C-band and used HH polarization that is optimum for ice monitoring, in a number of imaging modes with resolutions from 9 m to 100 m. Radarsat-2 was launched in December 2007 and has a finer resolution and a flexible selection of polarizations. The Japanese launched their own satellite in 1992, JERS-1, carrying an L-band radar with a resolution of 18 m, and an optical sensor and now have a SAR instrument on the ALOS satellite.

The latest ESA radar satellite, Envisat, was launched in 2002 and it has recently been decided to extend its life to 2013, after which many of its functions would be taken over by the Sentinel programme under GMES. This is a very large satellite, weighing over 8 tonnes and carrying a payload of ten instruments, including the Advanced Along-Track Scanning Radiometer (AATSR), the Medium Resolution Imaging Spectrometer (MERIS), as well as a microwave radiometer and a radar altimeter. The Advanced Synthetic Aperture Radar, ASAR operates at C-band in five selected polarization modes. It has a resolution of 28 m to about 1000 m depending on the imaging mode. The microwave instruments consume so much power that they are only switched on for about 20 min of the 100-min orbit.

In 2007, the German TerraSAR-X was launched. This is an X-band (3 cm) system that collects radar data in several different modes down to 1–2 m in images 10 × 10 km. The COSMO-Skymed is a series of four civilian/military satellites operated by the Italian Space Agency, the first of which was launched in 2007. They have a short revisit time and are equipped with SAR instruments that can operate in three different modes, which makes them ideal for agricultural mapping.

5.8.6 Laser systems

Most laser systems (Section 5.6) have been designed to fly on helicopters or aircraft. This is partly to avoid possible hazards from the laser beam itself, but also because of the types of small-scale, high-precision applications for which they are usually employed. One of the few satellite-borne laser systems is GLAS as discussed above. This was designed to measure ice-sheet topography, cloud heights, and aerosol vertical structure, but over land and water it provides along-track topography. It emits 40 short, 4-ns, pulses per second in the NIR at 1064 nm and in the visible green at 532 nm. The eye-safe energy levels of the two beams are 100 and 50 mJ, respectively. The IR beam is for surface altimetry, and the green for atmospheric measurements. As pointed out above, the narrow beam (divergence is 0.12 mrad) leads to only a very small fraction of the earth's surface being sampled during its 91-day repeat cycle.

5.8.7 Airborne systems

Air photographs are still a valuable source of remotely sensed data, and many mapping agencies routinely collect such data. We have already mentioned hyperspectral, Lidar, and microwave systems in the previous sections; most of these are not aircraft specific. Kramer (2002) gives the specifications for over 200 instruments, but as their availability changes rapidly, interested readers will be able to find up-to-date information on the appropriate instrument-specific websites. Here, we will just mention a few of the more widely used ones and give fuller specifications of some of these in Appendix 3. Other instruments used on balloons, helicopters, UAVs, etc., will be discussed in Chapter 11. Whereas up until recently satellite systems were usually produced and operated by agencies, many of the airborne systems have been designed and manufactured in the user's own laboratory. Access to these systems, and to their data, is usually limited to the individuals involved, though there are a number of commercial systems, some of which are flown by agencies such as JPL and NERC (UK).

Many types of camera, of varying degrees of sophistication are used, flown on specially adapted light aircraft. We shall here, however, mention just some of the many airborne scanning systems that have been used.

Airborne sensing is, in fact, used extensively for local and regional remote sensing, for example, and also for extending satellite remote sensing. It can provide support for instrument development and calibration, and has often been the proving ground for satellite systems. Airborne missions are often of short duration, flown in conjunction with fieldwork and for instrument calibration. Most airborne systems are not platform specific and can be flown on any type of aircraft, but which usually have to be specially adapted with mounts and windows. For technical reasons it is usual to have separate scanners or cameras to cover the visible/NIR and the short to mid-infrared (e.g. the Eagle and Hawk sensors from the Specim Company in Finland),

Lidar is one of the most common types of airborne sensing, and we have mentioned some laser systems currently in operation in the previous section, and also some microwave systems in Section 5.8.5. Many of the scanners used, though, are multispectral linescanners or hyperspectral instruments (section 5.8.3), although airborne thermal scanning is particularly useful, especially for vegetation monitoring and water-balance studies. One of the first was the Airborne Imaging Spectrometer (AIS), a pushbroom scanner operated by JPL, which was used for measuring vegetation stress amongst other applications. From this was developed the better-known Airborne Visible/InfraRed Imaging Spectrometer (AVIRIS), which became the first operational hyperspectral instrument in 1986. This has been frequently upgraded, and the present system, which consists of 4 spectrometers with a total of 224 spectral channels, is flown by JPL on an ER2 aircraft at a height of 20 km providing 20 m spatial resolution.

Daedalus Enterprises have developed a number of sensors, the best known probably being the AADS1268, the Airborne Thematic Mapper (ATM), which has been operated by many agencies worldwide. This is an 11-channel instrument built originally to emulate the Landsat MSS sensor bands, but the latest version now covers the channels of MSS, TM, and SPOT HRV, plus some additional ones. It is an opto-mechanical scanner providing 1.25 m spatial resolution from 1000 m, or 25 m resolution from 20 km on ER2, with over 700 pixels in a scan line. Daedalus also produces the Thermal Infrared Multispectral Scanner (TIMS), which has six spectral bands.

Another commonly used instrument is the Compact Airborne Spectrographic Imager (CASI). This was developed by ITRES from their Fluorescence Line Imager (FLI). It operates in the visible and near infrared, with 1.9 nm sampling intervals, providing a maximum of 288 spectral bands and 512 pixels per scan line. But because of the huge amount of data that such a system would generate, it is operated either as a multispectral imager in spatial mode in up to 19 operator-selected bands, or as a multipoint spectrometer (spectral mode) in up to 39 selected points over the full spectral range. The pixels are 1.5 m square from a height of 1200 m. It samples the downwelling radiation simultaneously that allows the conversion of the signal to spectral reflectance values on a pixel-by-pixel basis using standard atmospheric correction. The current system is CASI-2, which has several default bandsets including one corresponding to VEGETATION on SPOT and one corresponding to SeaWiFS.

Other commercial systems include the various versions of Digital Airborne Imaging Spectrometer (DAIS) produced by GER, and the Reflective Optics System Imaging Spectrometer (ROSIS).

These systems are continually being supplemented by new systems including multispectral, hyperspectral, Lidar, and radar sensors with the prospect of fluorescence sensors.

5.9 Data reception

To most users of remotely sensed data, the processes by which it is collected, pre-processed and disseminated are completely invisible with the process behaving as a 'black box'. Datasets at a number of different levels can now easily be ordered online from agencies or other providers and will arrive in a very short time via the internet, or occasionally on CD ROM. But there are a number of stages that will have been involved in turning the raw digital numbers transmitted from the satellite into images which can be processed on a computer in a laboratory.

Data are transmitted from the satellite encoded onto a high-frequency radio signal, often a microwave signal, which is received at a ground receiving station usually by means of a dish antenna. Data can only be received from a satellite when it is in direct line of sight of the receiver,[28] and the receiving antenna has to be pointing directly at the satellite. This poses no problem for a geostationary satellite that is effectively stationary, but for a polar-orbiting satellite an antenna that follows the path of the satellite overhead is required. To a ground observer, such a satellite would first appear above the southern horizon, following a path that would take it overhead and then down towards the northern horizon, at which point it will be lost from view (this is for a satellite making a northerly pass). A low-earth orbiter will be in view for about 20 min of its 100-min orbit. The receiving antenna is usually programmed to anticipate the point at which the satellite will first appear in view over the horizon and to lock onto it when the signal is first received. It will then follow the satellite until it loses sight of it again. This requires a sophisticated mount for the antenna that is able to rotate about both the vertical and horizontal axes. Low-resolution data that can be transmitted on a lower-frequency carrier, such as the AVHRR data from the NOAA satellites, can be collected using quite a small dish antenna, a metre or two in diameter, but the high-resolution data from Landsat, for example, and radar data need large antennae, several tens of metres in diameter, and very much more sophisticated electronic locking systems. Such data are usually only collected by the National agencies because of the expense involved, but low-resolution data (*direct broadcast data*) can easily be collected by a much simpler inexpensive off-the-shelf system (such as that manufactured by the University of Dundee, Cracknell and Hayes, 2007), and many universities and companies now have their own. The primary data from Meteosat is collected at EOSAT at its control centre in Germany where it is processed

28. Occasionally, the data may be received and retransmitted by another satellite, a data-relay satellite.

and retransmitted back to Meteosat that now acts as a data-relay satellite. This secondary, processed, analogue data can be received using a simple wire aerial and a radio receiver, and these are even available in some schools now. The large national ground stations will usually have many receiving antennae of different sizes, each dedicated to a particular satellite, because a number of satellites may be in view at any one time.

Once the data have been received, they can be quickly pre-processed to improve their geometric and radiometric fidelity and mounted on the ground station's website as a *quicklook*, often with a grid and coastline overlay. These can be accessed by users all over the world in order that they may select suitable images to purchase. They may then order images online and receive them in the same way, thus being able to undertake near-real-time analysis or to study natural disasters and other rapidly changing phenomena such as oil spills or wild fires. This was not possible in pre-internet days when users had to rely on the vagaries of the postal services to obtain both quicklooks and imagery.

In the early days there was no systematic policy on storing and archiving data. It was usually stored as raw data on large computer-compatible tapes at the ground station and, although a valuable research resource, there was no general policy on what should be kept and for how long. With the advent of more efficient storage devices, and particularly with the increase in sophistication of computers, most data are now archived not only in raw form but as a matter of routine processed to a number of different levels of corrected and geometrically rectified images. It is often also presented in the form of value-added products such as maps of geophysical quantities including sea/land-surface temperature, chlorophyll, or vegetation indices. We shall see in Chapter 6 how such data may be processed in order to ensure that the information is useful to the user.

Further reading

Most remote-sensing textbooks will cover systems to some extent. Particularly useful are *Remote Sensing and Image Interpretation* by Lillesand, Kiefer, and Chipman (2007), and *Introduction to Remote Sensing* by J.B. Campbell (2007). The relevant chapter in Campbell is excellent as an easily understood introduction to microwave systems, but more detailed information is to be found in *Introduction to Microwave Remote Sensing* by I.H. Woodhouse (2006). A very comprehensive survey of missions and sensors is given by H.J. Kramer (2002). This is an encyclopaedia containing almost every satellite, scanner and remote-sensing programme, including airborne sensors, and campaigns. Unfortunately, the most recent edition was published in 2002, and is by now already rather out of date. S. Liang (2004) covers calibration in some detail and Barrett and Curtis (1999) give a briefer overview. A very good series of tutorials on the basic technology of remote sensing is that published by the Canada Centre for Remote Sensing (*http://www.ccrs.nrcan.gc.ca/resource/index_e.php#tutor*), with the ones on radar being particularly useful.

Websites

Wan, Z. MODIS Land-surface temperature products users' guide, 2006: **http://www.icess.ucsb.edu/modis/LstUsrGuide/usrguide.html**

Tutorials on remote sensing from the Canada Centre for Remote Sensing: **http://www.ccrs.nrcan.gc.ca/resource/index_e.php#tutor**

Dundee satellite receiving station: **http://www.sat.dundee.ac.uk/**

Sample problems

5.1 In November 1984, the space shuttle *Discovery* was placed in circular orbit at an altitude of 315 km in order to catch up with a disabled satellite in a circular orbit at an altitude of 360 km. Suppose that these two objects were initially on opposite sides of the earth, how many orbits would the satellite have made for the shuttle to be immediately beneath the satellite?

5.2 An environmental satellite flies at an altitude of 700 km above the surface of the earth in a near-polar orbit. Calculate its orbital velocity. What is the velocity over the ground? Another satellite is planned to fly at a height of 1700 km. What would be the period of this satellite?

5.3 A rotating mirror cross-track scanner is flown at a speed of 720 km h^{-1} above the surface of the earth at an altitude of 10 km and each pixel corresponds to an area on the ground of 10 m × 10 m. At what speed must the mirror rotate if adjacent scan lines are neither to overlap nor to have gaps between them? What is the dwell time for each IFOV?

5.4 If the field of view of the Landsat ETM+ is 0.26 rad, what is the swathe width? If images are just to touch at the equator, how many orbits would be required to give complete earth coverage, and how many days will elapse between overhead passes at any point on the equator? State any assumptions made.

5.5 A pushbroom scanner produces a panchromatic image with pixel size 10 × 10 m and a swathe width of 50 km. How many individual detectors are there in the scanner array? If each detector is 10^{-5} m across, what is the physical size of the array?

5.6 A pulse of microwave radiation is emitted vertically downwards from an instrument onboard the ERS-2 satellite. Estimate the distance moved by the satellite by the time the reflected pulse returns. If the pulse is emitted from an antenna of diameter 1 m, what is the size of the footprint illuminated on the ground?

6 Preparation and manipulation of optical data

6.1 Introduction

In Chapter 5 we looked at some of the systems used to produce remotely sensed data. Here, we outline the necessary intermediate steps required to prepare the raw data for presentation and subsequent analysis, while discussion of methods for interpretation of the resulting images for the extraction of thematic information and to provide useful information about spatial variation of vegetation biophysical variables (such as leaf-area index or canopy photosynthesis) will be deferred to later chapters. The preliminary steps include geometric correction and radiometric correction as well as various forms of image enhancement. Even though most satellite and airborne data are supplied already corrected for the basic geometric, and sometimes radiometric, errors introduced by the collection system during acquisition (e.g. levels 1A, 1B, 2A, etc.), some further refinement by the analyst is often needed. There is frequently a requirement for some further image correction, especially for airborne data in which the distortions may be particularly severe. In any case, however, it is useful to understand the nature of the corrections applied and their potential impacts on data integrity. The manipulation of digital data is referred to as *image processing*, and comprehensive software to facilitate the various steps of image manipulation and interpretation is readily available. In many cases it is possible to purchase *value-added products* such as maps of *vegetation indices*, surface temperature, or chlorophyll that may need no further processing; nevertheless it is advisable to understand the derivation of these variables and especially the likely errors and uncertainty associated with them.

Most users will not need to know the mathematical processes that are actually being performed by image processing software, but it is useful to understand the basic principles involved so that intelligent choices can be made between different options offered, such as which resampling method to use, what type of stretch to use, or which bands to employ. The majority of users of remotely sensed data will only ever need to perform a few of the available operations and so we will, in this chapter, just briefly explain why certain procedures are required and what they can do. Modern packages such as ENVI or ERDAS™ Imagine provide sophisticated procedures for undertaking many of the required manipulations but the terminology used and the interfaces of each may appear different at first sight. For this reason, the reader is referred to the many textbooks on image analysis for remote sensing such as Mather (2004), Richards and Xia (2005), Jensen *et al.* (2008), or Russ (2006) for information about the principles involved before embarking on the tutorials and exercises included with the individual package being used.

We are all familiar with forms of image processing in our everyday lives. In photography we can make an enlargement, alter the contrast of a picture, or print an extract with some form of enhancement, such as soft focus. When we adjust the contrast, brightness, or colour balance on our television set or computer monitor we are simply changing the appearance of the visual display. These are all analogue processes, but if the image is in digital form we can perform many mathematical manipulations to individual digital numbers or sets of digital numbers, allocate any colour to any sets of digital numbers, or apply statistical manipulations to them. Most of these processes are performed to make the image more recognizable or more interpretable by the analyst – but it is critical to remember that no amount of image processing can increase the amount of information present. On the other hand, some transformations such as resampling or image compression do lose information so care is necessary in choosing both the types of manipulation and the order in which they are made.

We will consider a logical progression of processes, starting with the correction of geometric and radiometric distortions. Having obtained an image that bears some semblance to the scene being imaged, we will then consider different ways in which the image may be displayed, followed by methods of enhancing the appearance of the image to make it easier to interpret. The subsequent step of using spectral information for image interpretation in vegetation analysis will be addressed in Chapter 7, where we note that our ability to extract information about the underlying vegetation or other surfaces in an image increases as we increase the spectral information available. Hyperspectral imaging (where one may have more than 200 spectral bands) improves the capability of multispectral analysis, especially in the realm of extracting subpixel information. A further application of remote-sensing data is in what may be called *data assimilation*, where the remote-sensing measurements of variables such as surface temperature or vegetation cover are combined with models of underlying biophysical processes to estimate variables such as land-surface energy fluxes or soil-moisture content that are not directly observable. Examples of data assimilation will be encountered in the practical applications to be discussed in Chapter 11. The above division is, of course, subjective, and other authors may allocate the various processes differently between sections.

6.2 Image correction

Image processing is usually performed in a number of stages. Some form of *pre-processing* is almost always necessary as the raw data received from a satellite or airborne imager need to be transformed into a form that can be handled by the interpreter. Some of this takes place when the data are received at the ground station to correct them for known errors that may have occurred in the process of collection. Such distortions could be due to the motion of the satellite, the curvature of the earth's surface, non-linear scanning, banding/striping (due to faulty detectors), or non-linear response of the detectors. Other corrections may be required for atmospheric effects or geometric distortions due to changes in pixel size along the scan line (Section 5.4) and rotation of the earth. If a true geometric representation is required, then the image must be warped (*reprojected*) so that it will register exactly with map coordinates or with other images. In order to achieve this correction, the pixels must be *resampled* into a new grid (i.e. new pixels are formed from weighted proportions of the old ones).

6.2.1 Geometric correction

Geometric errors

The scale of an aerial photograph, that is the relationship between a distance measured on the photograph compared to the actual distance on the ground, varies across the photograph. This is because, as the angle from nadir increases towards the edge of the photo, so the distance from the camera to the point in question increases. Similar distortions arise when using ground-obtained images (e.g. from meteorological towers). When a photogrammetrist tries to match up points in two adjacent air photos, he needs continually to adjust the scale of one of them (using a *transferoscope*). Other forms of distortion may be introduced by variations in flying height produced either by the pilot or by changes in topography below, and by instability of the platform (pitch, roll, and yaw).

Similar distortions occur in scanned images and these may be more severe in satellite images because of the much greater distances involved. We can consider two types, systematic and non-systematic errors. The systematic ones are usually predictable and tend to occur in all images obtained by a particular system. These can usually be corrected from a knowledge of the orbital parameters of the platform and the characteristics of the scanner, and apply to all the images collected by that system. These are the corrections usually applied before the data are distributed. Non-systematic errors, however, are unpredictable and usually only apply to a particular image. They need to be corrected on an individual basis, often by the user.

The major sources of systematic geometric errors are:

Earth rotation – Earth rotation tends to skew scanned images, giving them the familiar rhombus shape (Fig. 6.1). As a polar-orbiting satellite proceeds along its track, the earth rotates eastwards beneath it. This means that the start of each scan line begins at a point slightly to the west of the previous one, and successive scan lines need to be shifted slightly to the west. Note that the surface velocity of the earth, and hence the amount of shift between lines, varies with latitude.

Platform motion – The forward movement of the satellite between the acquisition of successive pixels with scanning sensors causes an effect known as *scan skew*. The line of scanned pixels is now not perpendicular to the track of the satellite, giving it a slight upward curve. This effect does not occur with a push-broom scanner as the complete line is recorded at one time. For TM and ETM+ on Landsat, which use oscillating mirrors that scan in both directions, instead of rotating ones, scan-line compensation is incorporated into the instrument to produce parallel scanlines in both scan directions and thus reduce the effect of the forward motion on the image.

Non-linear scanning – The velocity of the scanning mirror may vary during the sweep across the field of view. This is particularly severe for an instrument that uses an oscillating mirror, rather than a rotating one, as the motion is simple harmonic. The reduction in velocity towards the extreme of the scan will increase the dwell time and elongate those pixels.

Panoramic distortion – The ground instantaneous field of view of a scanner increases from nadir to the extremity of the scan line due to the increasing angle of view (Fig. 6.2). The resulting compression at the edge of the raw image is very apparent (e.g. for

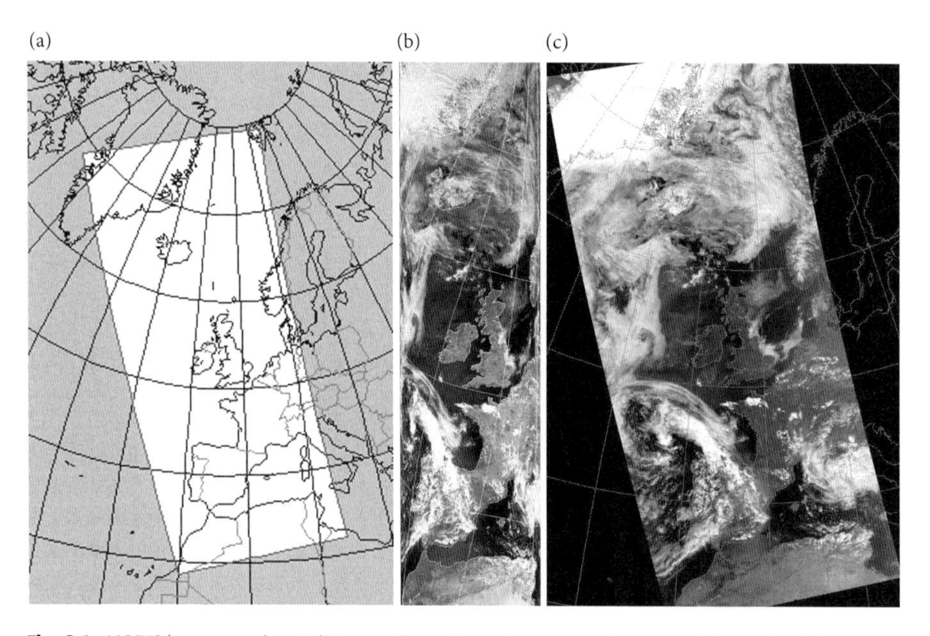

Fig. 6.1. MODIS image over the UK (13.15 UTC) Northbound over UK on 13 May, 2008, showing (a) the skewed path of the satellite (Aqua), (b) a raw image before geometric correction (near infrared (841–867 nm) showing geometric compression at the edges of the swath, and (c) a geocorrected RGB composite image (reproduced with permission from NEODAAS and University of Dundee). (See Plate 6.1)

(a)

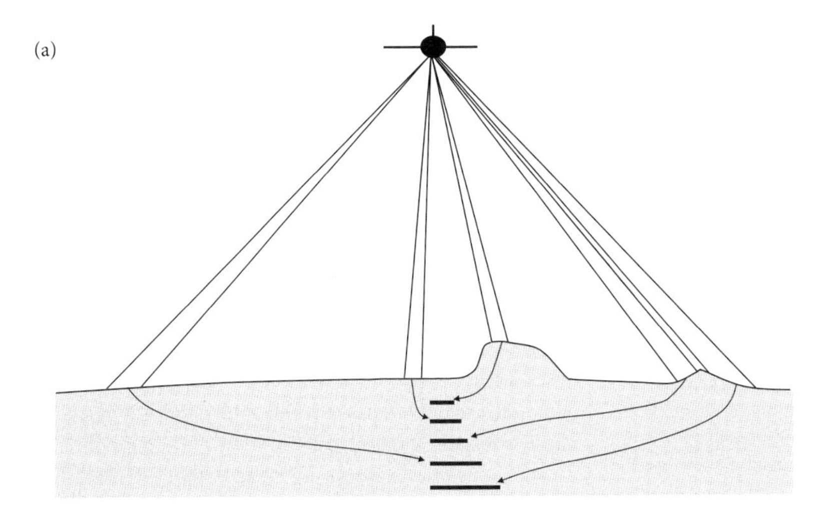

(b)

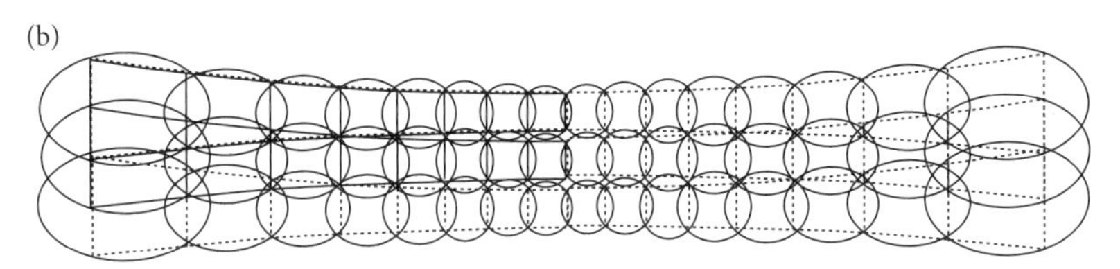

Fig. 6.2. (a) Illustration of how the area on the ground represented by pixels with a constant IFOV at the sensor varies with the angle of view and with the topography; each of the lines indicated represents the relative dimensions on the ground represented in the corresponding single pixel (not to scale), showing that the area viewed decreases with increasing height of the land, while it increases at the edge of the image because of the angle of view (and for high altitude images as a result of the curvature of the earth). The slope of the surface with respect to the view angle also has large effects. (b) Planar view of three successive scan lines showing the 'bowtie' effect, where the area viewed by each pixel increases as one moves from the nadir view (at the centre of the scan line) to the edge of the swath, with increasing overlap between pixels at the edge.

Scandinavia in Fig. 6.1b) and is exacerbated by the curvature of the earth's surface. This is particularly severe for instruments with a wide field of view such as the AVHRR (see Section 5.8.1) that has a viewing angle of up to 56° from nadir. Even HRV on SPOT can be rotated to view at an angle of up to 27° from nadir. A result of the image compression near the edges is a characteristic distortion in which what should be linear features appear curved. Another problem that can arise is the 'bowtie' effect (particularly apparent with MODIS) where the increasing area viewed near the edge of the swath results in pixels overlapping (Fig. 6.2(b)).

Aspect ratio – When a pixel is formed in an image, the signal received from the instantaneous field of view is allocated to a pixel representing an area of ground beneath (see Section 5.4). The size of the pixel is chosen so as to form a complete mosaic. For example, the size of a Landsat MSS pixel is 56 m across track by 79 m along track. This rectangular area is commonly displayed on a computer screen as a square pixel. The displayed image therefore needs to be reprojected in order to reproduce the geometric fidelity of the original scene.

Apart from distortions introduced by the scanning system, the geometric fidelity of the image can be affected by random variations in both the altitude of the platform and its attitude. The non-systematic errors include:

Platform instability – Airborne imagery is particularly sensitive to changes in attitude (yaw, roll, and pitch: see Fig. 6.3 and also the cover) as all of these can distort the image (especially with scanning systems) as well as moving the nadir point away from its expected position in an unpredictable way, necessitating *ad hoc* correction. An example showing the effect of substantial platform instability is presented in Fig. 6.4, which shows the large amount of transformation required to obtain the geocorrected image (Fig. 6.4(b)). Although not as severe as for airborne sensing, satellites also suffer from instabilities. Perturbations in the orbit (Section 5.2.3) may cause height or velocity variations and departures from the expected path, and slight changes in attitude may also occur.

Terrain effects – Again on a smaller scale than for airborne sensing, variations in the height of the surface below the satellite will have an effect on the scale of the

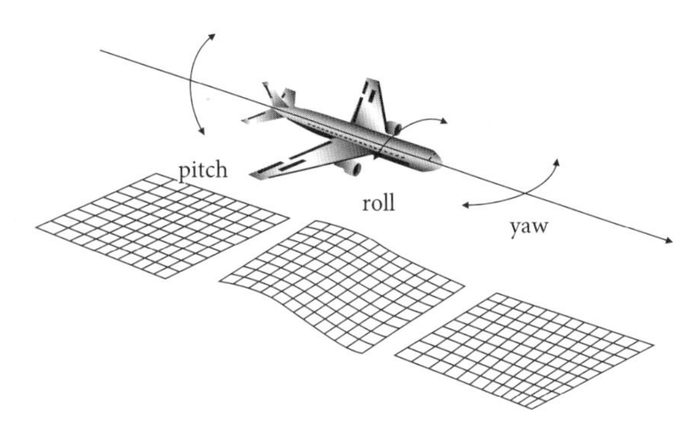

Fig. 6.3. Illustration of the geometric effects of aircraft instability showing rotations about the three axes; pitch (rotation about the horizontal axis perpendicular to the line of flight – leading to compression or expansion of whole scan lines), roll (rotation about the line of travel – leading to variation in the swath position), and yaw (rotation about a vertical axis – leading to skewing of the scanlines).

image at that point, and, over high-relief terrain, this may vary markedly across an image. In addition, the area viewed in a single pixel depends on the angle of the surface relative to the view angle (Fig. 6.2).

Correction methods

The objective of geometric correction is to generate an image that faithfully represents the spatial aspects of the scene itself. The process of geometric correction is illustrated in Fig. 6.5 where the original (distorted) image is transformed so that it now has a one-to-one relationship with a map of the area. This allows one to use the image as a map-like representation of the scene, or physically to overlay it over a map, another image, or an image product (e.g. a GIS). We will assume that first-order corrections for systematic errors have already been performed when we receive the data. Any remaining systematic errors and all non-systematic errors must be corrected on each individual image. Correction usually takes place in two stages. First, a rectangular grid is chosen, which is usually based on the relevant map grid, and a set of points is established to represent the new pixel centres. Secondly, new pixel values are calculated from the original ones to be associated with those points, thus forming a new image. Overall these two steps have the effect of warping the image to conform to this cartographic representation and the process is referred

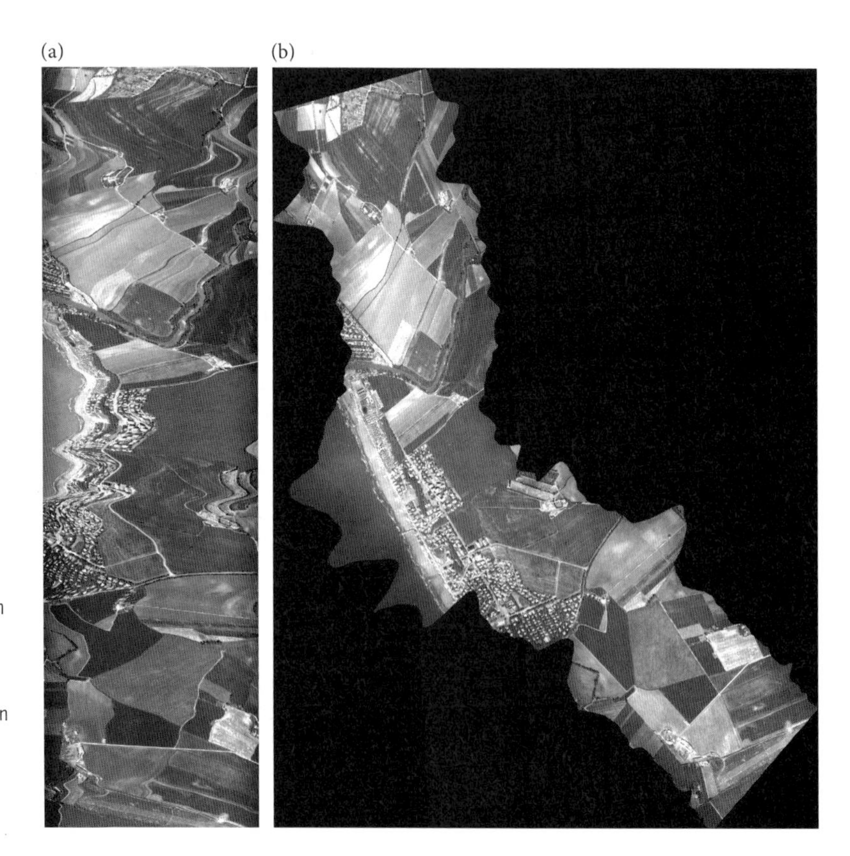

Fig. 6.4. Consequences of aircraft instability on the area imaged from an aircraft: (a) original greyscale image as collected by an Airborne Thematic Mapper (ATM) over Tarquinia in Italy, (b) the same image after geocorrection to correct for the effect of distortions caused by aircraft instability (image from a14093a – NERC ARSF project MC04/07). (See Plate 6.2)

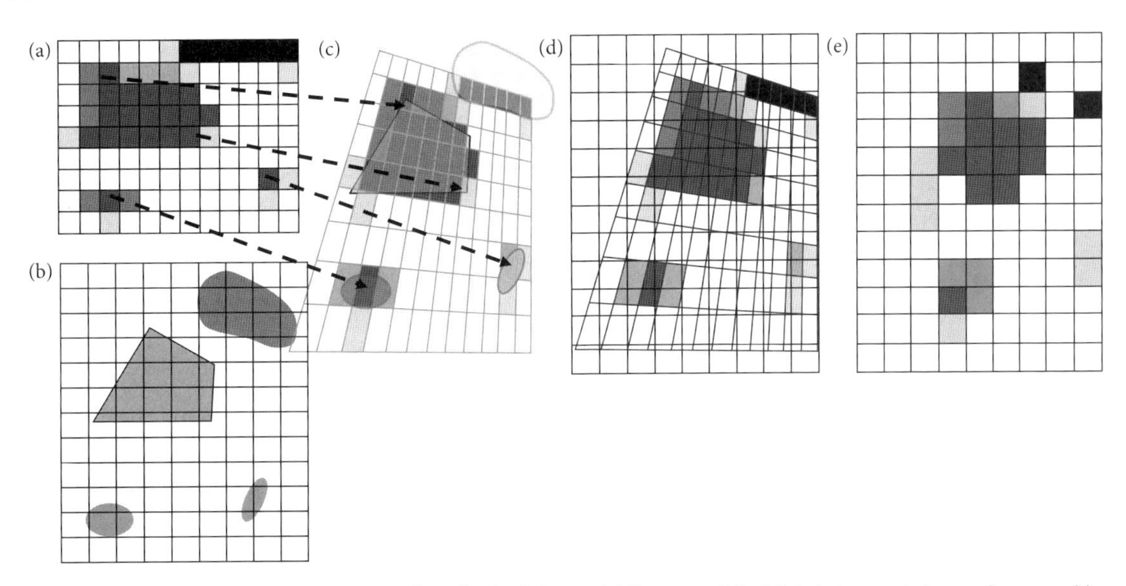

Fig. 6.5. Illustration of reprojection and resampling of a simple image. (a) Shows an original distorted greyscale image of an area with a number of recognizable objects, (b) shows a map of the corresponding real landscape, (c) illustrates the transformation (warping) of (a) so that it can overlay the map; this is achieved by using a series of corresponding ground control points (GCPs) in the image and on the map. The correspondence of GCPs in the image and map is shown using dashed lines. The warped image is then overlain by a rectangular grid in (d) and the new pixel values in the final georegistered image (e) are obtained using the nearest-neighbour approach where the value for any pixel takes that of the nearest neighbour in (d). (See Plate 6.3)

to as *reprojection* (see Fig. 6.5), while the calculation of the new pixel values is called *resampling*.

Often, an empirical method is used to carry out the geometric transform. This compares the differences between positions of common points that can be identified in both the image and on a map of suitable scale. The coordinates of easily identifiable features (such as road junctions or characteristic coastal features), either as grid references or as northings and eastings, are input into the image processing package together with the location of the corresponding image pixel, and the image is suitably warped to the map coordinates. This is referred to as registering the image to the map. The features used to tie down the image are called *ground control points* (GCPs). For the best results, especially with airborne images, a large number of GCPs should be used that are evenly distributed across the image. The accuracy of the registration will depend on the sophistication of the particular algorithm used in the transform, and it can be checked by choosing a number of other points (not the tie points) and calculating the root-mean-square value of the differences between their calculated and actual positions. It is also possible to register one image to another one, so that they can be overlain, if absolute georegistration is not required. For this application, as is common for ground-level or in-field sensing, conventional image processing software such as Adobe Photoshop provides a very handy transform tool for warping by direct visual matching in real time.

There are a number of algorithms available for resampling. The simplest is to calculate the distance of the centre of a new pixel from the centre of neighbouring old pixels and to substitute the value of the nearest old pixel (nearest-neighbour resampling) which is illustrated in Fig. 6.6. This method has the advantage that it involves just a movement of the original pixels without any change to their radiometric content so that the spectral distribution within the image is unchanged. Other methods such as bilinear interpolation and cubic convolution involve substituting values calculated from the neighbouring pixels on the basis of different weightings of their respective contributions in proportion to the distances of their centres from the centre of the new one. These more sophisticated resampling approaches can lead to better estimates of the new pixel values, and are especially useful when attempting to interpolate to a finer grid.

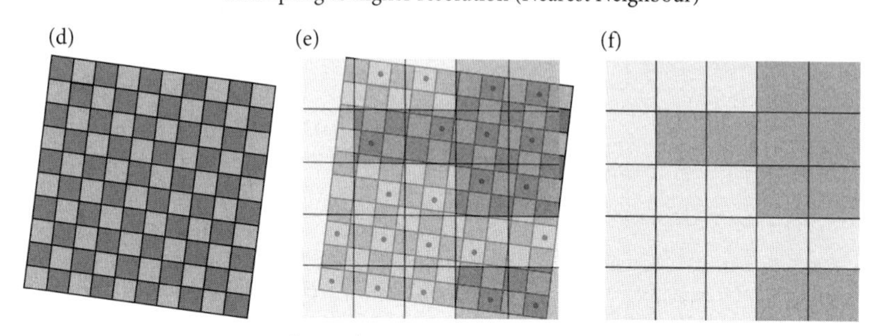

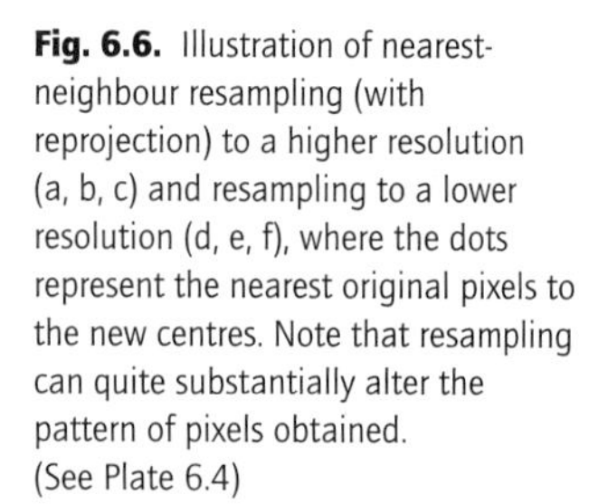

Fig. 6.6. Illustration of nearest-neighbour resampling (with reprojection) to a higher resolution (a, b, c) and resampling to a lower resolution (d, e, f), where the dots represent the nearest original pixels to the new centres. Note that resampling can quite substantially alter the pattern of pixels obtained. (See Plate 6.4)

Unfortunately, however, these sophisticated resampling procedures have to be used with care as they can drastically alter the data. It is preferable that some types of analysis (such as classification) are undertaken before any resampling takes place since resampling forms admixtures of the original pixels and thus their spectral integrity will have been lost.

6.2.2 Radiometric correction

Histograms

As well as errors in the location of the pixels, errors may also occur in the radiometric values of the pixels. These may be due to instrumental faults, resulting in incorrect values being recorded, or due to illumination or reflectance variations over the target, which will affect the accuracy of any quantitative measurements obtained from the data. A useful concept for the study of the radiometric properties of an image is the *histogram* (Fig. 6.7). This is a graph of the frequency of occurrence of every digital value in the image. The distribution of the values within an image band will be characteristic of the ground cover in the image. Water has a much lower reflectance than landcover and so would contribute lower digital numbers. An image containing both land and water will give a histogram with two peaks (*bimodal*). The distribution of digital values will depend on the illumination of the scene and on the reflectance of the surface.

Noise

Noise in an image can be either random or systematic. Random noise may be caused by electronic interference, errors in scanning, or intermittent problems with data transmission or recording. It can be reduced by smoothing the image (see *filters* later), usually by substituting the average value of the surrounding pixels for the rogue pixel value. Care must be taken, however, that legitimate variations are not also smoothed out by applying a too-severe filter. Systematic noise is due to malfunction of the detectors. *Striping* or *banding* may be caused by an imbalance between the detectors. In a whiskbroom scanner such as the Thematic Mapper, which scans sixteen lines in each sweep for each spectral band, one or more of the detectors in that band may have a different response from the rest, due possibly to aging at a different rate. This manifests itself as a distinctive horizontal banding pattern or periodic variation in the brightness of lines of data. Destriping algorithms are available that attempt to reduce this

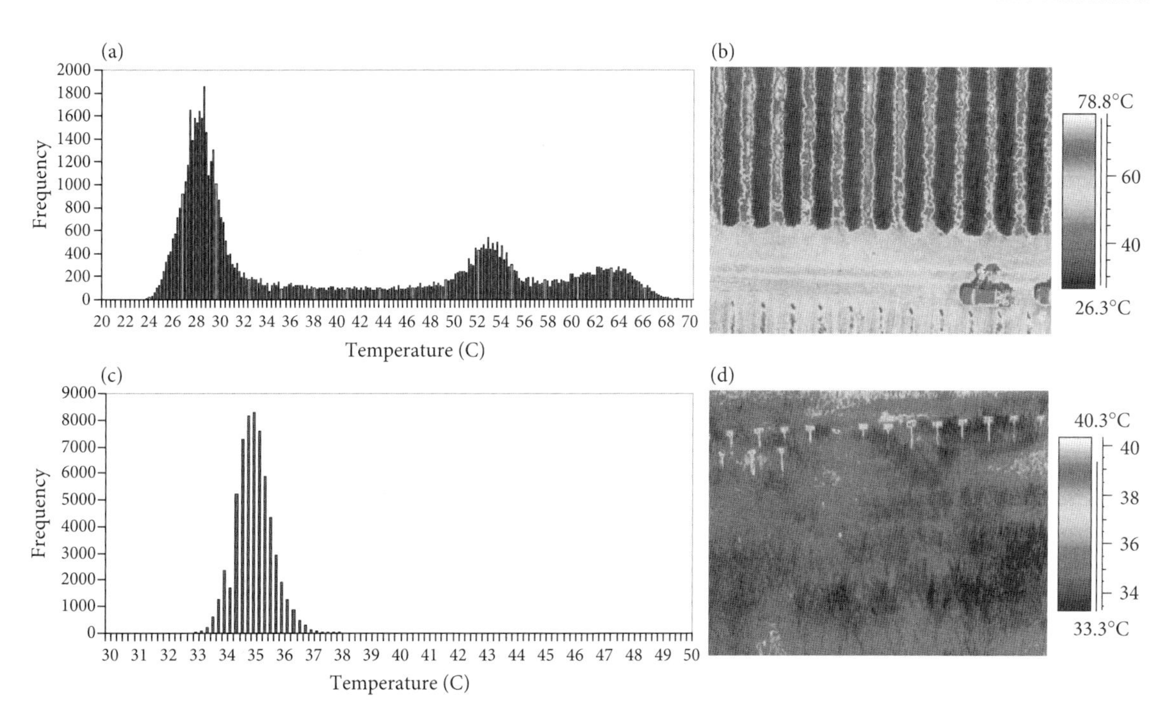

Fig. 6.7. Contrasting histograms showing the temperature distribution for the two thermal images shown at the right using a greyscale palette. (a) Histogram of temperatures in the aerial thermal image (b) of a vineyard in S. Australia (see Fig. 8.1) showing the contrast between the cool canopy and the hot soil, and (c) temperature histogram for a thermal image of a field of rice in Wuhan (d) showing a relatively homogeneous and narrow temperature distribution. (Photos A. Wheaton and H.G. Jones.) (See Plate 6.5)

effect by analysing the histograms from each detector and suitably adjusting the response of the rogue detector(s) so that its output is similar to the others. An extreme case of this fault is when one detector ceases to function completely, leading to a completely blank line in the image. Such *line dropout* can be corrected for by substituting the average value of the pixels immediately above and below for the missing ones, or a value based on the statistics of the surrounding pixel values. Note that these remedies are purely cosmetic in nature, producing a visually more pleasing image, but not restoring the lost data. Care must be taken when analysing such data as the correction procedure may have altered the radiometric balance of the image. A pushbroom scanner does not suffer from this problem, but may exhibit vertical banding instead.

A particularly powerful technique that is frequently used with hyperspectral data is to use the *minimum noise fraction* (MNF) transform.[29] The MNF transform is a data-reduction technique that involves two sequential principal components transforms (see Section 6.4.4) to identify the noise and then to segregate or remove it and to reduce the (redundant) dimensionality of an original hyperspectral dataset.

Although not strictly noise, a related phenomenon is *aliasing* that arises in some images where there is some regular feature on the ground at a spacing close to that of the pixels (or a multiple of this) (Fig. 6.8). The effect is to generate a series of spurious lines or fringes.

Sun angle and viewing geometry

The magnitude of the signal received at the sensor depends upon both the illumination (and its angular properties), and on the reflectance of the target, which itself varies with view angle. The angular variation of reflection will be discussed in more detail in Chapter 8, because it provides a particularly powerful tool for derivation of vegetation structural properties from optical RS. Here, it is appropriate just to summarize some of the main effects.

First, we need to take account of factors that can affect the incoming irradiance: these include atmospheric effects (such as cloud, haze, and other scatterers – see *atmospheric correction* below), variations related to solar

[29]. Frequently referred to as the *maximum* noise fraction transform as it aims primarily to segregate the maximum amount of noise.

Fig. 6.8. (a) High spatial resolution panchromatic image showing an area of vineyard with regular arrangement of the vines, (b) corresponding lower-resolution image from the ATM sensor (pixels approximately 1 m) showing aliasing and the appearance of spurious patterns in the vineyard (part of image a14093a – NERC ARSF project MC04/07).

angle variation with time of day and time of year, variation of earth–sun distance during the year (increasing by about 6% between January and July), and finally, variation of the angle between the incident beam and the surface. Various solar radiation calculators (e.g. SunAngle: http://susdesign.com/sunangle/) are available for easy calculation of solar angles at any place or time.

Differing solar angles have the consequence that two images of the same area taken on different dates or times will look different even if the spectral characteristics of the scene have not changed. It is for this reason that sun-synchronous satellites cross the equator at the same local time each orbit, thus ensuring approximately constant direction of illumination. This is particularly important when studying vegetation growing in rows or in high-relief terrain. The reflected radiation also depends on the topography of the surface in relation to the solar beam (Fig. 6.9) as a result both of topographical shading effects, smaller-scale shading (for example by trees in the image), and by variation in the angle between the surface and the solar beam.

Even where the true reflectance of the vegetation is constant over the scene, large differences in reflected radiance can result from surface angle and

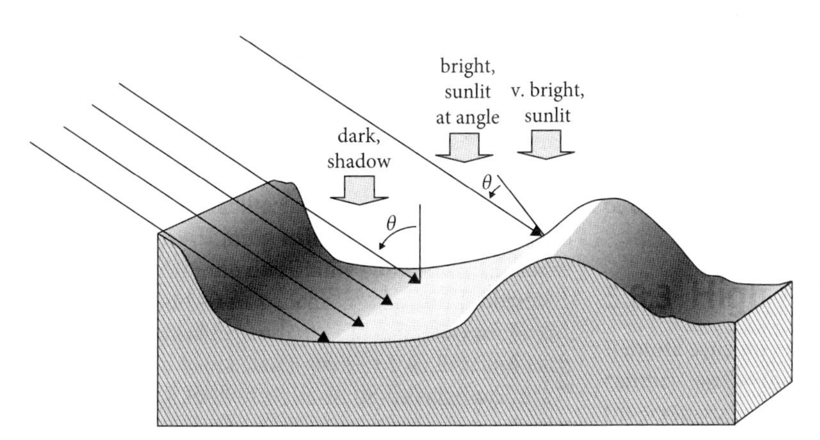

Fig. 6.9. Illumination and topographic effects showing that even for surfaces with a constant true reflectance over the scene, the radiance, and hence the apparent reflectance varies. For example, surfaces in shadow appear less bright than surfaces in full sunlight, which themselves depend on their angle with respect to the solar beam. Irradiance on the slope facing the sun is higher than on the horizontal and hence the apparent reflectance is also greater.

shading effects. Although measured radiances are usually converted to 'at satellite' or 'at surface' reflectances by satellite algorithms, it is important to recognize that these derived 'reflectances' are not necessarily good estimates of the true surface reflectance. A pixel on a sloping terrain will receive a different amount of illumination from one that is on a horizontal surface. Both solar angle and slope effects can be partially corrected for by dividing the observed pixel value by the cosine of the illumination angle with respect to the normal. The change in solar zenith angle will also affect the pathlength of the radiation through the atmosphere and consequently the amount of atmospheric correction. These effects will also change the amount of shadow that occurs in the scene.

Secondly, we have to consider the view angle, and its relationship to the angle of illumination, as this also has a large effect on the apparent reflectance. Variations in viewing angle across an image can be quite substantial, especially with instruments such as the AVHRR with its very wide swath and with instruments that can be pointed away from nadir, such as the HRV on SPOT or some of the new high spatial resolution instruments. We will look at such effects and how they may be utilized in Chapter 8.

Atmospheric correction

We saw in Chapter 3 that meaningful information about the surface target is contained in the physical properties of the radiation leaving the surface target, whereas the signal measured by a remote detector contains a combination of surface and atmospheric contributions. A proportion of the radiation received at the detector arises from scattering of incident radiation by the atmosphere without having interacted with the surface. This therefore carries no information about that surface, and so needs to be removed. Similarly, the radiation leaving the target is attenuated by scattering and absorption in the intervening atmosphere. The calculation of a reflectance at the surface also requires information on the incident radiation at the surface, which itself depends on atmospheric attenuation of the incoming radiation. These effects are summarized in Box 6.1.

It may be reasonable in some studies to ignore atmospheric effects, such as for a single supervised classification of a single image, or when looking for gross changes over a fairly small region (e.g. floods or coastal erosion). In most cases, however, appropriate correction is essential. It is particularly important for example when using multispectral and multitemporal data, when merging datasets from different instruments, when relating upwelling radiance with physically based models, or when attempting to make quantitative estimates of the properties of objects on the earth's surface. Inadequate atmospheric correction underlies the general lack of transferability of a classification from one image to another, while atmospheric correction is particularly important for the analysis of low-reflectance objects or when using physically based radiative-transfer models (see Sections 3.1.3 and 8.3).

There are many different approaches to atmospheric correction of varying complexity and the interested reader is referred to more specialist remote-sensing texts for further information (e.g. Jensen, 2005; Liang, 2004). Here, we outline the bases of the main methods available that can be grouped as follows: (1) calibration with *in situ* measurements of geophysical parameters, (2) the use of a model atmosphere

BOX 6.1 Calculation of surface reflectance ρ_s from a satellite sensor

Illustration of the contributions to the radiation received at a satellite sensor and their relationship to the target reflectance. In this figure, all terms refer to radiation within a particular spectral band, λ.

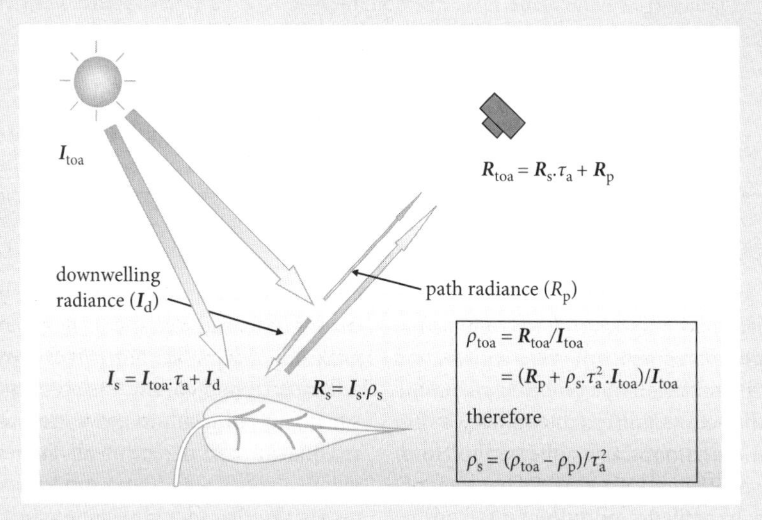

where I_{toa} is the spectral solar irradiance on a horizontal surface at the top of the atmosphere (as calculated from eqn (2.23)), I_S is the direct beam spectral solar irradiance at the surface, R_S is the reflected spectral radiant flux density at the surface, R_p is the path spectral radiant flux density, I_d is the downwelling solar irradiance due to scattering in the atmosphere (this is often ignored), R_{toa} is the spectral radiant flux density measured at the sensor, τ_a is the spectral single-transit atmospheric transmittance, ρ_s is the hemispherical surface spectral reflectance, ρ_p (= R_p/I_{toa}) is an effective path reflectance. Note that values for incoming and reflected radiation are expressed in terms of radiant flux densities (W m^{-2} µm^{-1}) absorbed or emitted by a horizontal surface and that it is assumed that scattered and reflected fluxes are isotropic, so that the observed radiant flux density at the sensor R_{toa} can be estimated from the measured radiance at the sensor, L_{toa} (W m^{-2} µm^{-1} sr^{-1}) as $\pi.L_{toa}$. It is also possible to refine the model to take account of the differing atmospheric pathlengths, and hence τ_a, for incoming and viewed radiation as a function of their respective zenith angles.

For a dark pixel (e.g. a shaded pixel or deep water) $\rho_s \cong 0$, so that $\rho_p \cong \rho_{toa\ (dark\ pixel)}$, therefore allowing one to eliminate the path radiance term, while the single path atmospheric transmittance term may be estimated either from surface pyranometer data or from other estimates of the optical air mass and radiative-transfer models (Section 2.3.2).

with parameters determined from historical climatological data, (3) a similar approach but using simultaneous meteorological data from the satellite or from ground stations, (4) the evaluation (compensation or elimination) of the atmospheric effects on a pixel-by-pixel basis using image-derived information. All these approaches can be classified as *direct* methods in which the atmospheric parameters are somehow obtained and used to correct the atmospheric effects to individual spectral bands. Another possibility is to use *indirect* methods, in which the atmospheric effects are not really quantified and the problem is instead overcome by avoiding (e.g. by using the *atmospherically resistant vegetation index – ARVI* – introduced in Chapter 7) or minimizing (e.g. *maximum value composite* – MVC) the effects. Correction of data is even more important over water surfaces where the size of the signal is so much smaller than over land due to the low reflectance of water in the optical region, but, on the other hand, the aerosol contribution is much more spatially variable over land, posing possibly more severe correction problems.

The direct methods are the most effective and the ones most commonly used. These generally make use of radiative transfer models, which consider both the extinction and the introduction of energy into the beam of radiation, for the assessment of the atmospheric contribution to the at-satellite measured signal by modelling the radiation pathway from source to sensor. A mathematical analysis of the effect of the atmosphere on electromagnetic radiation produces the *radiative transfer equation* (RTE). Even very simplified forms of the RTE are almost impossible to solve mathematically, so simplified models are adopted or it is common to use a *radiative transfer code* (RTC) such as 6S (Vermote *et al.*, 1997) or MODTRAN (Berk *et al.*, 1998). Such RTCs can usually provide very accurate estimates of the atmospheric effects and of surface–atmosphere interactions as long as the correct input parameters are used. *In situ* measurements must be made in representative areas throughout the image, noting that atmospheric conditions may vary substantially over the image, and the parameters used in the model atmospheres must be relevant to that geographical location. Such methods are more successful in large-area applications using low spatial resolution data, or when the correction is small compared with the surface signal. Tables of generalized parameters are available (e.g. within the MODTRAN package), specified by geographical location and time of year (e.g. for 'maritime tropical areas, summer', or 'temperate continental areas, winter'). These consist of vertical profiles of pressure, temperature, ozone, and water vapour, but it is much better, if at all possible, to use simultaneous meteorological data that are now sometimes obtained from the same satellite as is the image, or from specific spectral bands in more general sensors.

A particularly powerful approach is to use image-derived information to eliminate the effects of atmosphere, often using multilook or multispectral data. These approaches are particularly useful for correction of thermal data to derive surface temperatures as will be discussed in the next section. The multispectral method is widely used in optical remote sensing, such as the *darkest pixel* method, alternatively called *dark target* (DT), or *dark object subtraction* (DOS) method. This approach and the developments from it (Chavez, 1996; Moran *et al.*, 1992; Wu *et al.*, 2005) rely on the assumption that within an image there are some pixels in complete shadow, so that any radiance measured for these pixels must be attributable to the atmosphere (Box 6.1). Alternatively, since water absorbs near infrared radiation (NIR) very strongly, virtually nothing is reflected from water bodies at these wavelengths. Any signal received in a pixel that is mostly water or shadow can be assumed to have emanated from the atmosphere, and this signal indicates the level of correction that is needed in the other pixels, and, to a certain extent, at other wavelengths. Often, sufficiently large water bodies may not be present in an image, and so a variation on this method, the *dense dark vegetation* (DDV) method, is used. This makes use of the fact that vegetation, such as dense green forests, absorbs red light very strongly and so looks dark in that spectral band (they may only reflect 1–2% of the incident light). This is obviously not a very accurate method, but is quick and easy to apply and is often better than not using any correction at all. Another common approach makes use of the fact that shortwave radiation is preferentially scattered by atmospheric aerosols so that reflectance in the blue can be used to estimate aerosol concentrations for incorporation into radiative-transfer models.

For multitemporal studies it is useful to base the correction on ground targets which do not change significantly with time (*pseudo-invariant targets*, PIT). The PITs used are often man-made surfaces such as asphalt or concrete. A series of images of the same scene can then be normalized to the one with the least atmospheric interference, with any difference in observed reflectance for the PITs being assumed to be caused by atmospheric interference and to apply to all pixels. This difference is then subtracted from the observed values for all other pixels to give a corrected value. Because the reference image is likely to still have some atmospheric interference this only achieves a 'relative' correction. A better approach might be to use surface-obtained spectral reflectances for the PITs and to regress these against observed 'at-satellite' radiances using an empirical line approach as illustrated in Fig. 6.10 (Hadjimitsis *et al.*, 2009). The use of PITs depends on the assumption that the relationship between 'at-satellite' and surface reflectances is linear and that atmospheric effects are constant over the whole image.

The advantage of image-based methods is that they do not require information relating to the state of the atmosphere at the time of the data collection, which is essential for radiative-transfer modelling-based correction methods. The disadvantages are that the selection of dark values can be subjective, and it ignores

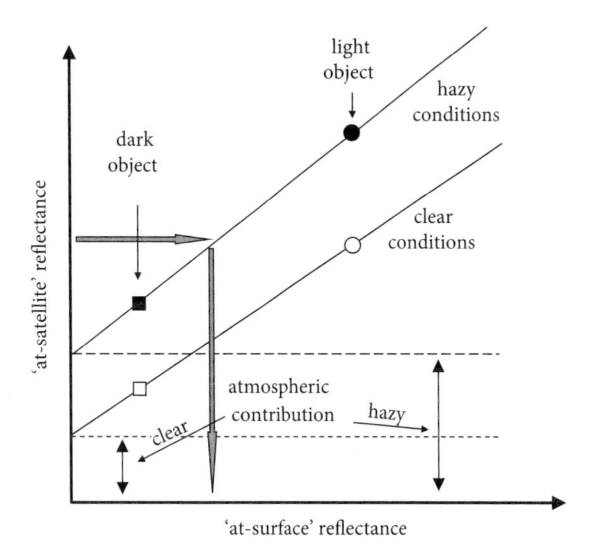

Fig. 6.10. Basis of the empirical line atmospheric correction method, showing how an observed 'at-satellite' reflectance is converted to a true 'at-surface' reflectance using known 'at-surface' reflectance values for contrasting dark and light objects and assuming a linear relationship between satellite and surface reflectances and a constant atmospheric contribution, obtained by extrapolating to zero surface reflectance. The broad arrows show the estimation of the ground (true) reflectance from the observed 'at-satellite' reflectance for a specific unknown pixel in hazy conditions. The light and dark objects should ideally have as contrasting reflectances as possible in each of the wavebands to be corrected.

the effects of atmospheric absorption. For applications where quantitative retrieval of surface parameters is required, a radiative transfer or other type of model must be used. Such a model should be capable of explaining the nature of a measured signal and relating it to the physical properties of the surface being observed.

Atmospheric effects are perhaps more important when using hyperspectral data in which individual absorption bands may affect some of the wavebands very severely. These variations would tend to be smoothed out over the broad bands of a multispectral image.

6.2.3 Thermal data and estimation of surface temperature

Surface temperature is estimated from the radiation received at the satellite in one or more thermal bands. As indicated in Fig. 3.23, the 'at-satellite' thermal radiance comprises not only the directly emitted radiation that varies as a function of temperature of the target, but also radiation from the environment reflected by the target towards the sensor, thermal radiation scattered by the atmosphere towards the sensor and any attenuation by absorption in the atmosphere. Estimation of the target surface temperature from the observed brightness temperature therefore requires the elimination or correction of these various confounding effects. The situation is complicated further by the fact that estimation of surface temperature requires a simultaneous estimate of the surface emissivity, which may vary with wavelength (Section 3.2). Unfortunately, even where we have brightness temperature measurements in n thermal wavebands, there remain more unknowns than we have observations (n emissivities + surface temperature), so all methods require some independent information (usually on emissivity, but sometimes including the state of the atmosphere) to allow estimation of the surface temperature from satellite images (Norman *et al.*, 1995a; Qin and Karnieli, 1999). The main approaches include the single-channel method, split-window algorithms, multiangle techniques, and combined methods.

A particular problem in measuring land-surface temperature, which is spatially rather variable on land, is the validation of estimates obtained at the usual scale of satellite imagery (e.g. 1 km for MODIS thermal data). There is usually no easy way of obtaining a representative estimate of the surface temperature for such pixels, unless they are very homogeneous. Furthermore, not only is the kinetic temperature not the same as the apparent temperature of the visible surfaces, but for surfaces such as the sea, bulk water temperatures differ from the remotely sensed temperature that detects only the top few mm (the *skin effect*).

Atmospheric correction for thermal data

(i) Single-channel methods. These are the only approaches available for many of the sensors with only a single thermal channel. The technique requires information on the distribution of water vapour and temperature in the atmosphere (it can be assumed that the atmospheric effects of other gases are constant); this information can be derived from vertical sounding instruments on satellites or from weather prediction models. A radiation-transfer model such as MODTRAN is then used to estimate the required atmospheric corrections. Accurate surface temperature estimation depends on the accuracy of emissivity estimates with (rather conservative) emissivity errors of ±0.025 leading to temperature errors of ±2 °C for typical mid-latitude conditions (Dash *et al.*, 2002),

(ii) Multilook approach. Another approach to atmospheric correction of remote estimates of surface temperature makes use of near-simultaneous measurements of surface temperature at different look angles, and hence different pathlengths through the atmosphere. Observations can be from different satellites or from a single satellite using sensors such as the Along Track Scanning Radiometer (ATSR) on ERS and Envisat that has a conical scan and so observes a feature both at near nadir (0–22°) and in a forward direction (up to 55°) almost simultaneously. Empirical equations can then be used to make use of the two different effective temperatures measured through two different pathlengths of atmosphere in order to estimate the 'true' temperature of the surface, though any angular variation in surface emissivity needs to be taken into account.

(iii) Multispectral approach. Temperature measurements can also make use of the differential absorption of thermal radiation by the atmosphere in different wavebands using the *split-window* technique. The general approach is to derive land-surface temperature from linear combinations of the brightness temperatures measured in the different channels using empirical coefficients to correct for the atmospheric effects. The AVHRR, for example, has two longwave thermal channels (channel 4: 10.3–11.3 μm, and channel 5: 11.5–12.5 μm) that are suitable for such split-window approaches. The main difficulty lies in derivation of the empirical coefficients. In its simplest form (for derivations and references see, e.g., Dash *et al.*, 2002; Qin and Karnieli, 1999), the split-window equation for AVHRR is

$$T_s = T_4 + a(T_4 - T_5) + b, \tag{6.1}$$

where T_4 and T_5 are the brightness temperatures in channels 4 and 5, and a and b, respectively, are parameters relating to atmospheric conditions and surface emissivity.

Estimation of emissivity

All the approaches to estimation of land-surface temperature described above require accurate estimates of the emissivity appropriate for the surface of interest. A range of approaches based on different assumptions can be used to estimate emissivity, mostly involving the use of ancillary visible data. Perhaps the simplest approach to the estimation of ε is to make use of the high correlation between *NDVI* and emissivity, with dense vegetation often having an emissivity as high as 0.994, while bare arid soil may have an emissivity as low as 0.925 (see Section 3.2.2). Van de Griend and Owe (1993) found a high correlation between ε and the logarithm of *NDVI*

$$\varepsilon = a + b \ln(NDVI), \tag{6.2}$$

where a = 1.0094 and b = 0.047 for semi-arid areas in Botswana. The precise constants vary with region (Valor and Caselles, 1996). An alternative approach is to estimate emissivity on the basis of an independent ground-cover classification (using multispectral data) and an associated emissivity knowledge base or 'look-up table' (Snyder *et al.*, 1998). This approach is an important component of the MODIS land-surface temperature algorithm where ε is tabulated for 14 landcover classes, together with corrections for seasonal changes (e.g. leaf fall in autumn) and for the angle of view (see Section 3.2.2 for further details). Other approaches to estimation of channel-specific emissivities and band ratios have been reviewed elsewhere (Dash *et al.*, 2002).

Also of interest are those approaches that attempt to estimate both emissivity and surface temperature together. Perhaps the best known combined approach is the temperature emissivity separation method (TES) proposed by Gillespie *et al.* (1998) that iteratively corrects for atmospheric effects. The MODIS algorithm also makes use of day and night images, where the surface emissivity is assumed not to change, to get further information in parameterization of the surface temperature retrieval algorithm (Wan, 1999).

6.3 Image display

Having obtained a reasonably true geometrical representation of the ground features with radiometric fidelity, the analyst usually needs to view the image on a screen or as hard copy to identify and check features, even though the analysis itself will probably be done automatically. There are a number of processes that can be applied to improve the visual impact and to help the operator to interpret the image. These can be separated into techniques for optimal display of the data, as outlined in this section, and those data transformations that enhance the appearance of the image, for example by manipulating the contrast, as described in Section 6.4. Note that in each case the processes are purely cosmetic and are not necessary for machine interpretation; indeed data transformation to improve visual appearance may even impede further quantitative analysis.

6.3.1 Resampling/windowing

A complete satellite image contains a huge number of pixels, nearly 40 million in the case of a TM image. To process such data is quite computer intensive, especially if many bands are used. It is common to degrade the resolution to allow quick-looks or overviews of the scene by aggregation of groups of pixels (e.g. by averaging 5 × 5 blocks of pixels). Another way to reduce the amount of processing required is to extract a portion of the image and to process only the pixels in this *window* and to discard the unwanted remainder of the image. In fact it is quite usual for a user to purchase only the portion of an image that he requires, particularly when the original is very large.

6.3.2 Density slicing

The human eye/brain system is not very efficient when it comes to quantitative analysis. It is difficult to recognize a particular shade of grey or colour if the surroundings change. This can be demonstrated in a number of 'optical illusions' where a grey level or shade of colour seems to change depending on the background. When studying a black and white photograph, it is not possible to carry the memory of an exact shade from one region to another, and this may make it difficult to identify similar features across the image. If only a single band is being used, then a particular grey level or range of levels can be identified by displaying the relevant digital numbers in a particular, distinctive colour, which is easily recognized by the interpreter wherever it occurs in the scene. Different colours can be allocated to different groups of contiguous digital numbers. Each group of numbers is called a 'slice' and the process is referred to as *density slicing* or *image segmentation* (Fig. 6.11). The intention is to improve the visual interpretability of the image, but this occurs at the expense of loss of spectral detail. Choice of colour is often done interactively by sliding the boundary up or down until the desired visual separation is obtained. A wide range of ready-made colour ramps are available

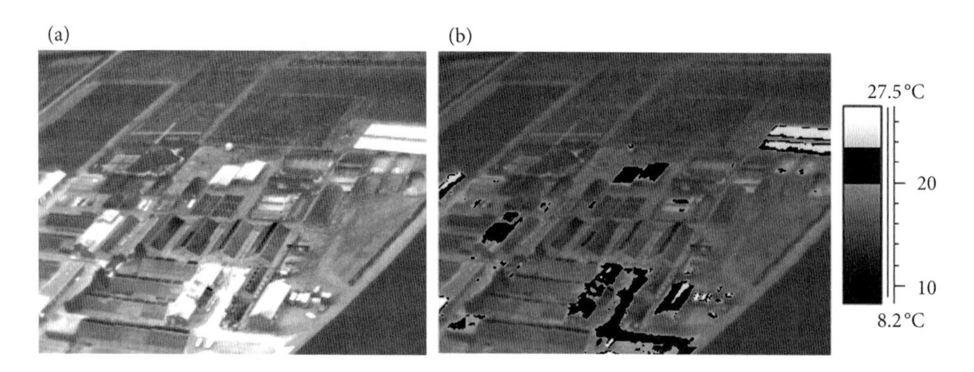

Fig. 6.11. Illustration of density slicing, showing (a) a greyscale thermal image of the farm buildings at the Scottish Crop Research Institute, Dundee (13 June, 2007), and (b) the same image with all pixels between 20 and 22 °C highlighted as a density slice in black (photos H.G. Jones). (See Plate 6.6)

that can be selected according to application; for thermal images it is common to use a scale from 'cooler' colours (blue and green) for low temperatures through yellow to a 'hotter' red and white. An extreme case of slicing is *thresholding*. In this, all digital numbers below (or above) a certain value are given a distinctive colour – for example water (low *DN*s) can be displayed in black, thereby allowing all the other 'land' pixels to be unambiguously identified and analysed. Sometimes it is difficult to unambiguously identify land/water boundaries in the visible channels – wet sand or mud may blur the actual boundary. As water has almost zero reflectance in the near infrared bands, it may be useful to identify all the near-zero values in such a band, and mask off these pixels in the visible bands (*masking*). Although density slicing is really just a form of visual display, it can sometimes be used as a crude form of *classification* (see Section 6.5).

6.3.3 Colour composites

Multispectral images (e.g. Fig. 6.12) are usually displayed in the form of colour composites. It is only feasible to view the information from three channels at once using conventional three-colour display systems. Therefore, three of the image bands are selected and each displayed on one of the colours of the monitor. There are a number of ways in which the bands may be chosen and in which colours may be allocated. For example, if a blue, a green, and a red band were allocated to the blue, green, and red displays, then this would be a *natural colour composite* and would look similar to the original scene (it may not look exactly like an actual colour photo as the result would depend on the specific spectral response of each band). Different combinations of bands may be used to highlight or identify specific features of the view area.

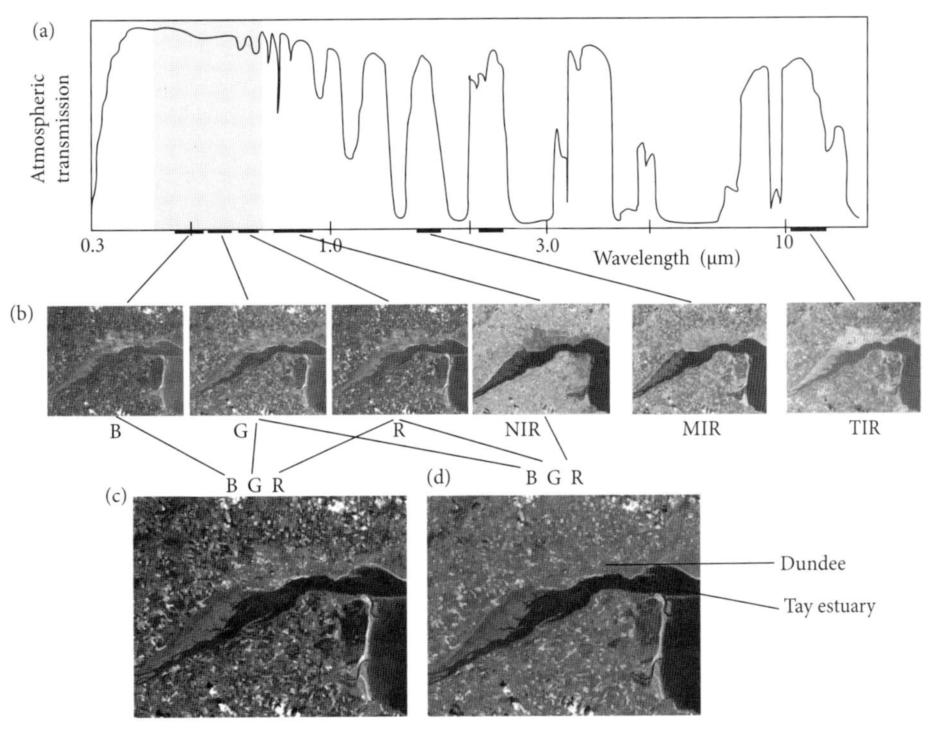

Fig. 6.12. (a) An atmospheric transmission spectrum, showing the positions of the seven Landsat ETM+ bands, (b) corresponding greyscale single-channel images for six of these bands for an image over part of eastern Scotland showing Dundee and the Tay estuary (path 205, row 21, for July 2000). These images can be combined in different combinations to highlight different aspects of the scene. The most obvious combination is to display the red, green, and blue channels in the R, G, and B screen colours as shown in (c) to give a natural colour composite. Alternatively different combinations may be combined to give one of several false-colour composites, including, (d) a standard false composite where R = NIR, G = Red, and B = Green (Landsat image courtesy of USGS). (See Plate 6.7)

For example the high reflectance of vegetation in the near infrared (Section 3.1) can readily be used to distinguish areas of vegetation from soil (Fig. 6.12) by allocating the green, red, and near infrared bands to the blue, green, and red displays, respectively. This produces a standard *false colour composite* (FCC) in which vegetation shows up in differing shades of red. This is particularly useful for studying the type and condition of vegetation. Other false colour composites can be formed using different combinations including the non-visible bands[30] – often the mid-infrared bands are useful in geological analysis. Although there may be good physical reasons for selecting particular wavebands, the choice is often made empirically, perhaps by using statistics to select bands with little correlation that contain different information. If correlated bands were used, the intensities of each gun would be similar for every pixel, leading to low saturation and an image of pastel shades, whereas the use of uncorrelated bands produces an image of high contrast and strong colours. Colour composites including false colour should be distinguished from *pseudocolour* images, where a palette of colours is applied to a single-channel image.

6.3.4 Other forms of display

If a *digital terrain model* (DTM) is available, some image processing software packages enable an image to be draped over the DTM producing a pseudo-3D effect. It may also be possible to view this image from different viewpoints, allowing a virtual 'flythrough' of the scene. Stereo images may be displayed using conventional red/green anaglyph techniques or polarizing computer screens and viewed using the appropriate spectacles.

6.4 Image enhancement

A wide range of data transforms is available to enhance the appearance of images. The eye is not very acute at differentiating between small intensity or colour variations, whereas a computer can easily distinguish between two adjacent digital numbers (see density slicing above). Enhancement techniques either increase these differences or in some cases reduce them to remove unwanted noise, in both cases making the features of interest more visible. The order of application of geometric correction and of other image enhancements can be critical; for example where processes such as classification (see below) are to be applied, this should be done before any spectral enhancement as that may upset spectral fidelity and the balance between bands.

6.4.1 Contrast stretching

Displayed images using raw band data often show poor contrast or are difficult to interpret, and a number of enhancement processes can be used to improve their interpretability. Note that these do not actually change the information content in the image, but are simply cosmetic in nature. If the computer is being used to analyse the image (e.g. using classification techniques or calculating temperature values), such enhancement is not necessary, and in fact is often inadvisable since it may upset the relative interband values. One such enhancement technique is *contrast stretching*. Within any particular image or area, there is usually only a restricted range of digital numbers producing an image with very poor contrast. In order to make the best use of the range of digital numbers available for display on the computer, the range actually used in the image can be spread out over the full range available, producing an image with much higher contrast. Since the statistics of each band may be different, each band of the image can be stretched independently to give the most effective result. There are different mathematical algorithms used to perform such stretches. The simplest is just a linear stretch in which the lowest pixel value in the image is mapped to zero and the highest to 255 (assuming a standard 8-bit display). All other intermediate values are calculated in proportion. Other ways might be to stretch only part of the original range, compressing all of the other values to maximum (or minimum). This is often used when an image contains both land and water and only one of these features is of interest (water has much lower reflectance than land

[30] There are six different ways of allocating three bands to the three colour guns. There are 120 different combinations of the six optical bands of TM data, and 720 if the thermal band were included.

and such an image usually displays a bimodal distribution – see *histograms* above). Apart from the linear stretch, others that are commonly used are histogram equalization and the Gaussian stretch. The effect of stretching is illustrated in Fig. 6.13.

6.4.2 Spectral indices

Various types of mathematical manipulations (addition, subtraction, multiplication, etc.) can be performed on the digital numbers, pixel-by-pixel, for two (or more) different spectral bands in an image or set of coregistered images of the same geographical area. This process generates a new variable or *spectral index* that can be displayed as a new image layer; this may not only look different but it may display information in a more subtle or informative way. As the resulting numbers may be non-integral or not lie within the range of the display system, some scaling is usually required for display. Although band addition is rarely beneficial, except perhaps for reducing the noise component by averaging, band subtraction and band division have many uses.

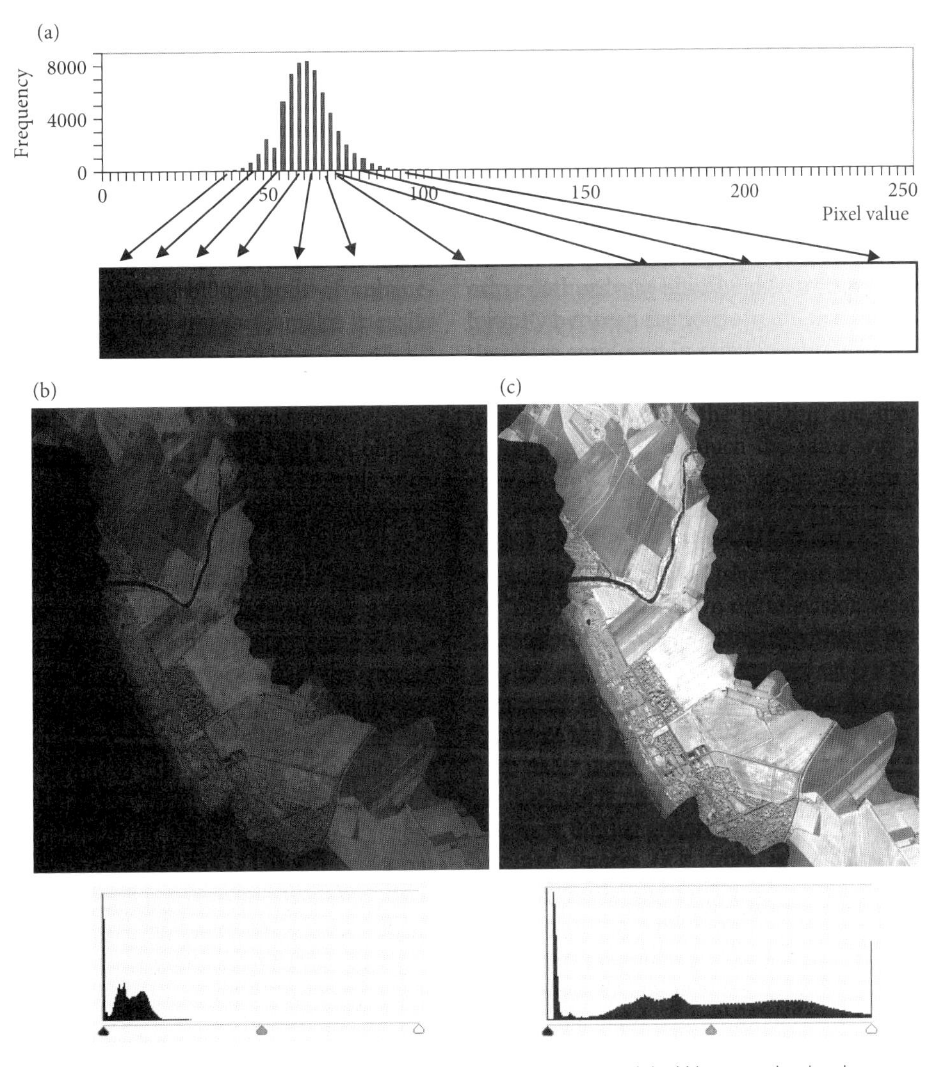

Fig. 6.13. (a) An illustration of the process of contrast stretching, where an original histogram showing the frequency distribution of brightness values for one channel of an image is spread out over the full brightness range available. (b) An original ATM image with its brightness histogram, and (c) the result of applying a contrast stretch where the brightness of pixels is distributed across the intensity range. Images are from an ATM dataset over Tarquinia in Italy (Image a140113a – NERC ARSF project MC04/07).

The most popular application of the technique is in the derivation of *vegetation indices* as measures of canopy biophysical properties (see Section 7.2 for a detailed discussion). Vegetation indices mostly make use of the fact that vegetation shows large differences in reflectance between the near infrared and the visible bands, while surfaces such as soil show comparatively small reflectance differences between these wavelengths. Although the use of spectral indices is a particularly powerful tool for visual enhancement of images, its main application is for extracting useful information on structure and functioning of vegetation, so we will defer to Chapter 7 a detailed discussion of the wide range of spectral indices that have been proposed.

6.4.3 Spatial filtering techniques

A number of filters can be used either to smooth an image or to identify and enhance features such as boundaries and edges. A filter is a regular matrix of numbers (a *kernel*) which operates successively on all the pixels in an image to produce new pixel values, and hence a modified image. The kernel is centred on one pixel in the original image and all pixels within the box are multiplied by the corresponding value in the kernel and added together. This sum is divided by the sum of the elements in the kernel, and the nearest integral value is used to replace the central pixel. The kernel passes along each row of pixels in the image, operating on them in turn, producing a new image.

A smoothing filter will average the digital values of a block of pixels and this average value will be substituted for the value of the central pixel of the group (see Fig. 6.14). If this procedure is carried out for all the pixels in the image it will have the effect of reducing the range of values in the image, particularly the random variations that may occur because of inhomogeneity of surfaces or noise. This has the effect of blurring the image and hence removing the finer detail. The larger the filter, the nearer the average values become to each other and the greater the blurring. It is therefore important to choose the optimum size of the filter to reduce the unwanted distractions yet to retain the wanted detail. A smoothing filter is also known as a *low-pass filter* because it suppresses the high-frequency variations between neighbouring pixels while retaining the slowly varying, low-frequency background component. One problem with this type of filter is that it applies the same amount of smoothing to all pixels, regardless of the amount of textural information there is in the image at that point. One type of filter designed to overcome this problem is an *adaptive filter*. In this, the weight of the filter changes as the filter passes across the image depending on the variance in grey level at that point. An example would be the *gamma filter* that is used to suppress speckle in SAR images without reducing the spatial resolution too much.

This type of spatial filtering is often referred to as *convolution* where the kernel is applied successively to each of the pixels. In general the value of a central pixel (V) is given by

$$V = \left[\frac{\sum_{i=1}^{n} \left(\sum_{j=1}^{n} d_{ij} c_{ij} \right)}{N} \right], \tag{6.3}$$

where d_{ij} are the values of each pixel in the kernel and c_{ij} are the values of the coefficients and N is the number of cells in the kernel (except for zero-sum filters where N is set = 1). We show the operation of a simple smoothing filter (n = 3) in Fig. 6.14. We can note that edge pixels are ignored (because there are not nine pixels within the window on which to operate) and replaced by zero in the final image (Fig. 6.14(d)); if a 5 × 5 kernel were used the two outer rows would be blank.

In contrast to smoothing filters, edge-detection and edge-enhancement filters enhance the high-frequency variations by increasing the magnitude of the difference in value between neighbouring pixels, thus making steps in values more noticeable. These are therefore referred to as *high-pass filters*. Edge enhancement is particularly useful in remote-sensing studies where the location of linear objects such as roads or the location of edges defining objects or areas of interest (e.g. fields) is of particular relevance. Some examples of high-pass filters and their effects are illustrated in Fig. 6.15. Of particular value are the gradient filters, used to detect edges. These edges may be field boundaries, hedges, roads, rivers, etc. Such filters are often used to detect lineaments and structural details in geological studies. Some gradient filters are directional in nature and can be used to detect linear features in specific directions. In general, asymmetry of the kernel across a horizontal direction would pick out horizontal features, while asymmetry about a diagonal would pick out diagonal features. A filter that will enhance the edges in an image (as distinct from just detecting them) is the *mean difference* filter. This is formed by subtracting

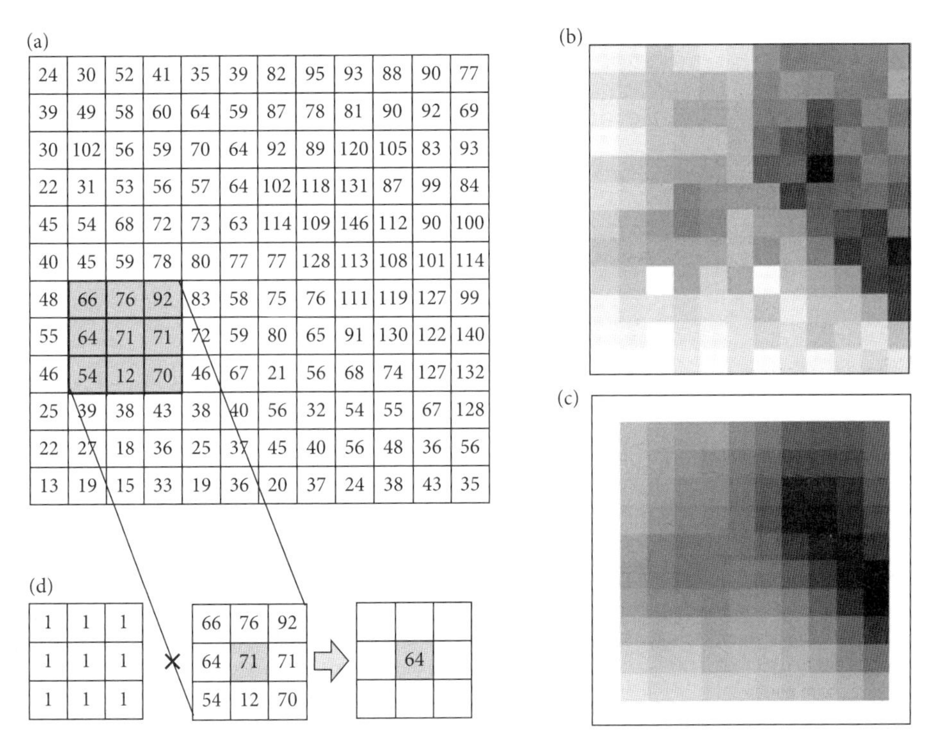

Fig. 6.14. The operation of a simple smoothing filter on an area from Fig. 6.13: (a) shows the pixel values for an area of an image, (b) shows the corresponding part of the image, and (c) the smoothed image after application of a 3×3 smoothing filter as shown in (d) operating on the 3rd pixel along the 8th row from the top. The new value (64) for the centre pixel of the highlighted area is the average of the nine pixels according to the operation of eqn (6.2) using the kernel shown at the left. Note that the two exceptionally light pixels are smoothed out by this filter, and that the edge values are not calculated and are replaced by zero.

the weightings of a mean (smoothing) filter from the unit filter (all elements zero apart from the central one, which is unity) This produces an image that is similar to the original except that edges and boundaries are more distinct.

Among other high-pass filters available are the *Laplacian* filters and the *Prewitt* filters; a wide range of high-pass filters is available in the usual image-processing softwares, and includes both linear and non-linear algorithms, each of which can produce characteristic effects. The most appropriate for any study is usually chosen by trial-and-error; some examples of high-pass filters and their effects are illustrated in Fig. 6.15.

6.4.4 Principal components

Information in a multispectral image is often distributed fairly uniformly between the different bands. Adjacent narrow bands are usually highly correlated and carry a lot of redundant information. The main use of *principal components* is to generate a new reduced set of bands in which the information content is concentrated and that have little correlation. Indeed, the first principal component can contain anything from about 92 to 98% of the total information, and typically over 99% is contained in the first three components. Principal components are linear combinations of the bands, chosen in such a way that they are uncorrelated with each other (Fig. 6.16). This figure illustrates the case for two bands, where we can see that the sample data largely fall on a line that is not parallel with any of the original axes; this would be defined as the first principal component (PC1). PC2 would be drawn perpendicular to PC1 and in the direction defining the second most variance.

If three spectral bands were used, a third feature axis (channel 3 brightness) would be formed normal

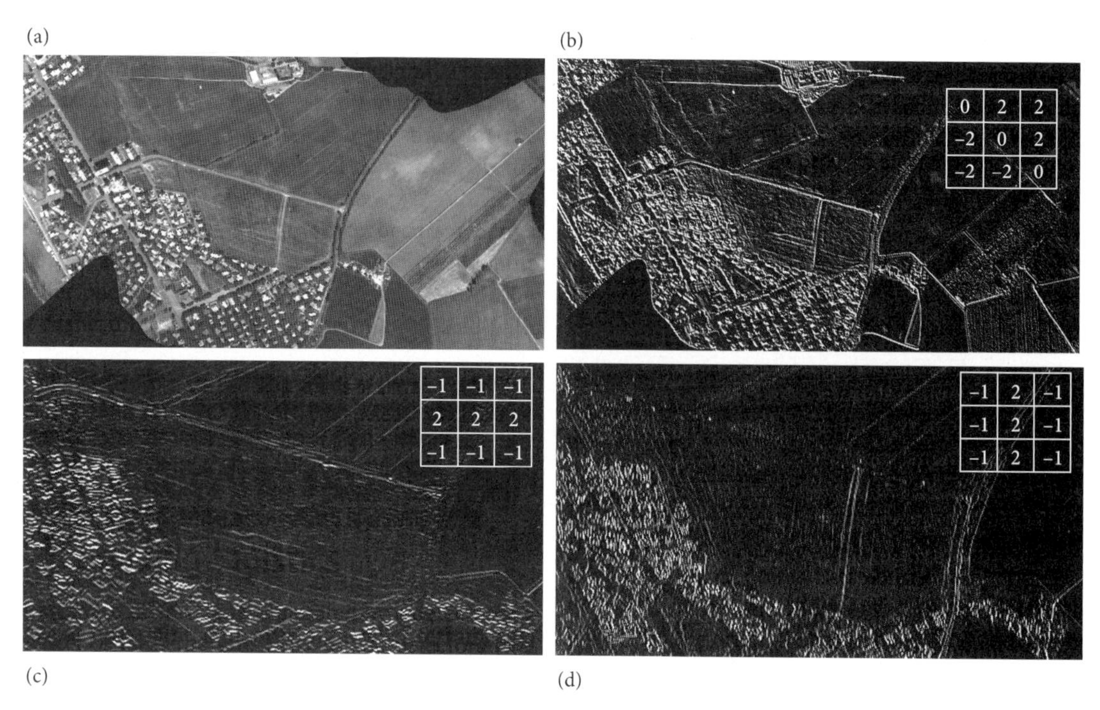

Fig. 6.15. Illustration of some 3 × 3 edge detection methods, and the corresponding kernels: (a) the raw image; (b) the effect of a Sobel filter that detects diagonal edges, (c) the effect of a horizontal filter (with the image enlarged 2× to illustrate the detection of horizontal edges more clearly), and (d) the effect of a vertical filter (again with the image enlarged). [Image a140093a – NERC ARSF project MC04/07].

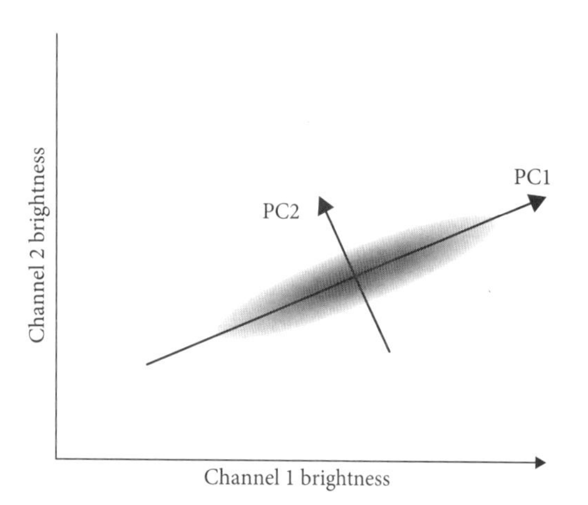

Fig. 6.16. Diagrammatic representation in two dimensions of the transformation from two spectral bands into new principal component axes. When the original channels 1 and 2 are plotted in two dimensions in feature space, we see the data for this example fall largely within a narrow diagonal ellipse. Note that this implies considerable correlation between the two original bands. The principal component transform rotates the coordinates to a new system (labelled PC1 and PC2) such that the majority of the variation in the original data is encompassed within the first principal component axis (PC1) which lies along the semi-major axis of the ellipse. The second axis, PC2, is perpendicular to the first so as to ensure that there is now no correlation between these new axes.

to the first two. Now PC1 would not necessarily lie in one of the planes formed by the axes and PC2 would be drawn perpendicular to PC1 and in the direction defining the second most variance. PC3 would then be drawn perpendicular to the first two. Although it is not possible to visualize, this procedure is not limited to just three bands, and many more can be used, say all six optical TM bands, producing six principal components. The lower-order components contain very little information, just that which is not contained in the bands of higher order, but nevertheless sometimes a particular feature may show up in one of these, although it is often not clear exactly what that feature may represent.

A linear contrast stretch (Section 6.4.1) performed on a band of multispectral data will stretch out the distribution along one of the main colour axes, and because three such multispectral bands may be fairly well correlated, a colour composite made from such data will show little saturation and few distinctive colours. If the stretch were to be performed along the principal axes in feature space, however, and a reverse transform applied in order to convert the data back into conventional RGB colour space, the resulting colour composite will consist of much brighter colours in which details of features show up much better. Because the bands have been decorrelated before the stretch is performed, such a procedure is referred to as *decorrelation stretching*.

6.4.5 Intensity, hue, and saturation transform

When any three bands of a multispectral image are displayed in the three colours (RGB) of a conventional colour monitor, they typically lack saturation and tend to appear as rather 'washed out' pastel shades because of the usual high degree of correlation between different spectral bands. The *intensity, hue, and saturation* (IHS) transform (see Box 6.2) provides a useful tool for enhancing the colour for visual-display purposes. The hue variable may also be better that the original RGB image for the identification of green vegetation. The IHS transform can also be used in data fusion (Section 6.4.6). In order to apply this transformation one uses the following steps:

(i) Transform the three bands from RGB colour space into the IHS system.

(ii) Apply contrast stretching (Section 6.4.1) to the intensity (I) and saturation (S) components (there is no sense in stretching the hue (H) as this would just change the colour).

(iii) Retransform back to RGB to give a more vivid image.

BOX 6.2 Colour space

We have seen that we can represent any spectral colour by the addition of appropriate amounts of the three primary colours red, green, and blue. These three colours can be used as axes in *colour space*, in which any colour can be designated by its position in three-dimensional space defined by the values of its three primary colour coordinates. White is found at the top rear of this cube where all three colours are at maximum brightness, and black is at the front bottom; hence any spectral colour can be uniquely quantified.

It is also possible to specify a colour in terms of three other qualities, namely *intensity* (I, *or value, or brightness*), *hue* (H), and *saturation* (S). Hue relates to the predominant colour, saturation to the purity of colours, and intensity to its overall lightness or darkness. The spectral colours can be represented around the perimeter of a circle; those around the perimeter are pure colours, called *saturated* colours. The central point represents white, formed by an admixture of all colours, and represents the most unsaturated colour. The saturation of any colour within the circle is determined by the distance from the perimeter to the centre. If an axis is drawn perpendicular to the circle through its midpoint, this can be used to specify the intensity, being maximum (white) at the centre of the circle and zero (black) at the base of the cone. The surface of the cone thus formed represents the saturated colours of different intensities. A point within the cone uniquely defines a colour in IHS space. Points along the axis of the cone ($S = 0$) represent shades of grey, while the purity (saturation) of the colour increases as you go outwards from the axis to the surface of the cone that represents the saturated colours of different intensities

(a)

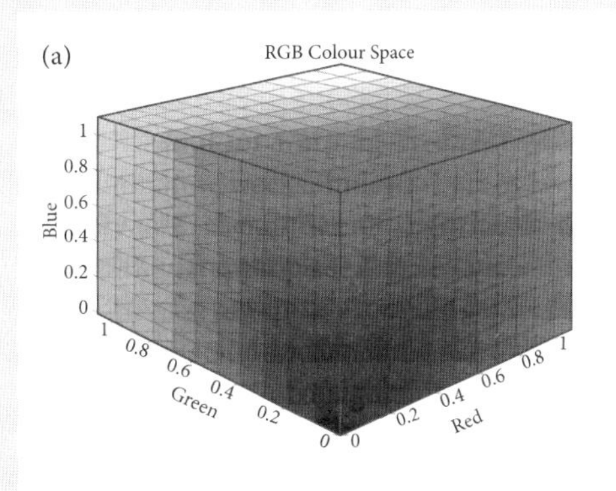

(b)

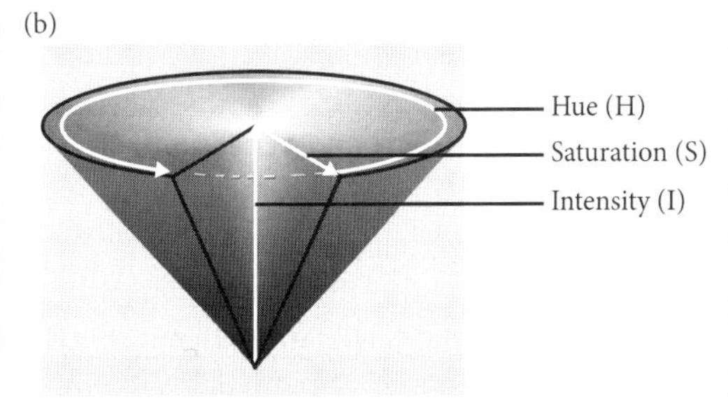

(See Plate 6.9)

6.4.6 Data fusion

It is frequently useful to be able to make use of data of different types from a range of spatial scales. The combination of such varying datasets, often obtained by different sensors, is usually referred to as *data fusion* (or *data merging*). The approach is particularly useful and most simply used for the enhancement of the effective spatial resolution where some information is only available at a coarse spatial resolution but there is some available at higher spatial resolution. For example many satellite sensors have some channels at higher spatial resolution than other channels and information from the higher-resolution channels can be used to resample the coarser channels to the high resolution. In such data fusion the spectral information in a low spatial resolution multispectral image is combined with the spatial structure in a higher-resolution (often panchromatic) image to 'synthesize' a multispectral image at the higher resolution.

Examples of where this might be useful occur wherever features of interest are smaller than pixel size, such as in precision farming. High spatial resolution imagery, with less than 1-m pixels, is coming onto the market, but at a price. Also the images usually cover only a restricted area. Researchers have been striving to find ways of improving the spatial resolution of images in order to get the best of all worlds, high spatial resolution over large areas at an affordable cost. There are a number of ways that have been developed to try to achieve this, but it must be borne in mind that it is never possible to get any more real information than that which was captured in the original images.

The simplest, lowest level of data fusion is simply to overlay say a SPOT panchromatic (pan) image (10 m) over an HRV multispectral image (20 m). This is done by displaying the pan band on one colour of the computer monitor and two MS bands on the other colours. The shade of colour of each quadrant of the MS pixels will be slightly modified by the intensity of the smaller pan pixel, which may make it possible to visualize the MS pixel as being subdivided.

More sophisticated methods involve algorithms that attempt to combine the data from the two images, rather than just overlaying them. The IHS transform described above is particularly useful for this. The procedure used for this type of data fusion (or *pan sharpening*) is similar to that outlined above for colour enhancement of a single image, though it is first necessary to resample the low spatial resolution image (MS in this case) to the same pixel size as the higher-resolution pan image, ensuring that they are carefully coregistered. Then contrast stretching is applied to ensure that each image has the same range and variance (*histogram matching*). The fused image is created by replacing the intensity component of the transformed MS image by the corresponding pan pixel value on a pixel-by-pixel basis, then reconstituting an RGB image. Now each band of the MS image has actually been modified by the overall panchromatic intensity in each small pixel, giving a much better approximation to a high-resolution MS image than the simple overlay method. It is also possible to use the first principal component, which roughly approximates to the average value of the pixel, to replace the intensity component. Sometimes, an edge-enhanced image may be fused with the multispectral data in order that small geometrical features, such as buildings or roads, are more easily discerned. Simpler than the IHS transform is the *Brovey fusion* method, which generates new intensities of pixels by forming linear combinations of the original intensities with those of the pan pixels. For example for band 1 we have

$$DN_{f1} = [DN_1/(DN_1 + DN_2 + DN_3)]DN_{pan}, \qquad (6.4)$$

where DN_{f1}, DN_1, DN_2, and DN_3, respectively are the digital numbers for the fused band 1, and the original values for bands 1 to 3, and DN_{pan} refers to the value in the panchromatic image. A disadvantage of the fusion techniques such as the Brovey and the IHS transform is that they can only be applied to three bands at a time.

When simply aiming to improve the clarity of images for viewing it can be acceptable to distort the colour information as occurs with standard image fusion approaches. Alternative fusion approaches, especially those based on the application of *wavelet transforms*, can be very effective in merging data while resulting in little spectral distortion (see, e.g., Amolins *et al.*, 2007). Wavelet transforms are essentially extensions of the concept of high-pass filtering, enhancing the high-frequency (short-distance) variation in the spatial domain while retaining local spatial information (unlike Fourier transforms that are applied equally to the whole image; see Section 6.5.2).

It is of course also possible to make quantitative use of any finer detail obtained from other types of information where we have it (e.g. with high-resolution data on say vegetation distribution), to infer the subpixel behaviour of another variable, e.g. temperature, on the basis of a linking model (in this case of emissivity and evaporation). These and other fusion techniques are implemented within most commercial software. A very comprehensive overview of the fusion of images of different spatial resolutions is given by Wald (2002).

6.4.7 Data assimilation

An extension of the data fusion approach is to combine remotely sensed variables such as surface temperature or vegetation cover with models of underlying biophysical processes in the process known as *data assimilation* in order to estimate variables such as land-surface energy fluxes or soil-moisture content that are not directly observable remotely (e.g. Reichle, 2008). It follows that data-assimilation techniques are particularly useful in studies of vegetation function including estimation of CO_2 and water-vapour fluxes. In general the combination of observation with functional models allows one to interpolate and extrapolate observations to temporal and spatial scales of relevance to any particular study. Applications could include the merging of low-resolution satellite data with models of evaporation to improve the spatial resolution of estimates of soil moisture, while merging of infrequent leaf-area estimates from Landsat data with a crop-growth model allows prediction of crop growth and yield at any time.

There are many approaches to data assimilation ranging from the use of simple deterministic models through to those that incorporate stochastic variation and that treat the errors both of the model and of the observation. The basic approach to the treatment of errors in data assimilation is illustrated in Fig. 6.17 for the case where we have a model that gives a prediction of the current state of a system (m) on the basis of a modelled extrapolation from the previous state, and we also have an observed value (o) estimated from the satellite data (again usually involving application of a model to derive the value for the variable of interest). From these we wish to obtain a best estimate ($\hat{x}$) of the true state of the system ($\bar{x}$). Given that both o and m have errors or uncertainty (σ_o^2 and σ_m^2) associated with them (see Chapter 10 for a more complete discussion of uncertainty and the meaning of the variance terms, σ^2), it is likely $\bar{x}$ lies somewhere between o and m. In fact it should be closer to the value with least error and the best estimate is given by

$$\hat{x} = (1 - K)\,m + K\,o, \qquad (6.5)$$

where K is the *Kalman gain* ($= \sigma_m^2/(\sigma_m^2 + \sigma_o^2)$) and is a weighting factor that determines how the best estimate relates to the modelled and observed values (Fig. 6.17); a value near 1 arises when $\sigma_m^2 \gg \sigma_o^2$ so that our best estimate of x is close to the observed value. For complex systems the approach can be extended to include the use of more sophisticated algorithms that combine the modelled and observed data, including the use of approaches such as the extended Kalman filter and particularly the ensemble Kalman filter (see, e.g., Bouttier and Courtier, 2002; Evensen, 2002). The latter approach particularly is suitable for the estimation of the large sets of parameters required in complex climate and ecosystem models where the system shows both chaotic and non-linear dynamics. Examples include the assimilation of coarse resolution satellite passive microwave data into a soil water and crop model to improve regional crop forecasts (e.g. de Wit and van Diepen, 2007), and in the estimation of CO_2 fluxes using ecosystem models (Quaife *et al.*, 2008).

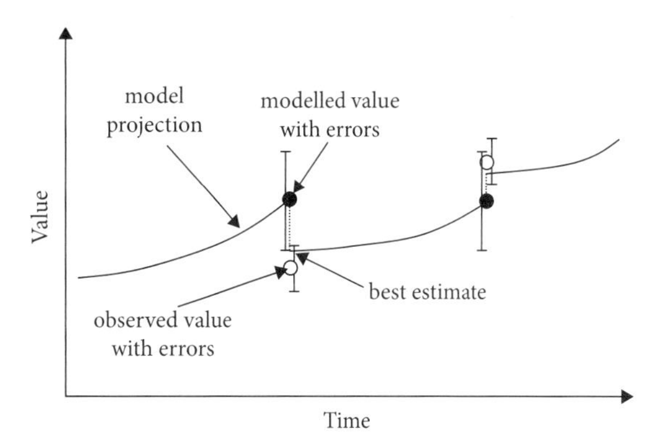

Fig. 6.17. Illustration of the use of data assimilation where a variable is predicted on the basis of forward projection of a model. The trajectory is then modified at intervals by observations, with the amount of adjustment depending on the relative confidence in the modelled and observed values.

6.5 Image interpretation

There is a wide range of techniques available for interpretation of remotely sensed images in terms of recognition and identification of specific features or their characteristics. Analogue photographs are interpreted (*photointerpretation*) visually by looking for key features and patterns within the image. A skilled interpreter can frequently deduce features such as topography from the arrangement of shadows, distinguish between roads and rivers, distinguish between crops, identify desert areas or built-up areas, and so on. With modern digital imagery an extensive range of tools is available to automate image analysis including the use of *pattern recognition* and spectral or textural classification approaches. The most straightforward methods for the extraction of thematic information about the areas imaged will be introduced here, with a more detailed discussion of classification deferred to Chapter 7. The most sophisticated techniques use artificial intelligence approaches or involve the use of training techniques such as artificial neural networks (ANNs) to recognize particular image features or regions.

6.5.1 Classification

Most of the previous procedures have been performed in order for the analyst to more easily extract the information he requires from the image. Some features are distinctive enough to be recognized in a panchromatic or single-band image, but usually it is the difference in reflectance of different surfaces in different wavebands that enables subtle variations to be noticed and thus to enable detailed identification to be made. We have already seen that vegetation reflects green light quite strongly, red light very little, and near infrared very strongly so that a false-colour composite of these three bands enables gross vegetation features to be distinguished, although not necessarily unambiguously. A careful study of small areas in each of the three image bands separately might also enable similar identification, but it would be tedious. What is needed therefore are automated techniques that can identify different surface types in an image and produce a thematic map with little or no user intervention. This process is called *classification* and usually depends on analysis of the spectral information on a pixel-by-pixel basis. In the usual approach the computer algorithm attempts to classify each pixel into one of a range of possible landcover types on the basis of its characteristic spectral reflectance, though other information can also be used. Spectral classification relies on the fact that very few surfaces reflect equally in all wavebands and a number of different algorithms have been developed to differentiate between different classes of landcover through their different spectral reflectivities, their *spectral signatures*. Since most classification procedures are based on spectral information we will leave a detailed discussion of this until Chapter 7.

6.5.2 Spatial and texture analysis

Another way of improving the analysis and increasing the amount of useful information is by considering spatial information such as *context* (the relationship of a pixel to nearby objects and features) and *texture* (a measure of the homogeneity of neighbouring pixels).

Context: When one looks at a picture, one does not look simply at one point, but observes that point in the context of its surroundings. A single piece of a jigsaw puzzle may be difficult to identify by itself, but when inserted in its proper place, the 'bigger picture' can be seen. Contextual analysis takes account of spatial relationships between the pixels in an image and the classes assigned to them. Usually, standard classification is performed first, but then a pixel can be reassigned on the basis of where it is and what is surrounding it. For example, it is unlikely that a patch of cabbages will occur in the middle of water. Some analysis may be undertaken visually, but it is also possible to use a moving-window algorithm (similar to a filter, see Section 6.4.3) to compare the class allocated to a particular pixel with the class of its neighbouring pixels. For example, the class of the central pixel in a 3×3 window is compared with the classes of the other eight pixels in the window. If it is not a member of the majority class (in this case five or more pixels) then it is reallocated to that class. This will smooth out the classification by weeding out rogue pixels due to signature variation as well as noise or speckle. Such smoothing needs to be applied with care, especially in highly heterogeneous images.

Texture: Classical statistics assumes that data are independent. For a regular grid of pixels such as in a remotely sensed image where there is usually a reasonable degree of spatial dependence, it is likely that neighbouring pixels will be more similar than distant ones. The variations in tonal patterns in a photograph will give some indication of the roughness or smoothness of an area or the distribution of features. A field of cabbages will 'look' different from a field of grass, even though the colour may be similar. The texture of the image contains important information about the spatial arrangement and structure of components and quantifies grey-level differences (contrast), the scale of changes, and any directionality in the variation. In general, texture measures are calculated on a pixel-by-pixel basis giving local values for the texture measure used and they can therefore be represented as a new image that can then be used as an extra band to aid classification of the image. It is worth noting that the observed texture depends both on the scale of the underlying variation we are concerned with in the real world (e.g. the pattern of forestry types at km scales, the pattern of individual trees at 10 m scales, and the patterns of individual leaves at cm scales) and on the spatial resolution of the images (one cannot detect leaf-scale, or even tree-scale variation in standard 250-m MODIS imagery). It follows that the spatial resolution of the imagery needs to be at an appropriate scale to detect the texture of interest.

Texture measures calculated in the spatial domain are often subgrouped into *first-order statistics* (occurrence) and *second-order* (co-occurrence) statistics (see Table 6.1). First-order measures are derived from the histogram of pixel intensities in a given neighbourhood (as defined by a moving window) and take no account of spatial relationships; these include values such as mean, range, variance, skewness, standard deviation, and entropy. Second-order statistics on the other hand are calculated from what is known as a *grey level co-occurrence matrix* (GLCM), which indicates the relationships between the grey levels of pixels in specified directions or distances (Haralick *et al.*, 1973). Each element of the GLCM is a measure of the probability of occurrence of two greyscale values separated by a given distance in a given direction. If one used the full 256 grey levels of an 8-bit image it would result in a 256 × 256 matrix representing all the possible combinations of levels, which would not only be computationally very intensive but would also contain many zeros (representing combinations that do not occur), thus limiting the statistical validity of the approach. Therefore it is common to reduce the number of grey levels used (e.g. from 8-bit (256 levels) to 5-bit (32 levels) or even to 4-bit).

Practical application of the method requires a symmetrical matrix so the estimation is always done in both a forward and a backward direction and the resulting matrix is normalized to express values in each cell as probabilities (*P*). In fact one needs to derive four separate GLCMs to describe the co-occurrence of adjacent pixels in each of the four directions (horizontal neighbours (0°), diagonally right (45°), vertically (90°), and diagonally left (135°)). Although a wide range of second-order statistics can be derived, only a few are commonly used and considerable skill is required to select the texture measure, window size, input channel (which may often be a derived variable such as *NDVI*), and spatial separation appropriate for any specific study. Examples of second-order statistics include the angular second moment, contrast and correlation; generally the texture value for any pixel is then derived as the average of the calculated texture values for each of the four directions. Some of these texture measures and their properties are summarized in Table 6.1. The higher-order statistics are particularly useful because of their ability to provide information about the spatial dependency between pixels.

An example showing textural characteristics of different vegetation surfaces is shown in Fig. 6.18. This figure shows that the spatial characteristics of different vegetation communities in the northern Chihuahuan Desert of New Mexico are clearly distinguishable visually. In this case, rather than calculation of a new texture image, results are expressed simply as box-plots of the means and ranges of a particular texture value for each of the vegetation types.

It is possible, at least in principle, to perform textural analysis on an image and to use the derived values as additional bands (analogous to the spectral bands) and then to use this additional information to enhance the power of classification procedures. This may be considered as effectively a form of context analysis. The use of *spatial statistics* (or *geospatial statistics*, since it is applied to the environment) provides a powerful tool for analysis of spatial variation in an image (Foody *et al.*, 2004). There are several ways in which texture can be assessed in addition to the first- and second-order statistics outlined above. These include the following.

1. *Semi-variogram* (sometimes just called a *variogram*). This is effectively a plot of variability against ground distance. The semi-variance is

Table 6.1 Some image texture measures (see, e.g., Jensen, 2005; St. Louis *et al.*, 2006; see also http://www.fp.ucalgary.ca/mhallbey/tutorial.htm). These are calculated on moving windows (which can be 3×3, 5×5, or larger sizes) centred on the pixel of interest.

	Formula	Comments
1st-Order measures[1]		
Arithmetic mean ($\bar{x}$)	$\bar{x} = \frac{1}{w}\sum_{i=1}^{w} x_i$	Where w is the number of pixels in the window; acts as a smoothing filter
Standard deviation (s)	$s = \sqrt{\frac{1}{w}\sum_{i=1}^{w}(x_i - \bar{x})^2}$	Indicates variability within the window
Range (r)	$r = \max(x_i) - \min(x_i)$	
2nd-order measures[2]		
Angular second moment (ASM)	$ASM = \sum_i \sum_j \{p_{i,j}\}^2$	A measure of homogeneity or orderliness
Contrast (CON)	$CON = \sum_i \sum_j p_{i,j}(i-j)^2$	Emphasizes cells away from the diagonal; matrices with most cells on the diagonal show little contrast
Entropy (ENT)	$H = -\sum_i \sum_j p_{i,j}\log(p_{i,j})$	Another measure of orderliness; sometimes classified as a first-order measure

[1] 1st-order measures take no account of the spatial relationships between pixels. In these, x_i is the grey level in the ith pixel in the moving window and summations are done across the w pixels in the window.

[2] 2nd-order measures are calculated on the GLCM co-occurrence matrix where $p_{i,j}$ is the (i,j)th entry of the normalized GLCM and indicates the probability in that cell (defined as the number of occurrences in that cell divided by the sum of all the values in the table); summations are across i and j. For an image reduced to 8 quantization levels, i and j would each vary between 0 and 7.

defined as half the average squared difference between the pixel values separated by a given distance (the *lag, h*), and, when plotted as a function of log h shows a characteristic shape for areas of different textures (such as urban and agricultural areas, for example). This can be done for several wavebands (e.g. red, green, infrared), and *NDVI*. Modelled curves can be fitted to such variograms in order to help analyse the image. Because larger pixels will tend to smooth out variability, different curves will be obtained using images of different spatial resolution, and it may be possible to use such curves to select the optimum pixel size for a particular purpose. For further information on the use of semi-variograms the reader should consult an appropriate geostatistical text (e.g. Foody *et al.*, 2004).

2. *Fractal analysis.* In the natural world, large-scale features often possess a certain degree of similarity in geometric form and complexity to small-scale features. Natural surfaces can therefore be considered to be *fractal* in nature and fractal analysis may be useful for their study. In Euclidean geometry, a two-dimensional figure is a plane and a three dimensional figure is a solid. Fractal geometry, however, permits non-integer dimensions.

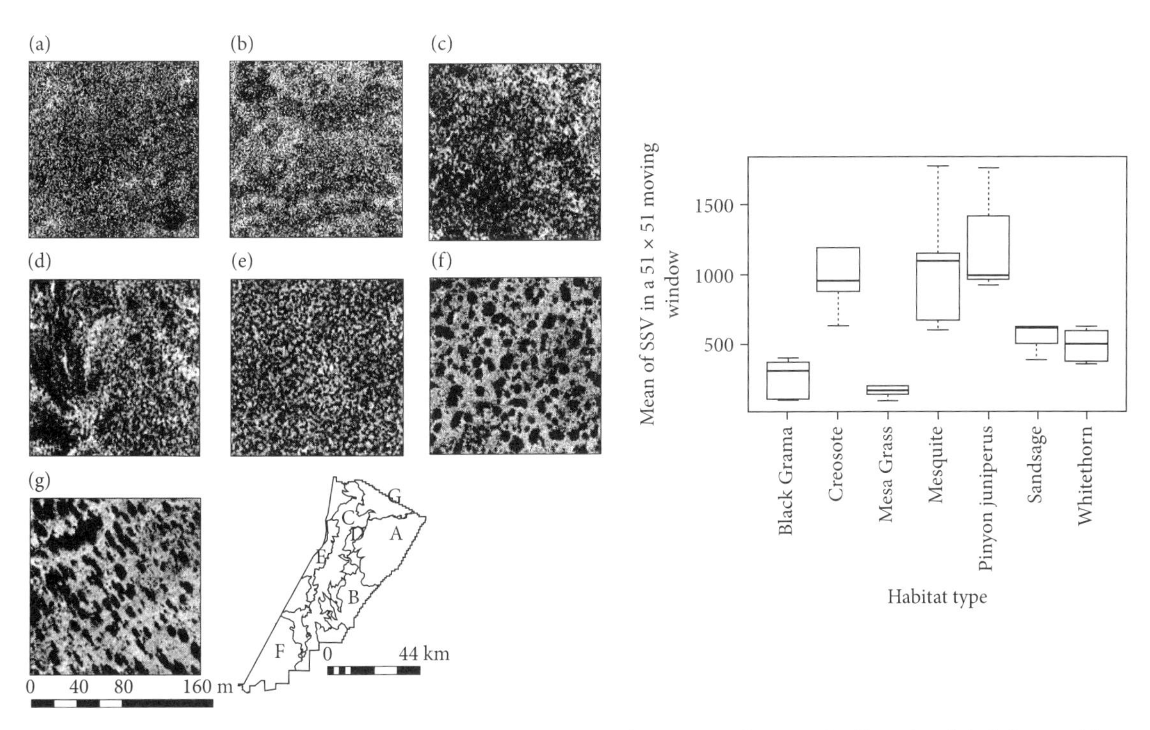

Fig. 6.18. (a) Areas of the seven main habitat types in the northern Chihuahuan Desert of New Mexico illustrating their distinct spatial patterning (textures): (A) black grama, (B) mesa grassland, (C) creosote, (D) whitethorn, (E) sand sage, (F) mesquite, and (G) pinyon-juniper. (b) An example of the average behaviour of one texture measure (sum of squares variance (SSV) across a 51 × 51 moving window) for the different vegetation types in (a). (With permission from St. Louis *et al.*, 2006.)

The Fractal dimension measures the degree of irregularity or complexity of an object. A dimension between 2 and 3 indicates the degree of space filling that occurs. A very rough surface fills more of the third dimension in space than does a smooth one. The fractal dimension may therefore be used to characterize surfaces with different textures. A lot of recent work has taken place to assess the use of fractals (actually multifractals that scale differently in different directions are a better measure of environmental texture) in image classification (Parrinello and Vaughan, 2002; Parrinello and Vaughan, 2006). Fractals can also be used for image compression.

3. *Fourier analysis.* The Fourier transform of a quantity that varies with time is a function of the frequency components that comprise that quantity, usually expressed as a sum of sine and cosine components having different amplitudes. Spatial variation within an image can also be considered in terms of frequencies; short-range variations occurring more rapidly with distance (therefore having a higher frequency) than long-range variations. The Fourier spectrum of an image will therefore be characteristic of the spatial scale components in the image and of the texture of the land surface. Frequency-domain filtering can be carried out using a *fast Fourier transform* either on the whole (single band) image or in a window of interest. As well as being used to analyse images by studying the distribution of spatial frequencies, this concept can also be used to design filters for special purposes.

4. *Wavelet analysis.* The use of Fourier analysis in remote sensing is limited since it assumes that the image repeats itself in all directions to infinity, that the mean and variance of the pixel values are constant over all regions of the image, and that the same frequency mix is present in all parts of the image. Wavelets are similar in concept to the frequency components in a Fourier transform, but a discrete wavelet transform decomposes the input signal in terms of space and scale simultaneously.

Larger window sizes are used in regions where there is low-frequency information and smaller ones where there is high-frequency information. This enables local analysis of large images to be performed. Hence, wavelet analysis is in many ways more suitable than Fourier analysis in the study of complex images that contain a wide range of frequency components. The spectrum of wavelets can again be used to analyse the texture at different scales. Wavelets have also been used in data compression and in noise filters.

6.5.3 Object detection and analysis

As an alternative to the essentially pixel-based analysis that is used in conventional spectral vegetation index and spectrally based classification studies one can attempt to identify groups of pixels that form discrete objects on the basis of characters that might include overall shape or texture as well as their spectral similarity. Sometimes called *geographic image retrieval* (GIR), or *object retrieval*, the approach seeks to retrieve regions in the image that have similar content to a known target object (Purves and Jones, 2006). Although analogous to the supervised classification approach (Section 7.4.2), it makes use of patterns of information relating to groups of related pixels. The features sought could include as well as, or indeed instead of, the spectral information, the shape, arrangement, or texture. The approach is particularly efficient when seeking to identify (usually repetitively) specific objects that occupy only small areas (e.g. trees in an orchard). This avoids the need for complete classification of the whole image where one has a specific interest in one component.

A particular advantage of the use of objects is that it is possible to make use of textural as well as spectral information in the analysis of the objects, thus giving much more information than is available from per-pixel spectral analysis alone. An example of such an approach would be to investigate the spatial relationships within and between the blocks of any class (e.g. the typical size or shape of homogeneous blocks or the distances between them). This analysis can be achieved by the use of *fragmentation statistics*; these can be particularly useful for landscape analysis and will be discussed in Chapter 11.

Further information on object identification and analysis may be found in appropriate image-analysis and machine-learning texts (e.g. Soille, 2002).

6.5.4 Scale and scaling up

Related to the above consideration of texture and geostatistics is the problem of *scale*. We all think we know what is meant by scale, but the usage of the term across disciplines, and even within a discipline, is largely tacit. 'Large scale' might imply fine detail say to a cartographer, but large area (and hence coarse detail) to a geographer. The particular definition being used is usually obvious from the context and we will not go into semantics here. In remote sensing, scale is inherent in the idea of resolution. Usually, large-scale phenomena occur over long temporal scales (in part because space and time are linked through transport mechanisms) and a similar covariance was seen in Section 5.7.3 where we were discussing trade-offs between spatial and temporal resolutions.

Scale is a real issue when relating the different datasets in a GIS, and relates to both the extent and the level of detail. A map is basically a single-scale product, whereas a GIS can operate at multiscales, and so must accommodate *multiscale* data. Natural phenomena almost always involve multiple processes, so we need multiscale observations. Scale affects both the extent and the level of detail. We must consider the characteristic scale, both spatial and temporal, of the earth process or phenomenon that we are studying. Different levels of information relate to different scales so we must decide on the best scale to use and how to relate data and information at different scales. Geostatistics can sometimes help to decide on the optimum scale. The shape of the variogram or the change in fractal dimension (which changes with scale because natural landscapes are not truly fractal over a range of scales) can be used to indicate the scale of a particular process in the environment. The texture within an image can be quantified by measuring the standard deviation within a 3×3 moving window. This produces an index that can effectively measure the spatial variability of the image data. The difference in texture of images at different scales and resolutions will indicate the heterogeneity of the area.

The second problem, that of the relationship of data at different scales, is central to the consideration of *scaling up* or *up-scaling* such as in relating modelled values of canopy reflectance to remotely sensed values.

This problem is particularly important when integrating different types of data as in a GIS (or in an integrated GIS – IGIS). How do you integrate point data (such as field measurements or rain gauge measurements) with spatial data (such as satellite images or climate), or spot heights from a DEM? There have been many attempts to do this (e.g. interpolation, kriging, etc.) over the years.

Of particular relevance to the discussion in Chapter 3 is the extension of models across a range of coarser scales. For example, the modelling of the reflectance of a single leaf needs to be related to that of a tree, and that in turn to a forest and on to the response of global vegetation. The relationships between spectra and biophysical variables at one scale are not the same as those at another scale. Relationships between say *NDVI*, cover types as those and elevation are scale dependent. Fine scale data may be aggregated to give coarser data (e.g. averaging $n \times n$ pixels in a SPOT image will give a lower-resolution image with less variance – the process has smoothed out the finer detail). The converse, attempting to obtain information at a smaller scale, is mentioned in the consideration of unmixing spectrally mixed pixels (Section 7.4.2) and above in data fusion. We will return in Chapter 10 to a more detailed discussion of scale and especially the problems related to transitions between different scales of observation.

6.5.5 The multiconcept

Above, we have considered several ways in which multispectral datasets can be combined so as to improve the success of data extraction. This is an example of *synergy*, where the results obtained from using several sets of data in combination are better than the sum of the results obtained from analysing each set individually. There are other ways of combining datasets:

Multispectral – The use of multispectral data has already been briefly mentioned earlier in this chapter, and its analysis is discussed in detail in Chapter 7. It is one of the main tools for measuring and monitoring vegetation type, health, and change.

Multitemporal – Multitemporal datasets (data obtained of an area at different times) can be used to analyse changes that may have taken place between observations using subtraction or by forming colour composites. Pixels that have similar values in both images will have differences of around zero, whereas pixels that have changed will have large differences. These data can be suitably scaled and used to provide a visual display of the areas in which change has occurred during that time period. Similarly, a colour composite in which one band from each of the time sequence of images is displayed on each colour gun will show regions of change as distinctive colours, whereas regions that have not altered will have the same intensity on each gun and therefore show up as pastel shades. Such sequences can also be used to detect motion. The displacement of features in successive images can permit the visualization of cloud motion or the motion of certain sea-surface temperature features. By measuring the displacement in successive images, and knowing the time interval between the acquisition of the images, quantitative estimates of velocities can be made. Pattern-recognition methods, sometimes using neural networks, are now employed to perform this automatically since they have the ability to track not only linear movements of the features, but also directional changes (e.g. rotation). Another use of time series is in the generation of maximum-value composites. It is unlikely that, in any one AVHRR image say of Europe or even just Britain, there will not be areas obscured by cloud. But over a week or ten-day period, it is probable that most areas will be cloud free in at least one image. A composite image (*maximum value composite* – MVC) is formed by selecting, on a pixel-by-pixel basis, the value that is best in all the images collected over that period.

Another feature of multitemporal imagery is its power for identifying species that have differing phenological trends (see Chapter 11). For example, winter cereals can be distinguished from spring cereals by their differing ground cover in winter, or different tree species may be identified on the basis of their characteristic seasonal developmental changes.

Multilook – Multilook or multiangular data, images of an area taken from different viewpoints, have been used in stereoscopy for many years, but can also be used for atmospheric correction (either from different orbits of the same satellite e.g. NOAA AVHRR, from data from different satellites, such as CZCS and AVHRR, or from the conically scanned ATSR on ERS and Envisat, see Section 6.2). Interferometric SAR (Section 6.6.4) also relies on the physical displacement of the receiving antennae between different images in order to construct an interferogram from which elevation data can be obtained or regions of change detected. The use of multiangular data for investigation of vegetation properties is developed further in Chapter 8.

Multi datasets – Lastly, different datasets can be combined, that is to say images from different instruments. These images must have the same spatial resolution, and so may need to be resampled, and must be coregistered (Section 6.2), so that there is a correspondence between the pixels in each image. These can then be combined say as colour composites, can be used in a classification procedure, or used for data fusion (see Section 6.4.5). A more general form of this type of combination of multi-datasets is in a geographic information system (GIS) in which many different types of spatial data, map, socio-economic, infrastructure, land use, classified images, etc., can be displayed for analysis.

6.6 Radar image interpretation

The properties of microwaves were covered in Chapter 2 and their interactions with soil and vegetation were discussed in Chapter 3. In this section we will see how radar images are corrected and interpreted (see Woodhouse (2006) and Lillesand *et al.*, (2007) for detailed information). SAR data are extremely complex and require much processing in order to produce an image-like representation that can be used for interpretation, which is highly computer intensive. Much of this pre-processing has already been carried out by the agency before most users receive their SAR 'images'. Raw SAR data consist of intensity values as a function of direction, frequency, and delay time. The delay time suffered by a pulse tells us about the distance travelled to the reflecting object, the Doppler shift in frequency gives information about the location of that target in the along-track direction, and the look angle gives information about the across-track position. This enables the data point to be located in a *ground resolution cell* (the equivalent of a pixel). But, due to the high obliquity of the image, further geometric processing is required to produce a vertical view of the area imaged. It also needs radiometric correction to relate the image intensity to the radar backscatter coefficient and to take account of the reduction in backscattered signal strength as the look angle increases in the far field (not only does the distance to the target increase, but backscattering is strongly dependent on the viewing angle). The final image can be treated like a conventional digital image and subjected to various processes to enhance and aid interpretation.

6.6.1 Radar geometry

A radar image looks superficially like a high oblique aerial photograph taken with low sun angle. In areas of high relief, there appear to be sharp shadows that highlight the structural details, but this analogy is an illusion. There are certain quirks in the geometry of the image caused by the side-looking view, particularly in high-relief terrain or when imaging tall buildings; these need to be understood before the image can be interpreted (Fig. 6.19). The area of ground immediately behind a tall object will reflect no signal back to the receiver. This is termed *radar shadow*. This is not like in optical images where 'shadow' means low level of illumination, but where information is present. In the case of radar, no microwaves are incident on this area and so there are none to be scattered – there is no information in this region of the image. Also, the leading side of a hill appears to be shorter than it should be – *foreshortening* – and the downward slope appears to be extended. This is because the difference in slant distances from the satellite to the foot and to the top of the hill are smaller the more oblique the view. If the hill is very steep, the distance to the top may be less than to the foot and the reflection from the peak is received before the reflection from the foot. This makes the hill appear to curve towards the viewer[31] and is termed *layover*. Another strange effect occurs when a moving object is viewed. Image synthesis is based on the relative motion of the platform and the ground to locate objects. It assumes that all points on the ground are stationary. If a moving target, such as a boat or train, is imaged, the relative motion between that object and the ground confuses the system and it may be imaged in the wrong place. Examples of this are ships that appear to be separated from their wakes, and trains appear to be travelling along in the field adjacent to the railway tracks.

31. In an aerial photograph, the converse happens where the effect of perspective is to make tall buildings appear to lean *outwards* towards the edge of the photo.

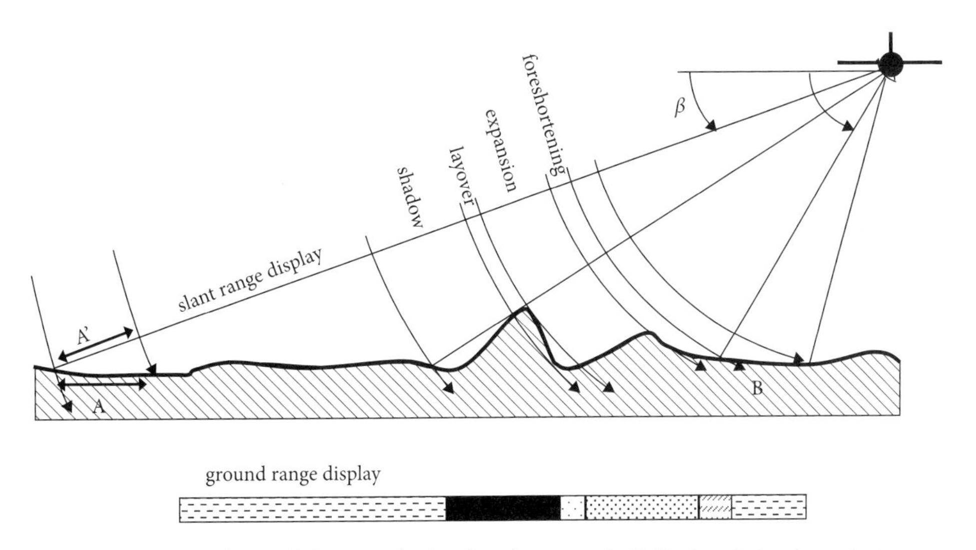

Fig. 6.19. Illustration of some of the geometric distortions that occur with SLAR when displayed on a slant range display, showing layover, foreshortening, and radar shadow. Also shown is the corresponding ground range display which differs from the slant range display.

6.6.2 Speckle

Due to the coherence of microwaves, interference effects occur between the signals returned from different scatterers within a resolution cell. This is due to different distances travelled by the component wave and produces a random variation in brightness, in other words, noise. The normal way to reduce noise in an image is to use a smoothing filter (Section 6.4.3), but this can reduce spatial resolution. For speckle reduction, *adaptive filters* are usually used in which the weights within the moving window are adjusted depending on the severity of the noise (normal non-adaptive filters use the same weights over the whole image that smoothes away high-frequency information as well as noise). The weights are calculated anew for each window position based on the mean and variance of the grey level within the window, or else by using a discrete wavelet transform. The degree of smoothing therefore depends on the local image statistics and so an adaptive filter is more likely to preserve subtle details such as edges or high texture areas. Another way to reduce speckle is to use *multilook* processing. This makes use of several independent images of the same area produced using different portions of the synthetic aperture (*looks*) that are averaged to give a smoother image. Since each look is produced by only a part of the synthetic aperture, the spatial resolution of the image is correspondingly reduced in relation to the number of looks.

6.6.3 Types of radar imagery

Radar images can be obtained in a number of different formats depending on the amount of processing to which they have been subjected:

Single-look complex (*SLC*) – This has been produced using the full length of the synthetic aperture and therefore has the highest resolution, but also has the maximum amount of speckle. It is complex in that it consists of both a real component (representing the amplitude of the signal) and an imaginary component (representing the phase, which is required for interferometry). This is the nearest to raw data that most users will get. It has not been reprojected onto a reference surface and remains in slant range coordinates, and, since the ground resolution cells are rectangular, it will appear distorted when displayed on a screen with square image pixels.

Multilook detected (*MLD*) – This has been processed using the multilook procedure described above. Usually between three and six looks have been used, which provides a good compromise between spatial resolution and speckle. Because the image has been produced by averaging, the phase information is no longer present. The data are in the form of real digital numbers that can be converted into a calibrated $\sigma°$ value using the calibration data provided. It can be in either ground range or slant range coordinates, but has not been fully geocorrected.

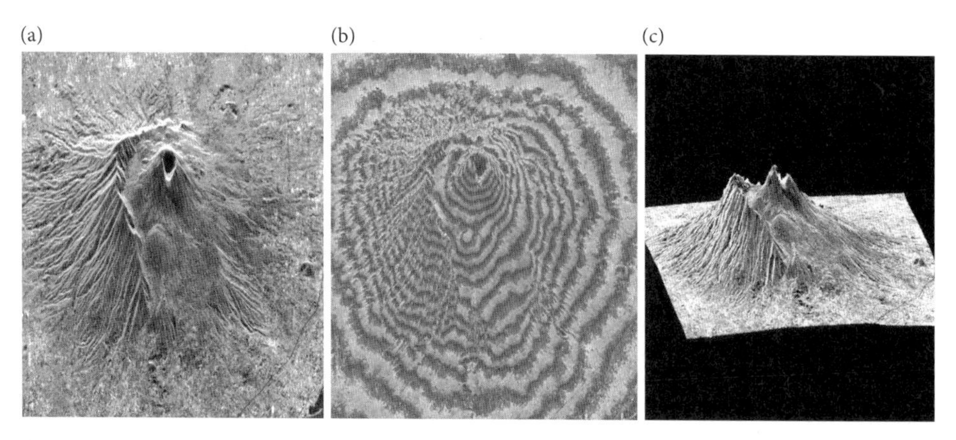

Fig. 6.20. (a) A SAR image of Mt. Vesuvius from ERS-1, (b) a corresponding interferogram, and (c) a 3D rendering of the derived Digital Elevation Model (images courtesy of ESA – European Space Agency). (See Plate 6.8)

Precision images (*PRI*[32]) – These are multilook images which have been resampled into square pixels, rotated to take account of the view direction of the detector, and warped so that they are properly georeferenced to some geographical reference system. They are not geocoded, so still show topographic distortions. Geocoded data will have made use of a digital elevation model (possibly produced by the radar system itself, see below).

6.6.4 Interferometry

The coherence and extreme electronic stability of radar systems permit the interference between different signals to be used to measure distances very accurately (see Section 2.6). This requires two measurements separated in either space or time (or both). From aircraft, a second receiver is placed as far as possible from the transmitting antenna, at the other end of the aircraft or on a boom, such that the signals received by the two receivers have travelled different distances and will therefore have a relative phase difference. From space, the stability of the systems is so great that data from different orbits, separated in time by hours, or even weeks, can provide the necessary baseline separation to produce interferograms. The tandem mission of ERS-1 and ERS-2 (Section 5.8.4) provided a baseline separation of about 1 km, but in general a baseline of about 200–300 m is optimum. Not only must the frequency be very stable, but the positions of the two receivers must be known very accurately. An added problem, particularly severe from space, is that an interferogram produces contours one wavelength apart, but the absolute distance travelled by the wave is unknown. A process of *phase unwrapping* is required to obtain height information. In this, taking one point as a baseline, since each fringe is separated by one wavelength (i.e. a phase difference of 2π) then by counting the number of fringes between two points their height difference can be obtained (Fig. 6.20). In order to obtain absolute heights, it is necessary to know very accurately the height of at least one point in the image. It is also worth noting that two radar images must be coregistered to within one-tenth of a pixel before 'interfering' them. This is a rather simplistic description, and the reader is referred to Woodhouse (2006) for more information.

6.6.5 Image interpretation

Various theoretical models have been developed to try to describe how objects reflect microwaves, but most information is still obtained from empirical observations. When considering the backscatter from various objects in the image, the characteristics of both the object (texture, electrical properties, moisture content) and of the system (wavelength, polarization, incidence angle) need to be considered. The unusual geometry and information content make radar images more difficult to interpret than conventional optical images. Multispectral classification is no longer possible – each image is strictly monochromatic (formed

[32] The acronym *PRI* is also used for *photochemical reflectance index* and, in electronics, for *pulse repetition interval.* It is usually obvious from the context as to which is being referred to.

using one single wavelength). Colour composites can, however, be produced by displaying images of different wavelengths (e.g. C band, S band, and P band) on each of the colour guns of a computer. It may then be possible to distinguish between different surfaces by their different scattering properties at different wavelengths. Similarly, their different responses to microwaves of different polarizations can be studied in the same way. Images of the same wavelength and polarization, but taken at different times will show changes that have taken place (flooding, coastline, deforestation) showing up in distinctive colours that can indicate the direction of change. In all these techniques, the sets of images must be carefully coregistered and, if possible, radiometrically matched. Colour composites can be formed using one radar image and two multispectral images, and a radar image can be included as one of the bands in conventional classification.

The most obvious type of information in a radar image is structural. Rivers, mountains, roads, and railways show up clearly, as do built up areas. The geometrical shapes of field boundaries are very distinctive and the textural differences of different landcover types (forests, pasture, scrubland) are usually very apparent. One of the major uses of interferometric SAR (InSAR) is in the production of digital elevation models. A special Shuttle mission (the Shuttle Radar Topography Mission, SRTM) in 2000 collected single-pass[33] radar interferometry data between about 60° N and S that includes about 80% of the land area of the world. The system collected 12 terabytes of raw data during the eleven day mission and has produced DEMs of much of the area with absolute horizontal and vertical accuracies of about 20 and 16 m, respectively. Optical satellite images can be draped over DEMs to produce perspective views of the area, allowing flythroughs to be generated. As well as for DEM creation, InSAR data can be used as input to 3D mapping programs such as NEXTMap® (http://www.intermap.com/nextmap-digital-mapping-program). Other uses of interferometric data are for studying changes in landform due to coastline changes, subsidence, or volcanic activity (bulges can sometimes be seen on the side of volcanoes when they are about to erupt).

Radar images can be used cartographically to produce maps in a process called *radargrammetry*, and the parallax between two images can be used to provide a stereoscopic view that can be used to calculate the heights of features (*stereo radargrammetry*).

Further reading

A very readable, but nevertheless thorough, account of image processing is given in 'Computer processing of remotely sensed images' by Paul Mather (2004), which he describes as a "relatively gentle introduction to a subject which can, at first sight, appear to be overwhelming to those lacking mathematical sophistication, statistical cunning or computer genius". As such it is descriptive rather than numerical, yet nevertheless it is certainly not superficial. Useful but briefer discussions may be found in Lillesand *et al.* (2007) or in Jensen (2005). A more complete coverage of the more mathematical concepts can be found in more specialized books such as those by Richards and Xia (2005), Russ (2006), and Tso and Mather (2001).

Websites

NEXTMap® 3D mapping programme: **http://www.intermap.com/nextmap-digital-mapping-program**

SunAngle solar radiation calculator: **http://susdesign.com/sunangle/**

Sample problems

6.1 How many different two-band ratios can be created from the six optical TM bands? How many false-colour composites can be obtained using all seven TM bands?

6.2 The following table shows the digital numbers in a TM image in three bands for deciduous and coniferous landcover under different illumination conditions. Find a band ratio (or a normalized index) that will uniquely separate the two tree types:

Landcover/ illumination	Band 2	Band 3	Band 4
Deciduous			
sunlit	48	50	90
shadow	18	19	23
Coniferous			
sunlit	31	45	54
shadow	11	16	24

33. Single-pass interferometry uses two receivers on the same platform, rather than using data from different orbits. In the case of SRTM, the second receiving antenna was located on the end of a 60-m long telescopic boom that was extended when the Shuttle was in orbit.

6.3 Below is a small extract of the pixel values observed within an image:

169	157	162	140	140	139
159	158	55	156	144	157
146	157	155	148	164	86
150	221	143	255	88	90
145	153	152	82	90	95
158	164	93	101	92	99

The values 55 and 221, shown shaded, appear to be erroneous. The other two shaded values, 164 and 90, straddle an edge. Apply a 3×3 filter that will correct the erroneous values, but maintain the genuine transitions due to edges.

7 Use of spectral information for sensing vegetation properties and for image classification

7.1 Introduction to multispectral and hyperspectral sensing and imaging

We have already described (Chapters 2 and 3) the ways in which electromagnetic radiation of different wavelengths can interact with natural surfaces and the resulting spectral properties of various components of vegetated surfaces. The various stages in the pre-processing of spectral images and their visualization were outlined in the previous chapter; here we will consider in more detail how one can interpret the spectral response, especially in the optical region (visible and near infrared), to derive useful information about the physical and biological characteristics of the vegetation. Further use of spectral reflectance combined with information on its angular variation to extract biophysical parameters will be addressed in Chapter 8, while some examples of specific applications will be developed further in Chapter 11.

The simplest sensors that enable one to determine the spectral responses of different surfaces are radiometers with a fixed field of view that can sample the average reflected radiation emanating from a defined area of surface; these can either operate over relatively broad spectral bands (e.g. red or green) or they may have high spectral definition as in a spectroradiometer (where each channel may have a bandwidth of a few nm). Such spectroradiometers are widely used for ground calibration of remote imagery, where the ability to average large areas can be an advantage.

For most remote-sensing applications, however, we require information on spatial variation so we collect images rather than field averages; the imagers used may range from simple broadband sensors (such as a conventional hand-held three-band red-green-blue visible camera or the TM on Landsat) to imagers with many higher spectral resolution channels, the so-called hyperspectral instruments. Conventionally, the terms superspectral and hyperspectral are used to refer to imagers that record more than about 10 or 50 separate spectral bands, respectively, though in practice there is no generally accepted distinction between multi-, super-, and hyperspectral imagery.

These multiple images (one for each waveband) can be stacked to obtain what is sometimes referred to as an 'image cube' (see Fig. 5.3). An advantage of hyperspectral over multispectral images is that they contain more information and can therefore be used to detect both narrow absorption peaks and quite subtle differences in the pattern of spectral reflectance between different surfaces. These differences can be used to discriminate rather sensitively between them. On the other hand, however, the considerable correlation that exists between the spectral responses of adjacent bands means that the amount of additional information contained does not increase in proportion to the number of bands, also the signal-to-noise ratio may be lower in hyperspectral data as a result of the reduced bandwidth.

Multispectral analysis is limited to the use of a few wavebands. Although it is possible to select a few of the narrower bands from hyperspectral sensors and analyse them in a similar way to multispectral images, a wider range of analytical approaches are available for such imagery. For example, where enough wavebands of data are used, it is possible to construct the spectral response (i.e. response as a function of wavelength) for each pixel. These spectra can then be compared with a library of spectra that have been obtained using normal laboratory spectrometers. This technique is widely used in geology for identifying minerals and can also be a powerful technique for vegetation analysis.

The use of simple visual analysis (e.g. using colour or false colour) was outlined in the previous chapter; here we discuss the main approaches available that use spectral data to extract information about the biophysical properties of the surface: these include (i) the use of simple empirical vegetation indices based on a few spectral bands only, (ii) the use of higher spectral resolution data for the extraction of relevant spectral features to quantify specific biochemical or biophysical characteristics (these can use both empirical statistical approaches and more physically based methods), (iii) the use of image segmentation or classification approaches to distinguish different surfaces, and (iv) the use of texture analysis and object-oriented image analysis that take account of patterns of variation between adjacent or nearby pixels. The more mechanistic approaches for information retrieval, based on the inversion of radiative-transfer models and the use of multiangular data, will be developed further in Chapter 8.

7.2 Spectral indices

As we saw in Chapter 3, different materials have characteristic spectra, often showing characteristic absorption maxima or minima at particular wavelengths. The complexity of these spectra means that there is a need to derive simplified approaches that can be used to determine key biophysical parameters of vegetation. *Spectral indices* are new variables generated by mathematical combination of two or more of the original spectral bands chosen in such a way that the new indices are more clearly related to biophysical parameters of interest such as canopy leaf-area index than are any of the original bands. The general spectral-index approach can be illustrated using the example of estimation of water content. Some typical spectra for soils and for leaves are shown in Fig. 7.1; this shows that both types of surface exhibit substantial dips in their spectral reflectance at around 1450 and 1940 nm. Water strongly absorbs radiation at these wavelengths and the magnitude of the observed depression depends on the water content of the target material and can therefore be used as an estimator of its water content. It follows that measurement of the absolute reflectance at a characteristic wavelength (e.g. 1450 nm for water) could, in principle, be used to estimate the water content in the target surface. Unfortunately, other components or structural features also contribute to the observed signal at any wavelength; for example for leaves, as was pointed out in Chapter 3, the presence of leaf hairs can substantially affect the observed reflectance over a wide range of wavelengths, while for canopies, interference by structural features including variation in leaf-area index (*LAI*) and leaf-angle distribution (*LAD*) substantially modifies the observed spectral signature.

In an attempt to correct for any such perturbing component it is common therefore to compare the reflectances at two fairly close wavelengths where there is a marked difference in reflectance for the material of

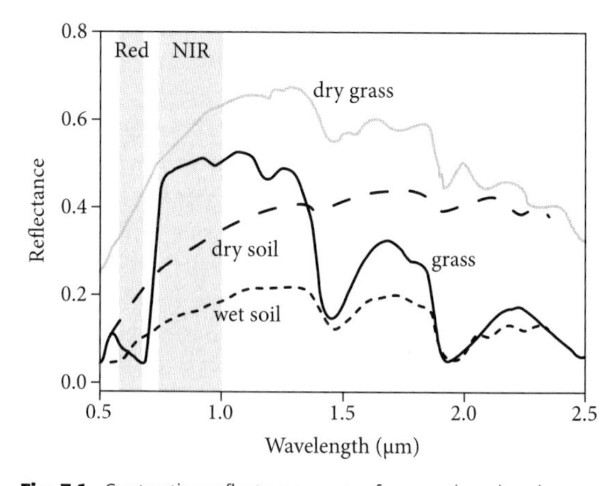

Fig. 7.1. Contrasting reflectance spectra for grass (——) and dry grass (grey line) (data from ASTER spectral library) and wet (- - - -; 0.1 MPa water tension) or dry (– – –) soils (data from Baumgardner *et al.*, 1985) showing both the sharp increase in reflectance from vegetation at c. 700 nm, the 'red-edge' (not found in soils), and the characteristic increasing depth of the water absorption features at 1450 and 1940 nm with increasing water in the tissue or the soil. The wavelengths of the red and NIR bands of the AVHRR sensor are shown as shaded areas.

interest, but where the reflectance due to other potentially interfering components does not differ substantially. Not only does this allow one to normalize the data for variation in other characters unrelated to the component of interest, but it also provides opportunities for enhancing the sensitivity of the measure by taking ratios or differences between the two readings. In the case of water, as we shall see below, it is common to compare the reflectances in a water band (1450 nm), and in a control band (e.g. at 1300 nm) where there is little radiation absorption by water. By taking ratios between these two reflectances one can derive a nearly linear relationship with leaf relative water content, while at the same time correcting for other factors such as the presence or absence of leaf hairs.

This principle of deriving spectral indices from measurements at two (or more) wavelengths is widely adopted in remote sensing, especially in the use of *vegetation indices* (*VI*) for studying vegetation cover. The use of just two wavebands greatly simplifies interpretation of what are otherwise rather complex spectra. Originally conceived as broadband indices that could be applied to the rather broad spectral channels available for example on satellites such as Landsat or NOAA-AVHRR, with the increasing availability of hyperspectral instruments, the principle has been increasingly extended to the development of more specific narrowband indices that can be used for the detection and quantification not only of vegetation cover but also of a wide range of specific leaf constituents such as pigments, proteins, and leaf water.

7.2.1 Vegetation indices and vegetation descriptors

Vegetation indices are usually dimensionless measures derived from radiometric data that are primarily used to indicate the amount of green vegetation present in a view. Most vegetation indices are based on the sharp increase in reflectance from vegetation that occurs around 700 nm (the *red-edge*), a change that is characteristic of green vegetation and not found for most other natural surfaces that show relatively slow changes of reflectance with wavelength over this region (compare the grass and soil reflectance spectra in Fig. 7.1). Over the years many vegetation indices have been developed to exploit this phenomenon; here we concentrate on the underlying principles driving the choice of vegetation index, outlining just some of the most popular indices (see Box 7.1). Further details of the range of such indices available and their applications may be found in the many detailed reviews on the subject (Baret and Guyot, 1991; Jensen, 2005; Lucas and Carter, 2008; Myneni *et al.*, 1995; Smith *et al.*, 2004; Steven, 1998; Tucker, 1979).

When one plots the reflectance in the infrared against the reflectance in the red for different surfaces for all the pixels in a typical image one obtains a cluster of points scattered over some or all of the shaded area indicated in Fig. 7.2. It is apparent that different surfaces cluster in different regions. Dense vegetation clusters at the top left-hand corner because of its high near infrared reflectance and low red reflectance, while pixels for bare soil cluster along the lower-right diagonal of this diagram. Indeed, for any soil there is usually an approximately linear relationship between ρ_{NIR} and ρ_R that has been termed the *soil line*, with dry soils being bright in both IR and R, and wet soils less reflective in both wavebands (cf. Fig. 7.1 and see also Baret *et al.*, 1993; Richardson and Wiegand, 1977).

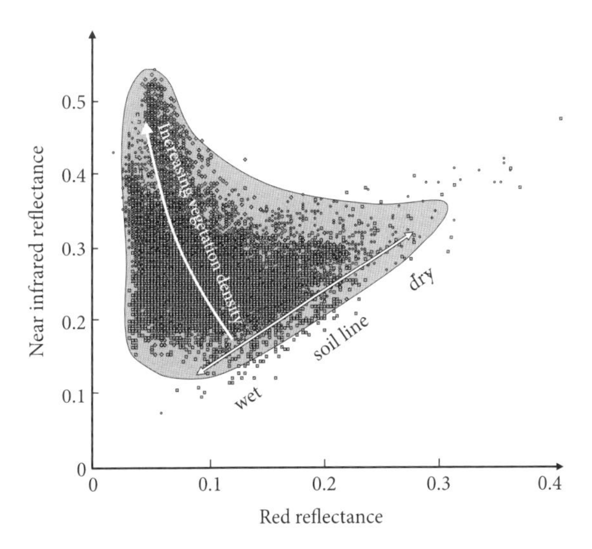

Fig. 7.2. A scatter diagram showing a typical relationship between near infrared and red reflectance for an area of mixed landcover (the grey area); points represent individual pixels from a Landsat ETM image (25 June 2003) near Evora, Portugal (path 204/row 33; courtesy US Geological Survey and Eurimage). Pixels at the top left-hand corner represent dense vegetation, while those along the diagonal at the bottom right-hand side represent soil. The 'soil line', given by the relation $\rho_N = a.\rho_R + b$ (where ρ_N is the near infrared reflectance and ρ_R is the red reflectance) indicates the range of values expected for bare soils varying between wet and dry as indicated.

Basic vegetation indices

There are many ways in which the differential response of vegetation and soil in the red and near infrared can be used to derive quantitative vegetation indices. Appropriate formulations can be used to minimize the sensitivity to irradiance and to other factors such as variation in atmospheric transmission:

(i) The simplest approach to the use of the sharp increase in reflectance of vegetation at around 700 nm might be to take the difference between the amount of radiation reflected (radiances or digital numbers obtained from the sensors) in the near infrared (NIR) and in the red (R). Because of the sharp change in reflection for vegetation between these wavelengths, the value of this difference would be large for surfaces that are predominantly vegetated and smaller for a bare soil surface. The use of reflectances rather than the raw radiances or *DN*s does much to help to reduce the sensitivity to illumination conditions. Using reflectances, the simplest vegetation index can be defined as the difference between reflectances in the infrared (ρ_{NIR}) and the red (ρ_R), giving a *difference vegetation index* ($DVI = \rho_{NIR} - \rho_R$).

(ii) Alternatively, one can take the ratio of the observed radiances, or better the ratio of the reflectances, at the two wavelengths to calculate a simple *ratio vegetation index* ($RVI = \rho_{NIR}/\rho_R$): this automatically compensates to some extent for differences in the lighting conditions, even when using radiances. The dashed lines in Fig. 7.3(a) illustrate lines of equal *RVI*, showing that they are related to the variation in *LAI*, though they are not quite parallel to *LAI*.

(iii) A third and even more powerful normalization that forms the basis for most current indices, applicable to both reflectances and radiances, is the *normalized difference vegetation index* (*NDVI*) that is obtained by dividing the difference index by the sum of the near infrared and red reflectances to give

$$NDVI = (\rho_{NIR} - \rho_R)/(\rho_{NIR} + \rho_R). \tag{7.1}$$

Note that the sum ($\rho_{NIR} + \rho_R$) represents (twice) the average reflectance in this wavelength range. Division by this factor reduces the effect of non-uniform illumination, such as that due to aspect, and thus helps to make for better comparability of the *VI* across an image. Although this index gives the same lines passing through the origin as *RVI* in NIR/R spectral space (Fig. 7.3), *NDVI* has the particular advantage that it is generally 'well behaved' and scales between 0 and 1 (except for

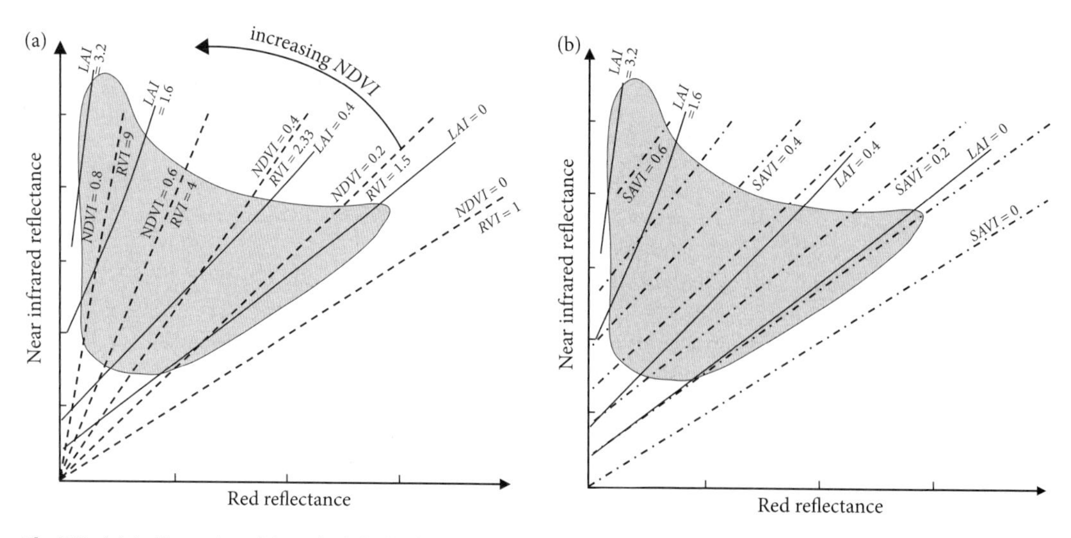

Fig. 7.3. (a) An illustration of the typical distribution of pixels (shaded area) in near infrared and red spectral reflectance space (from Fig. 7.2), where the dashed lines illustrate lines of equal *NDVI* (and equal *RVI*), while the solid lines indicate lines of equal leaf-area index as calculated for nadir viewing using the SAIL radiative-transfer model by Baret and Guyot (1991). (b) A similar diagram, but this time showing lines of equal *SAVI* (see Box 7.1 for definition).

cloud, snow, and water surfaces where $\rho_R > \rho_{NIR}$ so that negative numbers are obtained). It is worth noting that although radiances received at the detector (*DNs*) are often used in place of reflectances in this formula, the use of reflectances is strongly preferable as again this improves the correction for differences in incident radiation and for absorption and scattering in the atmosphere.

(iv) Since most vegetation indices do not tend to unity at high *LAI* (or to zero for bare soil), it is frequently useful to 'scale' the *NDVI* or other vegetation index to give a scaled *VI* (VI^*) according to

$$VI^* = (VI - VI_{min})/(VI_{max} - VI_{min}), \quad (7.2)$$

where the subscripts max and min refer, respectively, to the values for dense vegetation and for bare soil, sometimes estimated from images as the maximum and minimum positive values found within an image. The actual values for these extremes vary with index used, the particular wavebands used for R and NIR, and with canopy type and illumination conditions.

Input data (DN, radiance, or reflectance)

As should be apparent from the above discussion, satellite-derived *VI* should ideally always be calculated on the basis of corrected 'at-surface' reflectances if they are going to be truly representative of that surface. Unfortunately, many reported values in the literature only use 'at-sensor' reflectance, while some only use radiances or *DNs*. Not only will the values calculated on these different bases not be comparable, even for the same site on different days, but they will also in general lead to different relationships between the *VI* and variables such as crop canopy cover or *LAI*. Although the use of reflectances leads to *VI* that are in principle not sensitive to changes in illumination, differences in apparent reflectance that occur as view and illumination angles change can be significant; these effects are discussed in more detail in the next chapter.

Local variation in irradiance

An obvious, but little recognized difficulty in calculating reflectances arises from the fact that the amount of radiation reflected depends on the incident radiation (which itself varies locally within a scene as a function of factors such as cloud shadow and the slope of the surface in relation to the incident sunlight). Unfortunately, calculation of 'at-surface' reflectances may not adequately take account of this local variation in the input radiance, so in practice it is not always possible to derive a true reflectance for each pixel. For example with higher spatial resolution imagery, where illumination is from a different angle than the view one can often see areas within the field of view that are shaded (Fig. 7.4) and so even though their actual reflectance (defined as the ratio between reflected and incident radiance) may be the same as for a sunlit area, they will reflect less light and appear darker, so that the *apparent* reflectance for shaded areas is less than that for sunlit areas. Not only does this affect the results for whole pixels, but it also leads to errors where one has mixed pixels (as will be common even with lower spatial resolution satellite images). As shown in Fig. 7.4 this can lead to an underestimation of true reflectances. The implications of this 'high spatial resolution' effect that depends on the angle between the illumination and viewing angle are analysed further in Chapter 8.

Further developments of vegetation indices

There are many alternative formulations that have been proposed that correct for specific deficiencies of the basic *NDVI*. An extended range of the common indices and their relative advantages or disadvantages are summarized in Box 7.1. One example that aims to improve the sensitivity for dense vegetation with rather high leaf-area index, is the *green normalized difference vegetation index* ($GNDVI = (\rho_{NIR} - \rho_G)/(\rho_{NIR} + \rho_G)$), where the green band is substituted for the red band.

Other modifications attempt to minimize noise caused by variation in the underlying soil reflectance or by atmospheric absorption. For example, in low-density vegetation canopies the greater absorption of R than NIR by leaves means that the radiation reaching the soil will be progressively enriched in NIR. This results in the radiation reflected off the soil being relatively enriched in NIR with respect to R compared with the ratio observed if there were no vegetation present. This secondary signal increases with increasing canopy density. In order to correct for this effect and for the positive intercept of the 'soil line' relating to zero vegetation, Huete (1988) developed the *soil-adjusted vegetation index, SAVI* (see Box 7.1 for details). This index includes an empirical correction factor (L) which is often assumed equal to 0.5. *SAVI* attempts to account for the fact that with increasing canopy density

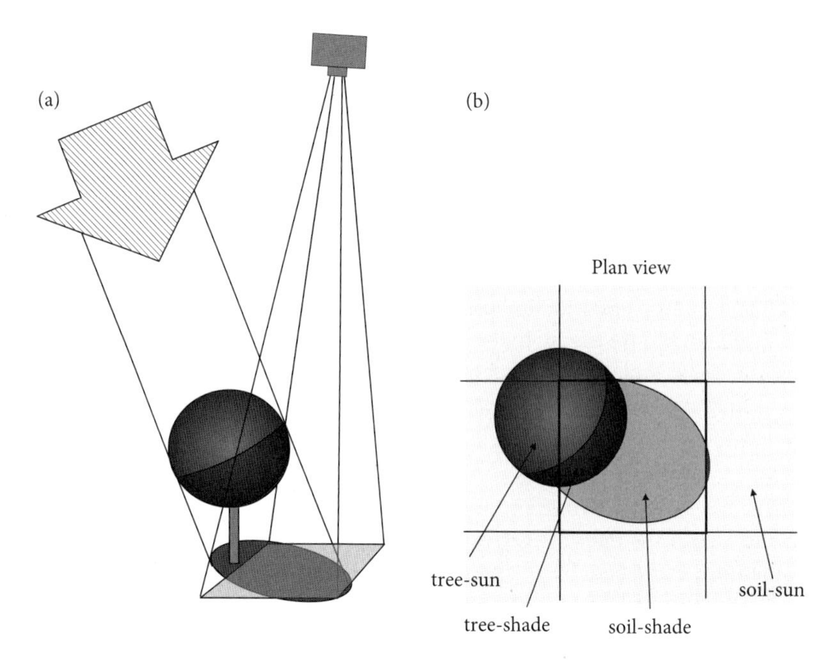

(c)

	fractional ground cover	true reflectance	I_o (W m^{-2})	reflected radiance (W m^{-2})	reflected radiance per total area	apparent reflectance
tree-sun	0.12	0.06	500	30	3.6	0.06
tree-shade	0.13	0.06	50	3	0.4	0.006
soil-sun	0.2	0.27	500	135	27.0	0.27
soil-shade	0.55	0.27	50	13.5	7.4	0.027
weighted reflectance		**0.22**				**0.077**

Fig. 7.4. (a) Schematic illustration of a view of a mixed pixel including a tree, some shaded soil, and some sunlit soil, (b) corresponding planar view showing 20% of the pixel as sunlit soil, 55% as shaded soil, and 25% as vegetation (13% shaded and 12% sunlit), (c) calculation of reflected radiances for the different areas in the above image, assuming an irradiance above the canopy of 500 W m^{-2}, and the consequent apparent reflectance of the whole pixel showing the impact of varying incident radiation on reflectance.

a greater proportion of NIR reaches the soil so that the apparent ρ_{NIR} increases relative to the apparent ρ_R. As can be seen from Fig. 7.3(b), *SAVI* is more closely related to leaf-area index over a range of soil reflectances than is *NDVI*, though the association is still not perfect and further improvements are possible (Baret and Guyot, 1991). In an alternative approach to correction for the soil line, Richardson and Wiegand (1977) defined a *perpendicular vegetation index* (*PVI*) as the perpendicular distance of a measured point from the soil line. However, as can be seen from Fig. 7.3 the perpendicular distance of an R/NIR measurement from the soil line is not linearly related to *LAI*.

A particular problem of determining vegetation indices from satellites is the variability introduced by atmospheric effects (e.g. varying aerosols or viewing angles) leading to differential attenuation of R and NIR. The *NDVI* calculated from satellite data without adequate atmospheric correction can be as low as 70% of the value that would be measured at ground level (e.g. Cracknell, 1997). A number of other modifications to the basic *NDVI*, such as the *atmospherically resistant vegetation index* (*ARVI*, see Box 7.1) are available that take account of differing transmission of different wavelengths in the atmosphere as conditions change. A common strategy for correcting for atmospheric aerosols is to make use of the fact that blue light is more scattered by aerosols than are red or near infrared. Addition of reflectance measurements in the blue, therefore, is used in indices such as *soil and atmospherically resistant vegetation index* (*SARVI*) and in the enhanced vegetation index (*EVI*) used for MODIS. An alternative strategy is to make use of the fact that atmospheric aerosols (other than large particles such as sand) are largely transparent

BOX 7.1 Vegetation indices (Note that in what follows, *N*, *R*, *G*, and *B* represent ρ_{NIR}, ρ_{Red}, ρ_{Green}, and ρ_{Blue}); *VI* represents vegetation index, *L*, *a*, *b*, β are constants. More extensive tabulations of *VI* may be found in many papers and texts including: Broge and Leblanc (2001), Jensen (2007); Lucas *et al.* (2008); Pontius *et al.* (2008).

	Formula	Advantages/disadvantages	References
Simple indices			
DVI (Difference *VI*)	$N-R$	Sensitive to illumination conditions, slope, etc.	Tucker (1979)
RVI (Ratio *VI*)	N/R	Partially corrects for variation in reflectance, illumination, especially if using reflectances	Birth and McVey (1968)
CI_{590} (Chlorophyll Index)	$(N_{880}/\mathrm{VIS}_{590}) - 1$	Apparently more sensitive to canopy N status than is *NDVI*, though see text (ρ_{590} is now used in preference to the original ρ_{540})	Gitelson and Merzlyak (1997)
Normalized indices			
NDVI (Normalized Difference *VI*)	$(N-R)/(N+R)$	Good for estimation of *LAI*; Clouds water and snow tend to negative values (when R > NIR)	Rouse *et al.* (1974)
GNDVI (Green *NDVI*)	$(N-G)/(N+G)$	Apparently better at higher *LAI*; particularly good at detecting chlorophyll as it increases over a much wider range of Chl than does *NDVI*, though the wavelength used varies from 470 nm up to amber (580 nm)	Gitelson *et al.* (1996)
SAVI (Soil-Adjusted *VI*)	$(1+L)\,(N-R)/(N+R+L)$	Corrects for varying soil reflectances, where *L* is a coefficient that varies from 0 at high *LAI* to 0 at low *LAI* (often assumed = 0.5)	Huete (1988); modifications, e.g. *TSAVI* / *OSAVI* Steven (1998)
TSAVI[1] (Transformed *SAVI*)	$a(N-aR-b)/(aN+R-ab+X(1+a^2))$	Corresponds well with *LAI*	Baret *et al.* (1989)
TVI (Transformed Vegetation Index)	$100\times((N-R)/(N+R)+0.5)^{0.5}$	Removes negative values; square root stabilizes variance	Deering *et al.* (1975)
PVI (Perpendicular *VI*)	$(N-aR-b)/\sqrt{(a^2+1)}$	Most effective at removing soil effect with low *LAI*	Richardson and Wiegand (1977)

(*Continued*)

BOX 7.1 (Continued)

	Formula	Advantages/disadvantages	References
ARVI (Atmospherically Resistant *VI*)	$(N-RB)/(N+RB)$ where $RB = R-\beta(B-R)$	Atmospherically resistant *VI*, corrects for changes in atmospheric transmission (β corrects R according to differences between R and B)	Kaufmann and Tanré – see Huete *et al.* (1997)
SARVI (Soil and Atmospherically Resistant *VI*)	$(N-RB)(1+L)/(N+RB+L)$ where $RB = R-\beta(B-R)$	Combines *ARVI* with *SAVI* (the constant β is normally 1 but can be varied to correct for aerosol (e.g. 0.5 for Sahel dust))	Kaufmann and Tanré – see Huete *et al.* (1997)
EVI (Enhanced *VI*)	$2.5(N-R)/(1+N+6R-7.5/B)$	Based on *SARVI*, and used as the operational index for MODIS products where the toa reflectances are atmospherically corrected	Huete *et al.* (2002)
AFVI (Aerosol-Free *VI*)	$(N-0.5\rho_{2.1})/(N+0.5\rho_{2.1})$	Insensitive to aerosols because the mid-IR (e.g. at 2.1 μm) is transparent to most aerosols except dust, but surfaces have similar reflectance to the visible (the factor 0.5 corrects for differences in ρ at 0.645 μm and 2.1 μm)	Karnieli *et al.* (2001)
WDRVI (Wide Dynamic Range *VI*)	$(\alpha^*N-R/(\alpha^*N+R)$ where $0.1 < \alpha < 0.2$	Reported to be more sensitive to high *LAI* than the standard *NDVI*, though see text	Gitelson (2004)
Multi-channel indices			
Kauth–Thomas transformation	For coefficients for Landsat 7[2]	Derives composite channels related to 'brightness' 'greenness', and 'wetness'	Kauth and Thomas (1976); Crist and Cicone (1984)
CAI (Cellulose Absorption Index)	$0.5^*(\rho_{2000} - \rho_{2200})/\rho_{2100}$	Reported to respond especially to plant dry matter	Daughtry *et al.* (2000)
MTCI (MERIS Terrestrial Chlorophyll Index)	$(\rho_{753.75}-\rho_{708.75})/(\rho_{708.75}-\rho_{661.25})$	The wave bands given are the centres of bands 10, 9, and 8 on MERIS	Curran and Dash (2005)
PPSG (Principal Polar Spectral Greenness Index)	$\tan^{-1}((PC_2-SF_2)/(PC_1-SF_1))$	PC_1 and PC_2 are the values on the 1st and 2nd principal components axes and SF_1 and SF_2 are the foci where lines of equal vegetation cover converge	Moffiet *et al.* (2010)
Hyperspectral indices			
REP (Red-Edge Position)		Sensitive to chlorophyll content	see Jago *et al.* (1999)

PRI (Photochemical Reflectance Index)	$(\rho_{570} - \rho_{531})/(\rho_{531} + \rho_{570})$	Estimates xanthophyll epoxidation as a measure of photosynthetic activity (N.B. sometimes subtract reference from sample λ)	Gamon *et al.* (1992)
PSSRa (Pigment-Specific Simple Ratio – *chla*)	(ρ_{800}/ρ_{680})	Good for estimating chlorophyll *a* at a leaf level	Blackburn (1998b)
PSNDa (Pigment-Specific Normalized Difference – *chla*)	$(\rho_{800}-\rho_{680})/(\rho_{800}+\rho_{680})$	Good for estimating chlorophyll *a* at a leaf level	Blackburn (1998b)
Ratio of derivatives at 725 and 702 nm	$d\rho/d\lambda(_{725})/d\rho/d\lambda(_{702})$	Responds to natural gas contamination of soils	Smith *et al.* (2004)
Water indices			
WBI (970 Water Band Index)	ρ_{900}/ρ_{970}	Rather variable between samples	Van Gaalen *et al.* (2007)
$NDWI_{1240}$ (Normalized Difference Water Index)	$(\rho_{980}-\rho_{1240})/(\rho_{980}+\rho_{1240})$	Use of shorter mid-IR wavelength less affected by atmospheric absorption	Gao (1996)
$NDWI_{1640}$	$(\rho_{858}-\rho_{1640})/(\rho_{858}+\rho_{1640})$	Shows less saturation than $NDWI_{1240}$ (as does using 2130)	Chen *et al.* (2005)
LWI (Leaf Water Index)	$(\rho_{1300}/\rho_{1450})$	A simple ratio water index	Seelig *et al.* (2008a)
NHI	$(\rho_{1100}-\rho_{1200})/(\rho_{1100}+\rho_{1200})$	Responds to heading in wheat, with values >0.18 indicating spike emergence	Pimstein *et al.* (2009)

[1] *a* and *b* are the constants in the soil line equation: $N = a.R + b$

[2] Coefficients for Landsat 7 ETM+ from Huang *et al.* (2002)

Index	Band 1	Band 2	Band 3	Band 4	Band 5	Band 7
Brightness	0.3561	0.3972	0.3904	0.6966	0.2286	0.1596
Greenness	–0.3344	–0.3544	–0.4556	0.6966	–0.0242	–0.2630
Wetness	0.2626	0.2141	0.0926	0.0656	–0.7629	–0.5388

Table 7.1 Examples of calculated vegetation indices using the data in Fig. 7.1 (assuming the wavebands of the AVHRR sensor; other sensors will give different values). The values of reflectances for the mixed surfaces are obtained by linear combination of the values for the components according to eqn (7.3).

	ρ_R	ρ_{NIR}	*DVI*	*RVI*	*NDVI*	*SAVI*
Pure surfaces						
Dry soil	**0.18**	**0.28**	0.10	1.556	0.217	0.156
Green grass	**0.08**	**0.49**	0.41	6.125	0.719	0.575
Dry grass	**0.36**	**0.56**	0.20	1.556	0.217	0.211
Mixed surfaces						
50% soil: 50% Gg	0.13	0.38	0.25	2.92	0.490	0.371
25% soil: 75% Gg	0.105	0.438	0.333	4.17	0.613	0.478

to radiation in the mid-IR, while surface reflectance in the mid-IR is generally proportional to visible reflectances (Karnieli *et al.*, 2001). Note that vegetation indices are not necessarily restricted to combinations of two bands; more sophisticated examples are outlined below (Section 7.3.3).

Relationship of vegetation indices to chlorophyll, LAI, fractional cover, and fAPAR

The popularity of *VIs*, especially of *NDVI*, arises because of their positive relationships with canopy density or vigour. Indeed they have been shown to be positively correlated with a wide range of functionally useful variables that all tend to vary together including chlorophyll content, leaf nitrogen, biomass, photosynthesis, productivity, leaf-area index (*LAI*), fractional vegetation cover (f_{veg}), and fraction of absorbed photosynthetically active radiation absorbed (*fAPAR*). Note that although *fAPAR* is defined as PAR absorbed, it is sometimes incorrectly used to refer to the fraction of PAR intercepted; a complete canopy will absorb about 94% of the total PAR intercepted. In view of the wide usage of vegetation indices to derive information about this extensive range of biological variables it is worth investigating some of these relationships in a little more detail.

Before considering specific relationships we should note that there are three somewhat distinct mechanisms that give rise to variation in *NDVI* and other *VIs*, though their effects can interact. The first, and most important, is the direct distinction between vegetation and soil so that changes in *NDVI*, at least at a remote sensing scale, are primarily related to the fraction of leaf or vegetation in a pixel. Superimposed on this effect, however, is a second influence that is based on the fact that even for full vegetation cover, or for an isolated leaf in a spectroradiometer, the calculated *NDVI* varies with the concentration of chlorophyll and other biochemical components and with structural factors that affect spectral reflectance (see Fig. 3.2). The third, but generally minor, effect relates to changing ratios between reflectance of R and NIR radiation as a function of changing view and illumination angles and variation in the fraction of shaded area visible (see Fig. 8.4 and Section 8.1.3 for further discussion).

(i) Fractional vegetation cover and fAPAR

The basic assumption is that for any area viewed by a spectroradiometer, or for a single pixel in an image, the spectral reflectance at any wavelength is a linear combination of the values for the background soil and for the vegetation. This is often referred to as the linear mixing model where the reflectances for the pure soil and for the pure vegetation are the 'end-members'. On the basis of this simplistic assumption we can derive the relationship between each *VI* and the *fractional ground cover of vegetation* (f_{veg}).[34] For example the overall red reflectance (ρ_R) would be given by

$$\rho_R = f_{veg}{}^{*}\rho_{R\text{-vegetation}} + (1-f_{veg})^{*}\rho_{R\text{-soil}}, \quad (7.3)$$

[34] This assumes no alteration of the spectral properties of radiation during penetration through the canopy, an assumption that may only be reasonable for large scales and clumped vegetation.

where $\rho_{R\text{-veg}}$ and $\rho_{R\text{-soil}}$ are the red reflectances of pure vegetation and soil end members. Table 7.1 shows some typical values of ρ_{NIR} and ρ_R for different surfaces ranging from pure grass to bare soil, and the values for mixed surfaces calculated using this equation, together with some corresponding calculated vegetation indices. For *NDVI*, for example, it is apparent that the value ranges in this particular case from 0.217 for bare soil to 0.719 for a completely vegetated pixel. Intermediate values for the vegetation index would therefore represent partially vegetated pixels.

Figure 7.5 shows how calculated values of the different basic indices vary as the green vegetation fraction increases from zero to full cover. It is apparent from this figure that only the *DVI* is linearly related to vegetation cover. The relationship is increasingly non-linear for *SAVI* and *NDVI*, and especially for *RVI*. The direction of curvature and the degree of non-linearity depends on the actual values for the pure component surface reflectances: it is possible to show that the condition for the response of *NDVI* to vegetation fraction in Fig. 7.5 to be convex is $(\rho_{NIR\text{-plant}} - \rho_{NIR\text{-soil}}) > (\rho_{R\text{-soil}} - \rho_{R\text{-plant}})$, while the response is concave where $(\rho_{NIR\text{-plant}} - \rho_{NIR\text{-soil}}) < (\rho_{R\text{-soil}} - \rho_{R\text{-plant}})$. The shape of the relationship between *VI* and fractional vegetation cover therefore depends on the values of the reflectances of the end-members (and hence on the exact wavebands used as well as on the spectral properties of the background soil and vegetation) as illustrated for *NDVI* in Fig. 7.5(b). A convex (saturation-type) curve is the most common; though in many cases linear relationships are adequate to estimate vegetation fraction from remote-sensing data. Quadratic or exponential functions are often used to approximate the saturation-type curves (see, e.g., Zhou *et al.*, 2009).

Unfortunately, the mixing model we have used thus far is overly simplistic, because shading of the soil in a canopy when the sun is at an oblique angle and observation is at nadir affects the absolute radiances of the reflected radiation. This means that the apparent reflectance of the soil end-member assumed in the linear mixing model does not remain constant as vegetation cover (and hence soil shading) increases. This effect, together with the corresponding shading of some of the leaves in the canopy, also changes the shape of the curve relating a *VI* to vegetation fraction. Figure 7.5(b) shows the potential magnitude of this shading effect if it is modelled assuming that the fraction of visible soil that is shaded is proportional to f_{veg}. More detailed modelling of these complex effects, including the preferential transmission of NIR as compared with R in canopies, requires the use of a more sophisticated canopy radiation-transfer model such as SAIL (see Chapter 8; Baret *et al.*, 1995).

Where the vegetation indices are linearly (or nearly so) related to fractional cover one can in principle estimate the fractional cover of vegetation (approximately equivalent to the fraction of radiation intercepted) from the observed *VI* by using the scaled vegetation

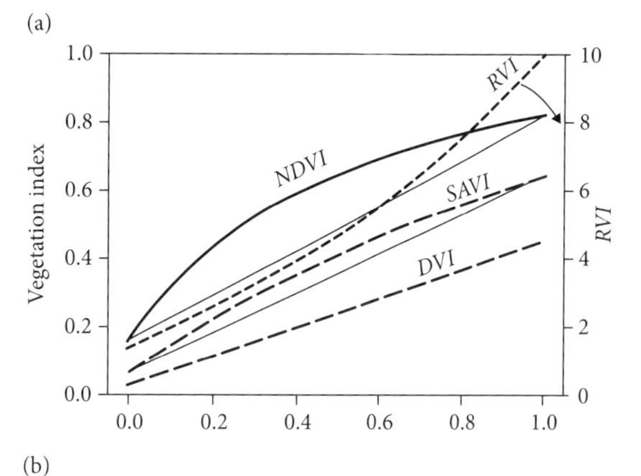

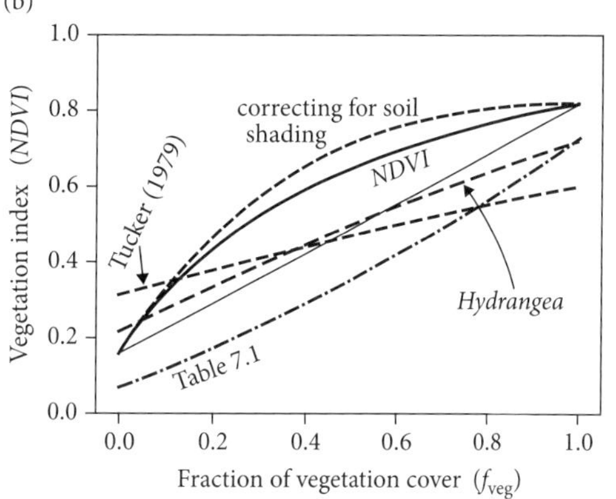

Fig. 7.5. (a) Calculated variation for vegetation indices (*NDVI*, *DVI*, *SAVI*, and *RVI*) as a function of fractional vegetation cover (f_{veg}), as calculated using eqn (7.3) and assuming typical component reflectances from Carlson and Ripley (1997) (vegetation: ρ_R = 0.05; ρ_N = 0.5, and soil: ρ_R = 0.08; ρ_N = 0.11) and neglecting any shading effect. Although *DVI* is linearly related to the vegetation fraction, *SAVI* and to an increasing extent *NDVI* and *RVI* are non-linearly related to f_{veg}. (b) Effect of differing reflectances and soil shading on the relationship between *NDVI* and f_{veg}. The solid line is the same *NDVI* line as in (a), while the short dashed line shows the effect of taking account of soil shading (assuming a 90% reduction in radiance for shaded soil), showing an earlier saturation. Other lines are for data from Fig. 7.1 (dots and dashes), and for 650 and 800 nm narrowband data either from Tucker (1979) (medium dashes) or for *Hydrangea* leaves over a dry loam soil (long dashes).

index (Choudhury *et al.*, 1994), VI^* (where the asterisk indicates a scaled index), according to:

$$f_{veg} \cong (VI - VI_{soil})/(VI_{veg} - VI_{soil}) = VI^*, \qquad (7.4)$$

where VI_{veg} ($=VI_{max}$) is the index for pure vegetation and VI_{soil} ($=VI_{min}$) is the index for pure background soil. For ground-based data it is usually straightforward to obtain reflectances for the extremes, while for a remotely sensed image these latter extremes may be estimated as the extremes in the appropriate area of the image if no better data are available. Rather than the relationship being linear, however, in some cases it has been found that $f_{veg} = (VI^*)^2$ (Carlson and Ripley, 1997) may be more appropriate, though a weaker power function (Baret *et al.*, 1995) can give a better fit on other occasions (Jiang *et al.*, 2006).

Because the fractional cover that is visible from the sensor varies with the view angle, it is most useful to correct any value obtained from a sensor at a particular view angle to the corresponding nadir value, which we define as the *true fractional cover* (f_{veg}). Unfortunately, as will be discussed further in Chapter 8, this conversion also depends on canopy structure and on leaf-angle distribution, though a reasonable approximation can often be obtained by assuming a random (−spherical) leaf-angle distribution. In this case we can make use of the shape factors derived in Section 3.1.3 to approximate f_{veg} as $f_{veg(\theta)}/\cos(\theta)$, where $f_{veg(\theta)}$ is the apparent fractional cover at a view zenith angle of θ. The NBARs (*nadir bidirectional reflectance distribution function adjusted reflectance*) data provided for MODIS attempts to approximate the nadir-view reflectances and hence provides a good starting point for estimation of *VI* and of vegetation cover.

For many ecological purposes a variable that is more directly related to canopy functioning, especially productivity, is the *fraction of absorbed photosynthetically active radiation* (*fAPAR*). This is approximately proportional to the fractional vegetation cover multiplied by the canopy absorptance ($\alpha_{vegetation}$) of PAR radiation (400–700 nm), though for accurate work it is necessary to take account of the absorption of radiation reflected from the soil and also the angular distribution of the incoming radiation. Although it is possible to define a unique value for f_{veg} for any canopy (referring to the nadir view), the fraction of radiation actually intercepted, and hence the fraction actually absorbed, varies with the angle of incident radiation (see Fig. 7.6), and with its distribution over the sky – which changes between sunny and cloudy days. The approach to saturation as *LAI* increases tends to be more rapid for a horizontal-leaved or planophile canopy than for one with randomly oriented leaves (spherical distribution) with its smaller extinction coefficient. As a result, any value derived for *fAPAR* is inevitably dependent on environmental conditions. Moreover, the index for pure background soil (VI_{min}) is in reality not a constant but varies with soil wetness and roughness.

Because of the importance of *fAPAR* in ecosystem functioning, effort has been devoted to the generalized derivation of instrument-specific indices that directly optimize the extraction of *fAPAR* from satellite data, correcting for variation in view angle, aerosols, and soil background (e.g. Gobron *et al.*, 2000). As we have seen, the addition of reflectance measurements in the blue, which is more sensitive to atmospheric scattering, to the conventional reflectances in the red and near infrared allows one empirically to correct for atmospheric effects. The sets of polynomial coefficients that give the optimal index are derived by numerical optimization over the range of conditions under which the index will be used. Up to 15 of these polynomial coefficients are required to give the optimal index (see, e.g., Gobron *et al.*, 2000; Liang, 2004 for examples for different sensors).

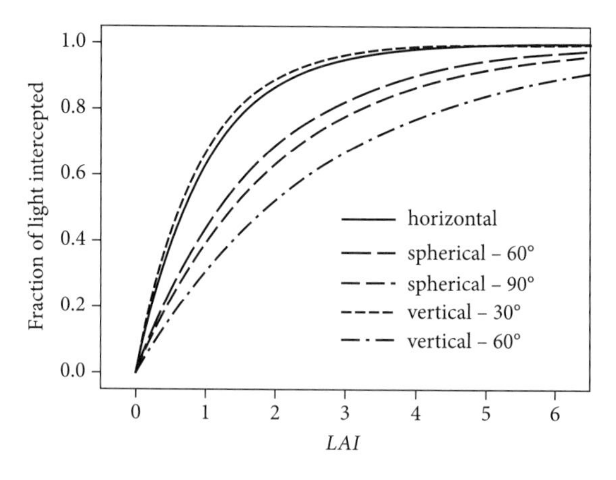

Fig. 7.6. Modelled fraction of radiation intercepted as a function of *LAI* for various elevations of the illumination source (expressed as solar elevation) and for various leaf-angle distributions (with horizontal being equivalent to a spherical (= random) distribution at an illumination elevation angle of 30°). The fraction of incident radiation intercepted clearly increases with increasing *LAI*. Note that the same calculations give the apparent fractions of ground cover for different view angles, with the curves for 90° view/illumination giving the true f_{veg}. Note that the results for the horizontal leaf-angle distribution are independent of view or illumination angle.

(ii) Relationship between vegetation indices and LAI

Although vegetation indices are most directly related to fractional vegetation cover (or *fAPAR*), they are commonly used as estimators of the *leaf-area index* (*LAI*; the one-sided leaf area per unit projected ground area). As was outlined in Chapter 3 and is clear from Fig. 7.6, the fraction of radiation intercepted by the canopy, and hence the *apparent* vegetation cover, depends not only on the *LAI* but also on the leaf-angle distribution and importantly on the angle of the incident radiation. The relationship between fractional vegetation cover or light interception and *LAI* is given by the integrated form of Beer's law introduced in Section 3.1.3:

$$f_{veg} \cong (1 - e^{-kL}), \tag{7.5}$$

where k is the extinction coefficient and L is the leaf-area index. The dependence of k on leaf-angle distribution and solar elevation was summarized for several leaf area distributions in Table 3.4. It is clear from Fig. 7.6 that a given change in *LAI* has less and less effect on radiation interception as canopy density increases, and that this relationship is even less linear than the relation between *NDVI* and f_{veg}. It is clear from this figure that remote sensing will be much less effective at distinguishing between canopies with high leaf-area indices than between sparser canopies. The effect of view angle on apparent radiation interception by canopies is discussed further in the next chapter.

Substitution from eqn (7.4) into eqn (7.5) and rearranging gives the following approximate relationship between L and VI^*:

$$L \cong -\ln(1 - VI^*)/k. \tag{7.6}$$

The consequence of this response is that *LAI* is very non-linearly related to most *VIs*, as shown in Fig. 7.7 for three contrasting *VIs*: *NDVI*, *TSAVI*, and *PVI*.

(iii) Conclusions on the relationship between vegetation indices, LAI, fAPAR, and other biophysical variables

As should be apparent from the above, although most scaled vegetation indices are related to *fAPAR*, the relationships are not usually linear; while relationships with *LAI* are usually very non-linear. Furthermore, in practice, substantial scatter in the relationship between any *VI* and any canopy biophysical parameter is to be expected in any realistic scene. Even though the relationship may be close for any one species in a given set of environmental conditions, a fundamental problem with all *VIs* is that the relationship between the *VI* and any biophysical parameter shows substantial scatter as a result of variation in canopy properties and illumination. Some modelled examples of the variation in the relationships between *LAI* and three *VI* are shown for a representative canopy in response to typical variation in chlorophyll per unit leaf area (Cab), average leaf angle (*ALA*), soil brightness (ρ_{soil}), and leaf clumping are illustrated in Fig. 7.7. In each case the effect of changing the parameter listed at the top of the column is shown by the grey area, while the outer lines indicate the additional variation introduced by varying the leaf clumping parameter (FracV) over a typical range. In many real situations all these factors may vary together, leading to potentially very great uncertainty in the estimation of biophysical properties from vegetation indices. It is worth noting that the various *VI* tested show different sensitivities to variation in Cab, *ALA*, and soil brightness, with *PVI*, for example, coping well with varying soil brightness but being very sensitive to varying chlorophyll concentration, while *NDVI* is less sensitive to variation in *ALA*.

It is common to go further and to use vegetation indices as proxies for biomass or canopy chlorophyll content; however, these applications involve even more assumptions than do their use for estimation of leaf-area index. It is clear that the standing biomass of different communities does not necessarily scale equally with *LAI*. For example forests, especially deciduous forests, may have a very high ratio of biomass to *LAI* in comparison with grassland. Therefore, any such use must use conversions appropriate for the type of vegetation being studied.

Indices such as *SARVI* have been reported to have less tendency to saturate at high canopy density and therefore to be more sensitive to canopy structure (Huete *et al.*, 1997) as has the *wide dynamic range vegetation index* (*WDRVI*; Gitelson, 2004). It has also been suggested that the *GNDVI* may be more sensitive to chlorophyll than is *NDVI* (Gitelson *et al.*, 1996). Unfortunately, these and the many other similar transformations that appear to increase the sensitivity to higher *LAI*, in fact just change the shape of the curve without revealing

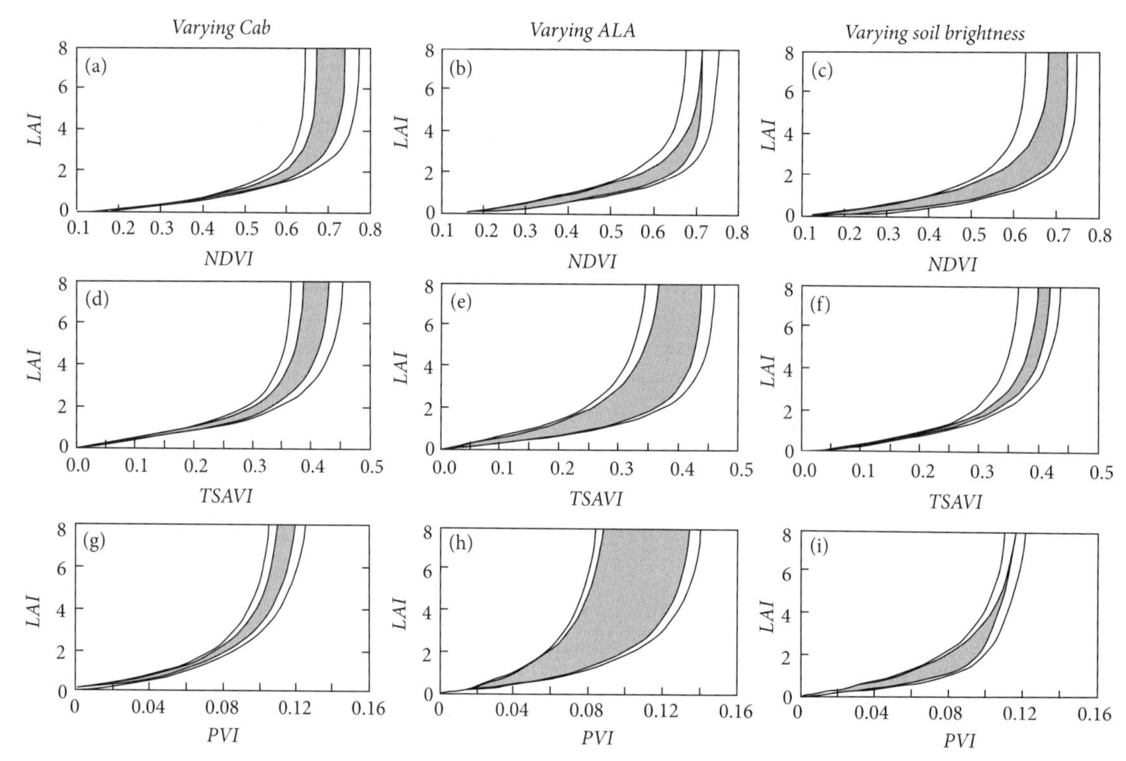

Fig. 7.7. Simulated sensitivity to extraneous factors of the relation between canopy leaf-area index (*LAI*) and different vegetation indices (*NDVI*, *TSAVI*, and *PVI* – see Box 7.1 for definitions). The grey areas indicate the sensitivity to variations in Cab (left), *ALA* (middle), and soil brightness (right). The outer lines give the additional variation introduced by leaf clumping at canopy level (FracV). Simulations were kindly provided by Clement Atzberger using the physically based PROSAIL model (see Chapter 8 ; and see Richter *et al.*, 2009 for the parameter meanings) with the following parameter setting: Cab: 45, Cm: 0.055, N: 1.5, *ALA*: 55, HotS: 0.5, RSOILred: 0.2, FracV: 0.95. For the sensitivity analysis *ALA* was varied between 35 to 55, RSOILred between 0.1 and 0.3, and Cab between 35 and 55. To assess the additional influence of leaf clumping at canopy level, FracV was varied between 0.9 and 1.0. All simulations were run for nadir view and a solar zenith angle of 30 degrees. The soil reflectance in the near infrared (RSOILnir) was linked to the soil reflectance in the red (RSOILred) assuming the following soil line: RSOILnir = 0.04 + 1.2 RSOILred.

more information. Indeed, though they may emphasize key parts of the response, they tend to be increasingly sensitive to small errors or noise in reflectance measurements over the critical range, so in practice do not generally result in real improvements in performance. A consequence of the many factors affecting the relationship between biophysical parameters and *VIs* is that there is inevitably substantial uncertainty in the derivation of such variables from satellite data.

Long-term continuity of vegetation indices

The absolute values of reflectances, and hence of the derived *VI*, differ between sensors and sample dates. This is because the particular wavebands used from different sensors are not all exactly the same in position or width and may not have the same sensitivity, while in addition the scale and the methods of calibration and the atmospheric correction procedures used may differ. This is why one should not, for example, try to compare the *NDVI* obtained from a Landsat image with that from the AVHRR. The sensitivity of *VIs* or derived variables such as *LAI* to the input reflectance data is well illustrated by an analysis of MODIS surface reflectance products (Yi *et al.*, 2008) where the changes in the blue reflectance calculated at different dates (due to changes in the atmospheric correction algorithm) led to substantial differences in the calculated *LAI* for wheat. Long-term change studies ideally require long time series of compatible imagery with a specific sensor. An approach for partially getting round this is the use of *spectral invariants* (e.g. Ganguly *et al.*, 2008; Lewis and Disney, 2007; see Chapter 8).

7.2.2 Narrowband indices

With the increasing availability of hyperspectral data and of spectroradiometers that have the power to resolve reflectance differences in rather narrow wavebands, down to 2 nm or less, there has been increasing interest in the use of narrowband spectral indices to study leaf pigments and other vegetation characteristics. By appropriate choice of wavelengths incorporated in the index it is possible to develop *pigment specific indices* based either on simple ratio measurements or expressed as normalized difference indices (e.g. Blackburn, 1998a; Box 7.1). Perhaps the most satisfactory approach to the selection of the appropriate wavelengths is to choose them using a rational approach based on a knowledge of the absorption spectra of the specific pigments. Otherwise, they can be chosen by using an empirical selection procedure whereby for a range of plant samples with known pigment composition, correlograms are constructed by sequentially regressing reflectance at each wavelength against pigment content and plotting the coefficient of determination (R^2) to identify the best wavelength combinations. This latter approach requires test samples with a good range of concentrations of the pigment of interest varying independently of other pigments; a criterion that is often difficult to achieve as concentrations of different pigments tend to covary. Such an approach has led to the identification of a wide range of empirically selected wavelength combinations (such as the ratios ρ_{695}/ρ_{420} or ρ_{695}/ρ_{760}) for distinguishing plant stresses (Carter, 1994, but see Chapter 11 for further discussion). Indeed one can extend the approach and evaluate all possible 2-band combinations within a given index type (e.g. normalized differences) and build 2D correlation plots that identify the optimum band combination for mapping the variable of interest (e.g. Fig. 7.8). In this example, as in many others, the ideal combinations of wavelengths varied with species, with the *VI* chosen, and with background soil. Application of this approach to Mediterranean pasture identified combinations of near infrared (770–930 nm) and red-edge (720–740 nm) wavebands as having most predictive power for *LAI*, biomass, and nitrogen content (Fava *et al.*, 2009), though other wavelength combinations have been found in other studies.

As a general rule, wavelengths where absorption is low are most appropriate for distinguishing high pigment concentrations, while spectral regions with high absorption are appropriate where pigment concentrations are low. A wide range of these pigment-specific indices have been developed, but their success tends to be rather dependent on the plant material and on environmental conditions and there is generally rather limited transferability across species or across scales. Various studies have derived narrowband spectral indices using three or even four spectral bands (see Blackburn, 2007) with most effort having been devoted to the development of chlorophyll indices. Although this approach can give improved precision

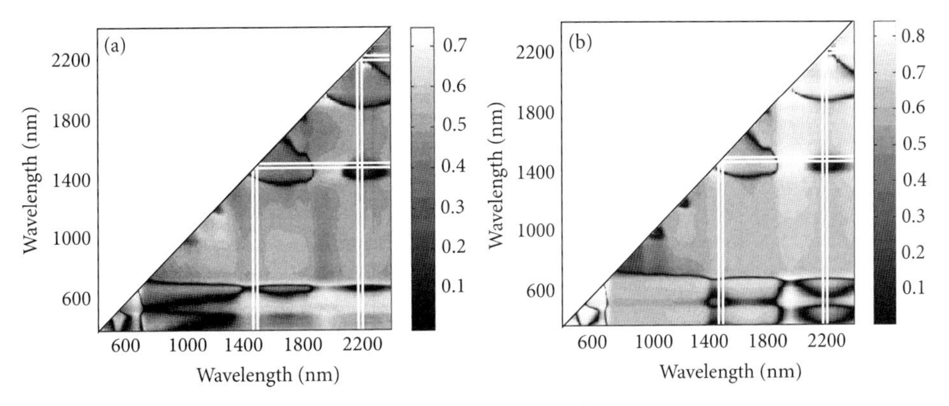

Fig. 7.8. Correlation plots showing the coefficient of determination (R^2) between *LAI* and all possible combinations of narrowband *NDVIs*, for (a) a dark soil background, and (b) with a light soil background. Light areas indicate the combinations of wavelengths with the highest correlations with *LAI*. The white lines indicate areas with no data (with permission from Darvishzadeh *et al.*, 2008).

over conventional *VIs* for defined experimental sets, they may become less useful when extrapolated to other biological systems. In practice, classic *VIs* based on broadband data (such as Landsat TM channels) tend to be less affected by external factors such as canopy structure, illumination, and atmospheric conditions than the corresponding narrowband indices when used as indicators of *LAI* or chlorophyll content. This effect may be a result of a higher signal-to-noise ratio for broadband indices. Various attempts have been made to target chlorophyll estimation by using bands such as those at 550 nm and 700 nm centred on chlorophyll absorption minima in combination with the chlorophyll absorption maximum at 670 nm, but these do not appear to be substantially better than conventional broadband indices for estimating vegetation cover or activity.

Note that information on leaf biochemical constituents can also be obtained using fluorescence studies as will be discussed in Section 11.2.

Photochemical reflectance index (PRI)

The so-called *photochemical reflectance index* ($PRI = (\rho_{570} - \rho_{531})/(\rho_{531} + \rho_{570})$) is a special case of a narrowband index that requires narrowband measurements of reflectance at 531 nm and 570 nm (Gamon *et al.*, 1992). This index aims to detect changes in the epoxidation state of the xanthophyll pigments in leaves, which is known to be associated with changes in the efficiency of the photosynthetic light reactions. The index is based on the slightly differing absorption spectra of epoxidated (violaxanthin) and de-epoxidated (zeaxanthin) xanthophylls centred at 531 nm (see Fig. 7.9). The measurement at 570 nm is used as the control region where both forms of the pigment have similar reflectances. The reason for interest in xanthophyll de-epoxidation is that these pigments in the chloroplasts play an important role in dissipating excess light energy that might otherwise damage the photosynthetic system, with an increased proportion of zeaxanthin occurring at high light when the efficiency of the system decreases so that, as is discussed further in Section 9.5.5, the ratio between them provides a useful indirect estimate of photosynthetic radiation-use efficiency and hence of photosynthesis rate. Unfortunately, the interpretation of spectral changes associated with xanthophyll epoxidation is complicated by other light-induced changes in leaf absorptance, scattering, and reflectance caused by processes such as chloroplast movement and the changes in light scattering associated with related membrane changes (centred on 535 nm). Interestingly, the xanthophyll-related change in absorptance (as opposed to reflectance) is actually greatest at 505 nm (Bilger *et al.*, 1989). Examples of the relationship between *PRI* and photosynthetic light-use efficiency at the leaf and the canopy scale are shown in Fig. 7.10 (see Section 9.7 for further discussion of photosynthetic light-use efficiency).

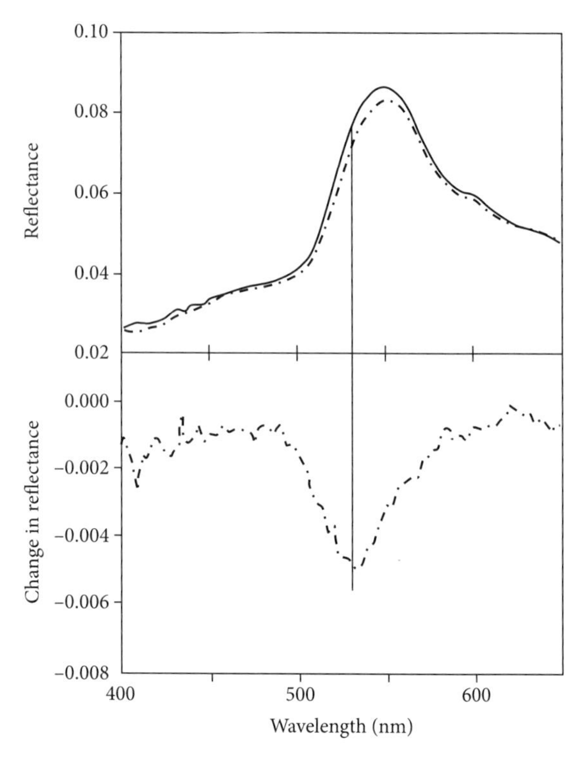

Fig. 7.9. Illustration of the change in reflectance at 531 nm when a shaded sunflower canopy was exposed to bright sun; the shaded spectral reflectance is shown by the solid line, while the corresponding curve after 40 min (when much of the xanthophyll is converted to zeaxanthin) is shown dashed. The difference between the two curves is shown in (b) (data from Gamon *et al.*, 1990). These authors demonstrated that this difference was related to a change in xanthophyll epoxidation (not shown) and to differences in photosynthesis.

7.2.3 Use of multiple wavelengths and derivative spectroscopy

There are a number of more complex vegetation indices that use the increased information available when using multiple wavelengths or hyperspectral data.

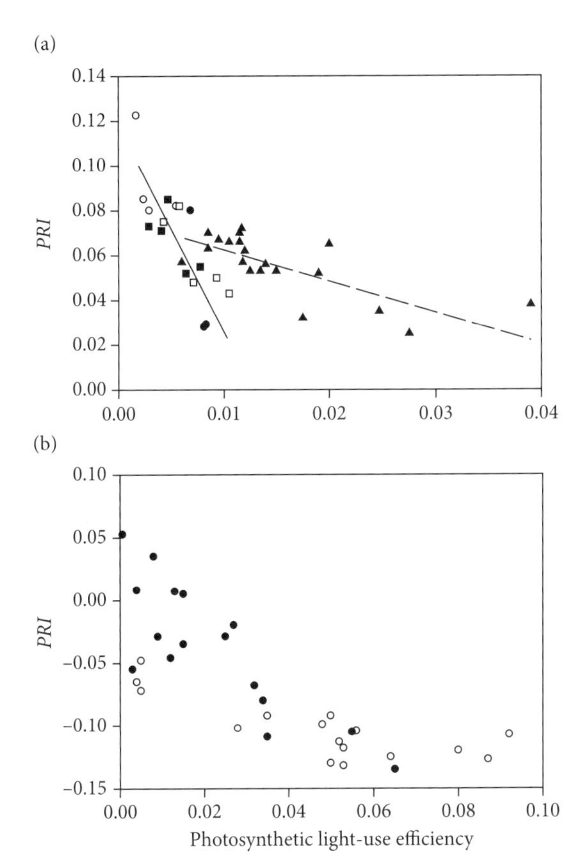

Fig. 7.10. Examples of the relationship between *PRI* (calculated as $(\rho_{570} - \rho_{531})/(\rho_{531} + \rho_{570})$) and photosynthetic light-use efficiency (*LUE*) for (a) single leaves of *Heteromeles arbutifolia* (○) and *Phaseolus vulgaris* (●) sampled in the field under varying irradiance (replotted from Peñuelas *et al.*, 1995), and (b) examples for tree canopies (using reflectances at 529 and 569 nm measured from a helicopter) where light-use efficiency was estimated from eddy covariance for boreal communities: fen (○), old aspen (●), old jack pine (■), black spruce (□); and for tropical mopane woodland (▲). Data from Nichol *et al.* (2000) and Grace *et al.* (2007).

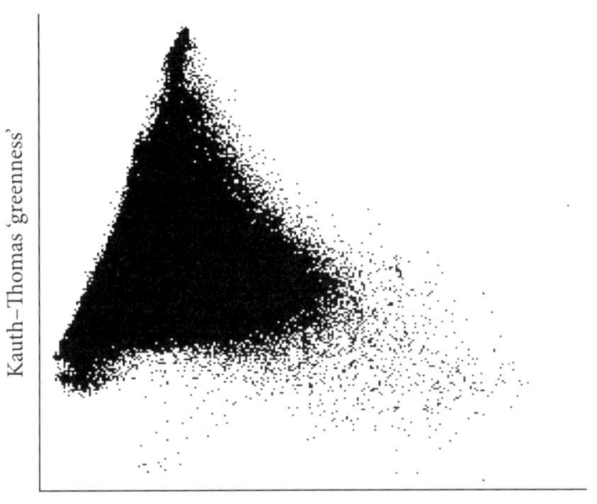

Fig. 7.11. The first two axes of the 'Kauth–Thomas' tasselled cap transform for Landsat TM data for 28th June 2001 for an area near Porto Alto, Portugal (data courtesy US Geological Survey).

In some, such as *SARVI* referred to above, correction of the basic *NDVI* is made by using an additional wavelength to isolate atmospheric effects.

An alternative approach is the reduction of multispectral or hyperspectral data to a limited number of derived variables that contain most of the information, through transformations such as the use of principal component analysis or through the optimization used in transformations such as the '*tasselled cap*' or '*Kauth–Thomas*' transformation (Kauth and Thomas, 1976). This transformation or its later variants extract from multiband sensors (e.g. the four wavebands in the Landsat MSS instrument or the six visible/NIR bands of the TM sensor) two principal variables, one called 'brightness' that relates primarily to soil reflectance and one called 'greenness' that relates primarily (but not exclusively) to plant cover; a third orthogonal variate has been related to 'yellowness' (in MSS) or to 'wetness' (TM) (Crist and Cicone, 1984). The transformation can be optimized for any sensor so that most data points fall in a pattern that resembles a 'tasselled cap', hence the name (Fig. 7.11). The related approach of extraction of a few principal components can be readily applied to the simplification of hyperspectral data. Various higher-order indices have been developed using polar transformation of the principal components to overcome the problem that characteristics such as green cover do not lie on a line parallel to a single principal components axis (e.g. *PPSB*; Moffiet *et al.*, 2010); this is somewhat analogous to the use of ratio *VIs*.

A wide range of statistical approaches are available to quantify pigment composition and other vegetation characteristics from reflectance in multiple spectral bands or indeed continuous spectra (Blackburn, 2007). These include, principal component analysis (PCA), factor analysis, artificial neural networks (ANNs), partial least squares regression (PLSR), support vector machine (SVM) regression, genetic algorithms, and stepwise linear multiple regression (SLMR). As with the optimization of two-band indices, it is again necessary to have a training set of spectra with a wide range of concentrations of the variable of interest in as wide as possible a range of background material. The usefulness of all these approaches always depends on the scope and quality of the training set available and, importantly, on the magnitude of the variation in perturbing factors such as canopy structure.

Geometry-based approaches or *angle indices* (*AI*) have also been developed to characterize the general shape of the reflectance spectrum at specific wavelengths. These make use of the reflectance triplets based on three contiguous bands and measuring the angle (e.g. Khanna *et al.*, 2007) formed at a vertex between the reflectances of these successive bands (λ_1, λ_2, and λ_3) as indicated in Fig. 7.12, or the perpendicular distance between the baseline and the peak (Borel, 1996). The approach is widely applicable to MODIS and other multispectral imagery where there are several contiguous bands.

Derivative spectroscopy and red-edge

Where we have spectroscopic information at multiple wavelengths, alternative approaches become available for the extraction of information on leaf pigments and other vegetation characteristics. The position of the sharp change in reflectance at 710–720 nm, known as the *red-edge position* (λ_{RE}, often abbreviated as *REP* or *REIP*, for red-edge inflection point) has long been known to be a particularly sensitive indicator of leaf chlorophyll content (see, e.g., Dawson and Curran, 1998; Jago *et al.*, 1999; Miller *et al.*, 1990). The sharp increase in reflectance at λ_{RE} indicates the boundary between the strong absorption by chlorophyll and the scattering of infrared wavelengths: increasing chlorophyll concentration results in a broadening of the chlorophyll absorption peak that moves the red-edge to longer wavelengths while losses of chlorophyll as in senescence lead to shorter wavelengths for the red-edge position. The red-edge can be detected using hyperspectral sensors as the wavelength at the point of inflection of the reflectance spectrum (or the wavelength at the maximum slope). In practice the position of the red-edge inflection point can be identified by fitting either a polynomial or an inverse Gaussian curve to the reflectance data around the red-edge at 710–720 nm or else it may be identified as the maximum of a curve fitted to the first derivative of the reflectance spectrum (see Figure 7.13). A difficulty is that the spectrum is often sampled only rather coarsely (e.g. AVIRIS only has a 10-nm channel separation) so it is difficult to estimate the λ_{RE} accurately in such cases. A three-point Lagrangian technique or other higher-order curve-fitting approaches may be particularly useful in such situations (Dawson and Curran, 1998).

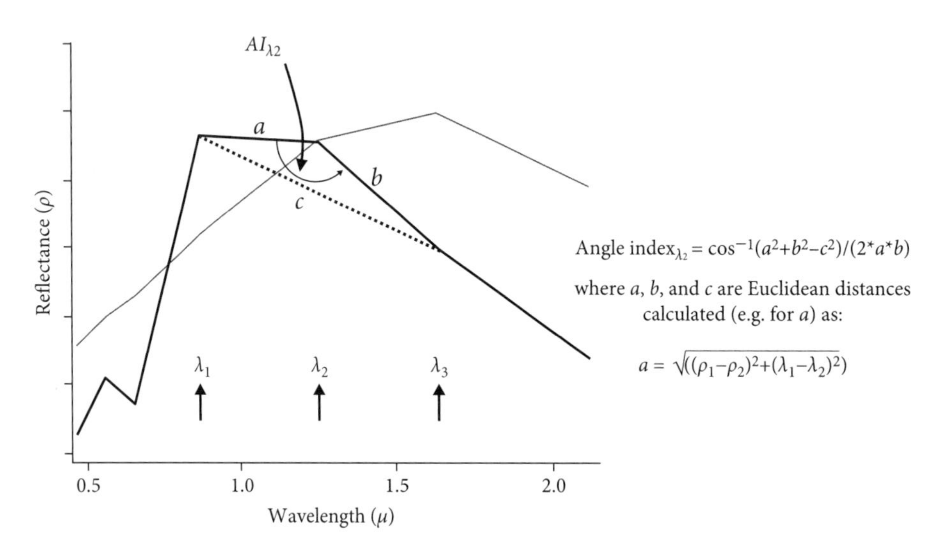

Fig. 7.12. Illustration of the calculation of angle indices (AI_λ) based on reflectances (ρ) in three contiguous wavebands (λ_1, λ_2, and λ_3). The thick-line spectrum is for vegetation and the thin-line one is for dry soil (normalized to λ_2) (modified following Khanna *et al.*, 2007).

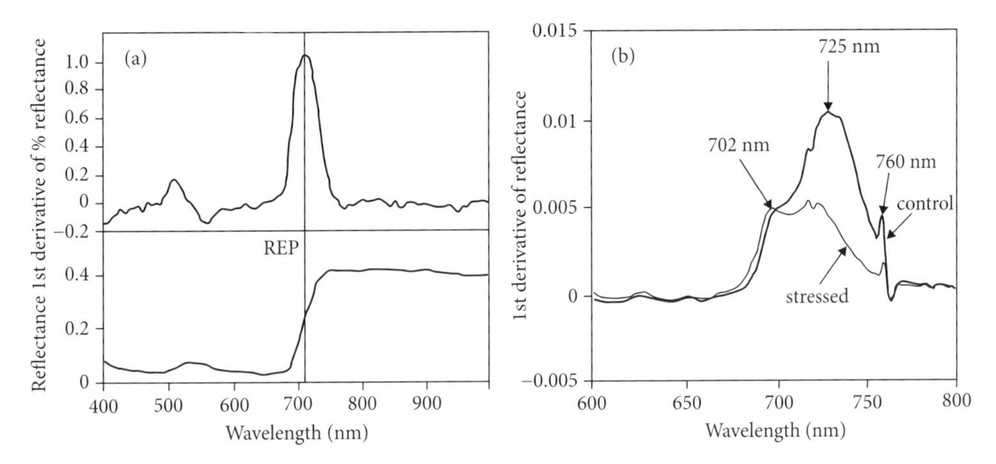

Fig. 7.13. (a) First derivative (i.e. the slope of the reflectance spectrum) of reflectance spectrum for a green leaf and (b) the corresponding reflectance spectrum (data from Blackburn, 2007a), (c) the variation of the first derivative of the reflectance spectrum with wavelength for control grass plots (heavy line) or plots treated by exposure to natural gas (light line) showing a marked shoulder at 702 nm and a peak at 725 nm (data from Smith *et al.*, 2004).

A particular advantage of the derivative spectroscopy approach is that the second derivative of the reflectance curve (i.e. the slope of the first derivative) has been shown to be very insensitive to the soil background (e.g. Demetriades-Shah *et al.*, 1990 and Broge and Leblanc, 2001). Unfortunately, in many cases there are two or more peaks apparent in the first derivative spectra, especially when the data are obtained using spectroradiometers with high spectral resolution. This often results in apparent jumps in the λ_{RE}, so alternative fitting approaches may be better (Cho and Skidmore, 2006; Cho *et al.*, 2008). The multiple peaks in the first derivative of the reflectance spectrum (e.g. Fig. 7.13) may themselves contain information of use for the sensitive detection of environmental stress effects on plants. Smith *et al.* (2004) have argued that the ratio between the peaks of the derivative spectrum observed at 725 nm and 702 nm may be a particularly useful approach to quantifying such variation. Other studies have demonstrated clear peaks and differences in the first derivative at other wavelengths (Darvishzadeh *et al.*, 2008). Further discussion on the detection of plant stress using spectral and other indices will be found in Chapter 11.

Hyperspectral derivatives (both first derivatives and second derivatives) have also been closely related to chlorophyll *a* content (Blackburn, 1998a), though the technique does not appear to distinguish well between the different chlorophylls (chl *a* and chl *b*). Similar derivative approaches are increasingly being adopted to discriminate plant stress responses remotely (see Chapter 11).

Laboratory techniques such as near infrared reflectance spectroscopy (NIRS), which are widely used in the food industry for chemical analysis, have potential as tools to identify specific chemicals in tissues. The reflectance spectrum in the range 1200–2400 nm depends on the differing content of C–H, N–H, and O–H bonds in organic material and can be used to estimate contents of specific organic materials and even total carbon or nitrogen, with some evidence of success at quantification of ecologically important isotopic concentrations ($\delta^{13}C$, and $\delta^{15}N$; Kleinebecker *et al.*, 2009). Unfortunately, multivariate statistical approaches using calibration samples are required for prediction of the composition in unknown samples (Workman, 2008). Although not currently used substantially as a remote-sensing technology, there is much interest in developing hyperspectral methods in the near to mid-infrared to obtain critical biochemical information on canopies in the field.

Spectral decomposition and wavelet analysis

Spectral decomposition holds promise for quantifying a range of leaf pigments from reflectance spectra. Wavelet analysis (see Section 6.5.2) is a particularly powerful approach that allows one to quantify concentrations of components in mixtures by decomposition of the overall reflectance spectra into different frequency components. This is rather similar to Fourier analysis that is also based on superposition of functions, but in this case the decomposition can represent local features of the spectrum. As a result it is possible to isolate the components due to the chemical of interest by removing the effects of interfering factors affecting leaf reflectance (e.g. chemical content, structure, thickness and water content), and canopy reflectance (structure, viewing, and illumination conditions). It is of course necessary to establish empirically the relationships between the individual wavelet coefficients and the concentrations of biochemicals of interest (see Blackburn and Ferwerda, 2008 for further information).

Use of spectral libraries

Spectral reflectance libraries have been used for many years for discriminating between different rock types in geology. Extensive libraries exist for mineral and rock types (see, e.g., the extensive series of publications by Hunt, Salisbury and colleagues – referred to by Clark, 1999), but the equivalent libraries for vegetation are relatively poorly developed (Rao, 2008), largely because of the inherent variability, so researchers often have to derive their own reference spectra for their field sites. Rather than use in quantification of pigments, however, their main use is in classification studies (see below).

Radiative transfer (RT) models and estimation of vegetation biophysical properties

Another important approach to retrieving canopy biophysical properties such as chlorophyll content or leaf-area index from spectral data, which does not use spectral indices, is based on the inversion of physically based radiative-transfer models. As was outlined in Section 3.1.1, inversion of coupled leaf and canopy radiative-transfer models such as SAIL combined with PROSPECT or LIBERTY provides a very powerful approach to the estimation of leaf biochemical composition, and one that is more mechanistically based than the simple correlations with spectral indices. Such models can be used in the 'forward' modelling mode to generate reflectance spectra for a wide range of biochemical and biophysical properties. The generated reflectance spectra can be used for developing *VI*s. Alternatively RTM can be used in the 'inverse' modelling mode to determine biochemical parameters such as chlorophyll content from observed canopy reflectances. The inversion of hyperspectral data provides a general approach to estimation of plant pigments and other vegetation characteristics, though, as will be discussed further in Chapter 8, inversion of RT models can be subject to substantial uncertainty as it requires a good model for canopy radiative transfer (Houborg *et al.*, 2007; Zhang *et al.*, 2008).

7.2.4 Water indices

Much effort has been expended over the years in attempts to estimate remotely the water content of tissues as a measure of water-deficit stress. We have already introduced the basics of plant–water relations and their control in Chapter 4, here we shall concentrate on the relationships between optical remote sensing signals and tissue water content. Microwave indices will be discussed in Section 7.3.2. The approach generally adopted for the study of vegetation water content is based on the presence of spectral bands in the near to mid-infrared where there are strong water-absorbing features. In theory any of the water absorption bands at 970 nm, 1200 nm, 1450 nm, 1930 nm, or 2500 nm (Fig. 7.14) could be used, but the accuracy of most sensors decreases at the longer NIR wavelengths where the depth of the absorption band is greatest. Absorption by atmospheric water vapour also restricts the amount of up- and downwelling solar radiation in the absorption troughs at the longer wavelengths. Therefore, use of the absorption band at 1450 nm is generally found to represent a good trade-off between sensor accuracy and depth of the absorption band (Seelig *et al.*, 2008b), but only for laboratory measurements. As with the use of other spectral indices, great improvements in sensitivity can be achieved by normalizing the results against values at nearby wavelengths where water absorbs more weakly if at all; for the 1450 nm band a suitable reference is 1300 nm. Using a simple ratio normalization this gives a *leaf water index* (*LWI*) as ($LWI = \rho_{1300}/\rho_{1450}$). A further complication (Carter, 1991) is that structural changes to the leaf as it dries out can also affect the calculated *LWI*. The problems caused by radiation absorption by atmospheric water vapour lead to the use of less strongly absorbing water bands, as used in the normalized difference water index $(\rho_{980}-\rho_{1240})/(\rho_{980}+\rho_{1240})$. Some examples of water indices are presented in Box 7.1.

In the laboratory one can also quantify water in leaves by the use of transmittance measurements with the same ratios used as for reflectance measurements. In principle, one would expect the transmission index to be exponentially related to leaf water content because of the exponential decay of radiation (especially in the water absorption bands) according to Beer's Law through the leaf (Seelig *et al.*, 2008a). For the transmission measurements one is primarily detecting the amount of water per unit leaf area or *leaf water content* (*LWC*; kg m^{-2}) as given by (fresh mass – dry mass)/*A* where *A* is the area. This can be expressed as an *equivalent leaf water thickness* (*EWT*, m) by dividing by the density of water; i.e. *EWT* = (fresh mass – dry mass)/ $\rho_w.A$, where ρ_w is the density of water.

Unfortunately, these measures of the absolute water content of leaves are not particularly useful physiological indicators of drought responses, as was pointed out in Section 4.2, so the more useful relative water content (*RWC*) is much more generally recommended. The response of the reflectance water index tends to saturate with increasing amounts of water for much the same reason as *NDVI* saturates with increasing *LAI* (shown as the increasing scatter at high water contents in Fig. 7.14(b)). Although the ratio water index used in this figure is well related to leaf relative water content over the full range of *RWC*, there is substantial scatter over the physiological range of water content so it only has a weak predictive ability.

A number of other water indices have been proposed, for example that based on the position of a pixel on an NIR:MIR scatter plot (Ghulam *et al.*, 2008). Because the remote-sensing indices of leaf water content primarily measure the amount of water in leaves or canopies, in most studies of canopy function it will be necessary to make the conversion to *RWC*, which will usually require calibration for the vegetation of interest. Furthermore, although the *EWT* sensed remotely may be related to the amount of water in the vegetation canopy, it is unlikely to be a reliable measure of plant water deficit, other than indirectly as a result of reduced plant growth or biomass, which is itself determined over the longer term by historic water availability for vegetation growth. Only when *EWT* varies at a constant canopy biomass or *LAI*, might one expect useful relationships. Generally, much more useful indices for use as indicators of water-deficit stress in vegetation are available using thermal sensing (see Section 8.5 and Chapter 11). Interestingly, variation in reflectance at the water absorption peak at 1200 nm has also been correlated with changes in crop development from vegetative growth to heading in cereals (Pimstein *et al.*, 2009).

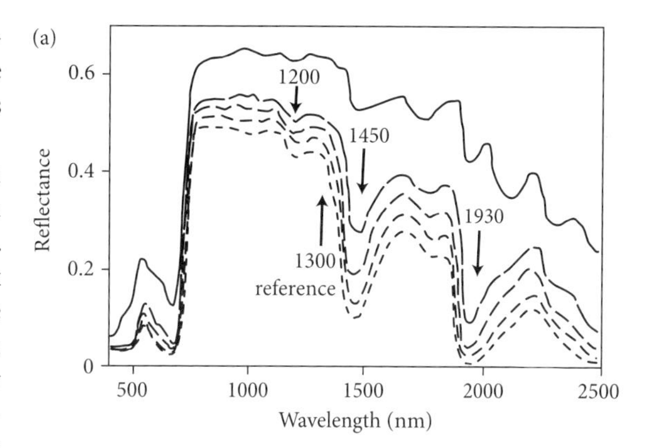

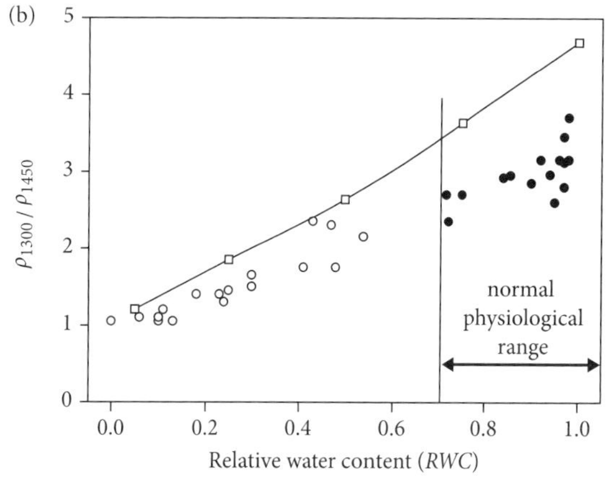

Fig. 7.14. (a) Reflectance spectra of *Magnolia grandiflora* leaves at relative water contents ranging from 5% (solid line) through 25%, 50%, 75%, to 100% (short dashes) showing the marked dips in reflectance with increasing water content around 1200, 1450, and 1930 nm and showing the reference wavelength used for *LWI* at 1300 nm (data from Carter, 1991), and (b) response of a reflectance water index ($LWI = \rho_{1300}/\rho_{1450}$) to leaf relative water content for the *Magnolia* data (solid line and open squares) and for *Spathyphyllum lynise* leaves (data replotted from Seelig *et al.*, 2008b) with values for the range of water contents expected for functioning plants (●), and for the full range of water contents (○).

7.3 Some other approaches to the estimation of vegetation cover

7.3.1 Estimation of fractional vegetation cover using Lidar

Small-footprint airborne laser scanning systems (ALS) can be powerful tools for estimation of fractional vegetation cover (see Section 5.6 for details). Even systems that only record first and last echoes can derive three classes of echo: first echo, last echo, and single echo. For vegetation, most single echoes will come from the ground. The fraction of cover (f_{veg}) for a forest can be estimated as $\Sigma E_{veg}/\Sigma E_{total}$ where the vegetation echoes (E_{veg}) are defined for forest as those above a threshold height (e.g. 1.25 m) above the ground. The calculated f_{veg} in this way has been found to be much higher when using first echoes than when using second or single echoes and approximations are needed to determine the true value for f_{veg} (e.g. Morsdorf *et al.*, 2006). The use of Lidar for studies of vegetation cover is developed further in Chapter 11. For close-range sensing (within a few m), ultrasound reflection using cheap proximity sensors can also potentially be used to provide an estimate of the canopy density, which is related to cover.

7.3.2 Microwave vegetation indices

Conventional optical vegetation indices can only monitor the surface of a canopy and they are not able directly to provide information about biomass or total woody material. There has therefore been interest in using passive microwave sensing, which at long wavelengths responds to relatively thick layers of canopies as alternative indicators of woody material and total plant biomass complementary to conventional *NDVI*. Advantages of using microwaves include the fact that they can be used from space at night or in cloudy conditions, though a severe limitation for many vegetation studies is the low spatial resolution of at best 10s of km. A number of possible approaches to the use of microwave emissions have been proposed. These include: passive microwave polarization difference temperatures (*MPDT*) or the corresponding normalized difference (*NPI*; see, e.g., Paloscia and Pampaloni, 1992) given by

$$NPI = C\,(T_{bv} - T_{bh})/(T_{bv} + T_{bh}), \tag{7.7}$$

where C is a scaling factor and T_{bv} and T_{bh} are the microwave brightness temperatures for vertically and horizontally polarized radiation. This type of index has been successfully correlated with *NDVI*, especially in semi-arid regions, and with plant water content per m^2. Although less sensitive to atmospheric transmission, microwaves are more sensitive to background emission from the soil. In order to correct for this soil effect, Shi *et al.* (2008) proposed an alternative approach applicable to short canopies where the microwave radiation can penetrate right through the canopy. They defined microwave vegetation indices (*MVI*) as the intercept (*A*) and slope (*B*) of the relationship between brightness temperatures at adjacent frequencies and showed that such indices are independent of the signals from the underlying soil. *MVI-A* tends to be positively related to *NDVI*, while *MVI-B* is negatively related, with the latter being more robust as it is not affected by surface temperature.

7.4 Image classification

Image classification is a fundamental tool for many remote-sensing applications where one is often interested in reducing the complexity of a remote image to a limited number of near-homogeneous classes that may represent, for example, different vegetation or landcover types. The classification of pixels into specific classes is most commonly based on recognition of their characteristic *spectral signatures*. Such a classification process produces a *thematic map* that displays the spatial distribution of a specific phenomenon such as the type of soil or agricultural crop. A process of calibration is required to convert the radiometric information in any pixel or identified area into thematic information such as vegetation type. Remote sensing software generally comes with the facility to classify areas of multi- or hyperspectral images into a number of such classes. The simplest and most commonly used classifiers (*point classifiers*) treat each pixel independently by assigning it to a class based on its own spectral properties alone; this contrasts with some more complex classification methods that consider the *patterns* of brightness (or *image texture*) within groups of adjacent pixels. The many approaches available for such pattern recognition or image simplification are summarized in Table 7.2 and

Table 7.2 Ways of grouping pattern-recognition/image-classification techniques (following Lu and Weng, 2007).

	Explanation	Examples
1. Context		
No context	Uses only single pixel/field data average	The majority of methods fall within this class
Contextual methods	Take account of the neighbourhood	
2. Use of supplementary information		
No supplementary information	Based solely on single image information	The majority of methods fall within this class
Use of additional information	Can include multitemporal observations, DEMs, mixed optical and radar, and data fusion and data assimilation approaches	Generally require non-parametric methods
3. Training data		
Unsupervised	Based solely on spectral information within the image: operator needs to label and merge resultant classes	ISODATA, *k*-means clustering, maximum likelihood, self-ordering maps (SOM), etc.
Supervised	Based on training dataset: signatures extracted from these and used to derive thematic classes	
4. Spatial requirement		
Point or per pixel	Each pixel classified independently; effectively classified according to its dominant component when not pure	
Per field/object	Uses the average spectrum of a parcel of adjacent pixels	
Mixed pixels	Several classes may be allocated as proportions of a pixel; assumes that the spectrum for a pixel is a combination (usually linear) of contributions of components	Fuzzy membership methods; spectral unmixing, ensemble methods
5. Statistical assumptions		
Parametric	Based on statistics of data	Max likelihood, min distance, discriminant analysis, etc.
Non-parametric	No assumptions about data structure; can readily incorporate various types of data	ANN, SOM, support vector machines (SVM), expert systems, *k*-nearest neighbours, ensemble methods and decision-tree approaches

some of the main methods are explained further in the following sections. A more complete coverage of the principles and practice of image classification and pattern recognition in remote sensing may be found in any one of the many texts on the subject such as those by Tso and Mather (2001), Mather (2004), Jensen (2005), Richards and Xia (2005) and Russ (2006). In addition, the most popular software packages such as ENVI and ERDAS™ Imagine both have quite extensive explanations of the techniques available. Here, we only aim to provide the essential information required to clarify what can and what cannot be achieved by the different approaches and to provide some guidance on the choice of an appropriate technique for specific applications.

The number of separate classes that can be distinguished successfully usually depends, amongst other things, on the number of uncorrelated spectral channels available; therefore hyperspectral images can be more useful than multispectral images.

7.4.1 Scatter diagrams

Multispectral classification is based on the fact mentioned in Section 5.4.1 that a digital image is spatially and spectrally discrete. Each pixel is represented by a set of radiance values (*DN*s), one for each of the wavelengths measured. The question to be answered is: is there a set of digital values that will uniquely identify the ground cover represented in that pixel? It is highly unlikely that the digital values for two distinct features in an image will be so similar in all wavelengths that they cannot be separated, especially if several wavelengths are used. Even if they have similar responses in two bands, often their response in a third band will be sufficiently different so as to differentiate between them.

One can explain the basis of most classification algorithms by looking at the two-dimensional case where we consider two wavebands only (Fig. 7.15, see also Figs 7.2 and 7.3). In this case we can plot a *scatter diagram* or *scatter plot* where, for each pixel, the value in one band is plotted on one axis of a graph and the value in the other band is plotted on the other axis. If this is done for all the pixels in the image, the points represent pairs of values for every pixel. Note that the spatial location of the pixels associated with that point is now lost. The points are plotted in a *feature space*,[35] not a physical space. Points representing pixels with similar spectral responses will tend to form *clusters*, hence the alternative name *cluster diagram*. It may be possible to relate a particular cluster to a type of landcover. For example, water has low reflectivity in all wavebands and hence all water pixels would form a cluster near the origin, whereas vegetation has low reflectivity in the red and high reflectivity in the infrared forming its own distinctive cluster, while soil does not have markedly different reflectances in these two wavebands, so clusters along the diagonal (Fig. 7.15(a)). For any pair of highly correlated wavebands, the points would tend to lie along a straight line through the origin. The objective of classification is to try to identify the clusters relating to different landcover types (*classes*). Other features can be plotted on one or more axes, such as height, aspect, soil type, etc., hence the alternative name *feature plot*. It is not easy to visualize a plot in more than three dimensions, but such analysis can readily be performed by a computer for all the bands in a multispectral or hyperspectral image.

The computing cost of classification increases with the number of wavebands (features) used to describe pixels. For some classifiers the cost only increases linearly with the number of features, while for others such as the maximum likelihood classifier discussed below, the cost increase with features can be quadratic. Therefore it is often useful, especially when using hyperspectral imagery, to optimize the choice of wavebands to use in the classification (in order to reduce redundancy between highly correlated wavebands). A common approach to such feature reduction is to analyse the spectral statistics of all the bands, and especially the relationships between different bands in order to select those wavebands that are least correlated. Another approach to the selection of bands to use in classification is to undertake a principal components transform (Section 6.4.4) that generates a new series of completely uncorrelated bands.

There are many approaches to the identification of clusters; these will be outlined in more detail in the following section. The simplest would be to specify the maximum and minimum values in each band for a pixel to belong to a particular class. This is equivalent to drawing a rectangle enclosing a cluster on a two-dimensional graph (or a 'box' in three or more

[35] The spectral bands (such as the seven ETM+ bands), together with other derived properties such as context or texture, are called the *features* that describe the properties of a particular pixel.

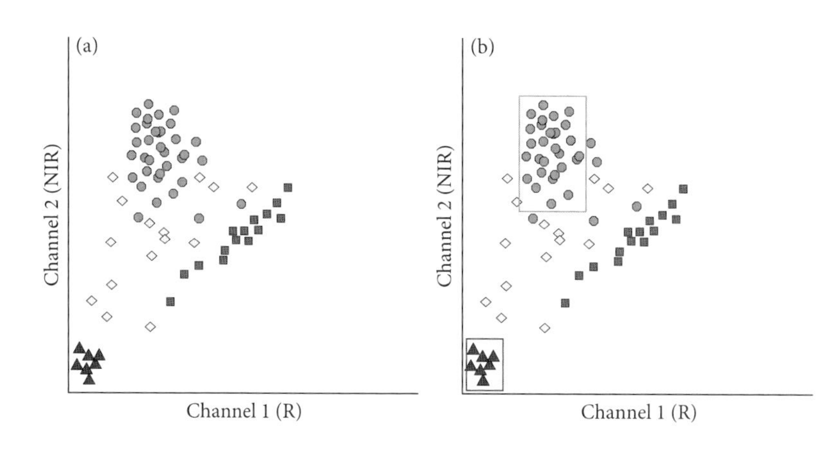

Fig. 7.15. (a) Illustrative scatter plot of pixel distribution in two spectral dimensions for an image where the value in the NIR is plotted against the corresponding value in the red for each pixel. Triangles represent water, squares represent soil, circles represent vegetation, and the diamonds represent mixed or unknown pixels. (b) The same data with boxes (parallelepipeds) drawn around the majority of water and vegetation pixels, defining the normal ranges for these classes.

dimensions), hence its name of *box* classifier, or more strictly a *parallelepiped* classifier (Fig. 7.15(b)). Because of the scatter of points within most real-life clusters the different boxes may overlap; in such cases a probability-based method can be useful.

The final stage of any classification process is to present the results. This is often done by means of a thematic map in which the different classes are colour coded (a *classification map*) or the results can be imported into a GIS. Decisions also have to be made as to whether every pixel is forced into membership of one of the specified clusters, or whether some pixels are allowed to remain 'unclassified'; alternatively 'fuzzy' classifications may be useful. Finally, it is usually necessary to assess the accuracy of the final result; this will be addressed in Chapter 10. Because results tend to be specific to the particular images used, it is not possible to specify 'the best' method, only the best method for a particular case.

7.4.2 Basic classification approaches

Unsupervised classification

In so-called *unsupervised classification*, the software itself classifies pixels based on their spectral statistics only, often grouping them into a pre-determined number of spectrally similar groups. The software then groups the individual points into the required number of clusters according to their separation in multidimensional space. Groups of points that fall close together will get allocated to one class. As it is only possible readily to visualize multidimensional remote-sensing data in two or three dimensions we illustrate the concept of distance between points in multidimensional feature space in Fig. 7.16 using a simple 3D example. A common measure of the spectral similarity of two pixels is the *Euclidean distance* (D_{ab}), which for two pixels (a and b) would be given, by analogy with Pythagoras' theorem, as the square root of the sum over all n channels of the squared differences between the brightness of the two points:

$$D_{ab} = \sqrt{\sum_{i=1}^{n} (a_i - b_i)^2}, \tag{7.8}$$

where a_i and b_i are the values for pixels a and b for the ith channel. As an illustration of the calculation, the results for three data points and four channels are summarized in Table 7.3 which shows that in this case a and b are much the closest pair of points. The data points are then each allocated into the nearest class.

Although the Euclidean distance is only one of a number of ways in which distance may be defined, the principle for all unsupervised classifications is similar, with the algorithm usually proceeding in an iterative fashion allocating pixels to groups until the clearest grouping is achieved, with the analyst stating either the number of classes or some *decision rules* for allocation of pixels to classes. Means are recalculated at each iteration and pixels reclassified, with classes being split or merged at each stage depending on the threshold rules used. Some pixels may remain unclassified if they do not meet the chosen criteria. Typical examples of unsupervised classification algorithms that are widely available include the ISODATA method and the k-means method (see references for further details of the various unsupervised classification approaches available). Unsupervised classification can also be achieved using

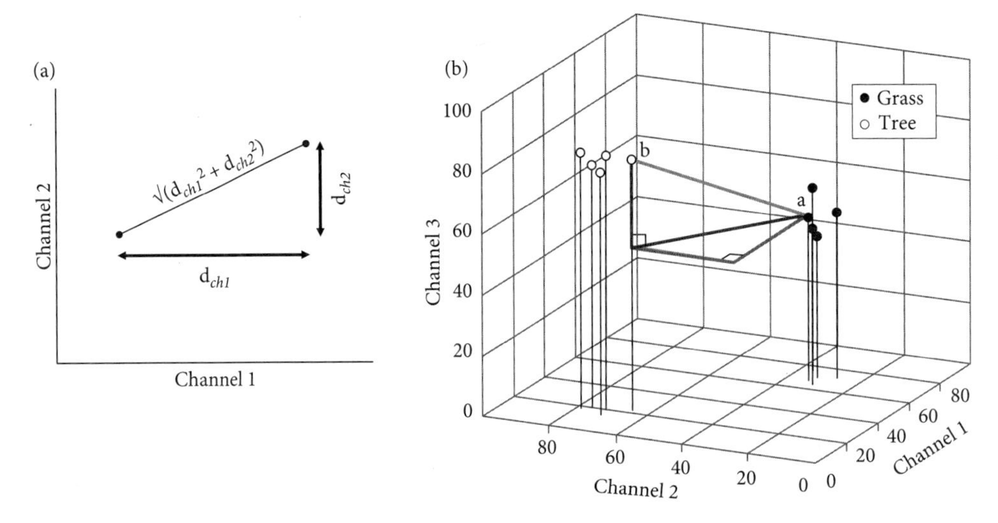

Fig. 7.16. (a) Calculation of the Euclidean distance between two sample points (a and b) for two channels using Pythagoras's theorem where the distance between them (D) is given by the square root of the sum of each channel difference squared. (b) Shows a three-dimensional scatter diagram illustrating the generalization of Pythagoras's theorem to multidimensions.

the *self-ordering map* (SOM) which is a neural network-based approach (de Smith *et al.*, 2009; http://www.spatialanalysisonline.com/index.html). A major problem with unsupervised classification approaches is that the classes that are output from such an automated procedure do not necessarily relate closely to the landcover classes on the ground. Nevertheless, in many situations it is feasible for the operator to assign the derived classes to physiologically meaningful features (soil, grass, water, etc.).

Supervised classification

In contrast to unsupervised classification, supervised classification methods start from an initial identification of certain areas or pixels from the image that are known (e.g. from ground survey) to comprise particular vegetation or other surface types of interest for the particular study. In this case the spectral characteristics of these *training pixels* are first measured, and, in 'conventional' approaches, a mean for each training class is calculated. The software then attempts to assign the remaining pixels to the most similar (nearest) training class. It is important to ensure that the training pixels chosen are as homogeneous as possible, and that each class is clearly separable (i.e. spectrally distinct). Because there is normally substantial spectral variation within any superficially homogeneous land type, it is usual to select a good number of samples of each type, well distributed around the image, in order to minimize any possible view-angle or other bias or spatial autocorrelation. When using the statistically based

Table 7.3 Calculation of Euclidean distances between two pixels: a and b in Fig. 7.16 having the following reflectances (DN) for three hypothetical channels.

	channel 1	channel 2	channel 3	$\Sigma(a-b)^2$	D_{ab}
a	65	32	50		
b	22	66	60		
a–b	43	–34	–10		
$(a-b)^2$	1849	1156	100	3105	55.7

classifiers described below it is strictly necessary that the training set contains $>n+1$ pixels (where n is the number of bands), though substantially larger numbers (of the order of $10n$) are preferable to allow good estimates of the variance and covariance properties of each class. Further discussion on the optimal selection of training areas may be found in remote-sensing texts, and will be discussed further in Chapter 10.

There are many ways in which the pixels in an image may be attributed to the different training classes. Each training class may be envisaged as consisting of a cluster of points in multidimensional space whose centre (or centroid) can be defined as its mean and any unclassified pixel is simply attributed to the nearest training class centroid, as in the *minimum distance to the mean* classifier. Depending on whether a distance threshold is set by the operator, all pixels may be attributed to a class or some may be left unclassified. In practice, such a simple minimum distance classifier neither takes account of the variability within classes and the fact that there is potential overlap between classes, nor does it take account of covariance between features that might give rise to elongated classes. A more robust classification, and the one most widely used in remote sensing, can be achieved by using methods that take account of the probabilities that a pixel is a member of each of the possible classes, rather than just the simple distance information. An example of this type of classifier is the *maximum-likelihood* classification that uses the training data to estimate means and variances of the different classes, and hence to calculate the probability of any pixel falling into a given class (see a simplified example for two dimensions in Fig. 7.17(a)). The computing requirement for maximum-likelihood classifiers is substantially greater than for simple distance measures, increasing rapidly with the number of features being used. Although this type of classifier is generally preferable, its parametric assumptions and requirement for covariance information means that it does not perform well where training information is limited or non-homogeneous so in such cases it can be better to use classifiers such as the minimum distance that only require positional information. Another commonly used classification method is the *spectral angle mapper*, which determines the spectral similarity between two spectra on the basis of a calculated 'angle' between the two spectra in multidimensional space. An advantage of this approach is that it can be relatively insensitive to illumination changes.

Classification results are notoriously sensitive to changing atmospheric conditions, use of different sensors, and time of year of image acquisition. Some examples of such effects are shown in Fig. 7.18, which also illustrates the use of a 'consensus' approach to classification where one takes the most common class for each pixel from a series of independently classified images.

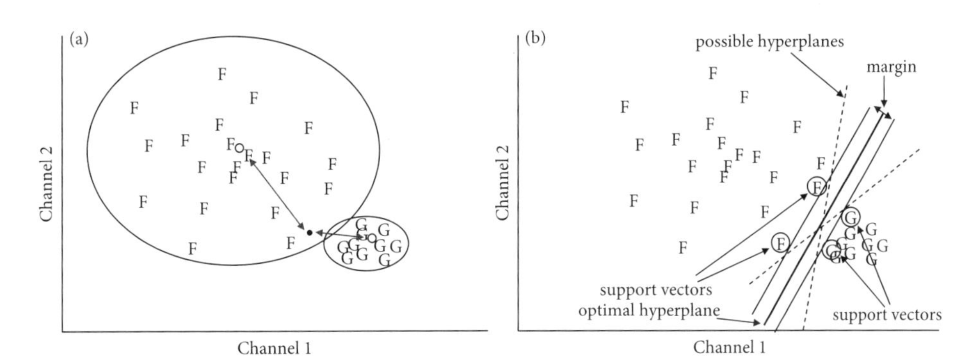

Fig. 7.17. (a) Maximum-likelihood classifier compared to the minimum distance classifier (simplified to two dimensions). The symbols (F = forest, G = grass) indicate the positions in two spectral dimensions of training pixels, together with the respective centroids (○) of each group, showing the greater variability of forest pixels. Although the unknown (•) point pixel is nearer to grass and would be classified as such by a minimum-distance classifier, it is clear that it has a higher probability of being a forest pixel, and would be likely to be classified as such using a maximum-likelihood classifier. (b) The same dataset illustrating the support vector machine approach and showing the optimal hyperplane separating the two groups, the support vectors (circled points) and the 'margin' determined by the position of the support vectors; the dashed lines indicate other non-optimal hyperplanes.

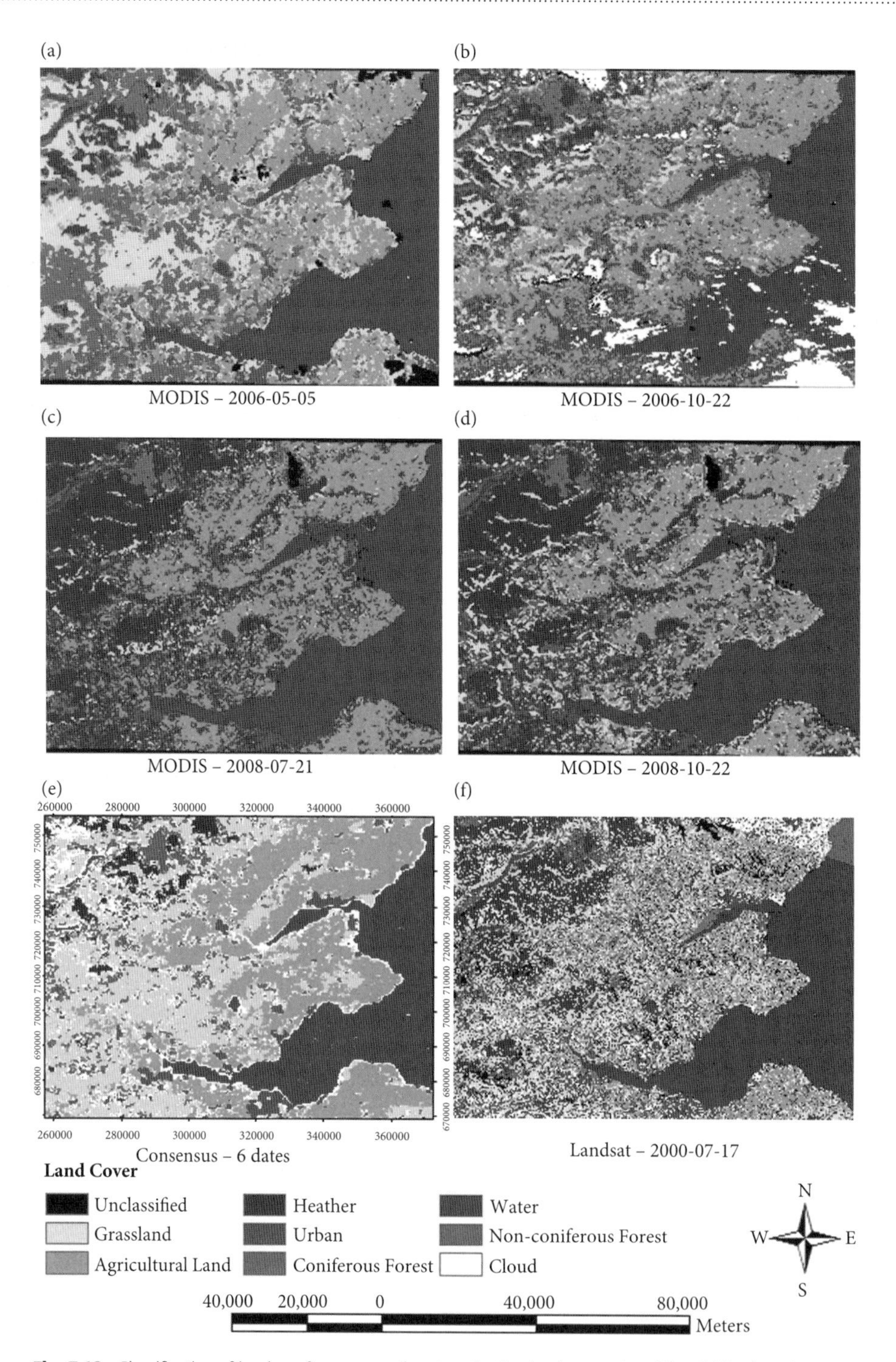

Fig. 7.18. Classification of land use for an area of eastern Scotland using a series of four fairly clear sky images from MODIS (a, b, c, d) as compared with Landsat TM (f) using a maximum-likelihood algorithm with the same parameters and the same training areas. In addition, a consensus image (e) for MODIS is shown (where colours are shown for five or more similar classifications using 11 NBARs images over the period day 43 to day 289 in 2006 and 2008) (MODIS data provided by NEODAAS at University of Dundee and Plymouth; Landsat data for 17th July 2000, courtesy US Geological Survey and Landsat.org). (See Plate 7.1)

Non-parametric methods

In many vegetation studies one is interested in both categorical variables such as might be obtained by a classification (species or landcover type) and in continuous variables such as biomass or tree density. In principle, mapping such attribute variables can be achieved using multivariate statistical modelling, but the approach can be invalidated by non-Gaussian variation of the attribute variables. Therefore, in such cases non-parametric approaches that do not assume a normal distribution of the data are preferable. Particularly useful are nearest-neighbour approaches such as the *k-nearest-neighbour* (*k-NN*) algorithm. This approach is distinguished from the minimum distance to the mean classifier outlined above by the fact that rather than relating each pixel to the nearest mean of a training class, it uses all the training pixels and classifies each pixel using the majority of *k* training pixels that are nearest in feature space (*k* is usually set to an odd number to ensure a single answer). This approach is being increasingly used in forestry applications (e.g. Tomppo *et al.*, 2009, and see Chapter 11).

Another powerful classification tool in supervised classification is the use of an *artificial neural network* (ANN). This and other non-parametric methods such as the use of *support vector machines* (SVM) make no assumptions about the data structure and can readily incorporate different types of data (see, e.g., Foody, 2008). An ANN is a computer program designed to simulate the human learning process by establishing linkages between input and output data, reinforced by repeated learning. The input to the ANN may be multispectral images from several dates (large volumes of data can be handled) and the output will be the classes (modest in number). Also, training data are needed to establish the linkages within the network, and often the process is two-way such that differences between expected and actual results can be used to adjust the weights involved in the analysis. SVMs are based on the use of statistical learning theory and aim to find the best hyperplane in multidimensional feature space that optimally separates classes (Fig. 7.17(b)); the points that constrain the width of the margin are known as support vectors. This approach, which only looks at the spectral boundaries between classes, is fundamentally different from the common nearest-neighbour and maximum-likelihood approaches that are based on the centroids of the different classes.

Each type of classification has its own advantages and disadvantages. The advantages of unsupervised classification primarily relate to the fact that no prior knowledge is required and natural groupings are brought out with little scope for subjective or human error as very little operator input is required. In such classification methods the computer will allocate pixels to one of a defined maximum number of classes, broadening the classes to accommodate outlying values and taking no account of mixed pixels, perhaps leading to misclassification. Unfortunately, the spectrally homogeneous classes identified may not correspond with classes of interest and importantly the classes identified are unlikely to be consistent over time when the approach is applied to different images. In supervised classification, however, only those pixels whose values lie within the pre-defined groupings that have been specified by the algorithm used will be classified as such, all other pixels remaining 'unclassified'. The accuracy of this classification therefore relies on the thoroughness of the analyst in selecting a representative range of training pixels.

In multispectral classification, the classes are assigned purely on the basis of the spectral signatures of individual pixels, but variability can be introduced due to such things as topographic and environmental effects. Accuracy can often be improved by incorporating additional features into the analysis, such as elevation, slope, aspect and soil, and geological maps, all of which might affect the specific signature for that pixel. Other features such as texture, field or forest boundaries, ponds and lakes, and buildings may be incorporated from a GIS.

Decision tree and ensemble classifiers. A useful approach to the improvement of classification is to apply several classifiers in combination in an ensemble approach (see, e.g., Doan and Foody, 2007). Yet another approach is the use of decision trees where ancillary information is built into the analysis. A *decision-tree classifier* involves a series of binary decisions to arrive at the classification of any pixel (Hansen *et al.*, 1996). A very simple example of a decision-tree approach is illustrated in Fig. 7.19. Particular advantages of these combined approaches are that they do not rely on any parametric assumptions and they can be applied in a very flexible manner to a wide range of problems, especially when these are related to mixed pixels.

Object-based classification. Where the end members are not very clear or are overlapping in spectral space,

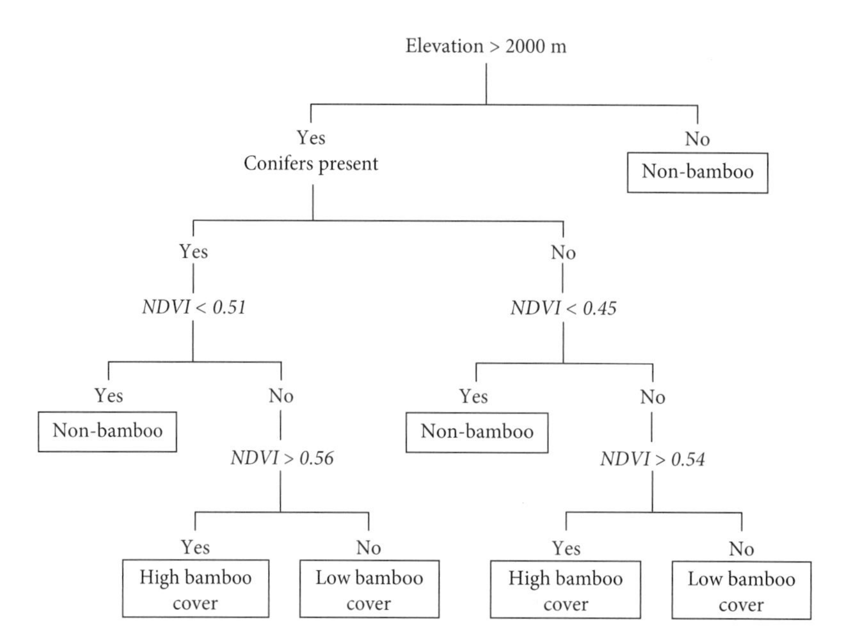

Fig. 7.19. Example of a simple decision-tree classifier (Wang *et al.*, 2009) where the density of understorey bamboo is estimated in one of three classes on the basis of successive decisions involving first an altitude threshold (obtained from a DEM), secondly the application of a landcover criterion (from a landcover map, or from image classification), and thirdly the application of a series of specific *NDVI* thresholds.

it may still be possible to obtain a classification using a different technique called *per parcel* or *per field* classification, as opposed to the usual *per pixel* classification. In this, instead of classifying each pixel in the image independently, the boundaries of homogeneous areas of landcover are drawn and all the pixels within such areas are forced into one class.

Simple per parcel classification, where the objects are defined in advance, has been enhanced by the development of more sophisticated object-based image analysis (OBIA) where objects can be determined by the software itself (see Blascke, 2010). This is a developing field that substantially enhances our ability to classify remotely sensed images (see for example, the GEOBIA website; http://wiki.ucalgary.ca/page/GEOBIA). A number of commercial softwares are becoming available that examine pixels in their context. One example is the Definiens suite of software (Definiens AG, Munich, Germany) which builds up a picture iteratively, recognizing groups of pixels as objects and using the colour, shape, texture, size, and context of the objects. This contrasts with pixel-based techniques that can only use the spectral properties of the single pixel. Object-based classification can be particularly valuable for higher spatial resolution imagery as the increased variability implicit within such imagery tends to confuse standard pixel-based classifiers resulting in lower classification accuracies than might be obtained with lower-resolution images where each pixel includes all components of a particular vegetation type (e.g. soil, trees, grass, and shadow). Data fusion (Section 6.4.6) is also an approach to obtaining subpixel information.

7.4.3 Mixed pixels

The above discussion of classification assumes that the ground area represented by each pixel contains one single class; this provides what is known as a *hard classification* of the scene. In general, however, it is rarely true that each pixel represents only one class (especially where the spatial resolution is coarse with respect to the variability of the landcover). Such pixels are called *mixed pixels* and are often the cause of misclassification, or at least failure to classify, using standard algorithms. The effect of mixed pixels on the results of classification will depend on which approach is taken. In supervised classification, this pixel may fortuitously have the same signature as one of the classes used as training samples and thus be misclassified. Alternatively, if its signature is different, then it will not be classified at all. In unsupervised classification, however, each mixture may form a class of its own.

For a mixed pixel, the intensity received by the detector in any waveband for that pixel will be the average response of each of the pure classes represented in the pixel (the *end members*) weighted in proportion to the fraction of the area occupied by that class. Therefore, the spectrum may not represent any of the classes that actually occur in the scene. Such mixed pixels often occur at the edge or border of an area, say a field boundary, or if another class is contained within the pixel (e.g. isolated trees in grassland).

A variety of techniques is available to handle mixed pixels. One approach to inhomogeneous areas is to use *fuzzy* or *soft classification* methods, rather than the above *hard classification*. Hard classification assumes that each pixel is pure (i.e. that the ground area is occupied by a single information class) and that pixels can belong to only one group – the choice is between 'yes' and 'no', with occupancy of one or zero. Fuzzy classification mimics human decision making that contains elements of subjectivity ('maybe'), and allows partial membership of several classes with occupancy between 0 and 1, all summing to unity. For example, a label of 0.4 (water) and 0.6 (forest) would mean that the probability that the area was forest was 60%. The typical output of a soft classification will be a set of fraction images, each describing the proportion of a particular landcover class within each pixel. It is possible to introduce fuzzy set theory into some of the standard 'hard' algorithms, such as maximum likelihood and ISODATA.

An alternative 'hard' approach makes use of the spectral signatures of a number of pure component classes. Each pixel is assumed to be a mixture of the *n* pure component materials, the so-called *end members*. In this case it is often possible to estimate the fraction of each of these end-member classes that occurs within the pixel using a process known as *linear spectral unmixing*. The maximum number of end members, and hence the number of component classes that can be identified, is limited by the number of spectral bands available, so that the approach is frequently most successful with hyperspectral data. The process of linear spectral unmixing is illustrated in Fig. 7.20 for the simplified case of two spectral bands. The first

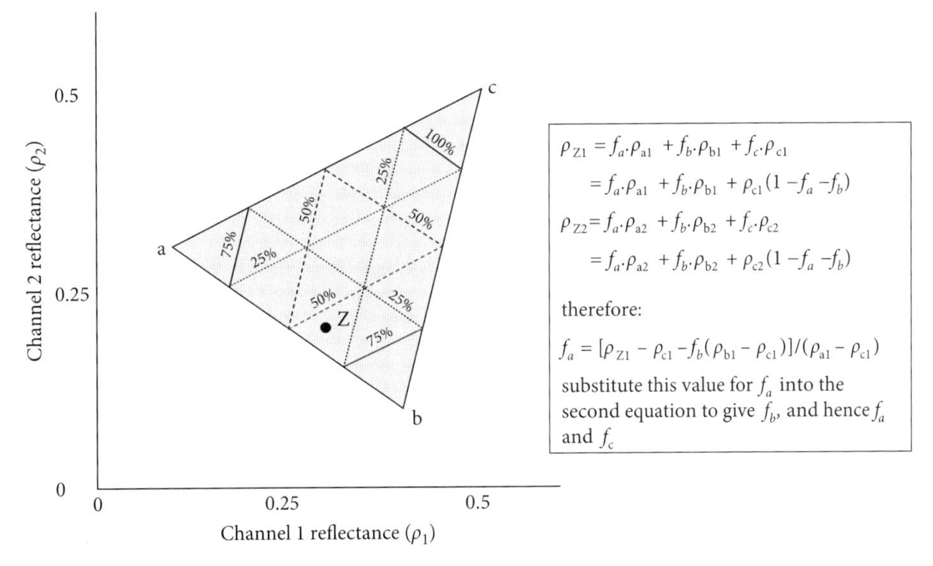

Fig. 7.20. An illustration in two dimensions of how linear spectral unmixing can derive the fractions (f_a, f_b, and f_c) of the pure end-members *a*, *b*, and *c* from the reflectance of a mixed pixel, Z. All possible mixtures of *a*, *b*, and *c* fall within the triangle defined by a, b, and c; the lines in this diagram show constant proportions of each component. One can estimate the fractions of end-members a, b, and c that are represented in the pixel whose spectral properties are given by the point Z from the two simultaneous equations shown. Since $f_a + f_b + f_c = 1$, there are only two independent fractions, f_a and f_b. The spectral reflectances in channel 1 are given by ρ_{a1}, ρ_{b1}, ρ_{c1}, and ρ_{Z1} and in channel 2 by ρ_{a2}, ρ_{b2}, ρ_{c2}, and ρ_{Z2}.

step in linear spectral unmixing is to determine the spectral properties of the n components either on the basis of pure training pixels from the image or sometimes from standard spectra.

The power of the technique and the number of components that can be resolved depends on the information content of the image and the distinctness of the spectral end members. Given perfect data the approach requires that the number of independent classes (i.e. the total number of classes minus one) cannot be greater than the number of wavebands. This is because with x simultaneous equations (one for each waveband) one can solve for at most x unknowns. A problem arises, however, whenever there remains some spectral variability within a single class. This is particularly likely where the spatial resolution of the detector is finer than the scale of the variability in the landcover. For example the spectral response for a particular crop may vary with stage of growth or degree of stress, resulting in a rather undefined end member. Linear unmixing also requires that the end members are as distinct as possible so that it is often difficult or impossible to distinguish different vegetation types in one image because they tend to be spectrally somewhat similar. A computationally more robust alternative technique can be to use a *subpixel classifier* approach, where one only specifies the spectral properties of one particular component material of interest to derive an image plane containing the fraction of that material in each pixel. This can be much more successful at separating spectrally similar materials as only the end member of interest needs specifying.

The use of ensemble classifiers is particularly useful for treating those cases where there may be a large proportion of mixed pixels and when soft classifiers are used. Using a suite of classifiers can substantially improve accuracy.

Further reading

Useful summaries of some of the main vegetation indices may be found in Jensen (2005) and Jensen (2007), while a good introduction to image classification methods may be found in Mather (2004).

Websites

GEOBIA website: **http://wiki.ucalgary.ca/page/GEOBIA**

http://www.spatialanalysisonline.com/index.html

Sample problems

7.1 Using the data in Table 7.1 calculate the expected *NDVI, SAVI, DVI, RVI* and the corresponding scaled values (*NDVI**, *SAVI**) for a target containing a homogeneous mixture of 40% dry soil, 20% green grass, and 40% dry grass. What would be the effect of re-greening of all the dry grass on these calculated indices?

7.2 Using the data from Table 7.1 derive the R and NIR radiances for the three surfaces if the incoming radiation in both R and NIR were (a) 100 W m^{-2} or (b) 350 W m^{-2}. Calculate for each surface the values of *DVI* and *NDVI* using these radiances, rather than the reflectances, Comment on any differences between these results and those shown in the table.

7.3 Given the following statistics for the three classes A, B, and C:

Class	Band X mean	Band Y mean
A	81	100
B	103	95
C	95	144

determine to which class a pixel with X = 85 and Y = 124 would be assigned using the minimum distance to the mean rule for classification.

7.4. Plot the dependence of fractional vegetation cover (f_{veg}) on leaf-area index assuming a homogeneous canopy with a horizontal leaf distribution. How would *NDVI* be expected to vary with *LAI* for this canopy assuming reflectances given in Table 7.1?

7.5 Below are small extracts of the pixel values observed within a remotely sensed image in the visible and the near infrared:

169	157	162
159	158	90
101	92	99

Visible

90	99	82
90	88	158
101	169	157

Near infrared

The extract represents two cover types, A and B. Apply an algorithm that will determine which pixels correspond to each cover type; comment on what the surfaces might be.

7.6 A satellite scanner is viewing an area of ground that, because of relief, is not uniformly illuminated. Digital values for two typical pixels in three bands are shown in the table. Determine (a) the best ratio that reduces the effect of illumination, and (b) a normalized difference vegetation index (*NDVI*) which is independent of illumination.

Illumination	Band 1 (red)	Band 2 (near IR)	Band 3 (mid IR)
sunlit	60	140	160
shadow	28	55	75

8 Multiangular sensing of vegetation structure and modelling of radiation-transfer properties

8.1 Introduction to multiangular sensing

So far in this book we have concentrated on the information provided to the remote sensing scientist in the solar spectral domain, though we have noted in passing that the spectral signal reflected from any surface depends both on the direction of illumination and on the direction of view. In this chapter we develop the treatment of anisotropy of reflectance as a tool that enables us to extract more information from remote sensing images than is possible from the use of spectral information alone. A useful introduction to the history of studies of anisotropy in the radiation field and multiangular sensing has been provided by Verstraete and Pinty (2001). Understanding and quantification of angular effects is critically dependent on our ability to model the angular variation in radiation-transfer properties of different plant canopies and their components, so perhaps somewhat unconventionally we combine our discussion of the measurement and analysis of multiangular effects with the development of radiation-transfer models that explicitly treat anisotropy in the radiation field. This is not meant to imply that these models are not also applicable to conventional nadir-viewing remote sensing; indeed the common assumptions of nadir-view remote sensing may actually be considered as a special, though limited, case of angular information. Particular advantages of an explicit treatment of angular effects include (i) the potential for greater accuracy than obtainable with simpler models and, (ii) when combined with appropriate multiangular observations, the potential for extraction of additional information on target structure.

In the first part of this chapter, therefore, we introduce the basic principles and terminology of multiangular sensing, followed by an outline of some of the approaches to the collection of multiangular data. We then extend the treatment of radiation-transfer modelling, introduced in Chapter 3, to incorporate explicit modelling of angular effects and then outline approaches to the inversion of radiation-transfer models for the extraction of canopy biophysical parameters.

8.1.1 Why use multiangular measurement?

Conventional passive remote sensing is based on the assumption that one measures the reflectance of the ground surface from one particular view angle, usually from vertically above (nadir viewing). In practice, however, the finite range of view angles of most sensors means that for many situations different parts of an image will view the surface at different angles, so that clear gradations in brightness may often be detected across the image (Fig 8.1 and Plate 8.1). In fact, the reflectance of complex three-dimensional surfaces such as plant canopies varies quite strongly as a function of both the angle of view and the angle of illumination, and also the angle separating these two directions. For many years this angular variation was treated as a confounding effect that needed to be corrected for in most remote sensing applications. More recently, however, there has been increasing recognition that this angular

Fig. 8.1. Aerial image of a vineyard from a balloon at 90 m showing the decreasing brightness of the image as one moves from the hotspot (the area around the shadow of the balloon on which the cameras are mounted). The changes in brightness largely arise as a result of changes in the proportion of shadow visible, with the proportion increasing as one moves away from the hotspot. (See Plate 8.1)

variation provides valuable further information about the surface biophysical properties. We have already touched on the consequences on such angular variation in our discussions of canopy reflectance (Chapter 3) and of spectral vegetation indices (Chapter 7). This chapter describes the principles underlying the measurement and utilization of the anisotropy in canopy reflectance as a tool for inferring canopy structure and related biophysical characteristics.

In spite of the well-known influences of view angle, there has been rather limited discussion of the problem in previous remote sensing texts except in more specialized volumes such as Hapke *et al.* (1993) and Liang (2004). What effort there has been has largely concentrated on the angular correction and normalization of images to correct for the effect of changing view angle (e.g. Roujean *et al.*, 1992) where the angular variation in reflectance is considered as noise rather than as an information source in itself. Traditional observing strategies have been designed to minimize problems caused by angular variation in reflectance, for example by flying linescan imagers along the solar plane, so that all parts of the image have the same angle of illumination (even though the angle of view differs along the scan line). Although the study of angular reflectance has not made it to many mainstream texts, this is a central area of modern remote sensing and there has been a substantial research effort into this complex and increasingly important subject over the last 20 years or so. Recent advances in computing power now allow us to run rapidly even quite complex models of canopy reflectance, but it is necessary to have a good understanding of the limitations and scope of such models if we are to optimize their use in remote sensing.

Historically the emphasis in the study of spectral properties of individual leaves has differed from that used in much remote sensing. The study of leaf optical properties has most frequently been achieved by the use of a spectrometer with the leaf enclosed in an integrating sphere, thus providing measurements of the diffuse reflectance and transmittance (see Chapter 3). Although this allows estimation of the leaf spectral absorptance that is, for example, of particular interest in studies of photosynthesis, knowledge of the directional reflectance properties is becoming increasingly important for incorporation into the newer and more sophisticated canopy radiative-transfer models.

It is a matter of common experience that when one views a body of smooth water such as a pond it appears brightest when it is viewed directly towards the sun, as a result of the direct (forward, largely specular) reflections of the sunlight. When one looks away from the sun, however, the water appears much less bright, because there is much less reflected light reaching the observer. In contrast to this example, many of us are also familiar with the *'hotspot'* phenomenon shown in Fig. 8.1 and also observed, for example, when viewing closely cut grass such as on a golf green, or when looking out of an aircraft window. Such a surface often looks brighter (has apparently greater reflectance due to backscattering) close to the observer's shadow (i.e. in this case looking away from the sun with the sun behind you) than when one views it looking towards the sun.

8.1.2 The basis for angular variation in reflectance

As was outlined in Chapter 2 (see Fig. 2.6), the reflection from a surface is a function of the angle of view and depends on the proportion of specular and diffuse reflectance. The mode of reflection from a surface varies from specular reflection, where the surface irregularities (roughness) are small in relation to the incident radiation wavelength to perfectly diffuse reflection, where the reflected light is scattered uniformly over the hemisphere above the surface (Lambertian scattering).

The origin of the anisotropic pattern of reflection from any plant canopy (Fig. 8.1), arises for a rather more interesting reason that relates directly to studies of vegetation structure by remote sensing. In its simplest form the basis of the effect is illustrated in Fig. 8.2, which shows that when an array of plants is illuminated by direct sunlight a proportion of the ground beneath is in direct sunlight, while a proportion is in shadow. Clearly the sunlit portion appears brighter to a viewer as it is receiving, and hence reflecting, more light than is the shaded area, but the proportions of sun and shade that make up any scene change with the view angle. When viewing the surface with the sun behind the observer almost all the canopy being viewed is sunlit – hence the bright apparent reflection when viewing from this angle – but when looking towards the sun much of the viewed surface is in shadow and hence looks dark (low reflectance). This leads to the apparent hotspot when the sun is immediately behind the observer (illustrated for the example of a grapevine canopy in Fig. 8.1). In fact, as is apparent also from this figure, not only does the effect relate to shadowed or sunlit soil but it also extends to the individual leaves in the canopy, so it clearly becomes necessary to have a good description of the shape and arrangement of the individual plants, including any clumping of leaves into branches, and of the distribution of leaf angles within each of those units within the canopy.

A complete treatment of the angular dependence of reflection from vegetation therefore requires a full understanding of radiative transfer processes within plant canopies. The basic radiative properties of plant leaves and of canopies were introduced for homogeneous canopies in Chapter 3; here we extend that treatment to consider in more detail the angular dependence of reflection, how it depends on canopy

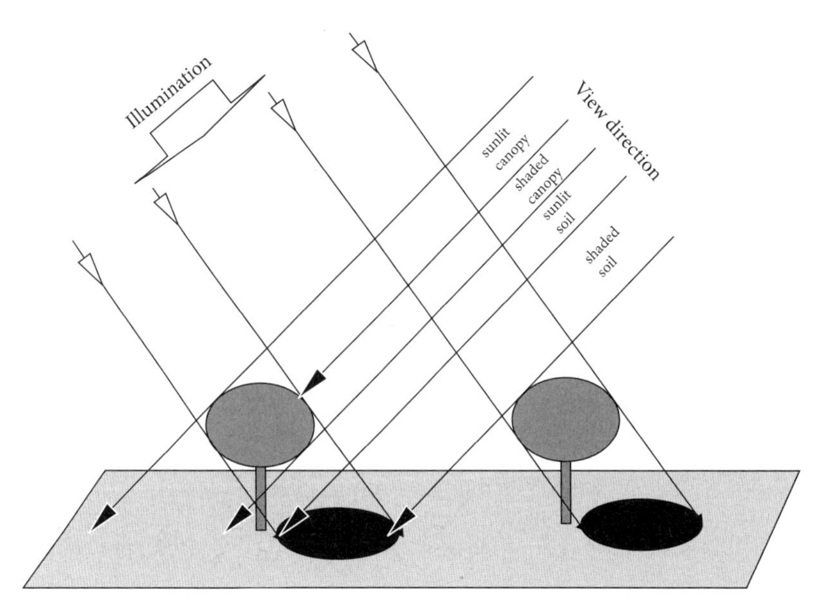

Fig. 8.2. Hotspot diagram: an illustration of how the proportion of shaded surfaces in a scene changes as the view angle changes. When the sparse tree canopy illustrated is viewed from the 'view' direction shown there is a substantial portion of shaded leaf and soil visible, but when viewed from the direction of illumination, only sunlit (and hence brighter) surfaces are visible. Therefore one sees a particularly bright image in the small area when viewing from closest to the illumination direction; this is called a 'hotspot' or bright spot. Compare with Fig. 8.1.

structure and canopy architecture, how it can be modelled, and how its variation can be used to infer canopy structure. Specifically, we outline the approaches available to determine key information about the canopy structure from multiangular measurements of either penetration or reflection.

8.1.3 Bidirectional reflectance definitions

When we measure radiation reflected from a surface we actually measure a *spectral radiance* (= the radiant flux density emanating from the surface; $W\ m^{-2}\ sr^{-1}\ nm^{-1}$); reflectance is the ratio of the reflected to the incoming radiation. The reflectance can be defined in a number of ways depending on the illumination and viewing angles and whether these are directional (restricted to a small solid angle) or hemispherical (integrating over the whole hemisphere). The terminology involved in description of reflectance is summarized in Box 8.1.

The full description of the reflectance properties of a surface is given by its *bidirectional reflectance distribution function* (*BRDF*; $f(\theta_i, \varphi_i; \theta_r, \varphi_r)$) which is defined for all possible illumination and view angles

$$f(\theta_i, \varphi_i; \theta_r, \varphi_r) = \mathrm{d}\boldsymbol{L}_r(\theta_r, \varphi_r)/\mathrm{d}\boldsymbol{I}_i(\theta_i, \varphi_i), \quad (sr^{-1}) \quad (8.1)$$

where θ_i and φ_i refer to illumination zenith and azimuth angles, θ_r and φ_r to view zenith and azimuth angles, $\mathrm{d}\boldsymbol{L}_r$ to reflected spectral radiance and $\mathrm{d}\boldsymbol{I}_i$ to incident directional spectral irradiance for the infinitesimally small view and incident angles. It is also usually a function of wavelength. The other terms for directional or hemispherical reflectance distributions can therefore be derived for the surface by integration over the appropriate angles. As sensors have finite acceptance angles one convenient terminology is to estimate instead the *bidirectional reflectance factor* (*BRF*; $R(\theta_i, \varphi_i; \theta_r, \varphi_r)$) that can be defined as the ratio of the radiance reflected by a surface in a given direction to that which would be reflected into the same reflected-beam geometry by an ideal (perfectly reflective, perfectly diffuse (i.e. Lambertian)) standard surface irradiated in exactly the same way as the target surface (Nicodemus *et al.*, 1977). Since an ideal Lambertian surface reflects the same radiance in all directions, its $BRDF = 1/\pi$, so that for a real surface its $BRF = \pi\ BRDF$. For field measurements we more correctly measure a *hemispherical (directional) reflectance factor* (*H(D)RF*), as light comes not only from the sun, but also from the whole sky hemisphere. As a result of this, the actual *HDRF* is not just a property of the surface but it also depends on illumination conditions.

BOX 8.1 Bidirectional reflectance terminology

We usually define directional reflectances by stating first the degree of collimation of the source followed by that of the detector (Schaepman-Strub *et al.*, 2006) so we have:

(*i*) *Directional-directional* (*bidirectional*) *reflectance* when both the illuminating and viewing angles are infinitesimally small.

(*ii*) *Directional-hemispherical reflectance*, with a small illuminating angle but the sensor view angle is so large that it integrates over the whole hemisphere above the surface. For a canopy this is also called 'black-sky albedo' and is used to refer to the scattering of direct solar radiation.

(*iii*) *Hemispherical-directional reflectance*, when the illumination comes from the hemisphere and the sensor has an infinitesimally small view angle. For canopies this is approximated by the scattering of the diffuse component of sky radiation to a sensor. Of course the value depends on the directional distribution of incoming radiation.

(*iv*) *Hemispherical-hemispherical reflectance*, when the illumination comes from the whole sky and the measurement integrates over the whole hemisphere. This is also often called albedo, though many studies use (ii) for this. In the absence of a direct component and when the diffuse component can be considered isotropic (such as when we have a uniformly overcast sky) we have the so-called 'white-sky albedo'.

All these reflectances can be further specified in terms of the wavelength of interest. As sensors and illuminators such as the sun generally have finite acceptance angles we should strictly replace the term 'directional' with 'conical' (giving, e.g., a biconical reflectance, see Schaepman-Strub *et al.*, 2006), but directional is in common use so we will retain that usage.

It is worth noting that some of these reflectance terms relate to other widely used terms. For example, the shortwave broadband bihemispherical reflectance from canopies is often termed the albedo (Rees, 2001), but of course, as we saw in Chapter 3, the actual value of albedo obtained for a given canopy depends both on the angle of sunlight and on the proportion of diffuse incoming radiation. The term 'black-sky albedo' is used to refer to the hemispherical (total) reflectance of shortwave radiation from the sun under clear skies with no diffuse sunlight component, though this never occurs in practice! 'White-sky albedo' approximates the situation with an overcast sky where the illumination is close to isotropic and there is no direct sunlight.

Some typical *BRF*s for different plant canopies are illustrated in Fig. 8.3, which shows the location of the high reflectance area or hotspot at the point where the view direction corresponds with the solar angle. Note also the fact that the *BRF* is typically symmetrical across the solar plane. The shape of the *BRF*, and especially the magnitude of the hotspot, varies markedly as a function of factors such as *LAI*, leaf size (or height of the canopy), and leaf-angle distribution (*LAD*). Measured *BRF*s can then be inverted to derive estimates of important biophysical canopy parameters. As the information content is greatest along the solar plane it is common to restrict measurements to data along the solar plane.

It is also worth noting here that the *BRF* measured in bright sunlight in the field is often assumed to approximate the true *BRF*, because the direct component of radiation dominates; nevertheless, it should be remembered that some error is necessarily involved in this assumption because of the effect of the diffuse sky irradiance. The *BRF* also varies as a function of wavelength (Fig. 8.4) because of the greater scattering and transmission of infrared radiation that will be most apparent in the shaded areas and the changing proportion of soil and leaf visible at different view angles. Figure 8.4(b) shows that there is greater view-angle dependence of reflectance in the infrared than in the visible. Although the absolute effect of view angle on spectral reflectance is quite small, there are substantial effects on ρ_{800}/ρ_{680} and hence on *NDVI* (Fig. 8.4(c)). Although some of the variation is related to changes in the fraction of soil visible, it is clear from Fig. 8.4(d) that there is an additional effect related to changing light quality. For any given fraction of soil visible, Fig. 8.4(d) indicates that the calculated *NDVI* is greater when viewing against the sun than when viewing from the sunlit side. This agrees with the fact that there is a greater proportion of shade when viewing against the sun, and shade tends to be enriched in the infrared (Chapter 7).

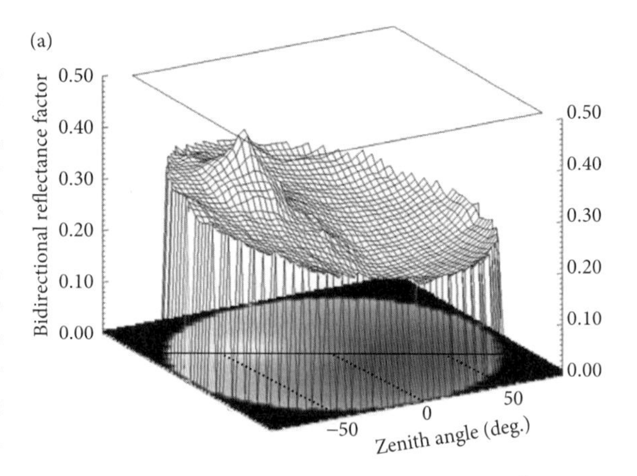

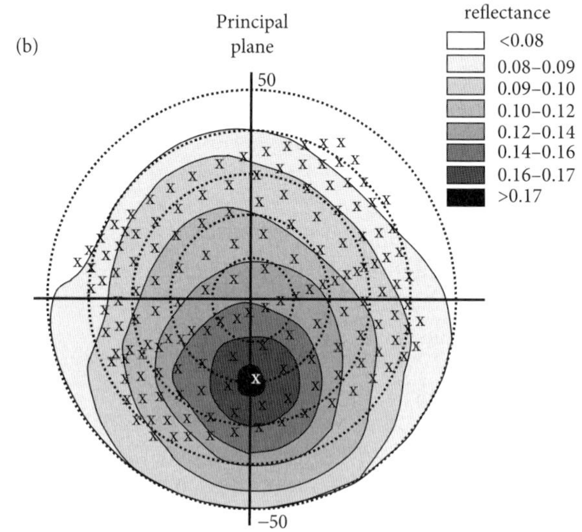

Fig. 8.3. Sample *BRDF* presentations showing (a) a typical 3D polar plot illustrating the hotspot of high reflectance at the solar angle (image from the late M. Barnsley), and (b) a simple polar plot of red reflectance for a shrub community (Lat. 28.92 N; Long. 116.75 E) derived from the POLDER-3/PARASOL database (http://postel.mediasfrance.org/en/BIOGEOPHYSICAL-PRODUCTS/BRDF/POLDER-3---BRDF-Data-Bases-/), showing the reflectance pattern, derived from satellite data at the marked points (data elaborated by the LSCE, and provided by the POSTEL Service Centre; the POLDER-3/PARASOL data are from CNES).

8.1.4 Principle of reciprocity

The principle of *reciprocity*, often referred to as Helmholtz reciprocity, is widely applicable and states

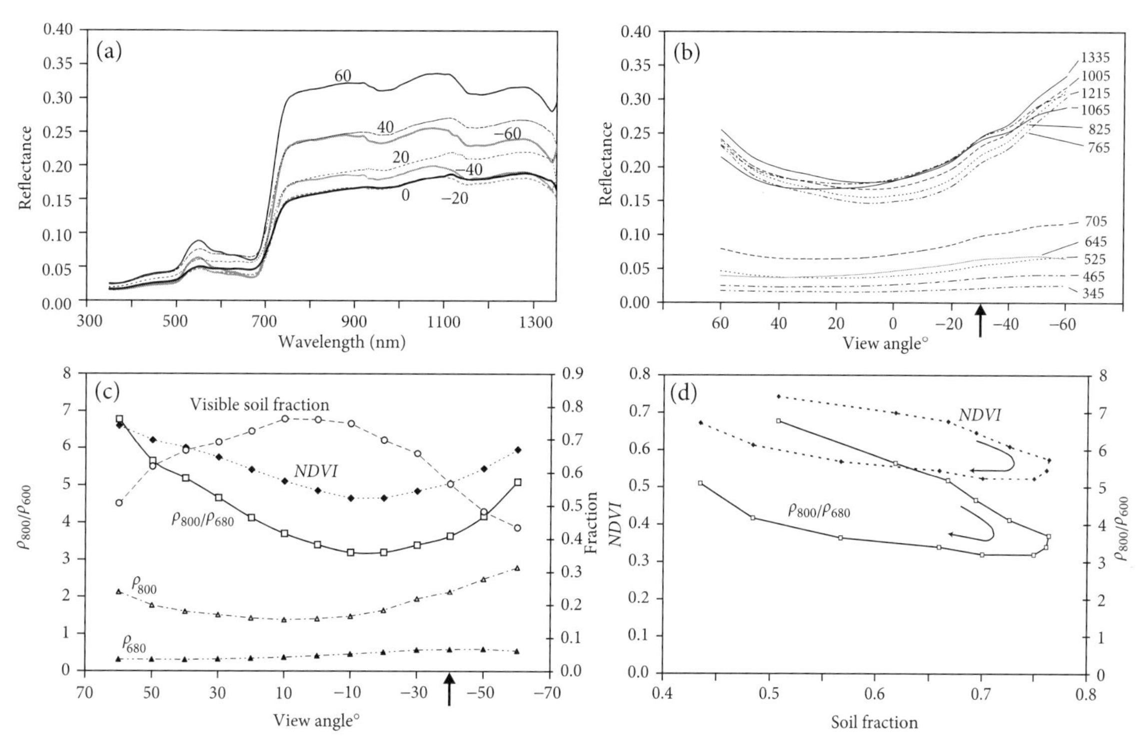

Fig. 8.4. Angular dependence of spectral reflectance for a maize canopy at Maccarese (Italy): (a) reflectance spectra for different view angles from −60° to +60° (where negative angles are for back reflectance), (b) angular dependence of reflectance at different wavelengths (average over 30 nm intervals), (c) variation with view angle of visible soil fraction, *NDVI*, the ratio of ρ_{800}/ρ_{680}, and the averaged red (400–700 nm) and infrared (701–1000 nm) reflectances. For any given fraction of soil visible, *NDVI* and NIR/R ratio are greater for the forward view (against the sun) than for the backwards view. (d) response of *NDVI* and ρ_{800}/ρ_{680} to visible soil fraction (with arrows indicating change from −60° to 60°. Spectral data obtained on 14th May 2008 with an ADC Fieldspec at 1 nm bandwidth; visible soil fraction estimated using image analysis; arrows mark the solar zenith angle; *LAI* = 0.69; data courtesy of R. Casa.

that 'in any linear physical system, the channels that lead from a cause at one point to an effect at another can be equally well traversed in the opposite direction'. It has been rigorously shown that directional reciprocity holds in remote sensing of homogeneous systems so that the emergent radiation in one direction due to incident radiation in another direction holds if the directions are reversed (see Leroy, 2001). This property is useful when modelling radiation transfer in canopies, and especially in the use of radiosity (Section 8.3.3). In heterogeneous canopy systems, however, reciprocity does not fully hold (Leroy, 2001). Departure from reciprocity arises in heterogeneous systems because of incomplete overlap between the volume of canopy viewed and the volume illuminated (see Fig. 8.5); in homogeneous systems the effects of areas outside the common viewed area would tend to cancel out. The deviation from reciprocity increases with increasing heterogeneity, increasing divergence between view and illumination angles, and with decreasing pixel size.

8.1.5 Extraction of canopy structure from *BRF* information

A major objective of much remote sensing of vegetation involves the estimation of canopy biophysical parameters such as *LAI*, *LAD*, radiation interception, or chlorophyll content. In the previous chapter we outlined the empirical statistical approach where canopy parameters are estimated from statistical relationships with remotely sensed spectral vegetation indices. Such approaches generally make no use of angular variation in reflectance. The use of multiangular information, however, provides a greatly enhanced tool for deriving information on the key canopy biophysical parameters. This information is of value both

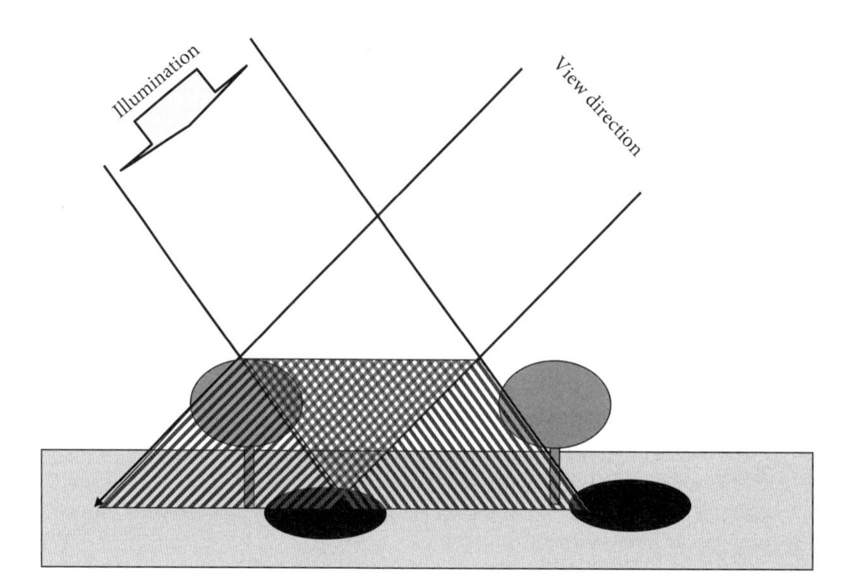

Fig. 8.5. The failure of reciprocity in heterogeneous systems arises because the volume of canopy viewed (▨) does not precisely correspond to the volume illuminated (▧) for the same surface area.

for interpretation of remotely sensed imagery and more importantly for incorporation into models of vegetation–atmosphere interactions such as for prediction of photosynthesis or hydrological feedbacks (see Chapter 9). The use of multiangular sensing for determination of canopy biophysical parameters and canopy structure is generally based on the inversion of canopy radiation-transfer models. Such inversion requires good models of radiative transfer in vegetation, and there has been a particularly intensive research effort in this area over the past few decades with the development of a wide range of more or less sophisticated radiation-transfer models. Useful more advanced reviews of the models and their inversion may be found in the following books and articles: Ross (1981); Myneni *et al.* (1989); Myneni and Ross (1991); Liang (2004).

A challenge in modelling the radiative transfer in plant canopies, and especially the spatial pattern of the reflection of sunlight back to space, is to simulate accurately all features of the *BRF* as functions of illumination and view angles. Not only do factors such as *LAI* and *LAD* affect the form of the *BRF*, but additional canopy characters such as the size, spacing, and arrangement of the elements (e.g. leaves and stems) and factors such as biochemical composition also contribute; they all also vary as a function of wavelength. The use of models to estimate radiative properties of a given canopy involves operation of the models in what is often termed the 'direct mode'. If, however, we want to use observed reflectance properties to infer the canopy structure parameters, which is the usual need in remote sensing, it is necessary to run the models in the opposite, or 'inverse mode'. Such inverse solution of canopy reflectance models will be discussed in detail below (Section 8.5) after we have introduced the main types of radiative transfer (RT) models available.

8.2 Measurement of the *BRF*

8.2.1 Basic approaches

The existence of effective tools for measurement of bidirectional reflectance properties of canopies is an essential pre-requisite for any modelling. Model inversion pre-supposes the ability to measure the *BRF*, or at least to approximate the full *BRF* by measuring the reflectance at a range of view angles. Measurements of the *BRF* of real surfaces have been attempted at scales where the target varies from a few cm up to a km. At the larger scales directional reflectance properties can be obtained from spaceborne and airborne multiangle spectroradiometers. Imaging sensors that achieve a measure of multiangular viewing solely as a result of their wide field of view (Fig. 8.6) provide rather limited information on *BRF* (limited especially by infrequent

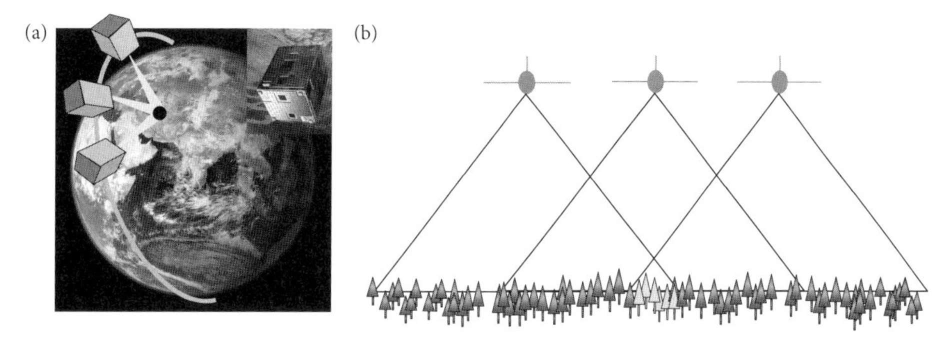

Fig. 8.6. (a) Illustration of the multiangle viewing capability of a pointable satellite detector such as CHRIS on the PROBA minisatellite (copyright European Space Agency), or from satellites with fixed along-track cameras at a range of view angles (e.g. MISR) which can take a series of images of one target from different directions during an orbit. Because such satellites tend to fly in near-polar orbits mid-day images collect information on reflectance variation along the solar plane, though cross-track information (at other angles) can also be obtained. (b) Multiple view angle imaging of vegetation using airborne sensors carried on a series of overlapping flightpaths (or satellite orbits) using wide field-of-view sensors to obtain cross-track data. The highlighted area can be viewed at a range of different angles depending on the spacing of the sequential flight paths. Image of globe with permission from NASA, downloaded from http://visibleEarth.nasa.gov/.

view of given ground points). As a result of the increasing recognition of the power of multiangular reflectance data an increasing number of missions have been flown to collect the necessary multiangle view data. For example, the Multiangle Imaging Spectroradiometer (MISR) on the Terra satellite has a series of nine fixed angle cameras with view angles of 0° (nadir) and fore- and aft-facing cameras at 26.1°, 45.6°, 60°, and 70.5°. As an alternative approach the Compact High-Resolution Imaging Spectrometer (CHRIS) mounted on the ESA minisatellite PROBA can be pointed to take a series of images of one target at different angles as the satellite passes overhead (Fig. 8.6(a)) and it also has some cross-track capability. Other examples include the Polarization and Directionality of the earth's Reflectances (POLDER) instrument that flew on the ADEOS satellite and the Advanced Along-Track Scanning Radiometer (AATSR). Multiangular airborne data can also be obtained from multiple flights over the target with overlapping swathes so that the sample site is visible at different angles each time (Fig. 8.6(b)) or by pointing the sensor.

For in-field studies and laboratory calibrations a range of strategies are available for obtaining multiangular reflectance data (Fig. 8.7): these involve either rotating the sensor on its own axis (Fig. 8.7(a)) thus viewing different areas of the target and requiring a large area of homogeneous target, or rotating the sensor around a given area of the target as in a goniometer (Fig. 8.7(b)). It is also possible to maintain the sensor at a given height above the target and to move the sensor; this corresponds with the standard approach for

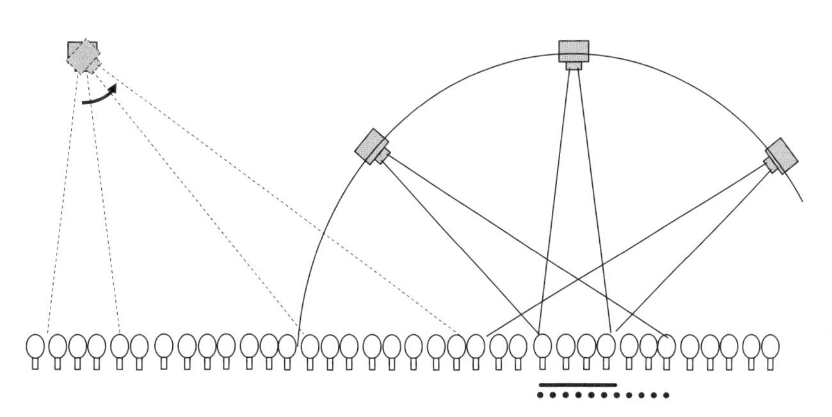

Fig. 8.7. (a) Illustration of rotation of the sensor about its own axis to obtain *BRDF* measurements: each image samples different parts of the vegetation, (b) rotation of the sensor around the target: each image samples the same spot (though the actual area viewed varies with view angle as indicated by the solid and dotted bars).

(a)

(b)

Fig. 8.8. Illustration of some goniometer systems: (a) a small-scale 3D field system (FIGOS – with permission from Schopfer *et al.* (2008); (b) a 2D bipod system for studying the *BRF* in the solar plane, where the downward-pointing imager is mounted in the middle of the cross-bar at the top (photo R. Casa).

satellite data as indicated above. A number of suitable instruments have been developed, small-scale examples of which are illustrated in Fig. 8.8. These small-scale instruments complement and supplement the data obtained with larger capital-intensive facilities including goniometers such as PARABOLA3 (Abdou *et al.*, 2000) or the European Goniometer Facility at Ispra (Meister *et al.*, 2000). In principle, all such instruments measure reflected radiation from a defined area of target at all possible angles, though in practice the hemisphere is usually subsampled at 5 or 10° intervals. Because the greatest amount of information is obtained from measurements in the solar (or *principal*) plane, full goniometric datasets are unnecessary for many purposes so simpler instruments such as the 'bipod' shown in Fig. 8.8(b) are often adequate (Casa and Jones, 2005).

An important feature to note from Fig. 8.7 is the fact that the target area varies with the view zenith angle, increasing as this angle increases. The viewed area increases with zenith angle even more in the situation where all views are from the same height than in a goniometric arrangement. This effect has important implications for sampling as it results in the nadir view observing smaller areas of the surface vegetation than the angled views, and hence there is less effective replication; this should ideally be compensated for in any sampling strategy chosen.

There is a significant difference between *BRDF* or *BRF* measurements in the laboratory, where it is possible to measure directly the reflected radiation for any given illumination angle, and in the field, where the reflected radiation includes reflection from direct (solar) illumination and reflection of the diffuse radiation coming from the sky. In the field it is easier to obtain the *HRDF* (the ratio of the radiance reflected in a specific direction to that reflected by a perfect Lambertian reflector under the same ambient irradiation). The latter is not solely a function of the surface properties but is also affected by the distribution of incoming radiation, however, under clear-sky conditions where diffuse radiation is less than 10% of the total incoming, *HRDF* is a good approximation to the surface *BRF*. Methods for correction of the observed *HRDF* to give the *BRF* are discussed by Abdou *et al.* (2000).

8.2.2 Some other practical considerations in *BRF* measurement

Reference surfaces: It is difficult to manufacture a reference surface whose reflectance approaches 100% over a wide spectral range, or one that shows a perfect Lambertian angular response. For a long time standard reference surfaces were made using $BaSO_4$; unfortunately such surfaces are rather unstable and variable in quality. More consistent results can now be obtained using sintered PTFE panels (e.g. Spectralon® panels from Labsphere, North Sutton, NH, USA – http://www.labsphere.com) that may reach a reflectance of around

99% over the solar spectrum, but still show significant deviations from Lambertian behaviour at low angles.

Dual-beam instrumentation versus sequential target and reference acquisitions: Because the incoming radiation changes continuously in the field, it is most accurate to measure both the incoming and reflected radiation simultaneously using a dual-sensor instrument with the sensors placed close to each other to avoid different illumination conditions. In situations where only one sensor is available, however, the spectra are acquired sequentially on the target (e.g. the canopy) and the reference (the Spectralon® panel). This approach only works when illumination conditions do not change between these acquisitions.

Scale and image component fractions: The information that is obtainable about a surface depends on the instrumentation and the scale of measurement. For example a simple radiometer only provides information on the average radiative flux emitted from a surface, at the scale of the viewed area or pixel; this average may, however, hide the variation in reflectance that occurs from the different surfaces that go to make up the viewed scene. The probability distribution of different radiances at smaller scales, as can be obtained through the use of imaging spectrometers, can provide substantial further information on canopy biophysical properties, beyond that which is obtained from the area average.

Although individual surface components are not readily apparent from satellite or even airborne imagery, measurements made from close above the canopy have the potential for separating out the various image component fractions. This gives the possibility of an alternative approach to the analysis of angular reflectance where one records the fractions of scene components, such as sunlit and shaded soil or leaves, visible at different view angles. Use of this information for the estimation of canopy variables such as *LAI* or *LAD* (Casa and Jones, 2005; Hall *et al.*, 1995; Peddle *et al.*, 1999) will be discussed in Section 8.6.3.

8.3 Radiation transfer and canopy reflectance models

In conventional problems in micrometeorology one is usually interested in determining the fluxes of radiation received or lost per unit area of horizontal surfaces. In such cases, whatever the actual orientation of the individual component fluxes (e.g. direct solar beam and diffuse sky irradiance) may be, they can be treated as a net vertical flux. More precise models of these fluxes can be obtained by explicit treatment of the angular dependence of the radiation field within vegetation canopies. Furthermore, for purposes such as the modelling of canopy photosynthesis it is more useful to know about the radiation actually intercepted by leaves and other canopy components and the distribution of magnitudes of these fluxes. As outlined in Chapter 3, the radiation actually intercepted by vegetation may be derived from this 'horizontal' irradiance by multiplying it by a 'shape factor' that depends on the directional properties of the beam and on the geometry of the surface.

In remote sensing, however, the usual requirement is to determine the surface reflectance and to make use of this in inferring canopy or surface biophysical characteristics. The use of bidirectional reflectance data greatly enhances the capacity to extract canopy biophysical information when coupled with appropriate radiative transfer (RT) modelling. The various radiative-transfer models available vary in the detail in which they treat the anisotropy of the radiation field in canopies. In such modelling it is often convenient to treat the radiation field as the sum of a number of components. These could include the unscattered irradiance ($\boldsymbol{I}^0$), radiation that has been scattered once ($\boldsymbol{I}^1$), and multiply scattered irradiance ($\boldsymbol{I}^{\mathrm{M}}$) that has been scattered several times (see Fig. 3.14). The simpler models ignore, or greatly simplify, the higher-order scattering. This is particularly the case when one uses radiation penetration through canopies to estimate canopy biophysical parameters such as *LAI* and *LAD* as outlined in Section 8.6, where gap-fraction approaches only treat the unscattered radiation. For canopy reflectance modelling, however, more sophisticated techniques are required. The main types of RT model can generally be classified into one of the four main conceptual classes described below, though there is an increasing number of hybrid models that combine aspects of more than one class. The classic text on radiation transfer in canopies was written by Ross (1981), while an advanced review of more recent work has been provided by Liang (2004).

8.3.1 Turbid-medium models

As outlined in Chapter 3, the simplest models are based on the theory developed by Kubelka and Munk (1931) to describe one-dimensional radiation propagation in a gas. These radiative transfer (RT) models assume that the vegetation canopy can be treated using the same theory as has been developed for radiation transfer in gases. They usually assume that the canopy can be represented by an infinite horizontal slab, within which the individual elements (leaves) are infinitely small (i.e. point scatterers) homogeneously distributed in uniform layers. In general, the positions of individual elements are only described by statistical distributions, with a random distribution being the most common, though some more sophisticated models allow for non-random distributions such as occur with clumped distributions. These models are most suited to homogeneous agricultural crops, and the simple Beer's law approach for radiation penetration introduced in Chapter 3 falls within this class. For the basic approach there is no lateral inhomogeneity so this can be treated as a one-dimensional transfer problem. Nevertheless, similar principles can be applied to more complex situations as outlined below.

The classical Kubelka–Munk (KM) treatment applies to a sparse medium with widely separated 'point' particles, and needs some amendment to account for vegetation canopies where the leaves are much larger proportionately. The basic one-dimensional radiative transfer equation describing the rate of change of irradiance with depth (eqn (3.3)) can be expressed as

$$\mathrm{d}\boldsymbol{I}(z)/\mathrm{d}z = -k(\boldsymbol{I} - \boldsymbol{J}), \qquad (8.2)$$

where $\boldsymbol{I}$ is the irradiance (or incident (downward) flux density) on a horizontal surface, z is height, and k is an extinction coefficient that depends on the leaf-angle distribution and on the direction of the radiation. The term $\boldsymbol{J}$ is a source function that represents any gain of flux resulting from scattering of radiation from other directions. This can be applied to the description of the attenuation of radiation through a canopy referring to the average flux in the vertical direction. Radiation from other directions is converted to the equivalent amount that is absorbed by a horizontal surface. The extinction coefficient depends on radiative losses by absorption and by reflection.

The basic KM theory is what is known as a two-stream model in that it uses two differential equations to describe the fluxes in downward and upward directions, but still essentially relates to the integrated vertical fluxes and so is not directly suitable for derivation of directional reflectances. Nevertheless, the coefficients in the equations can be related to solar and viewing angles and the basic equations may be generalized to describe the change in irradiance in any specific direction (see, e.g., Liang, 2004). The different models that have been developed to derive the directional reflectance properties of canopies differ in the detail with which they take account of primary and secondary reflections and the assumptions made about the proportion of specular and Lambertian reflection. A key development of the two-stream theory was made by Suits (1972) who added equations relating to the directional solar radiation flux and to the flux towards the viewing angle, thus making this a four-stream model. This model assumed randomly distributed leaves and treated the angular radiative properties of the canopy by using the projection of finite canopy components (e.g. leaves) onto horizontal and vertical surfaces to approximate the transmission and reflection from angled surfaces as the incidence angle changed. The approach has been extended to allow for specific leaf-angle distributions in the SAIL model (scattering by arbitrarily inclined leaves; Verhoef, 1984), or by more recent models with greater flexibility in the definition of leaf-angle distributions (e.g. Goel and Thompson, 1984). The basic parameters required by this class of models include the reflectance and transmittance of the individual leaves, and a background (soil) reflectance, while the canopy structure is defined by the parameters required to define the leaf-angle distribution (often two depending on the actual function used), the leaf-area index, and the canopy height. It is also necessary to define the fraction of diffuse radiation incident on the canopy, giving a total of seven parameters.

Although this type of model generates a bidirectional reflectance it cannot directly simulate the 'hotspot' effect, because it is based on point scatterers that cannot give shadows. Various approaches are available to incorporate such an effect. For example, Suits (1972) recognized that the probability of viewing a sunlit area at a particular depth in the canopy was not independent of the angle between illumination and view, so incorporated an empirical correction into the calculation. A key development in more recent models has been the movement from turbid-medium models to ones that explicitly account for the statistical distribution of finite leaf sizes within the canopy (Gobron *et al.*,

1997; Nilson and Kuusk, 1989; Verstraete *et al.*, 1990) so that the hotspot effect can be simulated directly. Additional parameters are required for this simulation of the hotspot. In general, these models treat the first orders of scattering explicitly and model the higher orders of scattering using standard turbid-medium models. Information on the shape of the *BRF* near the hotspot can be used to retrieve key canopy biophysical parameters. This inversion process is discussed in Section 8.5.

A complementary advance has been the inclusion of submodels that can allow for the variation of radiation transmission and reflection by canopy elements as a function of radiation wavelength. The PROSPECT model (Section 3.1.1: Feret *et al.*, 2008; Jacquemoud and Baret, 1990) simulates reflectance and transmittance spectra between 400 and 2500 nm and can be combined with the SAIL model (Jacquemoud, 1993; the model is available as PROSAIL at http://teledetection.ipgp.jussieu.fr/prosail/; Vapnik, 1995) or with other models (Kuusk, 1995). In a similar way, submodels have been described that mimic the spectral-directional behaviour of the underlying soil as incorporated in more recent extended versions of SAIL (e.g. Verhoef and Bach, 2007).

8.3.2 Geometrical-optical models

In contrast to turbid-medium models, geometrical-optical models have been developed particularly to describe forests and other discontinuous canopies. Geometrical-optical models operate by assuming that the canopy can be described by an array of geometrical objects of defined shape and optical properties arranged in space according to some statistical distribution (Fig. 8.9). The light interception and reflection can then be calculated analytically from geometrical considerations of the interception of light and shadowing by these objects. The overall reflectance at any angle is then calculated as a weighted average by summing the area fractions visible at that angle of each class of object (e.g. sunlit or shaded soil and sunlit or shaded leaf) for which a reflectance has been assigned. Examples of this type of model include those developed by Otterman and Weiss (1984) and by Li and Strahler (1985) who modelled a forest canopy as Lambertian cones. The approach has been extended (Li and Strahler, 1992) to incorporate various tree shapes (e.g. ellipsoids on sticks or combinations of cones and cylinders) and to fill the geometrical shapes with leaves

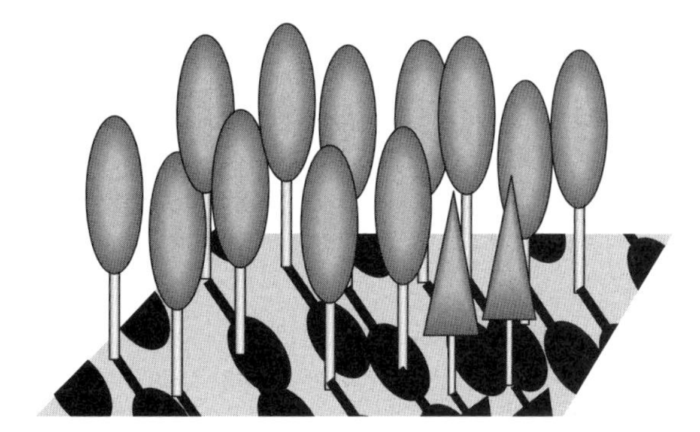

Fig 8.9. Illustration of the canopy as used for a geometrical-optical model where there is a combination of discrete plants (or plant parts) with the volume of each being treated as a homogeneous absorber. The proportion of sunlit canopy or soil at any view angle can be derived and used to estimate overall reflectance.

that obey the turbid-medium assumptions, or even the population of the larger shapes (trees) by branches as in the 'four-scale' model of Chen and Leblanc (1997).

The overall reflectance is obtained from

$$\rho = \rho_{g} \cdot f_{g} + \rho_{c} \cdot f_{c} + \rho_{c\text{-}sh} \cdot f_{c\text{-}sh} + \rho_{g\text{-}sh} \cdot f_{g\text{-}sh}, \tag{8.3}$$

where ρ is reflectance, and f is the fraction of the area covered for sunlit ground (g), shaded ground (g-sh), sunlit canopy (c), and shaded canopy (c-sh). The fractions f_g, f_c, etc., are obtained from simple geometric considerations. Simple treatments assume that shadows are perfectly black eliminating the last two terms, and even that sunlit canopy and ground are equally bright. The probability of viewing any of these classes can be obtained by geometry, though the calculations are quite complex for realistic canopies. For example, the probability of seeing sunlit ground is the joint probability of a ray in the solar direction and a ray in the viewing direction reaching the ground. When these two directions are far apart they can usually be assumed to be uncorrelated, but as the two angles come closer together, the probability of seeing sunlit soil is clearly also a function of the angle between the solar beam and the view angle with the hotspot occurring where this angle is small and there is a high probability of both rays traversing the same canopy gaps.

For sparse canopies, where interference between adjacent 'trees' is minimal, relatively simple formulae can be derived, based on the horizontal radius of the ellipsoid (r), the vertical radius of the ellipsoid (b), and the height to the centre (h). These formulae need adjustment when there are dense canopies and there is interference between the shadows of different ellipsoids (see Box 8.2). Geometrical-optical models are particularly appropriate for the description of forest canopies.

8.3.3 Monte-Carlo ray-tracing and radiosity models

Ray-tracing – Ray-tracing is a rendering technique that calculates an image of a scene by simulating the way rays of light travel in the real world. Conceptually a large number of 'rays' are randomly fired from the light source(s) into the canopy and probability density functions are specified for each possible interaction with canopy elements defining how rays are absorbed, transmitted, or reflected and the angles of reflection (e.g. Govaerts *et al.*, 1996; Govaerts and Verstraete, 1998). Each ray continues to bounce within the canopy until it either escapes from the top of the canopy or its energy level falls below a minimum threshold. A comparison between those rays that reach a specified sensor and those originally fired into the canopy is a measure of reflectance. Because the vast majority of rays never hit an observer, it would take forever to trace a scene, so ray-tracers frequently operate in reverse: rays start at the sensor and are traced backwards to the light source. A range of computing strategies can be used to speed up even forward ray-tracing. For every pixel in the final image one or more 'viewing rays' are shot from the camera, into the scene to see if it intersects with any of the objects in the scene. These viewing rays pass through the viewing window (representing the final image). Every time an object is hit by one of these viewing rays, the colour of the surface at that point is calculated. For this purpose rays are sent backwards to each light source to determine the amount of light coming from that source. These 'shadow rays' are tested to tell whether the surface point lies in shadow or not. If the surface is reflective or transparent new rays are set up and traced in order to determine the contribution of the reflected and refracted light to the final surface colour. Because of the computational intensity involved, ray-tracers are best at simulating the reflection of direct radiation; for diffuse irradiance the calculations can become excessively long.

BOX 8.2 Calculation of proportions of sunlit and shaded canopy and ground

For a sparse canopy of ellipsoids (Fig. 8.9), where h is the stick height, r is the horizontal radius and b is the vertical radius, one can calculate the proportions of sunlit (f_c) and shaded ($f_{c\text{-sh}}$) crown and sunlit (f_g) and shaded ($f_{g\text{-sh}}$) background that can be viewed from nadir for any angle of illumination (θ_i) (see Li and Strahler (1992) and Liang (2004)). The fraction of the viewed area which is canopy is given by $N\pi r^2$, where N is the average number of trees per unit area. This can be partitioned into viewed sunlit and shaded canopy by using a geometric factor B,

$$f_c \cong N\pi r^2(1+B)/2, \text{ and } f_{c\text{-sh}} \cong N\pi r^2(1-B)/2.$$

where, $B = \cos(\tan^{-1}[-1/(4KL^2)]).\ \sqrt{[(1+16K^2L^4)/(4L^2+16K^2L^4)]}$,

and in which $L = r/(2b)$, and $K = \tan(\pi/2+\theta_i)$.

The fraction of soil that is shaded is $N\pi r^2/\cos(\theta_i')$ where the cosine corrects for the increasing area as solar zenith angle increases, and θ_i' ($= \tan^{-1}((b/r)\tan\theta_i)$) is the zenith angle that would generate the same shadow area for a sphere. The fraction of the shadow that is visible, is obtained using a correction term A_0, defined as

$$A_0 = (\beta - ((\sin 2\beta)/2))(1 + 1/\cos\theta_i')/\pi \text{ if } (b+h)\tan\theta_i < r(1 + 1/\cos\theta_i'), \text{ otherwise } A_0 = 0,$$

and in which β is given by

$$\beta = \cos^{-1}[(1 + h/b)(1 - \cos(\theta_i'))/\sin(\theta_i')].$$

Using these corrections the visible shade and sunlit soil fractions are given by

$$f_{g\text{-sh}} \cong N.r^2\pi((1/\cos(\theta_i')) - A_0),\quad f_g \cong 1 - (f_c + f_{c\text{-sh}} + f_{g\text{-sh}}).$$

Further modification is needed for other view angles, dense canopies where there is mutual shading, for low illumination angles and for sloping surfaces (Liang, 2004). Wanner *et al.* (1995) discusses derivation of the corresponding geometric kernels and constants for BRDF models.

A particular advantage of ray-tracers is that they do not require analytical solutions of the radiation-transfer equations of canopies, which even for relatively simple canopy models can become exceedingly complex, or more often impossible. A good free ray-tracer is POV-Ray (Section 8.4.1) which also incorporates a programming language for generation of a 3D representation of the scene, or canopy, to be studied. Its use is illustrated in Section 8.3.6.

Radiosity – Somewhat related is the *radiosity* approach (Borel *et al.*, 1991; Goel *et al.*, 1991) which has been widely used in computer graphics applications and is particularly valuable for studies of long-wave radiation exchanges in canopies (Rotenberg *et al.*, 1998). As indicated in Fig. 8.10 the radiosity (B_i) of a surface *i* is defined as the total amount of energy leaving the surface per unit area per unit time (and hence is a flux density; W m^{-2}). This radiation includes all the leaving radiation whether emitted (though this is not generally relevant at optical wavelengths), transmitted, or reflected and sums all the contributions emanating from each of the *n* surfaces in the 3D canopy. Note that the radiosity approach assumes that all surfaces behave as diffuse reflectors/transmitters.

The radiosity approach provides a general method for calculating all the exchanges of light between all elements in the canopy with the radiation exchange between any two of the surfaces being calculated using 'view factors' (Fig. 8.10) that describe the view of the receiving surface from the emitter. The view factor describes the fraction of radiative energy emanating from the infinitesimal surface dA_j that reaches the surface dA_i. Only pairs of surfaces that are not obscured by intervening objects are treated. The spatially explicit treatment of all the individual radiative interactions between each of the canopy elements contrasts with the more conventional radiative transfer approach where volume-averaged properties of the canopy are used. Radiosity calculations are relatively efficient for diffuse radiation as the approach deals efficiently with radiation from all directions. In contrast to backward ray-tracing, the specification of the unknowns is independent of the viewer position, and hence most of the computational effort occurs before the selection of the viewer parameters.

The *view factor* ($F_{dAi \to dAj}$) of surface *j* from surface *i* is a purely geometric function depending on only the size, distance, visibility, and orientation of the elements (not on their optical properties), and is defined as

$$F_{dAi \to dAj} = \cos\theta_i \, dA_j \cos\theta_j / \pi r^2, \tag{8.4}$$

where θ_i and θ_j are the angles between the two surfaces A_i and A_j and their respective normals, and *r* is the distance between them (Fig. 8.10(b)).

A major limitation of the radiosity approach is the initial effort required to set up the view factor matrix describing all the possible view factors and to solve this for radiative transfer. This initialization effort can be very substantial, and the resultant matrices very

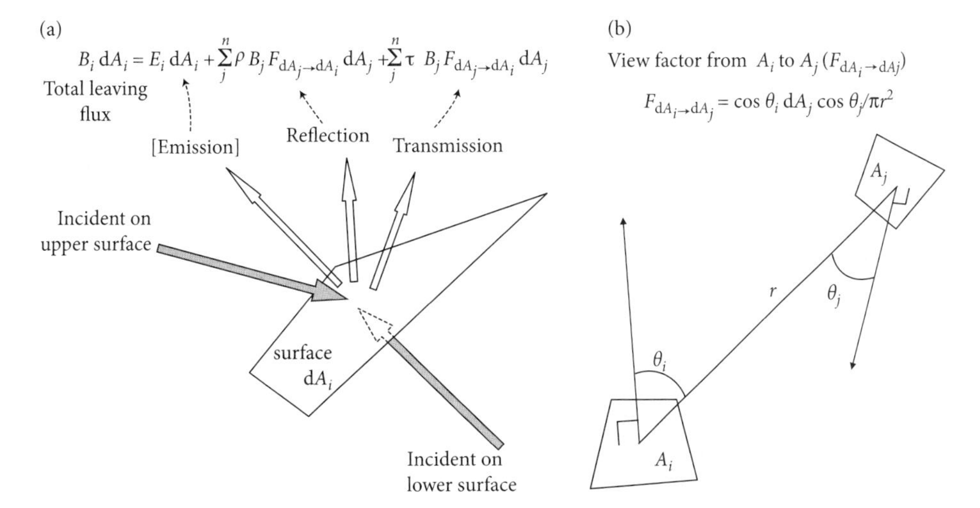

Fig. 8.10. (a) Radiosity (B_i) leaving a surface A_i is the sum of the emitted, reflected, and transmitted radiation, each of which sums the contributions from incident radiation from all other visible surfaces within the canopy. (b) Illustration of the calculation of the view factor between the surfaces A_i and A_j.

large, for complex scenes involving a large number of scatterers. However, once this has been achieved it is then straightforward to simulate canopy reflectance from any view angle. Chelle *et al.* (1998) and others have simplified the approach somewhat using a *nested radiosity* approach by treating the irradiance on any polygon in the scene as the sum of two components: the irradiance due to nearby scatterers and the irradiance from distant parts of the canopy. The near-field interactions are treated using the full-matrix radiosity method, while far-field interactions are treated using a volume scattering model such as SAIL.

8.3.4 Kernel-driven and empirical models

The geometrical-optical and radiosity models can be rather complex so various attempts have been made to develop simpler, largely empirical, models that are faster to use and easier to invert, yet can adequately simulate the directional reflectance of vegetation or soil (e.g. Nilson and Kuusk, 1989; Roujean *et al.*, 1992; Walthall *et al.*, 1985). The basis of these semi-empirical models is that reflectance is modelled as a sum of three 'kernels' representing the three main types of scattering: isotropic scattering, volumetric scattering from homogeneous canopies, and a geometrical term representing scattering from three-dimensional objects that cast shadows and can be mutually obscured as angles change. These models have the advantage over more detailed mechanistic models of more rapid inversion for estimation of canopy biophysical parameters. Such models can be written in the form

$$BRDF = f_{iso} + f_{vol} * k_{vol}(\theta_i, \theta_v, \varphi) + f_{geo} * k_{geo}(\theta_i, \theta_v, \varphi), \quad (8.5)$$

where the kernels, k_{vol} and k_{geo} that are semi-empirical functions of view zenith (θ_v), illumination zenith (θ_i), and relative azimuth (φ) and describe the volumetric and surface scattering properties, respectively. The f_{iso}, f_{vol}, and f_{geo} are weighting factors, with the first being a constant relating to isotropic reflectance, while the volume-scattering term refers to the bidirectional scattering function derived from a single-scattering solution for vegetation with a random (spherical) leaf-angle distribution (in many cases this does not account for the hotspot), while the geometric kernel specifically allows for the hotspot function by incorporating shading effects between the individual crowns.

Suitable expressions for the k_{vol} and k_{geo} functions have been given by Roujean *et al.* (1992) and Wanner *et al.* (1992), with the 'RossThick' kernel being appropriate to describe k_{vol} as a function of the incident solar angle (θ_s), the view angle, and the angle between the illumination and view azimuths:

$$k_{vol} = ((0.5\pi - \xi)\cos\xi + \sin\xi)/(\cos\theta_s + \cos\theta_v) - \pi/4, \quad (8.6)$$

where the phase angle ξ is given by

$$\cos\xi = \cos\theta_s \cos\theta_v + \sin\theta_s \sin\theta_v \cos\varphi. \quad (8.7)$$

Various options have been proposed for the k_{geo} kernel, of which the most popular have been derived from the Li and Strahler shadowing model (Box 8.2).

Further empiricism has been introduced in the empirical models of the type exemplified by the Rahman–Pinty–Verstraete model (Rahman *et al.*, 1993) which are not based directly on the physics of radiation transfer, but rather they approximate the shape of the *BRF* and its anisotropy by a limited number of parameters (usually four). These parameters include a scalar overall reflectance and parameters relating to the shape of the reflectance 'bowl', the degree of forward versus backscattering, and the magnitude of the hotspot. Although this type of model is non-linear, fast and effective inversions are available (Lavergne *et al.*, 2007).

8.3.5 Treatment of heterogeneous canopies

More complex models are required if we are successfully to simulate the bidirectional and spectral reflectance of more complex heterogeneous landscapes. One approach to this is that adopted by the 3D models such as the discrete anisotropic radiation-transfer model (DART; Gastellu-Etchegorry *et al.*, 2004) which simulates the reflectance and radiation absorption by landscapes using a combination of radiative kernels and representation of natural scenes as matrices of parallelepiped cells containing trees, shrubs, soil, water, etc. (Fig. 8.11). Each cell has an appropriate set of scattering, absorption, and transmission properties largely defined by *LAI* and *LAD* in that cell. The model operates by following the propagation of a discrete number of possible radiation vectors, including those originating from the sun and from the sky, through the canopy, rather like a simplified radiosity approach.

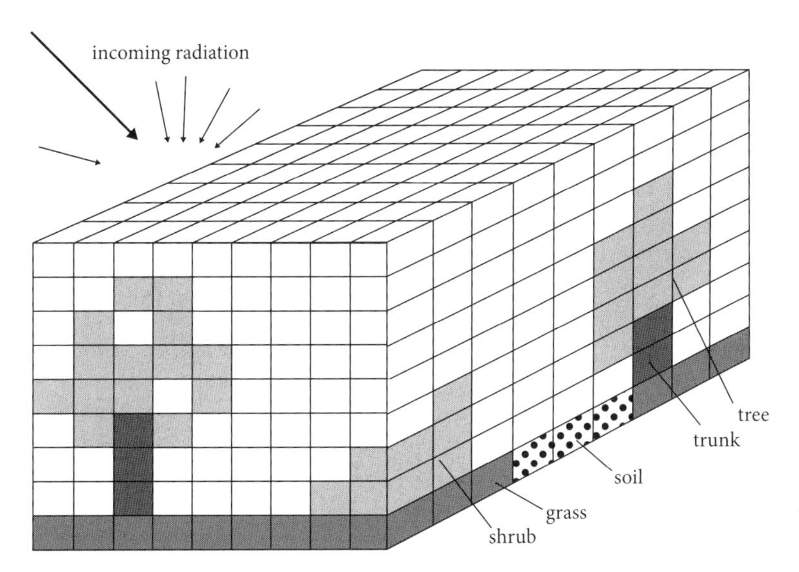

Fig. 8.11. Illustration of the landscape structure in the DART 3D model (see Gastellu-Etchegorry *et al.*, 2004) to simulate bidirectional canopy reflectance. Each cell has appropriate properties of *LAI*, leaf-angle distribution, reflectance, etc., defined and scattering, reflectances, and transmission are calculated for each cell for each incident radiation vector.

8.3.6 Spectral invariants

An important tool for the study of canopies is the use of *spectral invariants* as introduced by Knyazikin *et al.* (1998) and developed further by workers such as Lewis and Disney (2007). General canopy properties such as reflectance and transmission vary with both the angle of view and the wavelength of the radiation. Although the probability of a scattering event is a function of wavelength, where the objects are large compared to the wavelength (as with leaves) the photon free path between two successive interactions is independent of wavelength, and hence spectrally invariant. Therefore interaction probabilities for photons in vegetation are determined by canopy structure rather than wavelength. The recollision probability, p, is the probability that a photon once scattered will interact with another foliage element and is spectrally invariant but is a function of canopy structure. Interestingly it can be shown (Sternberg, 2007) that an average recollision probability ($\hat{p}$) can be calculated from uncollided transmission, that is from canopy gap fraction ($F(\theta)$) at different view angles (see Section 3.1.3) using the relationship

$$\hat{p} = 1 - \frac{1 - DIFN}{L}, \tag{8.8}$$

where L is the hemispheric leaf-area index and $DIFN$ is the so-called '*diffuse non-interference*' measured as the integral over all zenith angles of the canopy gap fraction (including a cosine correction), i.e.

$$DIFN = 2 \int_0^{\pi/2} F(\theta) \cos(\theta) \sin(\theta) \, d\theta. \tag{8.9}$$

Conveniently, $DIFN$ may readily be measured for any canopy using a simple canopy analyser of the LiCor 2000 type (see Section 8.6.2). Note that both $DIFN$ and L are independent parameters in eqn (8.8), so that this expression provides a correction for any clumping of the leaves. The proportion of sunlit or shaded soil or canopy, as used by Casa and Jones (2005), are other spectral invariants (see Smolander and Stenberg, 2005).

8.4 Canopy generators

The development of radiosity methods and efficient ray-tracers has opened up the possibilities of accurate characterisation of both real and model canopies where the distribution and orientation of all components are known explicitly. There are a number of ways in which one can generate models of canopies ('virtual canopies') for use in ray-tracer models. The most straightforward approach is to replicate an actual 3D canopy measured using stereoscopic imagers or 3D digitizing systems (Moulia and Sinoquet, 1993) to determine the spatial distribution of all the canopy components. More commonly, however, such measurements are used to derive the necessary parameters for input to canopy generators that use a set of simple rules to generate a canopy from component modules.

Canopy generators largely fall into one of two types: those that produce a population of leaves or plants on the basis of simple statistical distributions (e.g. the scene generator component in POV-Ray), and those that incorporate developmental or botanically relevant relational information based on the relationships between specific modules (e.g. leaves, stems, and apices) (e.g. L-systems and Y-plant). We discuss the application of these two model approaches below.

8.4.1 POV-Ray

The Persistence of Vision Ray-Tracer (POV-Ray™, http://www.povray.org) is a package that includes a powerful scene description language for defining the distributions and shapes of objects within the scene and the view and illumination angles. Although it is particularly easy to generate homogeneous canopies with regular or randomly distributed components, it is simple to extend the system to generate canopies of row crops, isolated trees, or more complex structures. The canopy can be built up by repeatedly inserting leaves into the defined canopy volume. Once the canopy has been constructed using the scene description language in POV-Ray or by any other canopy modelling software, it is then possible to run the ray-tracer part of the programme to visualize the scene and describe its reflectance (Fig. 8.12).

A fundamental requirement for the simulation of real canopies using simple biophysical parameters, as is required by the scene generator in POV-Ray, is the need for simple descriptions of leaf-angle distributions. Of the various possibilities, perhaps the most general is the use of Campbell's ellipsoidal model (1986) where the extent to which leaves may be randomly oriented, predominantly horizontal or predominantly vertical, can be simulated by the variation of a single parameter (Campbell's x parameter). Varying the value of x in the algorithm used in POV-Ray to simulate the positioning of leaves allows one to mimic effectively rather different canopies, especially when information on the average leaf shape and size is also included.

8.4.2 L-systems

Lindenmayer systems (L-systems: Lindenmayer, 1968) provide an effective modelling tool for generating 'virtual plants' (Prusinkiewicz, 2004). In this approach the plant can be regarded as a population of semi-autonomous modules. All modules of one type (e.g. leaves) behave according to a given rule, independent of where in the plant they occur. This allows concise model specifications to generate complex yet realistic architectures by repeated application of simple rules.

Development is considered as a sequence of events where parent modules are replaced by successor modules according to rules of replacement, somewhat in the manner of fractals. An L-system is simply a specification of the set of all module types that can be found in a given organism, the set of productions that apply to these modules, and the initial configuration of modules (the axiom), from which the development begins.

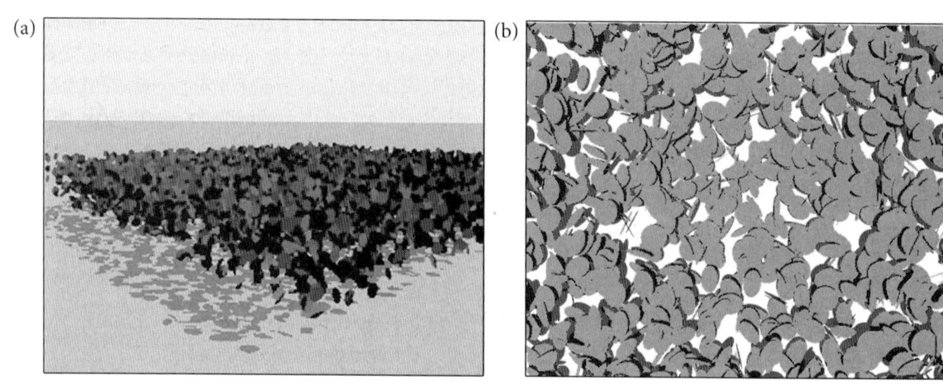

Fig. 8.12. (a) An image of a model canopy with randomly oriented and distributed circular leaves, illuminated from the right, generated by POV-Ray. Shaded areas are indicated as darker areas on either the leaf or soil backgrounds. (b) View into the hotspot at +45°, showing increasing proportion of shadow as the angle of view moves away from the precise hotspot. (Images with permission from R. Casa.) (See Plate 8.2)

Fig. 8.13. An illustration showing the generation of a compound leaf using a recurrent series of instructions where the leaf is modelled as a series of apices (thin lines) and internodes (thick lines). The inset illustrates the production rules that produce this development, with an apex giving rise to the branched structure shown, and the internode extending at each time interval (after Prusinkiewicz, 1998).

The synthetic process involved is illustrated for the generation of a compound leaf in Fig. 8.13. Various languages for implementation of L-systems have been developed, of which the most useful for plant studies is probably L-studio or its variants (Prusinkiewicz, 2004).

An example of a system that uses this approach is the Botanical Plant Modelling System (BPMS) developed by Lewis (1999), where a limited number of plant structure measurements can be generalized up to a canopy using L-systems. An incorporated ray-tracer can then be used to simulate the canopy radiation and hence its reflectance distribution.

8.4.3 Y-plant

Y-plant is an example of a three-dimensional crown architecture model for the analysis of light capture and carbon gain of plants where the geometry of the crown and self-shading are important. Again it is an example of one of those models that builds a canopy from its essential components (Fig. 8.14). The model is described in Pearcy and Yang (1996) and has been used to examine optimality of allocation in the crown architecture of an understory herb (Pearcy and Yang, 1996), and for studying the consequences of architectural plasticity and trade-offs between light capture and avoidance of photoinhibitory stress in a chaparral shrub species (Valladares and Pearcy, 1999; Valladares and Pugnaire, 1999).

The fundamental unit in Y-plant is the node, and plants and canopies can be constructed from a series of nodes according to some rule. A node can have three objects attached to it: (a) a leaf (with a petiole), (b) a stem, or (c) a branch. Note that the stem or branch segments defined at any node are those extending from the node, not the connection to its mother node. Designation of a particular segment as 'stem' or 'branch' is arbitrary. It only serves to identify which node is connected to the mother node by which segment. Therefore, a 'branch' at one node can (and usually will) become a 'stem' at subsequent nodes along the same path.

8.5 Model inversion

8.5.1 Principles

The methods available for retrieving canopy surface characteristics from remote sensing range from rather simple empirical calibration over experimental areas with known canopy characters to the inversion of more or less complex radiative-transfer models that predict reflectance properties as a function of canopy structure.

Given a set of reflectance measurements or an estimated *BRF* for a particular canopy, the inversion problem is to find the set of model parameters (*LAI*, *LAD*, etc.) that best fits the observed *BRF* (Fig. 8.15). The most suitable approach to achieve this solution of the inverse problem depends on the underlying model that is used to describe the radiative-transfer properties of the canopy. Unfortunately, only very few simple models can be

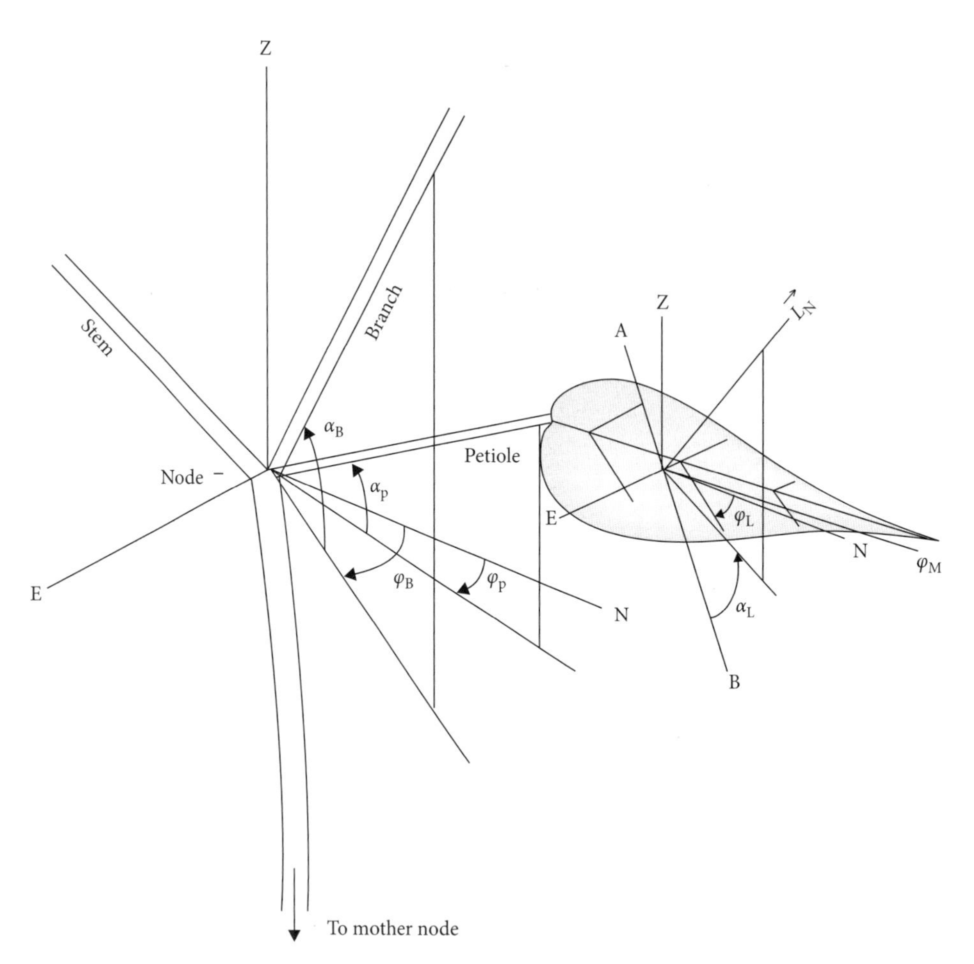

Fig. 8.14. Definitions of the angles required for Y-plant: E, N, and Z are the east, north and zenith directions, respectively, α is an angle from horizontal, φ is an azimuth (0–360°), the subscripts B, P, L, M refer to branch, petiole, leaf, and midrib, respectively, L_N is the leaf surface normal, the leaf angle is measured along the line A–B. Note: φ_M is the leaf orientation; φ_L is the leaf azimuth (with permission from Pearcy and Yang, 1996).

inverted analytically, and these generally do not simulate the behaviour of real canopies very effectively, so most practical solutions require the use of numerical or approximate methods. A particular difficulty is that the inversion problem is usually 'ill-posed' (Baret and Buis, 2008; Combal *et al.*, 2003) in the sense that the solution may not necessarily be unique with more than one set of fitted values potentially giving an equally good fit to the data. The potentially valid solutions are not necessarily close to each other in the parameter space, meaning that jumps and leaps may be observed. The ill-posedness of the solution is further amplified by measurement errors and model uncertainties. The latter refers for example to incorrect assumptions on canopy architecture (e.g. non-random leaf azimuth or non-Lambertian scattering by leaves). Together these can lead to unstable and inaccurate inversion results. When observing vegetation canopies, the main confusion arises between *LAI* and *LAD* (Atzberger, 2004). In fact, a sparse planophile canopy may have a spectral reflectance very similar to that of a dense erectophile canopy. The ill-posed nature of the inversion problem underlies the uncertainty apparent in biophysical variables derived from the use of vegetation indices that was discussed in the previous chapter.

Since in practice the modelled fit is unlikely to be perfect it is commonly necessary to define some 'merit function' that is optimized to obtain the best fit. A simple merit function (*M*) would be to minimize the sum of the squared differences between the observed and modelled values. Since, however, the range of reflectance at different view angles (and/or at different

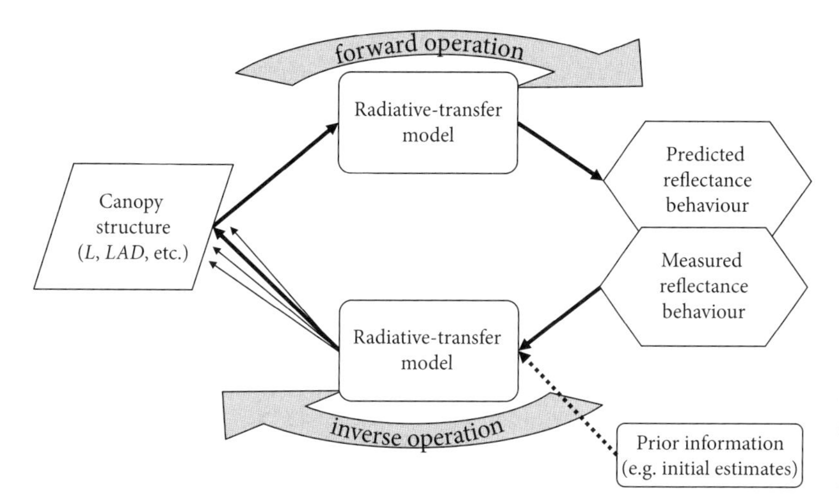

Fig. 8.15. An illustration of the operation of forward and inverse modelling in the retrieval of canopy structure information in remote sensing. In the forward operation the radiative-transfer model is used to predict reflectance behaviour, while in the inverse mode measured reflectances are used to estimate canopy structure parameters. The extra outputs in running in the inverse mode indicate different equally possible combinations of the structure parameters. Prior information, such as knowledge of the crop and its leaf-angle distribution or an initial guess as to the value of *LAI* can be used.

wavelengths) may vary several-fold, it is sensible to weight the observations, by using a weighting factor (w_j) such as $w_j = 1/\rho_j$. This gives

$$M = \Sigma\,(w_j\,(\rho_{oj} - \rho_{mj})^2), \qquad (8.10)$$

(for $j = 1 - n$) where ρ_{oj} and ρ_{mj}, respectively, are the observed and modelled reflectances for the *j*th observations. During optimization it may also be necessary to eliminate physically impossible values using some form of 'penalty function'. The use of prior information on the likely range of biophysical parameters can help to constrain the inversion process and to make it more robust. Such ancillary data could include prior knowledge of the canopy type and hence likely structural properties (for example from a cropping map) or its state of development (for example from a series of observations or a growth model). These *a priori* estimates may be used, for example, as starting values in an iterative inversion algorithm to focus on the likely true values and to avoid physically unlikely solutions. Alternatively, the prior information may be used in a modified cost function (similar to the mentioned penalty function), where not only the sum of squared difference between the observed and modelled value is minimized but at the same time the difference between estimated and prior values is also minimized (Combal *et al.*, 2003).

8.5.2 Direct or 'numerical' inversion

Most canopy radiation-transfer models are complex and non-linear so it is usually necessary to resort to numerical optimization, using one of the standard numerical approaches such as Nelder and Mead's downhill simplex method or other more sophisticated algorithms. Further information on model inversion methods may be found elsewhere (Liang, 2004; Myneni *et al.*, 1995). Where more than one set of parameters fit the observed data equally well (i.e. there is more than one distinct minimum), one needs to use auxiliary or a priori information to determine the correct solution (see below for further discussion). For example, Li *et al.* (2001) have shown how Bayesian inference can be used in the enhancement of inversion algorithms, where one makes use of pre-existing probability distributions for the parameters to be extracted to enhance the inversion efficiency. Other practical aspects of inversion include the need to take account of the inevitable correlations between parameters, the need for careful choice of starting values, a need for optimization of the number of sample angles used, and the need for recognition of the consequences of experimental error and of the possibility of local not global minima. For example, introduction of only 1% error can lead to an error in estimated *LAI* of up to 73% when inverting the SAIL model (Goel, 1989).

8.5.3 Look-up tables (LUTs)

The 'look-up-table' (LUT) approach (Casa and Jones, 2005; Combal *et al.*, 2003; Weiss *et al.*, 2000) has become widely used for model inversion, largely because of its simplicity. The method involves running the model in the forward mode many times, covering the full range of possible parameter values, to produce a multidimensional table of outputs. This pre-calculated set of results can then be searched rapidly to find the best fit parameters for any *BRF* dataset.

This means that the inversion is reduced to searching the LUT for the best fit with the measured data, usually by means of some weighted least-squares fitting procedure. The precision can sometimes be enhanced further by interpolating between points in the table using geostatistical software to get a more precise estimate of the best-fit parameters.

8.5.4 Machine intelligence

A wide range of alternative approaches are available that make use of artificial intelligence approaches such as artificial neural networks (ANN), genetic algorithms (GA), and support vector machines (SVM). As with the LUT approach it is necessary to train the ANN using large numbers of forward runs of the model; this set of pre-calculated results is used to learn the relation between the canopy spectral-directional signature and the parameter(s) of interest. Hence, the signature space is directly mapped onto the parameter space.

The intrinsic disadvantage of LUT and ANN approaches is that the forward runs on which they are based are inevitably incomplete, leaving a sparse and often unevenly scattered set of known values. This results in there being an infinite number of possible functions that could fit these points – an example of the ill-posed nature of the inversion problem. In such cases one can attempt to 'regularize' the problem to find a stable and reliable solution; this involves imposition of a smoothness constraint that trades off the precision of the fit to the data points against the smoothness of the function. The SVM and the related support vector regression (SVR) are based on statistical learning theory (Vapnik, 1995) and provide a particularly efficient approach to model inversion (Durbha *et al.*, 2007).

The most effective inversion approaches for large images where there are many independent pixels can be to combine components of conventional numerical or analytical inversions with artificial neural networks (e.g. Sedano *et al.*, 2008). In this approach the full inversion of the radiative-transfer model (in this case the Rahman–Pinty–Verstraete model; see Lavergne *et al.*, 2007) was achieved for a limited number of training pixels and then the others estimated using the ANN.

8.5.5 Further aspects of inversion

Incorporation of spectral information. The power of *BRDF* inversion methods can be enhanced by incorporating spectral information. This makes use of the fact that the *BRDF* can vary strongly as a function of wavelength, especially between visible and NIR wavelengths and hence discriminate well between canopies with differing amounts of soil visible. Examples of such an approach include the combination of the PROSPECT (see Chapter 3) and SAIL models (Jacquemoud, 1993; Jacquemoud *et al.*, 2009) These approaches combine those used in Chapter 7 with those dependent on angular responses.

Use of a priori knowledge to constrain inversion. As has been pointed out above the inversion of canopy radiation-transfer models to derive biophysical parameters is a difficult or ill-posed problem that can lead to rather unstable estimates of the requisite parameters enhanced by any errors in either the model or in the *BRDF* measurements. This noise makes the inversion subject to substantial error and may even give rise to erroneous local (as opposed to global) minima or false solutions. The types of *a priori* knowledge that can be used to regularize the solution range from general or 'global' knowledge about the land surface or physical limits to parameters (e.g. a negative *LAI* is impossible), to target-specific knowledge such as type of crop or the stage of crop development. In some cases we can set hard-bounded ranges for certain parameters, in others we may have good information about the parameters, or we know they may be insensitive in our systems and may be fixed, while in others we may be able to set an a priori joint probability density function for pairs of parameters. Bayesian inference can be used in a formal way to incorporate *a priori* knowledge into the inversion process (Li *et al.*, 2001).

The use of growth models provides an example where the inversion may be constrained by prior knowledge derived from the combination of a canopy growth model (especially for homogeneous agricultural crops) with either data from a sequence of remote images or from general knowledge of the seasonal development of the crop.

Object-based retrieval of biophysical variables. It has been suggested (Atzberger, 2004) that the accuracy of the inversion can also be improved by using an object-based approach where one takes account not only of the spectral signature of the pixel of interest, but also of the neighbouring pixels. The relevant neighbouring pixels can be grouped into objects by combining appropriate pixels by processes such as image segmentation or classification (Chapter 6) or the use of digitized field boundaries. In this way one takes advantage of the spatial autocorrelation of spectrally relevant vegetation characteristics in highly managed landscapes. One can derive a wide range of additional statistical properties to describe the object signature (e.g. variances, covariances between spectral bands, and so on) so there is extra information available for guiding or regularizing the model inversion. The rather complex information available when using such an object-based approach is most easily inverted by use of an ANN.

Taking advantage of temporal information. Generally, vegetation growth follows a smooth pattern that can be described by some curve (e.g. the *LAI* of crops can readily be described using a double logistic function). When the same target is observed several times during a growing season one may take advantage of this smooth development for regularizing the inverse problem (Baret and Buis, 2008). In a first run, the data from individual scenes are inverted using some standard approach. The retrieved *LAI* values are next subjected to some curve-fitting techniques to retrieve a smooth *LAI* development. These smoothed values may then serve as prior information for the following model inversion. The process may be repeated until convergence is achieved.

8.5.6 Radiative model comparisons

It is worth considering the relative advantages of the different approaches used for modelling of the

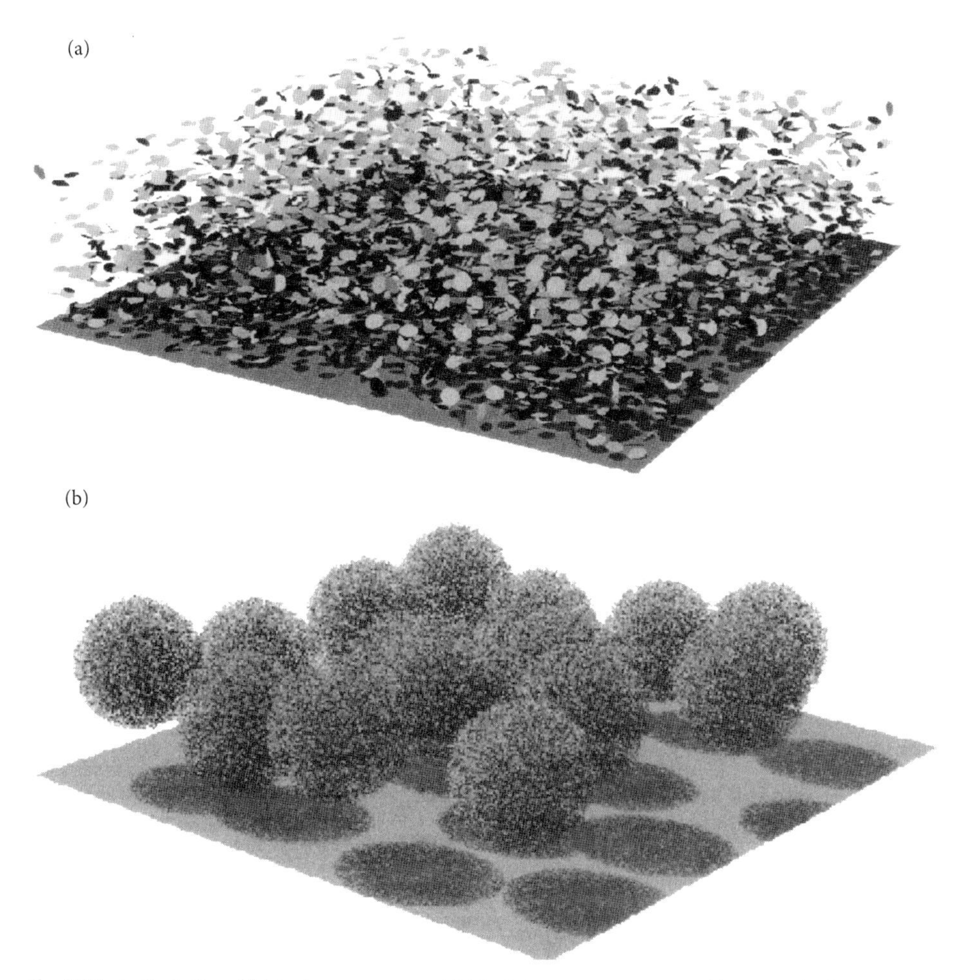

Fig. 8.16. An illustration of (a) the homogeneous and (b) the heterogeneous scenes used for comparing the performance of different radiative-transfer models in RAMI1 (with permission from Pinty *et al.*, 2001). (See Plate 8.2(a))

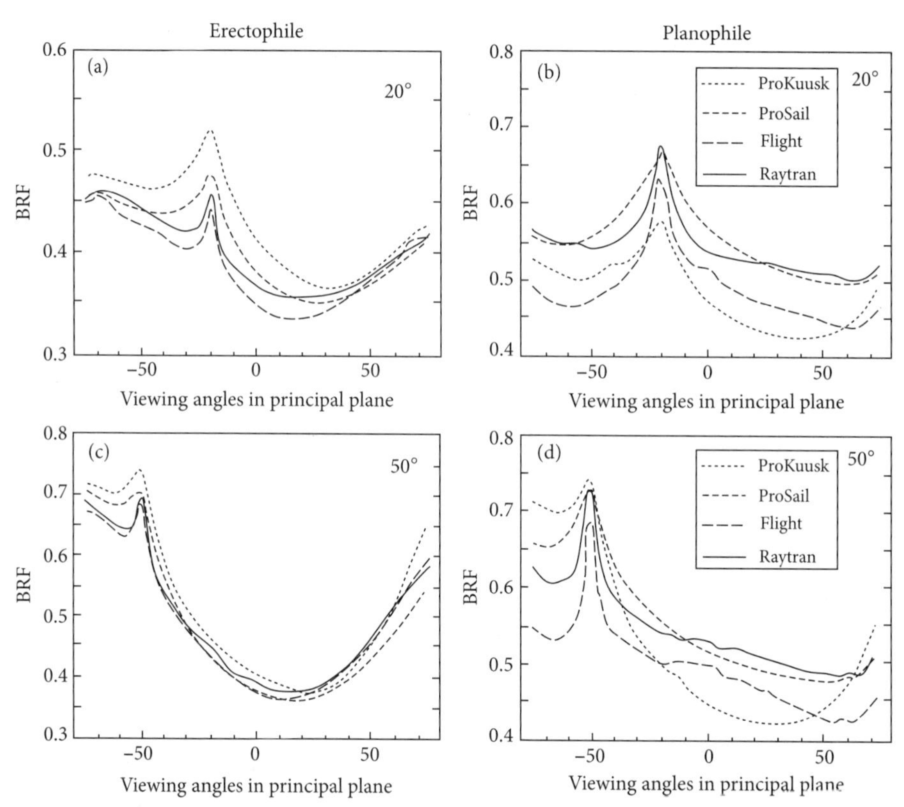

Fig. 8.17. Figure showing examples of the comparative performance of some of the different radiative-transfer models tested in the RAMI 1 (see Fig. 8.16). Results shown are the *BRF*s in the principal plane for simulations in the near infrared at illumination zenith angles of 20° (a, b) and 50° (c, d) for erectophile (a, c) and planophile (b, d) test canopies (see Pinty *et al.*, 2001, for details of the different models and the tests undertaken).

radiative transfer of plant canopies and especially for the simulation of the *BRDF* together with their relative ease of inversion. The major distinction between radiative-transfer modelling and ray-tracer or radiosity approaches is that the former tend to be based on (volume) average properties within any cell or layer, while the radiosity approaches are based on the actual reflection/transmission by discrete and spatially and orientationally explicit objects. The latter approach allows one to simulate spatial correlations, canopy gaps, and clumping directly, but requires much more information.

Because of the varying assumptions of different models used to simulate the anisotropy of land surfaces and the different assumptions used in the inversion of these models it is not surprising that different models can give rather different results (see Figs. 8.16 and 8.17 for a comparison of the performance of some representative models). The greatest problems arise in simulation of heterogeneous systems; nevertheless, even in the relatively simple case of a homogeneous canopy there are surprisingly large deviations between the predictions of different models, with the average local angular deviations being of the order of 5–10%. This has obvious implications when one is concerned to run the models in the inverse mode to extract canopy structure parameters.

In order to overcome the difficulty of quality control of different models there has been substantial effort over the years (e.g. Pinty *et al.*, 2001; Pinty *et al.*, 2004; Widlowski *et al.*, 2008) at developing a system that can test the performance of any model against a standard. As a result of this Radiation-Transfer Model Intercomparison (RAMI) exercise, a model checker has been developed that is now available on line (Widlowski *et al.*, 2008) at (http://rami-benchmark.jrc.ec.europa.eu/).

8.6 'In-field' estimation of canopy biophysical parameters

For validation of radiative-transfer models and extraction of canopy biophysical parameters from remote-sensing systems it is necessary to have independent measures of canopy characteristics such as leaf-area index or leaf-angle distribution and often, their distribution with height in the canopy. This can only be achieved rigorously by direct measurement, but this is generally extremely labour intensive, so we tend to have to rely heavily on indirect methods, frequently themselves based on measurements of radiative-transfer properties of the canopy.

8.6.1 Direct measurement

The fundamental check on all radiative-transfer-based approaches for estimation of canopy biophysical parameters is provided by direct measurement. *LAI* (usually used to refer to only the green leaf area) can readily be estimated from complete harvests of the plant material above known areas of ground, though this is often only feasible for herbaceous canopies and agricultural crops. After harvest of the sample area, the plant material is separated into stems/branches (or other components such as flowers) and leaves, if necessary after separating into component species for mixed canopies. The total leaf area of sample can then be obtained either by direct measurement (e.g. by tracing outlines on squared paper and counting squares), or by photographing leaves on a contrasting (e.g. white) background and estimating leaf area as a fraction of the image using approaches such as the 'magic wand' tool in Adobe Photoshop or other imaging software. Otherwise, the leaf areas may be measured by feeding the leaves through an automatic leaf area meter. In the more sophisticated leaf area meters (e.g. Licor 3000C or 3100C, from LiCor Inc., Lincoln, Nebraska) or Delta-T WinDIAS from Delta-T devices, Burwell, Cambridge, UK) the leaves are fed past the sensing head between two transparent belts, and an image projected to a camera that is set up automatically to accumulate leaf area. Automated systems are clearly preferable for the large numbers of samples that are required to get truly representative estimates for canopies, however, where automated systems are not available it is also possible to measure the area of a small subsample of leaves and scale this up to the whole sample on the basis of the ratio of the weight of the subsample to the weight of the whole sample. Stratified sampling can be used to provide information on the area distribution by height in a canopy. Information on leaf-angle distribution (*LAD*) within canopies is even more difficult or tedious to obtain by direct measurement; the range of instruments and approaches available for this have been outlined in Section 3.1.4.

For trees, total harvest is even more difficult than for herbaceous canopies and often totally impractical, so scaling up using subsampling strategies is most common. For example one can count leaves on individual branches and estimate leaf area per branch from the size of typical leaves and then scale up to whole trees on the basis of the number of branches per tree (e.g. Čermák *et al.*, 2007). Because of the difficulty of direct measurement of *LAI* and *LAD*, however, and their destructive nature, most studies rely on indirect estimates of leaf area from radiative properties of canopies based on the inversion of radiative-transfer models; these are outlined in the two following sections.

8.6.2 Within-canopy indirect methods

Indirect estimates of canopy biophysical parameters can be obtained using analogous approaches and similar radiation-transfer theory to those used in conventional remote sensing, but based on measurements of radiation attenuation below the canopy (for approaches using above-canopy measurements see Section 8.6.3). The below-canopy approaches are usually based on the theory of gap-fractions (see Chapter 3) and therefore only require models that treat the unscattered radiation beams. These approaches include a range of techniques that estimate some aspect of radiation attenuation through vegetation canopies.

Multisensor arrays: At their simplest, simple multisensor arrays (e.g. AccuPar80, Decagon Devices, Lincoln, Nebraska or SunScan SS1 from Delta-T

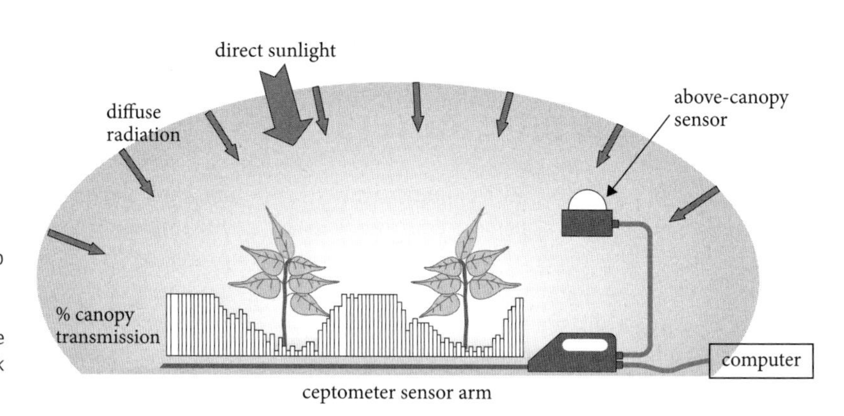

Fig. 8.18. Illustration showing the application of a simple multisensor transmission meter (SunScan SS1) for estimation of leaf-area index and leaf-angle distribution. The above-canopy sensor is used to estimate the incident irradiance and the fraction of direct radiation. The instrument can be set up either to estimate the average fraction of radiation transmitted through the canopy to the level of the sensor, or else it can be set up to measure the sunfleck fraction for the direct component.

Devices, Burwell, UK) or mobile sensors are used to measure irradiance at a large number of points below the canopy and to compare these with the above-canopy irradiance so as to estimate extinction coefficients of global radiation and hence to derive estimates of *LAI* (Fig. 8.18). It is necessary to measure transmission at a large number of points to allow for the substantial heterogeneity and hence sampling errors for most real canopies. The approach can be refined to take account of variation in radiation transmission at different sun zenith angles to infer not only the canopy *LAI* but also the leaf-angle distribution. It is possible to use such radiation measurements under the canopy in two closely related but distinct ways: the *radiation measurement approach* measures the attenuation of mean irradiance with depth in the canopy, the *'gap-fraction'* approach measures the proportion of area below the canopy that is illuminated by sunflecks, or the gap fraction. In principle the radiation approach involves solving the radiation attenuation equation (eqn (3.6)) for *L*, giving $L = \ln(\boldsymbol{I}/\boldsymbol{I}_0)/k$, where k is the appropriate extinction coefficient for the canopy and $\boldsymbol{I}/\boldsymbol{I}_0$ is the irradiance below the canopy as a fraction of that above. As indicated in Chapter 3, the value of k depends on the leaf-angle distribution and the angle of the incident light, being 1.0 for a horizontal-leaved canopy but varying for other canopies as outlined in Table 3.4. By substituting a known value of the Campbell ellipsoidal parameter (x) for the canopy in question, or by estimating it as 1.6 times the ratio of the number of leaves with angles of greater than 45° divided by the number of leaves with angles less than 45° one can then estimate L with some accuracy, at least for reasonably homogeneous canopies. The actual algorithm used takes account of measurements of the ratio of direct to diffuse radiation and knowledge of the solar zenith angle so that the canopy transmission corrects for the differing transmission for radiation arriving from different angles.

The alternative approach using gap fractions at specific angles is rather simpler, since it involves only the unscattered direct radiation using the theory derived by Nilson (1971) and outlined below. This angular information may be obtained by using hemispherical photography (Anderson, 1981; Evans and Coombe, 1959; Leblanc *et al.*, 2005), or by sensors with angular capacity (see next section). It can be shown (by substitution into the equations in Table 3.4) that radiation transmission and the extinction coefficient is particularly insensitive to assumptions about leaf-angle distribution for solar elevations around 33°. Both approaches can be adapted to obtain information on the variation of radiation penetration with zenith angle by using measurements at different solar angles (i.e. at different times of day). Software is available to help with analysis of such information (e.g. ter Steeg, 1993).

Although multisensor arrays measure total transmission by the canopy without any explicit angular information, angular information can be obtained by measuring the gap fraction from the percentage of sensors in sunflecks on bright sunny days at known solar angles. By collection of data at a range of solar angles it is possible to estimate not only *L*, but also the *LAD*, by using the inversion techniques outlined above.

Sensors with angular capacity: One approach to the estimation of canopy parameters from radiation transfer is based on hemispherical photography, where a hemispherical camera is set up underneath the canopy to give a fish-eye image such as that from under a maize canopy shown in Fig. 8.19(b). Such an image is then enhanced to separate leaf from canopy and to quantify the fraction of sky visible (= gap fraction, *F*) in different angle classes. Information on the angular distribution of radiation transmission can also be collected by the use of specialized sensors that are set up to measure canopy transmission in different angle classes (e.g. LiCor LAI-2000; Fig. 8.19(a)). In either case the aim is usually to determine the angular distribution of canopy gaps, and invert this information to derive *LAI* and *LAD*. For the gap-fraction approach, assuming azimuthal symmetry, F at any zenith angle, θ, is the probability ($P(\theta)$) of viewing the sky as given by (Nilson, 1971)

$$P(\theta) = \exp(-G(\theta)\, L_{\text{eff}}/\cos\theta) = \exp(-k(\theta)\, L_{\text{eff}}), \qquad (8.11)$$

where $G(\theta)$ is the mean projection of a unit area of foliage in the direction θ, θ is the zenith angle, $k(\theta)$ is the extinction coefficient at an angle θ (and equals $G(\theta)/\cos\theta$), L_{eff} is the effective (apparent) leaf-area index, and the term $\cos\theta$ corrects for the increasing pathlength through the canopy at larger zenith angles. The extinction coefficient is derived as explained in Chapter 3 as the ratio of the projected area of the canopy to the actual area, while the effective leaf-area index corrects for the fact that in practice leaves are not randomly distributed.

A disadvantage of all radiation transfer-based methods is that it is difficult to distinguish between radiation interception by leaves and by woody tissues such as stems and branches. An approach to overcoming this problem has been proposed by Kucharik *et al.* (1998; 1997) using a multiband vegetation imager (MVI). This is basically a wide-angle camera set up beneath the canopy but collecting images in the NIR and the R so that it can separate leaf material in the image from woody material based on their differing reflectances in the two wavebands (see Chapter 7).

Whatever method is adopted, however, the spatial heterogeneity in almost any natural canopy requires a sensing technique that averages data for as much canopy as possible with large numbers of replicate samples in order to obtain useful data. The second key problem with radiation-penetration methods based

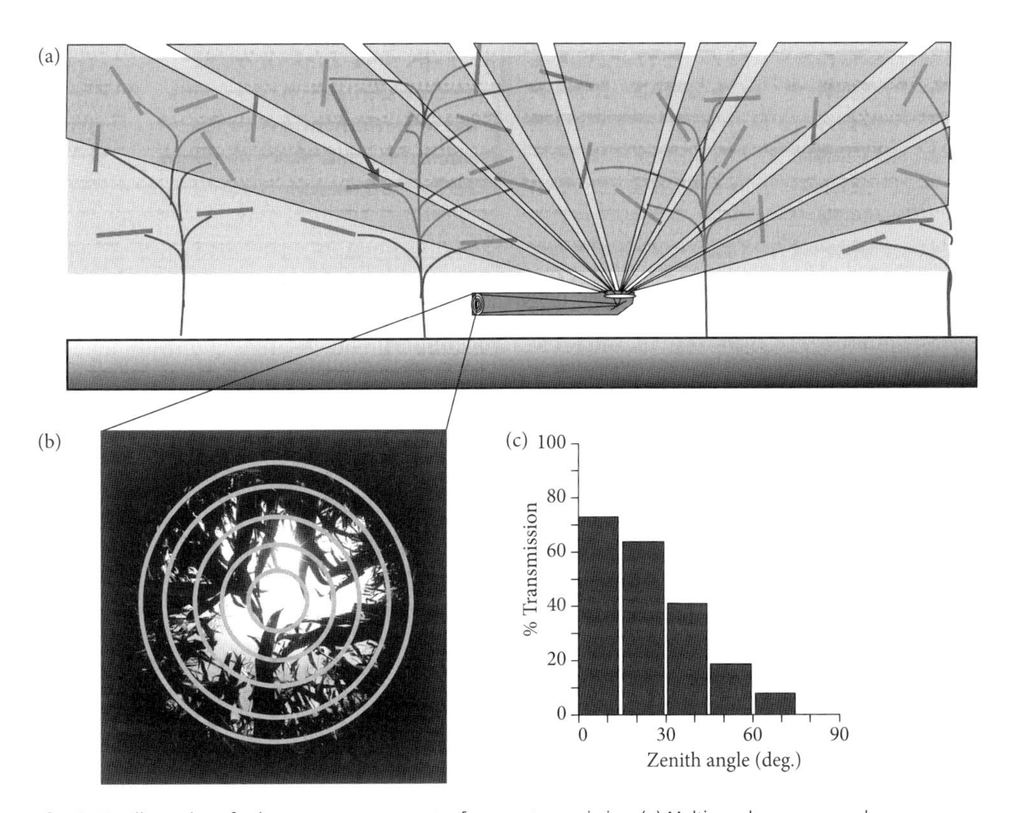

Fig. 8.19. Illustration of subcanopy measurements of canopy transmission. (a) Multiangular canopy analyser (LAI-2000) under canopy showing view angle classes, (b) corresponding hemispherical photograph showing the angle classes, (c) histogram showing radiation transmission (or gap fraction) in each class.

on the inversion of random leaf distribution models is that the radiation attenuation is substantially affected by clumping of leaves, with greater radiation penetration for a given leaf-area index as the degree of clumping increases. The observed value of L (L_{eff}) is related to the true value (L) by a correction factor or '*clumping index*' (λ_0 that is normally <1.0 for clumped canopies, though can be >1.0 for very regular canopies), where

$$L = L_{eff}/\lambda_0. \tag{8.12}$$

There have been suggestions as to how the clumping index can be estimated from the frequency distribution of the gap sizes (e.g. Leblanc *et al.*, 2005). This clumping error may give rise to uncorrected equations underestimating the true leaf-area index by as much as 50% (van Gardingen *et al.*, 1999). The correction may be further extended to allow for non-green parts of the canopy (branches and trunks also intercepting radiation) to estimate the true foliage area index as

$$L = L_{eff}(1 - \alpha)/\lambda_0, \tag{8.13}$$

where α is the ratio between the woody (hemisurface) area index and the total plant (hemisurface) area index (Chen, 1996). Substituting eqn (8.13) into eqn (8.11) gives

$$P(\theta) = \exp(-k(\theta)\, L\, \lambda_0/(1 - \alpha)). \tag{8.14}$$

Detailed equations and the methods of application, together with essential precautions to ensure adequate sampling of the canopy of interest may be found in the instrument manufacturers' handbooks. There are many useful reviews of methods for extracting canopy parameters from subcanopy radiation measurements (Bréda, 2003; Garrigues *et al.*, 2008; Hyer and Goetz, 2004; Jonckheere *et al.*, 2004; Leblanc *et al.*, 2005; Weiss *et al.*, 2004), while España *et al.* (2008) have extended the use of hemispherical photography to the analysis of canopies on slopes.

8.6.3 Above-canopy indirect methods

As well as it being possible to do conventional bidirectional reflectance modelling and inversion for measurements made in-field using, for example goniometry, it is also possible to use hemispherical photography from above canopies, especially for short canopies such as grassland or cereal crops. The radiation-transfer theory for above-canopy photography is the same as for below-canopy measurements. Although subcanopy hemispherical photography has the advantage that it is particularly easy to distinguish leaf and background when viewing the sky as background, the use of R and NIR channel data from above canopies make it straightforward to distinguish canopy from soil background using an *NDVI*-based approach.

Use of component fractions

A rather different approach to model inversion for near-canopy images uses not the total reflectance but the proportion of component image fractions such as shaded leaves or soil and sunlit leaves or soil (Casa and Jones, 2005).

Following Campbell and Norman (1998), for the simple case of nadir view, the fraction of the view that is soil ($f_{s.v}$) is $e^{-k(0)L}$ (see Box 3.1) and therefore the fraction that is vegetation ($f_{veg.v}$) is $1 - e^{-k(0)L}$, where $k(0)$ is the extinction coefficient for a zenith angle of 0°. For any depth, L, the fraction of leaves that is sunlit for a beam zenith angle of θ is $e^{-k(\theta)L}$, therefore the fraction of sunlit leaves at that depth in the field of view is simply the joint probability of a leaf being directly sunlit when below a leaf area index L ($=e^{-k(\theta)L}$) and of it being visible from the sensor at nadir ($= e^{-k(0)L}$), i.e. the product of these two exponentials. Integration over the full canopy gives the sunlit leaf area index that is visible (L_v^*) as

$$L_v^* = \frac{1-e^{-(k(0)+k(\theta))L}}{k(0)+k(\theta)}. \tag{8.15}$$

This can be converted to the fraction of sunlit leaves in the sensor view ($f_{sl.v}$) by multiplying by the appropriate shape factor ($k(\theta)$)

$$f_{sl.v} = k(\theta)\, L_v^*, \tag{8.16}$$

therefore the fraction of shaded leaves in the field of view ($f_{shl.v}$) will be given by

$$f_{shl.v} = 1 - e^{-k(0)L} - k(\theta)\, L_v^*. \tag{8.17}$$

The fractions of sunlit and shaded soil can be derived similarly. This approach uses only the first interception of direct radiation and avoids the complexities introduced by scattering as this usually only has a minor effect on the reflected radiation. Figure 8.20 shows how

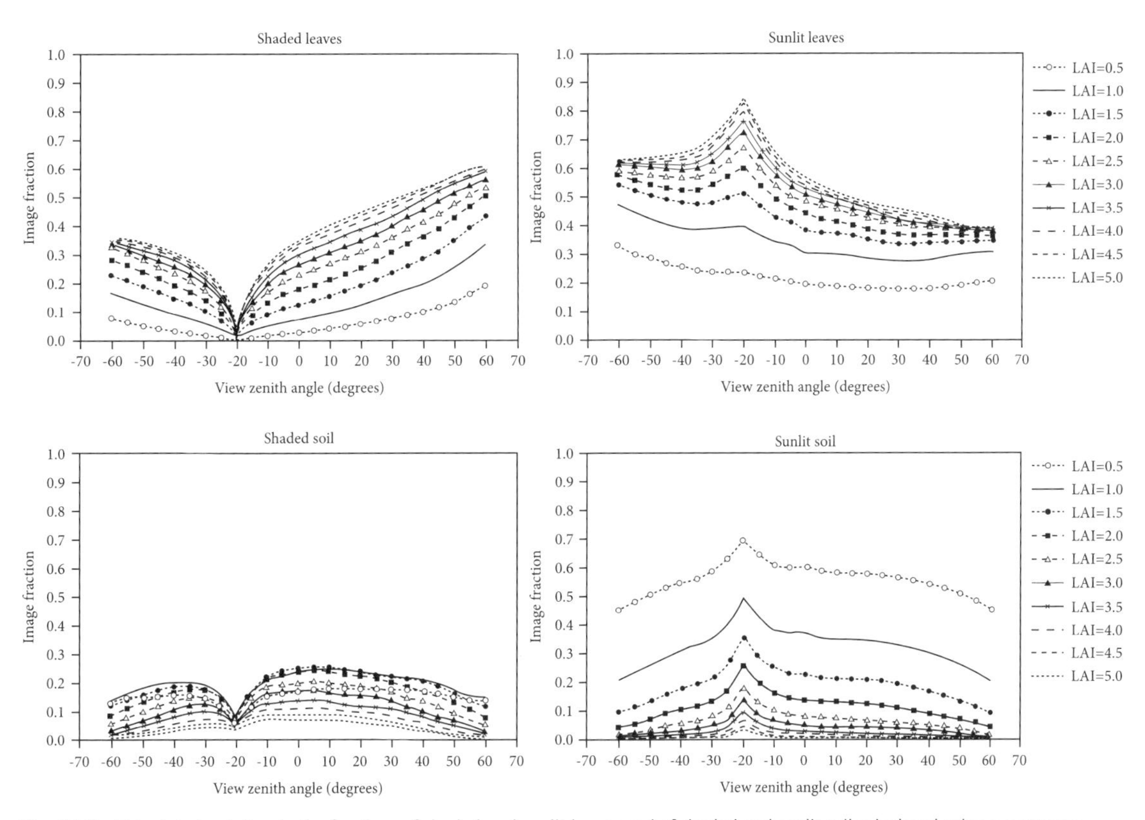

Fig. 8.20. Calculated variation in the fractions of shaded and sunlit leaves and of shaded and sunlit soil calculated using a ray-tracer approach when viewing a canopy with spherical leaf-angle distribution from different angles in the solar plane. Negative zenith angles correspond to the backscatter direction (from Casa and Jones, 2005).

Fig. 8.21. (a) A false colour (Red = red, NIR = green) image at +10° view angle of a potato plot using a two channel (Red/NIR) camera (ADC, Dycam Inc., Chatshereworth, CA), together with (b) the image classified into sunlit leaves, shaded leaves, sunlit soil, and shaded soil using a supervised classification in the ENVI image processing software (Casa, 2003). (See Plate 8.3).

these fractions are predicted to vary, using POV-Ray, for a range of view angles and different values of leaf-area index (L).

This approach is only applicable when using high spatial resolution imagery, as is possible for in-field calibration studies. We illustrate in Fig. 8.21 the application of this approach using close-to-canopy images of a potato crop, where a standard supervised classification procedure is used to classify an image of a potato canopy into sunlit and shaded soil or leaves.

An alternative simplified approach using only a set of brightness thresholds is presented in Fig. 8.22 to

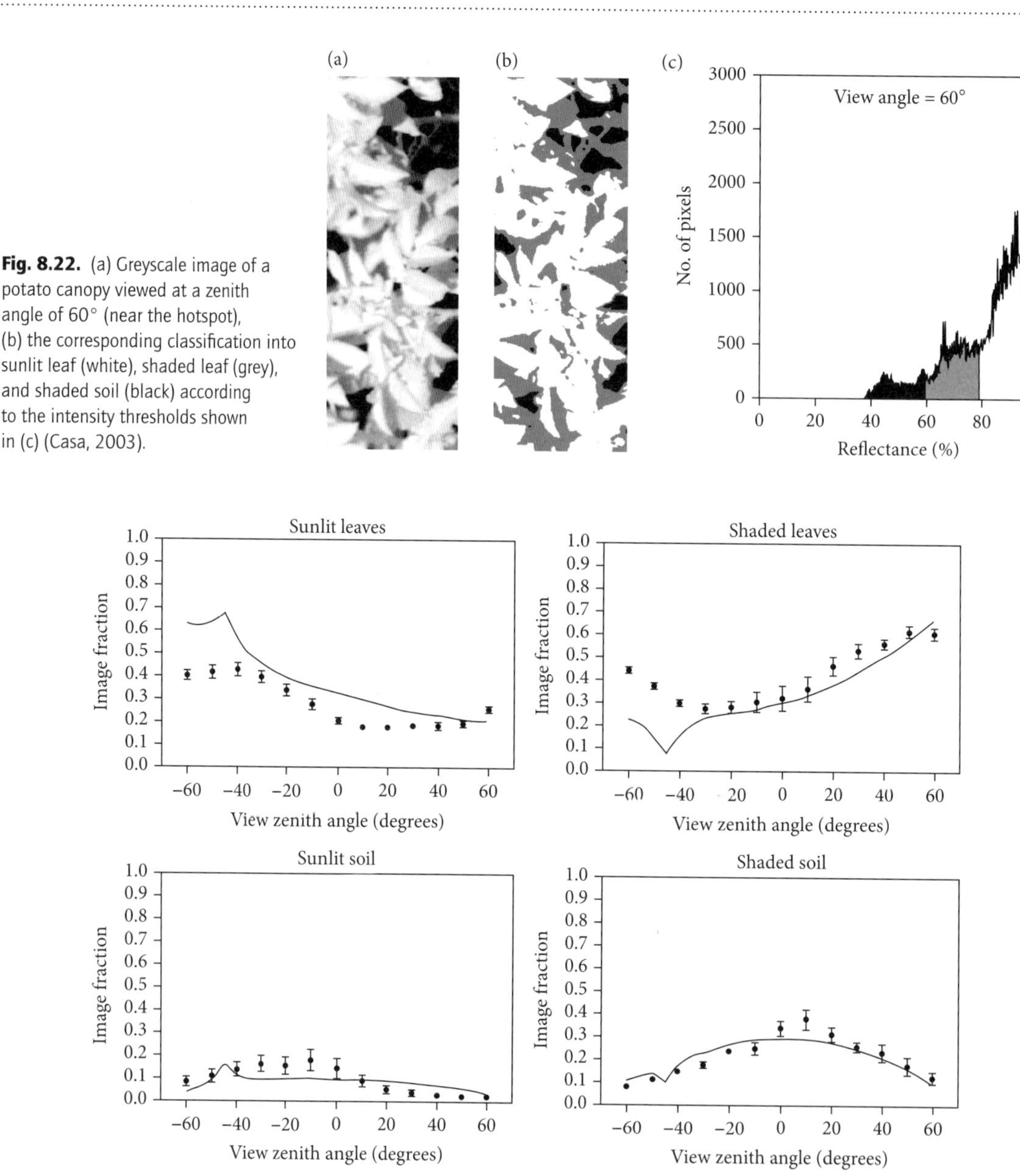

Fig. 8.22. (a) Greyscale image of a potato canopy viewed at a zenith angle of 60° (near the hotspot), (b) the corresponding classification into sunlit leaf (white), shaded leaf (grey), and shaded soil (black) according to the intensity thresholds shown in (c) (Casa, 2003).

Fig. 8.23. Multiangular image fraction components obtained from classification of ADC data acquired in 2003 on a potato plot with *LAI* of 1.9. Points are the means of four datasets acquired consecutively on the same plot with error bars representing the standard deviations. Lines are model outputs using canopies simulated using the POV-Ray system and retrieved as single best fit from a LUT search procedure (Casa and Jones, 2005). This dataset illustrates the sort of errors that may be found with real datasets, often as a result of limited field replication, especially near the zenith where the area viewed tends to be small.

illustrate the principles involved (when applied to a dense canopy) where the brightest pixels are assigned to a sunlit leaf, and less bright pixels to shaded leaf or soil. Figure 8.23 illustrates the results from a classification such as that in Fig. 8.21 (results from a series of measurements at 10° intervals using the method shown in Fig. 8.21), together with the best fit *scene component fraction* curves (Casa and Jones, 2005). The derivation of the LUT can be either, as in this case, from ray-tracer analysis of a simulated 'crop' as used by Casa and Jones (2005), or from any of the available radiative-transfer models discussed above that provide information on shade/sunlit fractions, noting that only the first-order scattering needs to be considered.

Further reading

There are few texts that cover the material in this chapter, though a useful but fairly advanced treatment may be found in Liang (2004). Useful introductions to radiation transfer in plant canopies may be found in texts such as Campbell and Norman (1998), Jones (1992), and Monteith and Unsworth (2008).

Websites

PARASOL/POLDER database: **http://postel.mediasfrance.org/en/BIOGEOPHYSICAL-PRODUCTS/BRDF/POLDER-3---BRDF-Data-Bases-/**

RAMI online model checker: **http://rami-benchmark.jrc.ec.europa.eu/**

SAIL and PROSPECT software downloads: **http://teledetection.ipgp.jussieu.fr/prosail/**

Sample problems

8.1 Assume a sensor with an 8° field of view and at a height of 2 m above a short grass surface, calculate the linear dimension of the viewed surface for nadir view, and for views at 30° and 60°. What would the corresponding values be if the sensor rotated about the viewed area as in a goniometer?

8.2 Calculate the probability of viewing the sky at 30°, 45°, 60°, and 75° elevation for (a) a horizontal leaved canopy with an *LAI* of 3.2, and (b) a canopy with a random leaf-angle distribution and *LAI* of 2.2 and a vertical-leaved canopy with an *LAI* of 1.5. What effect would a leaf clumping index of 0.5 have on your answers?

8.3 A simple multisensor transmission meter (SunScan) is used for estimation of leaf-area index under two contrasting canopies growing at sea level on the Greenwich meridian. In each case a series of measurements were made at 08.00, 10.00, and 12.00 h (GMT) on 1st July and the average percentage of the sensors that were shaded was recorded. For crop (a) the percentages of sensors shaded were, respectively, 79%, 64%, and 60% while for crop (b) the corresponding values were 60%, 58%, and 59%. What can you conclude about the leaf-area indices and leaf-angle distributions of the two canopies?

9 Remote sensing of canopy mass and heat exchanges

9.1 Remote estimation of energy-balance components and mass fluxes

As we have already indicated in Chapter 4, remote sensing is not generally at its best for the study of vegetation-related functional processes and fluxes; it is much more suited to answering questions about slowly changing structural features such as the distribution of vegetation type or its properties (e.g. leaf-area index, chlorophyll content, or even biomass). A particular difficulty is that neither heat nor mass fluxes between canopies and the atmosphere can be directly measured, though it is possible to derive useful information at least about the radiative flux components of the surface energy balance, though even some of these, such as the downward longwave radiation, are less easy to estimate than others. Remote sensing can also be used to provide information on the atmospheric humidity (though the overall atmospheric water content is not necessarily a good measure of humidity at the surface), atmospheric transmission, cloud cover (which, together with atmospheric transmissivity, is a key requirement for modelling of radiation inputs over time), and most importantly, the radiative surface temperature.[36] In fact, temperature is the key linking variable in the surface energy balance. Although surface wind speed would be useful for estimation of canopy transfer properties, this is not readily obtained from remote sensing over land (though useful estimates of surface wind-shear over the sea can be obtained from wave structure using microwave altimeters and scatterometers; see Section 4.5). Together, the measureable variables allow one to estimate indirectly the energy partitioning at the surface between radiation, sensible heat, and latent heat fluxes.

A second and critical deficiency of most remote-sensing studies is the fact that images are snapshots, usually separated by daily or longer time intervals; so we rarely get anything approaching continuous measurements (except at very low spatial resolution from some geostationary meteorological satellites). Of course *in situ* monitoring can allow continuous monitoring for the study of response dynamics. Furthermore, cloud cover may restrict further the availability of good optical data. The development of methods to interpolate between observations and to integrate fluxes over time is largely dependent on the use of models, both for interpolation and for estimation of variables that are not directly measurable from the variables or canopy biophysical parameters that we can sense remotely. Much use is made, therefore, of surface-vegetation–atmosphere transfer (SVAT) models or atmosphere models to refine the predictions from the raw remote sensing. Parameter estimation from RS data may also be aided by the use of growth models that incorporate previous information to derive good initial estimates.

[36] It is worth noting that the radiative temperature of a surface is not necessarily the same as the effective aerodynamic surface temperature for substitution into mass and heat-transfer equations of the type outlined in Chapter 4.

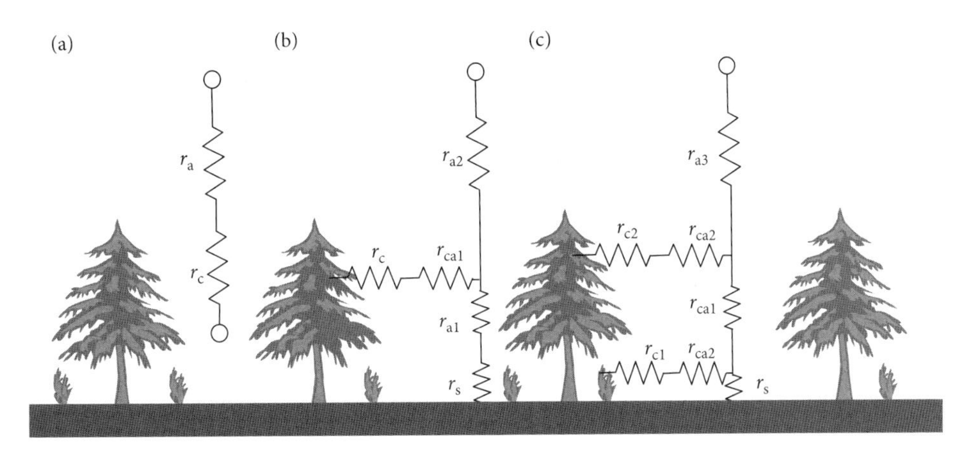

Fig. 9.1. Resistance network models to describe heat and mass fluxes between plant canopies and the atmosphere: (a) Simple 'big-leaf' or single-source model for canopy energy balance, where the whole plant–soil system is lumped together, (b) two-source models treating soil and canopy separately, (c) more complex models that treat fluxes from three or more sources. The various resistances involved include those in the canopy (r_c), the boundary layer (r_a), and at the soil interface (r_s).

Although 'in-field' sensing of fluxes can be valuable for some applications, such as precision agriculture, a great strength of remote sensing is its ability to derive spatially continuous estimates of fluxes over large areas including to the regional scale, and therefore as a source of information for global climate and climate change modelling. The models used to derive the key fluxes range from simple radiation balance and 'big leaf' approximations to more complex multilayer models (Fig. 9.1). In what follows we will first consider each type of energy flux in turn, culminating in a treatment of evaporation, then moving on to the estimation of CO_2 fluxes.

9.2 Radiation fluxes and surface temperature

We have discussed earlier the approaches available for estimating radiative fluxes at the earth's surface from remotely sensed short- or longwave data. Information on these fluxes is essential for estimation of the surface energy balance. Derivation of these fluxes requires correction of the observed 'at-sensor' radiances to the corresponding values at the surface. As was illustrated in Fig. 3.1 this requires correction both for attenuation and scattering of incoming solar radiation in the atmosphere and correction for losses (and gains) on the outward path to the sensor. The methods for atmospheric correction have already been touched on in Chapter 6; further details are largely beyond the scope of this book so readers are referred to appropriate texts and reviews (Bisht *et al.*, 2005; Jensen, 2005; Liang, 2004; Norman *et al.*, 1995a; Sellers *et al.*, 1990; Whitlock *et al.*, 1993) for further information.

9.2.1 Shortwave radiation

The above-atmosphere incoming shortwave fluxes can be derived from simple geometric considerations and the solar constant (see Section 2.7.1). Under clear-sky conditions it is possible to estimate the atmospheric attenuation and scattering (using the standard methods for atmospheric correction outlined in Section 6.2) and hence the surface irradiance, but in the presence of clouds, surface irradiance must be estimated from largely empirical relationships between transmission and cloud cover/type combined with the known information on the irradiance at the top of the atmosphere (I_{toa}). Key steps in the estimation of solar irradiance include (i) calculation of the top of atmosphere irradiance, (ii) derivation of information on the variation of cloud type and density at high frequency from meteorological satellites such as the MSG-SEVIRI instrument on Meteosat-8 (see Appendix 3) that provides data at up to 1 km spatial resolution every 15 min, (iii) conversion to an atmospheric transmission using an atmospheric radiative-transfer model that incorporates cloud and atmospheric properties and optical thicknesses, using appropriate supplementary information on total precipitable water in the atmosphere (e.g. from the ATOVS sounder) and surface albedo (from MODIS). Further details of the approach for estimating solar irradiance at the surface may be found for example in Deneke *et al.* (2008). Estimates of surface radiances are commonly obtained using radiative-transfer models such as MODTRAN and ATCOR, while a good review of the basic satellite algorithms for estimating surface fluxes was provided by Sellers *et al.* (1990). Further developments for estimation of the radiative components of the surface energy balance have been outlined by Whitlock *et al.* (1993) who used the Pinker radiative-transfer calculations and the Staylor (parameterized physical model using ISCCP radiances and TOVS meteorology) algorithms.

Although there is a tendency to concentrate on derivation of instantaneous values at the time of observation, for processes such as photosynthesis or evaporation it is generally much more important to understand the diurnal variation in these process and the integral values over time. Therefore the key problem for most functional studies is the need to fill gaps between valid satellite observations and to derive full diurnal trends; this has been a central activity of much recent research.

9.2.2 Longwave radiation and surface-temperature estimation

Airborne or satellite radiometers can be used to estimate both upward and downward fluxes of longwave radiation (see Norman *et al.* 1995a), but since satellites and airborne sensors do not measure surface radiation fluxes directly, models are required to link satellite observations and surface radiation. Measurements of upward longwave radiation, as well as being an important part of the surface energy balance, provide the basis for estimation of surface temperatures. The relationship between emitted longwave radiation and surface skin temperature (T_S) is given by the Stefan–Boltzmann equation (2.3), though allowance has to be made for the fact that most sensors operate over only the parts of the thermal band that fall within the atmospheric windows. As discussed below, there can be substantial errors in estimation of surface longwave fluxes from satellite observations. The most accurate values are obtained for sensors with several thermal bands that allow the use of split-window algorithms to allow correction for atmospheric effects by making use of known differences in transmissivity of the atmosphere at different wavelengths. The various approaches to the estimation of land-surface temperatures from satellites, including the use of split-window algorithms, were discussed in Chapter 6. Here, we summarize some of the main considerations when estimating fluxes or surface temperatures remotely. As outlined by Jones *et al.* (2004) problems include:

(*i*) *Effect of intervening atmosphere.* It is difficult to correct for the effect of the intervening atmosphere on both upward and downward fluxes. Indeed, as pointed out by Monteith and Unsworth (2008), even with clear skies 85% of the downwelling longwave at the surface is determined in the lowest several hundred m of the atmosphere, which is least accessible to satellite sensors. These errors are larger over land than over ocean (Ellingson, 1995) but can be much reduced from the c. 5 °C often found with GOES satellite data by the use of multiple thermal channels (Gu *et al.*, 1997).

(*ii*) *Anisotropy of the radiation field.* It is widely recognized that the incoming atmospheric radiation and the outgoing vegetation radiation, both in the shortwave and longwave, are anisotropic and depend on the view elevation and azimuth (Chapter 8). These effects are dependent on the angular radiative properties of soil surfaces, leaf-area index, canopy structure, and the proportion of soil likely to be seen by the sensor (Norman *et al.*, 1995b) and can lead to apparent temperature variation with view azimuth as great as 9.3 °C (Paw U *et al.*, 1989) or even 13 °C (Kimes, 1980) though an apparent temperature variation with view elevation of 4–6 °C may be more typical (Lagouarde *et al.*, 1995; McGuire *et al.*, 1989).

It is therefore difficult to obtain the full hemispherical fluxes from the directional radiances usually measured by remote detectors, though Otterman *et al.* (1995), for example, have developed a rather simple model that indicates that measurements made at an angle of 50° from the zenith should estimate the hemispherical emission within a few per cent.

(*iii*) *Emissivity estimation.* Although it is not strictly possible to estimate emissivity and temperature independently from thermal radiation received at satellites, a number of useful approximations are available (e.g. Norman *et al.*, 1995a; Qin and Karnieli, 1999). One simple approach is to estimate ε using empirically determined relationships to *NDVI* (Van de Griend and Owe, 1993). Crucially, errors in emissivity of around 1% can lead to around 0.6 K errors in the estimated surface temperature.

(*iv*) *Radiative versus aerodynamic temperature.* A further complication that makes validation of remote sensing difficult relates to differences in the sources and sinks for different forms of energy transfer. This results in the temperature measured at ground stations not necessarily being equal either to the effective radiative surface temperature for substitution into the Stefan–Boltzmann equation or to the effective aerodynamic temperature for other forms of energy exchange. This discrepancy arises because the emitted thermal radiation comes largely from the upper surfaces of the canopy (which are what is visible at the sensor) rather than from throughout the canopy profile, which may be more appropriate for other energy exchanges. Moreover, standard ground stations measure air temperature at a given height above the surface where temperatures may differ by several degrees from that of the skin temperature. Note that variation in radiative temperature from multiangular views may allow improved separation of soil and vegetation temperatures that are critical for incorporation in advanced models of energy partitioning in canopies.

(*v*) *Heterogeneous surface temperatures.* A less obvious error can arise when the viewed surface in the field of view is composed of an ensemble of surfaces at different temperatures (e.g. soil and leaves), because even if each has an emissivity of close to unity the wavelength distribution of the emitted thermal radiation from the ensemble will not correspond exactly to that of a black body. A temperature difference of more than 10 K between soil and canopy can lead to errors in radiometric estimates of surface temperature of the order of 1 K (Norman *et al.*, 1995a) while even larger variations can occur in some pixels.

(*vi*) *Spatial scale.* Unfortunately most available thermal satellite data are either infrequent (e.g. Landsat at 16-day intervals) or else obtained at rather coarse spatial resolution (AVHRR, MODIS, GOES, or Meteosat). For the higher frequency but low spatial resolution satellites it is possible to estimate subpixel-scale temperatures and hence energy fluxes at more useful spatial and frequency scales by disaggregation of the large pixels on the basis of independent higher spatial resolution estimates of vegetation indices and assumption of a simple relationship between temperature and the vegetation index (e.g. Kustas *et al.*, 2003; see also Chapter 6).

(*vii*) *Temporal integration.* The limited and potentially biased (e.g. as a result of cloud cover) temporal sampling by satellites can also lead to difficulties in estimating true mean surface temperatures or the corresponding mean longwave fluxes, certainly over long periods. Appropriate long-term integration or the estimation of diurnal changes usually relies on the application of some model for smoothing the data.

9.2.3 Net radiation

The instantaneous net radiation at the surface ($\boldsymbol{R}_{\mathrm{n}}$) can be calculated as

$$\begin{aligned}\boldsymbol{R}_{\mathrm{n}} &= (1-\rho_{\mathrm{S}})\boldsymbol{I}_{\mathrm{S}} + \boldsymbol{R}_{\mathrm{Ld}} - \boldsymbol{R}_{\mathrm{Lu}} \\ &= (1-\rho_{\mathrm{S}})\boldsymbol{I}_{\mathrm{S}} + \boldsymbol{R}_{\mathrm{Ln}},\end{aligned} \tag{9.1}$$

where ρ_{S} is the hemispherical surface reflectance, $\boldsymbol{I}_{\mathrm{S}}$ is the shortwave irradiance at the surface and $\boldsymbol{R}_{\mathrm{Ld}}$ and $\boldsymbol{R}_{\mathrm{Lu}}$ are, respectively, the downward and upward longwave radiation at the surface. As outlined in Section 2.7.2 $\boldsymbol{R}_{\mathrm{Ld}}$ depends on the effective sky temperature as seen from the surface and $\boldsymbol{R}_{\mathrm{Lu}}$ can be estimated from surface temperature using the Stefan–Boltzmann equation. As mentioned above, however, one is often interested in the diurnal variation and the longer-term integrated values, for which we are usually dependent on modelling.

A key parameter in eqn (9.1) is the hemispherical surface reflectance, which can be obtained using *in situ* measurements or else it can be estimated from either a knowledge of the canopy type and the typical expected reflectances (see Chapter 3), by the use of one of the available atmospheric correction algorithms (see below).

As an alternative to estimation of the net longwave radiation ($\mathbf{R}_{Ln}$) from all the component fluxes, it can conveniently be approximated as a simple function of daily shortwave energy receipt expressed as a fraction of the potential maximum for that day ($\mathbf{I}_{S-24h}/\mathbf{I}_S^*$) using a relationship such as that illustrated in Fig. 9.2.

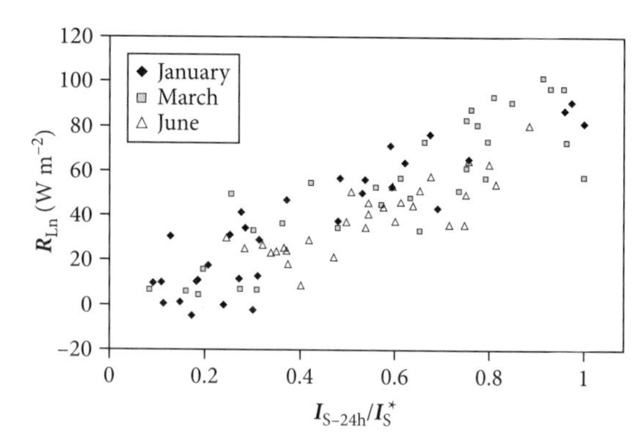

Fig. 9.2. Relationship between daily values of the net longwave radiation ($\mathbf{R}_{Ln}$), and the fraction of possible shortwave radiation ($\mathbf{I}_{S-24h}$) for a Douglas Fir forest at Cabauw in the Netherlands during 1987. Daily shortwave radiation integrals are expressed as a fraction of the maximum for that time of year and location ($\mathbf{I}_S^*$). Data kindly provided by F.C. Bosveld and analysed by Dr E. Rotenberg; details of measurements and the site are described by Beljars and Bosveld (1997).

9.3 Soil heat flux

The fraction of the incident net radiation energy that goes into the soil ($\mathbf{G}/\mathbf{R}_n = \Gamma$) has been shown empirically to be highly predictable from the canopy surface characteristics, and especially the value of *LAI*. As outlined in Section 4.4.6 the energy partitioning into the soil depends on energy partitioning at the surface (Γ') and the fraction of the above-canopy $\mathbf{R}_n$ that reaches the soil surface (Γ''). With bare soil, $\Gamma = \Gamma' \cong 0.4$ indicating that approximately 40% of the incident energy goes into heating the soil at midday. The attenuation of radiation through the canopy is approximately exponential (Section 3.1.3) so we can write (Choudhury, 1994):

$$\mathbf{G}/\mathbf{R}_n = \Gamma = \Gamma'.\Gamma'' \cong 0.4 \exp(-k\,L), \tag{9.2}$$

where k is an extinction coefficient for $\mathbf{R}_n$ and L is the leaf-area index. For more precise formulations see for example Bastiaanssen *et al.* (1998b). The extinction term can be estimated remotely by use of vegetation indices (Section 6.3.1), and where *SAVI* is used, eqn (9.2) can be approximated (Baret and Guyot, 1991) by

$$\mathbf{G}/\mathbf{R}_n \approx 0.4\,[(SAVI_{max} - SAVI)/(SAVI_{max} - SAVI_{min})], \tag{9.3}$$

where $SAVI_{max}$ and $SAVI_{min}$ are, respectively, the maximum (c. 0.814) and minimum values of *SAVI*.

9.4 Sensible heat flux

The remote estimation of sensible heat flux ($\mathbf{C}$) is usually based on the heat-transfer equation (4.8), which can be written as

$$\mathbf{C} = \hat{c}_p\,\hat{g}_H\,(T_s - T_a) = \hat{c}_p\,(T_s - T_a)/\,\hat{r}_H, \tag{9.4}$$

which requires information on both air (T_a) and surface (T_s) temperatures and on the transfer coefficient ($\hat{g}_H$) that controls the rate of heat transfer between the canopy and the bulk atmosphere. When this simplistic equation is used to describe heat exchange by vegetation canopies it is important to recognize that it assumes that the canopy temperature can be approximated by a single average value (T_s) even though in reality it includes an ensemble of surfaces (soil and leaves) at different temperatures. Strictly, one should use the aerodynamic surface temperature rather than the radiometrically measured surface temperature, though with remote sensing only the latter is directly obtained. The transfer coefficient or boundary-layer conductance depends on variables such as wind speed, canopy roughness (itself a function of height and canopy structure) and also on the size of the area of homogeneous vegetation as that determines the thickness of the boundary layer. Of the required variables only the radiometric surface temperature is readily measurable by remote sensing so a range of alternative approaches and approximations have been used.

9.4.1 Empirical approaches

A simple empirical procedure for remote estimation of **C** was proposed by Jackson *et al.* (1977), whereby the daily integral of **C** (C_{24h}) could be estimated for clear days from the instantaneous value of the temperature difference ($\delta T = T_s - T_a$) near midday according to

$$C_{24h} = B\,(T_s - T_a)^n, \tag{9.5}$$

where the exponent (n) is usually close to one and the constant B is an effective exchange coefficient and depends on both the canopy and the meteorological conditions. In practice, better fits are usually obtained by adding an intercept (A):

$$C_{24h} = A + B\,(T_s - T_a)^n, \tag{9.6}$$

where the required value of T_a can either be obtained from ground measurements, or it can be estimated remotely using pixels with known flux conditions as indicated below. By analogy with eqn (9.4) it is clear that the effective transfer coefficient, B (proportional to $1/r_{aH}$), is related to wind speed, atmospheric stability, and canopy roughness (Seguin and Itier, 1983). The value of B generally increases with increasing vegetation cover, varying from about 0.25 to 1.6 over the full range of vegetative cover (Carlson *et al.*, 1995) when C_{24h} is expressed in MJ m^{-2} d^{-1}. A significant problem with use of these equations with remote-sensing data is that the remotely sensed surface temperature, which is weighted to the upper leaves of a canopy, may be a poor estimator of the effective aerodynamic temperature that drives evaporation and sensible heat transfer, whilst there are often problems associated with view angle and estimation of emissivity (Chapter 8).

9.4.2 Endpoints with known flux conditions

A remaining problem is the remote estimation of T_a because air temperature can deviate substantially from the surface temperature. One useful approximation involves the assumption that the coolest pixels represent water surfaces or freely evaporating vegetation. For such pixels it is usually assumed that all the absorbed radiant energy is lost as latent heat and that $T_s \cong T_a$. This value of T_a is then assumed for neighbouring pixels. A recently irrigated field is generally a better choice than a lake as its aerodynamic properties will more closely simulate other vegetated pixels. In cases where no good wet pixels are available it is possible to use the same argument and extrapolate a plot of T_s against *NDVI* to a maximum limiting *NDVI* to obtain an estimate of T_a. In extremely arid areas there may be some advection of heat to oases of vegetation resulting in negative actual local values for **C**; where this is likely some correction needs to be made.

Another useful endpoint is given by dry pixels (non-transpiring vegetation or dry soil); in this case there is no latent heat flux so **C** can be assumed equal to $\mathbf{R}_n - \mathbf{G}$. In this case we can substitute into eqn (9.4) to estimate the transfer resistance as

$$\hat{r}_H = \hat{c}_p\,(T_{dry} - T_a)/(\mathbf{R}_n - \mathbf{G}). \tag{9.7}$$

Alternative approaches to the estimation of $\hat{r}_H$ are available where surface data on wind speed and canopy aerodynamic properties are available (Monteith and Unsworth, 2008). The assumption that estimates of T_a and $\hat{r}_H$ estimated from extreme wet and dry pixels in an image can be extrapolated to neighbouring pixels involves several serious approximations. These include errors induced by local advection, differences in radiation absorbed by vegetated and unvegetated pixels, variations in atmospheric stability as a function of surface temperature, and so on; improved algorithms that aim to overcome these problems will be discussed further in Section 9.5.

9.5 Evaporation

Evaporation is the most difficult component of the surface energy balance to estimate by remote sensing. As was pointed out in Section 4.5.2, for agricultural purposes **E** has most usually been estimated as the product of a reference or potential evapotranspiration rate ($\mathbf{E}_o$) and a crop coefficient (K_c) that is largely determined by canopy cover. Although K_c may be reasonably closely related to vegetative cover as estimated by the use of vegetation indices (at least for well-watered canopies), the estimation of E_0 remotely is more difficult. For example, although remotely sensed canopy temperatures can be substituted into the Penman–Monteith combination equation for the estimation of E or E_0 (Section 4.4.4) this equation also requires information such as wind speed, air vapour pressure deficit, surface resistance, and boundary layer resistance, which are all difficult to estimate remotely.

Many alternative approximations have been used in the estimation of E; the main methods are outlined below. These approaches differ in the scale of application (from small areas of agricultural land or forestry to regional applications where a variety of landcovers are involved), the type, frequency, and spatial resolution of the remote-sensing data, and the necessity for ground data. There are also differences in the approximations made, the degree to which mechanistic models are incorporated in the calculation, and the degree to which dynamic feedbacks between the atmosphere and the vegetation (e.g. stomatal behaviour and albedo) are incorporated.

Most remote-sensing approaches to the estimation of E depend on remote estimation of canopy temperature by thermal sensors, with E itself most commonly being determined as a residual after all other terms in the energy balance have been estimated (Courault *et al.*, 1996; Kalma *et al.*, 2008). As we shall see, there are particular difficulties in up-scaling local estimates to longer timescales and to regions, though remote sensing is a particularly useful tool for such up-scaling as it provides fully distributed data over the region of interest and in combination with an interpolation algorithm allows interpolation of point data. Alternative approaches to the estimation of E over longer time intervals from remote-sensing data include the use of water-balance studies based on microwave sensing of soil water.

9.5.1 Estimation as the residual term in the energy balance

The most usual strategy is to estimate evaporation as the residual after all other components of the energy balance have been estimated. Any of the methods described in Sections 9.2, 9.3 and 9.4 for estimation of R_n, G, and C may be combined and λE obtained by difference using the energy balance (eqn (4.7))

$$\lambda E = R_n - C - G, \tag{9.8}$$

noting that R_n is positive for a gain to the surface, while other components are positive for a loss from the surface. Estimation as a residual can lead to particularly large fractional errors in estimates of E when evaporation is small. Errors clearly depend on the accuracy of the component estimates with estimates of the sensible heat flux being particularly subject to error, because this depends on good estimates of the canopy to air temperature difference, which is often rather hard to obtain accurately remotely. Indeed errors in surface temperature estimation frequently limit the accuracy of remote estimates of E. A particular difficulty with many of the simple algorithms is that they tend not to be suitable for extrapolation to composite terrains at larger scales as a result of non-linear scaling problems. Furthermore, even when ground data are available they are usually only point measurements.

SEBAL and METRIC

A major step forward was the development of the SEBAL (Surface Energy Balance Algorithm for Land) algorithm (Bastiaanssen *et al.*, 1998a; 1998b) and its successors such as METRIC (Mapping EvapoTranspiration at high Resolution with Internalized Calibration; Allen *et al.*, 2007b). Because of the importance and widespread use of the approach it is worth considering this and its derivatives in some detail. These are *one-source* models as opposed to more sophisticated *two-source models* that treat the vegetated and non-vegetated fractions of landscape separately (Kustas *et al.*, 1989). The general approach involves the use of visible, near infrared and thermal satellite radiances to derive surface biophysical parameters such as albedo and vegetation indices that are then used to estimate the various components of the surface energy balance in eqn (9.8) and the necessary heat- and mass-transfer coefficients. An important aspect of the approach is the use of spatial variation in temperature and *NDVI* to provide internal calibrations. Empirical relationships with the *NDVI* play an important role in estimation of key parameters such as surface emissivity and soil heat flux (see Sections 9.2 and 9.3). For further details of the SEBAL algorithm the reader is referred to original papers by Bastiaanssen and colleagues (Bastiaanssen *et al.*, 1998a; 1998b; 2005).

A particular difficulty for remote sensing is the estimation of heat and mass transfer in the atmospheric boundary layer and especially the key transfer coefficients for substitution into eqn (9.4) and other flux equations. For this step the SEBAL approach makes use of the analogy between fluxes of sensible heat, latent heat, and momentum (or wind shear) in the atmospheric boundary layer above a canopy and the empirically observed (and theoretically justified) relationship between instantaneous surface temperature (T_s) and the hemispherical surface reflectance (ρ_s) as shown in Fig. 9.3. In this particular example T_s decreases as ρ_s decreases below c.0.23 as a result of increasing $\boldsymbol{E}$, and also decreases as ρ_s increases above c.0.23, but in this case because of the reduction in absorbed solar radiation. The algorithm in SEBAL essentially takes account of the slope of the radiatively controlled arm of the $T_s : \rho_s$ relationship to estimate the area-effective momentum transfer properties and hence the boundary-layer transfer resistance and how it varies with vegetation cover by using an iterative procedure to estimate $\boldsymbol{C}$ and hence $\lambda \boldsymbol{E}$ for each pixel. The approach also takes account of the variation in the differential between the remotely sensed radiometric surface temperature (T_s) and the near-surface air temperature (T_a) and assumes a linear relation between these variables and thus allows estimation of $T_s - T_a$ from the observed T_s for any pixel. These refinements can lead to a substantial improvement in estimates of $\lambda \boldsymbol{E}$ in comparison with the use of parameters derived simply from eqn (9.4). The parameterization of the near-surface temperature difference used in SEBAL as a function of the observed radiometric surface temperature avoids the need for absolute estimates of the aerodynamic temperature and the need for accurate air-temperature estimates. The METRIC algorithm is similar to SEBAL but is internally calibrated at two ground locations using locally based weather data (Allen *et al.*, 2007a; 2007b). A key step in METRIC is the expression of the calculated $\boldsymbol{E}$ as a fraction of a reference $\boldsymbol{E}$ for a well-watered alfalfa crop normalized for each pixel to estimates at the two ground weather stations.

Although SEBAL and METRIC take some account of the feedbacks between the fluxes and their driving forces, there remains a need to adequately account for feedbacks on the transfer processes (e.g. feedback control of stomatal conductance). It is also important to note that the feedbacks are themselves scale dependent and should ideally take account of the local heterogeneity (see, e.g., Timmermans *et al.*, 2008).

9.5.2 Estimation from net radiation

Estimates of $\lambda \boldsymbol{E}$ may be usefully bounded as the value should fall between zero for a dry non-evaporating surface to a maximum set by the amount of available

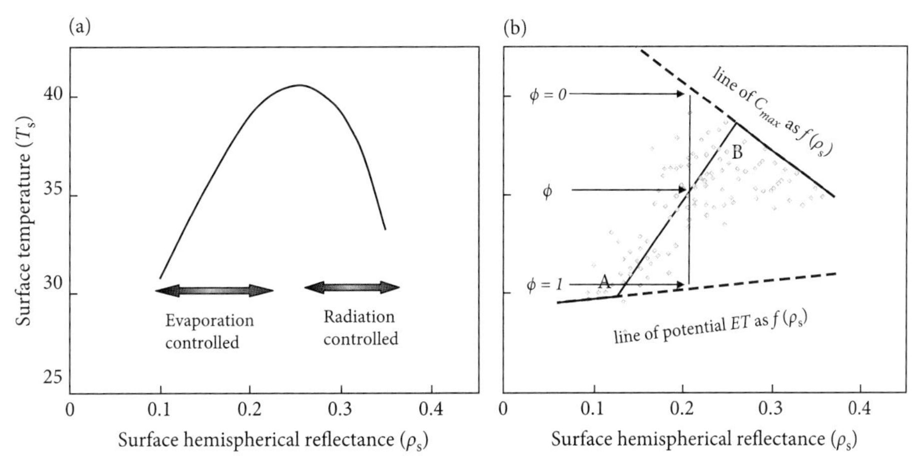

Fig. 9.3. (a) The typical relationship between surface temperature and surface reflectance, showing that for low surface reflectances, which correspond to marshes and otherwise wet or highly vegetated surfaces, temperature is controlled by the evaporation rate, while for surfaces with higher reflectances (drier and non-transpiring), the surface temperature is controlled by the radiation absorbed. Data derived from TM data for the Eastern Qattara depression in August 1986, after Bastiaanssen *et al.* (1998a); (b) Illustration of the method for determining the boundary lines and evaporative fraction (ϕ; as a proportion of the distance between the boundary lines) for a particular pixel from a scatter plot of T_s against ρ_s (based on the approach of Roerink *et al.*, 2000).

energy, which in the absence of advection between pixels is equal to $\boldsymbol{R}_n - \boldsymbol{G}$. For these purposes it is common to assume that the soil heat flux is negligible over 24 h periods. In this case the available energy ($\approx \boldsymbol{R}_n - \boldsymbol{G}$) determines the rate of evaporation according to the equilibrium evaporation rate (McNaughton and Jarvis, 1983; Monteith and Unsworth, 2008):

$$\lambda \boldsymbol{E}_{eq} = \alpha\,(\boldsymbol{R}_n - \boldsymbol{G})\,(s/(s+\gamma)), \tag{9.9}$$

where $\boldsymbol{E}_{eq}$ is the equilibrium evaporation rate, s is the slope of the curve relating saturation vapour pressure to temperature, γ is the psychrometer constant, and α, the so called Priestley–Taylor constant, is an empirically determined constant that is used to describe evaporation from large areas of homogeneous vegetation where often $\alpha \cong 1.26$ (Brutsaert and Sugita, 1996; Priestley and Taylor, 1972). This relationship holds reasonably well, and gives reasonably reliable estimates of $\lambda \boldsymbol{E}$ for substantial areas of vegetation where ground cover is nearly complete (high *NDVI*).

In practice, however, an image will frequently include large areas of less well-vegetated surface evaporating at anything from zero (for dry bare soil) to the maximum as determined by the available energy. Therefore, this equation can be written as

$$\lambda \boldsymbol{E} = \phi\,(\boldsymbol{R}_n - \boldsymbol{G}), \tag{9.10}$$

where ϕ is a bounded multiplier, otherwise known as the *evaporative fraction*, that can vary from 0 (for a non-evaporating surface such as dry soil) to 1 (for a freely evaporating surface such as a lake) and assuming no advection of energy from adjacent pixels. The question remains, however, as to how to determine the value of ϕ. One approach, used in the S-SEBI model (Roerink *et al.*, 2000; Sobrino *et al.*, 2007), is based on the same assumptions as used in Fig. 9.3(a). The rationale is explained in more detail in Fig. 9.3(b), where for wet pixels there is assumed to be little change in observed temperature as reflectance increases. Pixels with higher reflectance largely arise because of reductions in moisture availability and decreasing vegetative cover or transpiration, so that temperature increases from A to B as the evaporative fraction decreases to zero ($(\boldsymbol{R}_n - \boldsymbol{G}) = \boldsymbol{C}$) at B, then from B onwards, temperature decreases with further increases in reflectance reducing the absorbed net radiation. The evaporative fraction (ϕ) for any pixel can therefore be estimated from its position relative to the extrapolated extremes as shown in Fig. 9.3(b). These extreme lines are estimated from scatter plots of T_s against ρ. Some other approaches based on the use of information on canopy temperature are outlined below.

9.5.3 Estimation from the relationship between surface temperature and evaporation

Conveniently it can readily be shown that for similar surfaces in one environment, surface temperature is linearly related to $\boldsymbol{E}$. For a dry surface we can write

$$(\boldsymbol{R}_n - \boldsymbol{G})_{dry} = \hat{g}_H\,\hat{c}_p\,(T_{dry} - T_a), \tag{9.11}$$

while for any evaporating surface we can write

$$(\boldsymbol{R}_n - \boldsymbol{G})_s = \hat{g}_H\,\hat{c}_p\,(T_s - T_a) + \lambda \boldsymbol{E}_s. \tag{9.12}$$

Subtracting eqn (9.11) from eqn (9.12) and rearranging gives

$$\lambda \boldsymbol{E}_s = (\boldsymbol{R}_n - \boldsymbol{G})_s - (\boldsymbol{R}_n - \boldsymbol{G})_{dry} + \hat{g}_H\,\hat{c}_p\,(T_{dry} - T_s). \tag{9.13}$$

Where the radiative (and aerodynamic) properties of any two surfaces are similar the first two terms drop out and it follows that temperature difference between them is linearly related to differences in the evaporation rate. Alternatively it is possible to eliminate the rather difficult to estimate $\hat{g}_H$ from eqns (9.11) and (9.13), to give

$$\lambda \boldsymbol{E}_s = (\boldsymbol{R}_n - \boldsymbol{G})_s - (\boldsymbol{R}_n - \boldsymbol{G})_{dry} \cdot (T_s - T_a)/(T_{dry} - T_a), \tag{9.14}$$

where T_{dry} and T_a ($\approx T_{wet}$) can be estimated from extreme nearby pixels as above.

This principle has been widely adopted to utilize satellite imagery over a heterogeneous area by using the properties of a scatter plot of T_s : *VI*, or better, a plot of $T_s : f_{veg}$ (see Chapter 7). In such a plot one commonly finds that the points fall in a generally triangular or trapezoidal area as illustrated in Fig. 9.4. Since we know from the above discussion that temperature decreases linearly with increasing evaporation (eqn (9.13)) this implies that the lower-temperature boundary-line represents areas evaporating at the maximal potential rate determined by incident energy (as given by eqn (9.9)), while the highest temperatures observed relate to non-evaporating dry pixels. In principle one can obtain improved estimates by replacing measured surface radiometric temperatures by δT obtained as we discussed before for SEBAL. The fraction of available energy used in evaporation (ϕ) is therefore linearly related to temperature as shown. There is also a close relation to fractional vegetation cover, with fully vegetated pixels having the lowest temperatures and maximal evaporation and dry unvegetated pixels clustering at the top left-hand corner with the highest temperatures and the lowest evaporation rates. Scatter at the high vegetation cover end can largely be interpreted as variation in evaporation rate due to a short-term stomatal limitation for those areas of dense vegetation where water is limiting. The diagonal limiting line can be thought of as representing the result of linear mixing of dry soil and fully evaporating vegetation, so an f_{veg} of 0.3 would give rise to an evaporative fraction of 0.3 for these conditions. Points within the main lower triangle in Fig. 9.4(b) represent areas where the soil is not dry and there is some soil evaporation contributing to the total; for example the solid point shows an evaporative fraction, $\phi = x/y$.

For any point on this diagram we see that

$$1 - \phi = 1 - \boldsymbol{E}/\boldsymbol{E}_o = (T_s - T_{wet})/(T_{dry} - T_{wet}). \quad (9.15)$$

This is identical to what is known as the *crop water-stress index* (*CWSI*) that was developed by Idso and Jackson (Idso *et al.*, 1981; Jackson *et al.*, 1981). We discuss application of this to 'stress' sensing in Section 9.6.

This can be better expressed on a trapezoid (Fig. 9.4(b)). At the left-hand (relatively unvegetated) part of the plot there is rather greater scatter but a shortage of fully transpiring points, indicating few areas of free water in the image. A reasonable approximation is to assume that the energy absorbed in any pixel is absorbed by vegetation in proportion to its ground-cover fraction (and largely dissipated as latent heat), with the rest being absorbed by the bare soil, and that is largely dissipated as sensible heat when the soil is dry. Where there is a limited population of pixels with high vegetation cover, the upper edge of the T_s:f_{veg} line can be extrapolated to maximal f_{veg} to estimate the vegetation temperature and hence the potential evaporation (Boegh *et al.*, 1999). In this approach it is assumed that the observed temperature (T_s) is a linear sum of the vegetation (T_v) and soil (T_{soil}) temperatures according to

$$T_s = f_v T_v + (1 - f_v) T_{soil}. \quad (9.16)$$

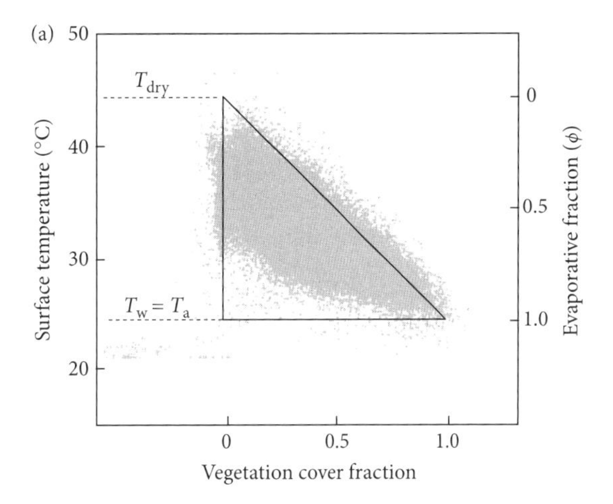

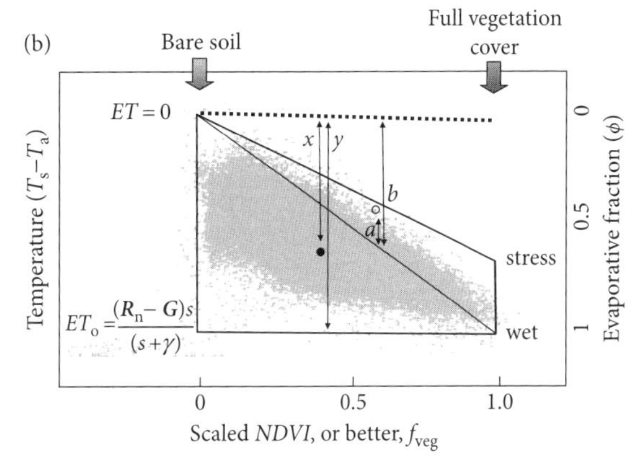

Fig. 9.4. (a) Relation between T_s and vegetation cover fraction (approximated by scaled *NDVI*). The interpretation is based on the assumption that ground cover is proportional to the scaled *NDVI*, and the assumption that the lower horizontal bound represents full evaporation from wet surfaces. At any point the evaporation, and hence the temperature, is made up of the linear sum of that from the soil and that from the vegetation weighted by the vegetation fraction. Points at low *NDVI* represent bare soil, lakes, etc., with higher temperatures occurring where the soil is dry. Data points from Gillies *et al.* (1997) for Walnut gulch, 9 August 1990.
(b) Illustration of the trapezoidal relationship showing the increasing temperature above the diagonal as stomatal closure occurs (typical of short-term drought stress). The indicated point (•) applies to a pixel with some soil evaporation, while the pixel (○) shows some stomatal closure. Note that this diagram is only an approximation as both the upper and the lower limiting lines can diverge from the horizontal, primarily because the net absorbed energy varies with f_{veg} because of changing albedo and also the changing fraction of soil heat flux. Note also that the upper line of the trapezoid does coincide with the line for $\boldsymbol{E} = 0$, because few vegetation types reach this point without the vegetation dying off. The open circle (○) illustrates the calculation of a cover-weighted corrected stress index. For explanation see text; a detailed analysis of this diagram is presented by Moran *et al.* (1994).

One problem with this general approach is that the amount of available energy can vary substantially between pixels with desert soil, for example, often being more reflective than vegetation, so that the position of pixels should also be corrected for absorbed radiation, and for soil heat flux. The approach can be enhanced by the incorporation of thermal inertia data (Anderson *et al.*, 1997; Stisen *et al.*, 2008). Although total $\boldsymbol{E}$ for a pixel can be assumed to be proportional to $T - T_{dry}$, it is worth noting that there is not necessarily any unique solution for centre pixels that would determine the relative contribution of soil or canopy components according to the dryness of the soil and the stomatal conductance.

Although various workers (e.g. Choudhury *et al.*, 1994) have used an expression of the form

$$\boldsymbol{E} = \boldsymbol{E}_o(1 - (VI_{max} - VI)/(VI_{max} - VI_{min})), \quad (9.17)$$

as an estimator of $\boldsymbol{E}$, this equation represents only the upper boundary line in Fig. 9.3(a) and will not work well where the soil is wet, or where evaporation from the vegetation is less than the optimum as a result of stress-induced stomatal closure. Nevertheless, this type of relationship may have some value as a predictor when combined empirically with temperature measurements (Nagler *et al.*, 2005).

Empirical application of the triangle/trapezoidal approach is sometimes limited by the fact that the extreme wet and the extreme dry pixels may not always be available in an image, therefore it is useful to have alternative estimates of these extremes. McVicar and Jupp (2002), for example, have estimated theoretical extreme temperatures corresponding to an infinite surface conductance to water vapour (a wet surface) and a zero surface conductance (a dry surface) by inversion of an energy balance model run using both local meteorological data and remotely sensed temperature, vegetative cover and $\boldsymbol{R}_n$. These extreme values were incorporated into the complement of eqn (9.15) to give what they call a *normalized difference temperature index* (*NDTI*):

$$NDTI = (T_{g=0} - T_s)/(T_{g=0} - T_{g=\infty}) = \boldsymbol{E}/\boldsymbol{E}_o, \quad (9.18)$$

where $T_{g=0}$ is the temperature calculated for $g_s = 0$, and $T_{g=\infty}$ is the temperature calculated for a wet surface. A key step is the interpolation across large areas using the RS data.

9.5.4 Data assimilation and the use of SVAT models

Complex soil–vegetation–atmosphere-transfer (SVAT) models that include a detailed parameterization of the surface energy balance are widely used to estimate water and energy fluxes. These models usually represent deterministic approaches where the plant–atmosphere exchanges are generally described with fine time steps (hour or less) and these may range from single-layer models to multilayer models where energy budgets are calculated for each layer (e.g. Olioso *et al.*, 2002). These numerical models generally require detailed parameterization and knowledge of many crop and environmental parameters, which as we have seen, are often difficult to obtain by remote sensing. Most of these models are driven by meteorological data that again limits their spatial applicability, especially in areas with few meteorological sites, though remotely sensed canopy temperature is often an important driving input. Frequently these models may be run using a sequential assimilation approach where key state variables in the model are set each time remote-sensing data are available. The approaches described for SEBAL and other models above can provide tools for the necessary spatial interpolation. One approach to the calibration of such models can be to use the diurnal cycles of temperature (Coudert *et al.*, 2008).

The fact that CO_2 and water-vapour fluxes are jointly controlled by stomatal conductance means that the two fluxes are correlated at the landscape scale. Coupled modelling of the two fluxes with their separate constraints can lead to improved estimates of each that can be used in the up-scaling of data from flux-tower networks such as FLUXNET (Anderson *et al.*, 2008). By coupling a model that separates observed surface temperature into soil and canopy components and that provides useful information on evaporation and on soil moisture status, with an analytical model of canopy light-use efficiency, Anderson *et al.* (2008) were able successfully to map carbon, water, and energy fluxes at a range of scales.

One-source, two-source, and more complex models

The simplest SVAT models assume a single source (Fig. 9.1), with the surface temperature representing an appropriate mean of leaf and soil temperatures, as in the application of the Penman–Monteith model to canopies. This assumption can be quite good for canopies such as homogeneous short grassland where almost all radiation is absorbed by a thin layer of canopy and most evaporation originates in the same layer. It is much less appropriate for sparse canopies with substantial exposure of soil. In such situations the large difference between soil and canopy temperature can lead to substantial errors in estimates of ***C***. To overcome this problem, especially in discontinuous canopies, a range of two-source models (Fig. 9.1) have been developed (e.g. Norman *et al.*, 1995b; Shuttleworth and Wallace, 1985), where ***C*** and ***E*** are partitioned between soil and canopy sources using the separate temperatures of the soil and canopy. Data on variation of apparent temperature with view angle, if available, can be inverted to estimate canopy and soil temperatures separately. For large areas of vegetation the approach has been further extended by combining a two-source model with information on the rate of temperature change (Anderson *et al.*, 1997), making use of models of the behaviour of the planetary boundary layer during the day. This combination approach has the advantage that air-temperature measurements are not required and also that absolute accuracy of temperature measurement is not critical.

Some other approaches and conclusions

It has also been suggested that evaporative fraction may be estimated using microwaves (Li *et al.*, 2009; Min and Lin, 2006), since there is a semi-empirical relationship between optical depth at microwave wavelengths and vegetation water content, with the shorter-wavelength emission being more determined by the upper canopy layers. Unfortunately this is based on rather loose empirical relationships between the microwave indices and surface transfer processes. Other possible approaches include the use of the complementarity relationship (see, e.g., Venturini *et al.*, 2008), which takes account of the fact that as ***E*** increases, and hence humidifies the atmospheric boundary layer, the potential evaporation decreases to match, according to: $\boldsymbol{E} + \boldsymbol{E}_o = 2\,\boldsymbol{E}_w$, where $\boldsymbol{E}_w$ is the evaporation rate from an extended area of water.

9.5.5 Conversion from instantaneous to 24-h and seasonal values

The conversion from instantaneous estimates of ***E*** to 24-h estimates is a key step in the application of remote sensing data. A reasonable estimate of the daily total is to assume that the instantaneously derived fraction of available energy that goes into evaporation ($\lambda \boldsymbol{E}/(\boldsymbol{R}_n - \boldsymbol{G})$) applies over the full 24-h period. Since over a 24-h period ***G*** normally is close to zero, the available energy is $\boldsymbol{R}_{n-24}$, which can be approximated by difference between the absorbed shortwave at the surface ($= \boldsymbol{I}_{toa-24} \cdot \tau_a \cdot (1 - \rho_s)$) and the net outgoing longwave at the surface. Longwave fluxes are large in relation to other components of the surface energy balance, yet the net longwave radiation can often be approximated by a linear function of incoming shortwave radiation expressed as a fraction of the maximum potential incoming shortwave for that location (Fig. 9.2). Allen *et al.* (2007b) provide good evidence that ***E*** expressed as a fraction of the reference $\boldsymbol{E}_o$ for a well-watered 0.5-m high alfalfa crop is approximately constant over the day and that this relation appears to be generally valid for agricultural crops, though for some stressed stands of vegetation subject to afternoon stomatal closure there may be exceptions to this. This relationship allows one to estimate the daily ***E*** for any pixel from the interpolated reference $\boldsymbol{E}_o$. The seasonal ***E*** may be calculated by integrating the 24-h estimates over time, interpolating as necessary between observations.

9.5.6 Conclusions on estimation of *E*

Remote sensing has proved valuable for estimation of ***E*** at all scales from single plants to regions for a wide range of applications. Interestingly, a meta-analysis of a wide range of 30 validations of remote-sensing approaches for estimation of ***E*** (Kalma *et al.*, 2001) suggested substantial errors with an average RMSE of c. 50 W m^{-2} with the more complex models often not performing better than the simpler empirical models. As with other applications of RS, data tend to be biased to cloud-free days except where microwave techniques are used, so RS is of less use in humid climates.

9.6 Thermal sensing for the detection of water-deficit stress and stomatal closure

Ever since the principles underlying the processes that control the leaf energy balance were clarified in the 1950s (Raschke, 1956), there has been interest in using canopy temperature as a remote indicator of plant water stress (or more correctly of the stomatal closure response to water deficits). The development of *stress indices* based on canopy temperature for the study of plant water relations will be discussed in detail in Chapter 11; here we outline the principles involved in regulating canopy temperature and how they impact on estimation of stomatal conductance from temperature measurements.

As we have seen (eqn (9.13)) canopy temperature is linearly related to the rate of water loss from the canopy. In turn the rate of water loss is closely related to stomatal conductance, though the exact relationship depends on the degree to which the plant canopy is 'coupled' to the atmosphere (Jarvis and McNaughton, 1986). Where the boundary-layer transfer process is efficient (small leaves, high wind speeds, rough canopies, etc., giving a high boundary-layer conductance) the plant is said to be well coupled to the atmosphere as heat and mass transfer are rapid. In such cases transpiration is approximately proportional to stomatal or leaf conductance (g_s). For large areas of homogeneous vegetation, however, the canopy is said to be poorly coupled to the atmosphere and in such cases the total evaporation rate is largely determined by the availability of energy (eqn (9.9)), in which case changes in stomatal aperture have little effect. These differences are illustrated in Fig. 9.5, which shows that forest (with typically high boundary-layer conductance) is much more sensitive to changes in stomatal conductance than are short smooth canopies such as grass. The relative sensitivity of transpiration ($d\boldsymbol{E}/\boldsymbol{E}$) to a relative change in stomatal conductance (dg_s/g_s) can be shown (Jarvis and McNaughton, 1986) to increase with increases in the ratio between boundary layer and canopy conductances according to

$$
\begin{aligned}
(d\boldsymbol{E}/\boldsymbol{E})/(dg_s/g_s) &= (g_a/g_s)/(s/\gamma + 1 + g_a/g_s) \\
&= 1 - \boldsymbol{\Omega},
\end{aligned}
\tag{9.19}
$$

where $\boldsymbol{\Omega}$ is known as a decoupling coefficient.

The way in which environmental and plant factors interact to determine leaf or canopy temperature was developed in Section 4.4.4 (see eqn (4.14)). This equation can be expressed in a slightly more accurate form by utilizing the concept of isothermal net radiation (Jones, 1992; Leinonen *et al.*, 2006). Expressing this equation in terms of the more familiar mass units for resistance rather than molar units this then becomes (Jones, 1992; see also Appendix 1)

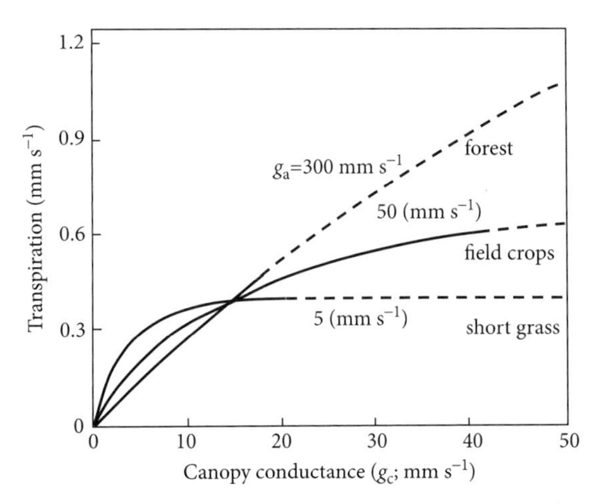

Fig. 9.5. Calculated relationships between transpiration rate and canopy conductance at different boundary-layer conductances (calculated for 400 W m^{-2} available energy, 1 kPa water vapour pressure deficit, and 15 °C). The solid lines represent the probable range of values of g_c for different crops (based on Jarvis, 1981; Jones, 1992).

$$
T_\ell - T_a = \frac{[r_{HR}(r_{aW} + r_s)\gamma \boldsymbol{R}_{ni} - \rho c_p\, r_{HR}\, D]}{[\rho c_p(\gamma\,(r_{aW} + r_s) + s\, r_{HR})]}, \tag{9.20}
$$

where T_ℓ is the leaf temperature, $\boldsymbol{R}_{ni}$ is the net isothermal radiation (the net radiation that would be received by an identical surface if it were at air temperature), D is the water vapour pressure deficit, r_{aW} is the boundary layer resistance to water-vapour transfer, r_{HR} is the parallel resistance to heat and radiative transfer, and other symbols have been previously defined. The equation can be rearranged to give the following expression for stomatal resistance as a function of the relevant variables (Guilioni *et al.*, 2008)

$$
r_s = \frac{-\rho c_p\, r_{HR}\,(s\,(T_\ell - T_a) + D)}{\gamma\,((T_\ell - T_a)\,\rho c_p - r_{HR}\,\boldsymbol{R}_{ni})} - r_{aW} \tag{9.21}
$$

Because stomatal closure is one of the first responses of plants to drought as plants act homeostatically to conserve water, the consequent effect on canopy temperature has been one of the most widely used means for remotely detecting plant stress (or more correctly the stomatal closure in response to plant water-deficit stress). However, as is apparent from Fig. 9.6, many factors other than stomatal conductance also affect leaf temperature. In particular, it is worth noting from Fig. 9.6(c) that canopies exposed to high wind speeds (or rough canopies such as forests) show smaller temperature responses to stomatal closure than do smoother canopies.

Stress indices

The major problem with canopy temperature as a measure of stomatal conductance (or 'stress') in the field is that the instantaneous temperature is very sensitive to environmental factors. It is therefore necessary to normalize the observed canopy temperature in some way to get a useful general indicator of stress. A key milestone in the development of *stress indices* was the normalization of canopy temperature against air temperature when Jackson *et al.* (1977) defined a stress degree day (SDD) as the difference between canopy temperature and air temperature at a specified time. Successive daily values obtained near midday could be integrated over time to give a measure of crop stress. The next advance was the definition of a *crop water stress index* (*CWSI*) by Idso and colleagues (Idso *et al.*, 1981; Jackson *et al.*, 1981). This involved further normalization to take account of both the effects of atmospheric humidity and the expected temperature of a well-watered crop (Fig. 9.7) according to

$$CWSI = (T_{canopy} - T_{nwsb})/(T_{upper} - T_{nwsb}), \quad (9.22)$$

where temperatures are expressed as differences from air temperature, T_{nwsb} is the temperature of the *non-water-stressed baseline* (nwsb) at the atmospheric humidity recorded and T_{upper} is the corresponding temperature of the upper limit at the given humidity and air temperature. The non-water-stressed baseline was obtained empirically by measuring the leaf–air temperature difference for a well-watered crop in the same experimental environment and indicates the lowest temperature difference likely in that environment. The value of the upper temperature limit represents the expected temperature of a non-transpiring crop. It is constructed by extrapolating the nwsb to a zero vapour pressure deficit and estimating the leaf–air vapour-pressure difference (D) from the temperature

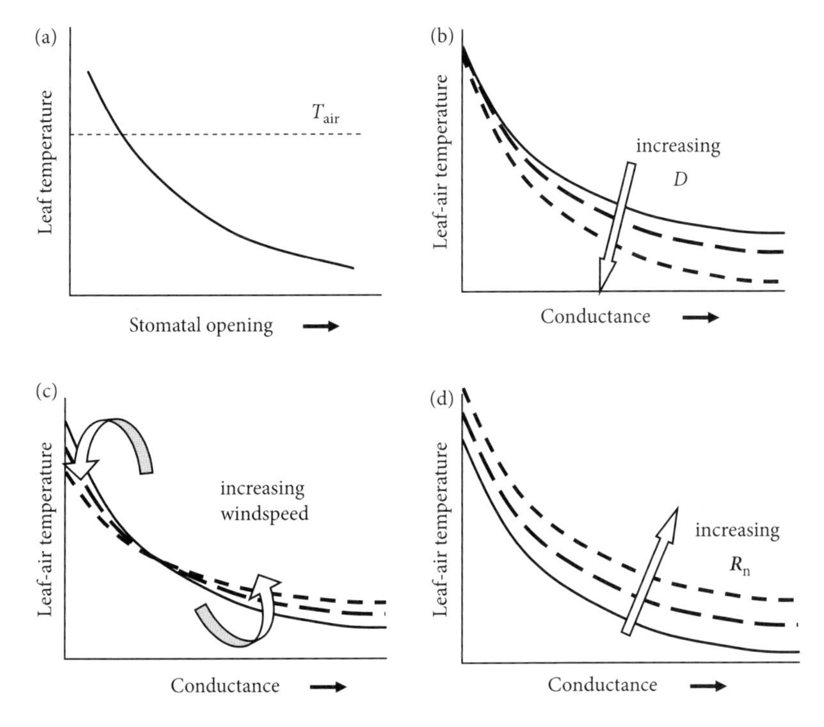

Fig. 9.6. A schematic diagram showing the dependence of the relationship between temperature and stomatal conductance on environmental factors as calculated by using eqn (9.20). (a) The general relationship between leaf temperature and stomatal opening. Variation with environmental conditions can be partly eliminated by normalizing against air temperature, as in the following diagrams; (b) the dependence on vapour pressure deficit of the atmosphere (D), (c) the dependence on windspeed, and (d) the dependence on net radiation absorbed (R_n).

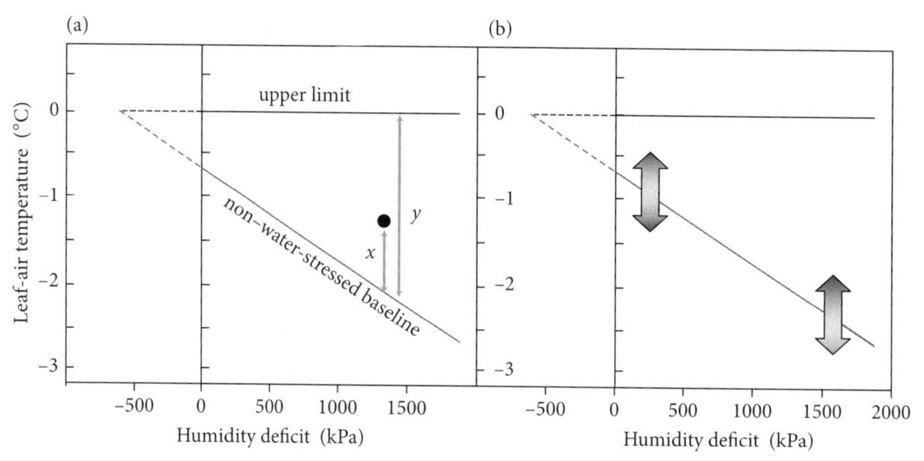

Fig. 9.7. Illustration of calculation of the Idso–Jackson crop water-stress index (*CWSI*). (a) Plot of foliage-air temperature against atmospheric humidity deficit, showing the non-water-stressed baseline representing a crop transpiring at the potential rate (no stomatal limitation) and the upper temperature limit representing a non-transpiring crop. *CWSI* for any point, such as that indicated, is then defined as x/y. (b) Illustrates that for a given error in temperature measurement the relative error in calculated *CWSI* is likely to be much greater in situations (e.g. temperate climates) with low atmospheric humidity deficits. (Lines calculated using eqn (9.20), assuming that the well-watered g_s = 10 mm s^{-1}, R_n = 500 W m^{-2}, T_a = 20 °C, leaf width = 0.1 m, windspeed = 0.8 m s^{-1}).

intercept on the ordinate, then extrapolating the line further to the negative humidity deficit that balances this value, and taking the horizontal (upper limit) line from that point.

As can be seen in Fig. 9.7(b), it is clear that the method is most useful in hot, dry climates where large atmospheric vapour pressure deficits occur. Experimental errors resulting from both temperature measurement errors and variation in other environmental factors (especially incident radiation and wind speed) will reduce the signal-to-noise ratio in more humid climates. Even in Arizona where the index was derived, it was found essential to restrict measurements to clear sunny days.

For general application, therefore, there has been much interest in deriving more robust stress indices. Although in principle one can replace the use of stress indices by using direct estimates of stomatal resistance or conductance obtained by substitution of canopy temperature into the energy balance solution (eqn (9.21); see also Guilioni *et al.*, 2008; Leinonen *et al.*, 2006) this requires data on all the relevant variables (e.g. $\mathbf{R}_{ni}$, D, T_a, r_{aH}) that are not necessarily easy to measure in the field. Estimation of the net radiation absorbed by individual leaves is particularly difficult. Therefore, the alternative approach to correction for environmental variation is to use wet and dry 'imitation' leaf references that behave similarly to real leaves in all respects except for their conductance to water vapour (Jones, 1999). Using the temperatures of the wet (T_{wet}) and dry (T_{dry}) surfaces Jones (1999) defined a stress index analogous to the Idso–Jackson *CWSI* as

$$SI_{CWSI} = (T_{leaf} - T_{wet})/(T_{dry} - T_{wet}). \qquad (9.23)$$

This differs from *CWSI* in that it uses a wet surface having an infinite surface conductance, rather than the finite value for a well-watered crop. Of various alternative formulations proposed is one that is proportional to stomatal conductance

$$SI_{gs} = (T_{dry} - T_{leaf})/(T_{leaf} - T_{wet}) = \beta\, g_s, \qquad (9.24)$$

where β $(=(c_p\, g_{HR} + g_{aH}\, \lambda\, s/p_a)/(g_{aH}\, c_p\, g_{HR})$ (see Jones, 1999)) depends only on wind speed and to a lesser extent, temperature. This form of index tends to be more stable than its reciprocal. When using such wet and dry references it is essential that the references have the same aerodynamic and optical properties as real leaves. To this end some workers have used real leaves either sprayed with water or covered in petroleum jelly to stop water loss (Jones *et al.*, 2002), while others have resorted to various artificial surfaces (e.g. Möller *et al.*, 2007). The latter approach becomes essential as the scale of measurement increases to airborne or satellite sensing. Conveniently, it has been shown (Leinonen *et al.*, 2006) that accuracy is not substantially compromised by using just a dry reference (omitting the more difficult to maintain wet reference).

Use of temperature variability

An alternative approach to the detection of water deficits might be to make use of temperature variability. Aston and van Bavel (1972) suggested that drought might lead to increased temperature variability within a field plot resulting from soil heterogeneity; such a response was confirmed by Gardner *et al.* (1981) for maize. There may be substantial opportunities for measurement of temperature variance across fields. At a leaf scale, Fuchs (1990) showed that temperature variance should increase as stomata close for a canopy with randomly oriented leaves. This effect arises because the different radiation intercepted by shaded and sunlit leaves would have a greater effect on leaf temperature where latent heat loss is a small component of the energy balance (closed stomata) than where stomata are open. Although there are canopies where temperature variance increases as stomata close, this is not always the case (Grant *et al.*, 2007), presumably where the assumption of random leaf orientation does not hold.

Interpretation of stress indices

The *CWSI* or other related stress indices strictly only give a measure of stomatal conductance for single leaves or for full canopies with no visible soil. As the fraction of background soil in the pixel increases, the calculated index becomes more and more dominated by the soil temperature. This means that when calculated for a mixed (vegetation/soil) pixel, the value of the calculated *SI* may be largely determined by the fraction of soil visible rather than by the stomatal closure. Of course, such a reduction in leaf area is often also a useful indicator of stress because the leaf area, and hence f_{veg}, of plant canopies decreases with decreasing water availability. Therefore, for most remote-sensing applications it is useful to combine the *CWSI* with a vegetation index (as an estimator of f_{veg}). For the quite common situation, at least in dry climates, of dry soil, one can interpret the middle diagonal line in Fig. 9.3(b) as representing fully transpiring (unstressed) vegetation (e.g. Moran *et al.*, 1994). Therefore, the degree of stomatal closure or a *cover-weighted corrected stress index* can be derived as a/b in this figure. Practical aspects of the application of stress indices in precision agriculture and irrigation management will be discussed in Chapter 11.

9.7 CO_2 fluxes and primary production

Although one might consider estimating photosynthetic rates (and respiration rates) from the sensing of heat generated or released, the metabolic heat fluxes are usually at least two orders of magnitude smaller than other components of the vegetation energy balance. For example, the total energy fixed in plant growth is c. 24.5 kJ/MJ intercepted (i.e. 2.5%; Jones, 1992) while the heat released by leaf respiration is likely to be less than 1.5 mW (g fresh weight)$^{-1}$ (Briedenbach *et al.*, 1997), which for a typical leaf (c. 200 μm thick) is less than 3 W m^{-2}. It follows that remote sensing is not well suited for detection of the energy fluxes directly associated with CO_2 exchange between the vegetation and the atmosphere. Only in very rare situations, such as in the thermogenic respiration shown by the flowers of some arum lilies does respiratory metabolism reach a rate at which the associated heat generation is detectable; in such cases the spadix may heat up to as much as 15 °C above ambient (Seymour, 1999).

Therefore, remote estimates of photosynthetic rate are most usually derived from remotely sensed estimates of the absorption of solar radiation (Grace *et al.*, 2007). Although a constant efficiency of light conversion was assumed in early work it is now accepted that the basic calculation needs to be corrected for variation in the *light-use efficiency* (*LUE*) of the photosynthetic tissues which results from the effects of environmental stresses such as drought. A general symbol for efficiency is ε; this provides a convenient symbol for distinguishing the different ways in that light-use efficiency can be expressed. Here, we outline the basis of the estimation of photosynthesis from radiation interception and compare this with the potential offered by studies of the xanthophyll cycle and chlorophyll fluorescence.

9.7.1 Photosynthesis estimation from radiation interception

Although neither the fluxes of CO_2 nor the direct energy fluxes associated with photosynthetic metabolism are detectable remotely, there is, however, a very close association between the amount of solar radiation absorbed by a plant canopy and photosynthesis. This provides the basis for the most popular remote-sensing approach to estimation of photosynthetic productivity. As was pointed out by Monteith (1977; see Fig. 9.8), the conversion efficiency of absorbed solar radiation to plant growth is surprisingly constant across a range of vegetation systems and equates to an efficiency of utilization of intercepted solar energy (ε_V) for C3 plants of about 2.5%.[37] Figure 9.8 shows that about 1.4 g dry matter are fixed on average by C3 crops per MJ of intercepted solar radiation. If we assume that dry matter has an energy content of about 17.5 kJ g^{-1}, this equates on an energy basis to an efficiency of approximately 2.45% for conversion of solar radiation to net primary production and c. 5% for conversion of the photosynthetically active radiation component (Jones, 1992). In practice, as shown in Table 9.1, conversion efficiencies vary with crops and the way in which they are grown, with C4 crops potentially more efficient at conversion of incident energy than are C3 crops (but not always in practice).

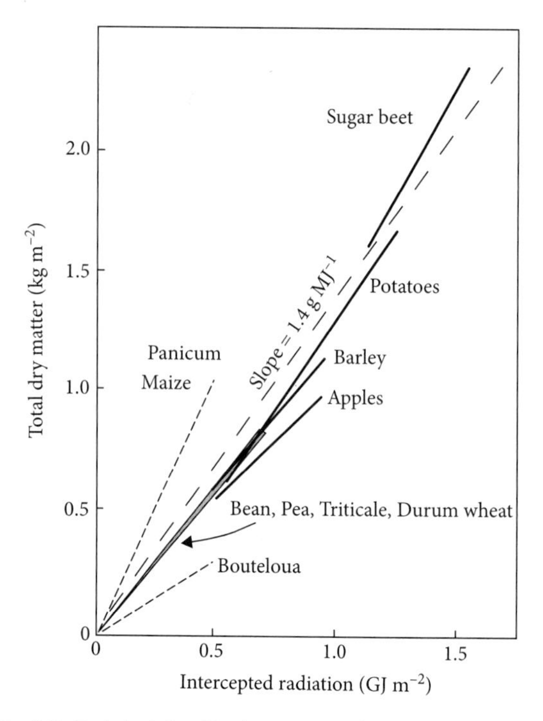

Fig. 9.8. Typical relationships between net primary production (measured as dry matter production) and intercepted solar radiation for a range of crops. Data for solid lines from Monteith (1977), the dashed lines indicate an approximate range of values for C4 grasses (converted from Kiniry *et al.*, 1989, 1999) and the shaded area indicates some other typical values for C3 crops (bean, pea, triticale, and durum wheat) (Giunta *et al.*, 2008). Note that values for intercepted PAR were divided by two to give the lines shown.

In order to estimate photosynthesis and ecosystem productivity by remote sensing we can consider the net photosynthesis or net primary productivity (*NPP*) as the product of the incident radiation and a number of independent efficiency terms:

$$\begin{aligned} NPP &= \boldsymbol{I}_S \times \varepsilon_c \times \varepsilon_f \times \varepsilon_A \times \varepsilon_V' \\ &= \boldsymbol{I}_{PAR} \times fAPAR \times \varepsilon_V' \\ &\cong \boldsymbol{I}_S \times fAPAR \times \varepsilon_V, \end{aligned} \quad (9.25)$$

where $\boldsymbol{I}_S$ is the incident solar irradiance at the surface, ε_c is the fraction of incident radiation in the photosynthetically active region (PAR; 400–700 nm), ε_f is the fraction of PAR that is intercepted by the canopy, ε_A is the absorbed fraction of the intercepted PAR that is absorbed by the leaves (often assumed equal to 0.85 (though as shown by Campbell and van Evert (1994) for $LAI > 3$ a better estimate would be 1.0), and ε_V' is the conversion efficiency of absorbed PAR to dry matter (often assumed to be $2 \times \varepsilon_V$). The second equivalent equation shows that this can also be written in terms of *fAPAR*, which is the fraction of photosynthetically active radiation that is absorbed by the canopy (commonly derived from *NDVI* or other spectral indices as described in Section 7.2.1). The third equivalent

[37] Although the maximum potential efficiency of photosynthesis (ε) is close to c. 8–10 quanta/CO_2, in practice values of about 15–22 quanta/CO_2 are required (taking account of the various losses in the system) which is equivalent to an energy efficiency of 11–16% in terms of absorbed photosynthetically active radiation (Jones, 1992). The fraction of solar radiation that is not in the photosynthetically active region and the subsequent respiratory losses involved in synthesis of complex organic molecules, growth, maintenance, and tissue losses/cycling reduce this to the 2.45% reported for what might be called the vegetation efficiency (ε_V) by Monteith (1977).

Table 9.1 Values of the conversion coefficient (ε_V; g MJ^{-1}) for solar radiation intercepted for different crops grown under 'optimal' conditions and the corresponding values when experiments were subject to water or nutrient deficiencies (ε_V^*; g MJ^{-1}) as reported by Azam-Ali *et al.* (1994).

Crop	ε_V	ε_V^*
C4 species (maize, sorghum, and millet)	2.43–2.69	0.57–1.30
C3 species (wheat, barley, potato, groundnut, sugar beet)	1.20–1.84	0.47–1.10
C3 species (rice)	2.05	1.00
C3 species (soybean)	1.30	0.23

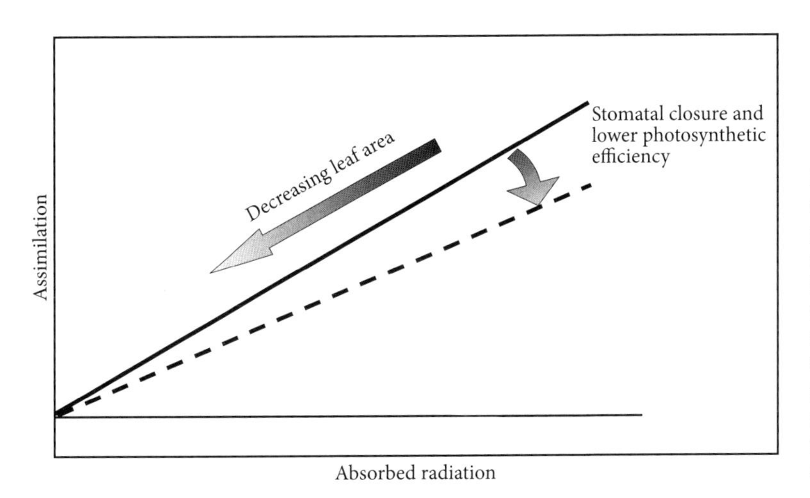

Fig. 9.9. The solid line shows an expected relationship between seasonal photosynthesis and total absorbed radiation for a typical crop. The arrows illustrate the typical effects of drought or other environmental stresses on the reduction in total seasonal assimilation, with the largest contribution usually being from the reduced leaf area (and hence reduced radiation absorption) rather than from reductions in photosynthetic efficiency.

arises because, as a useful approximation it is usual to assume that I_{PAR} is approximately half of the incident shortwave (I_s) at the surface.

The key variable in this equation is ε_V', which, as we have seen, is commonly referred to as the *radiation-use efficiency* or *light-use efficiency* (*LUE*);[38] this varies with factors such as vegetation type and with stress factors such as low temperature or drought, as indicated in Fig. 9.9. It is notable, however, that a large proportion of the inhibition of canopy photosynthesis by stresses such as drought is usually caused by the changes in leaf area that occur in response to stress (indicated by moving along the lines indicated in Fig. 9.9), rather than by the changes in slope or ε_V' that generally tend to be of secondary importance. Nevertheless, many crops do fall substantially below their optimal efficiency when subject to nutrient or water deficiency, as is common in many parts of the world, and there may be diurnal variation in *LUE* as stomata close in the afternoon. The generally good relationship between production and intercepted radiation, even though instantaneous conversion efficiencies can vary substantially, is to some extent an artefact of the induced correlation that occurs when taking cumulative sums.

We discuss below approaches to the estimation of *LUE* including the use of *PRI* and chlorophyll fluorescence. Although the infrequent observations characteristic of most remote-sensing systems are useful for estimation of the rather slowly varying *fAPAR*, they are of little use for direct estimation of I_{PAR} that changes

38. Care is needed with this term as some authors use different bases for its calculations involving either absorbed or intercepted radiation and PAR or total energy.

rapidly and continuously. Therefore, for estimation of *NPP* for any particular vegetation type using this equation, it is more usual and more useful to estimate I_{PAR} and its diurnal variation on the basis of an appropriate radiation or climate model for the environment being studied (e.g. Singh *et al.*, 2008).

9.7.2 Photosynthesis estimation from xanthophyll epoxidation

An alternative approach to the remote estimation of photosynthesis, and especially the efficiency of the process, is based on the observation that the photochemical efficiency of photosynthesis varies with the xanthophyll de-epoxidation state, which as we saw in Chapter 7, can be remotely estimated where appropriate hyperspectral reflectance data are available by means of the photochemical reflectance index (*PRI*; $(\rho_{531} - \rho_{731})/(\rho_{531} + \rho_{731})$). This index indicates the proportion of absorbed radiation that is dissipated as heat and not used for electron transport in photosynthesis. It therefore indicates the decreasing efficiency, both as radiation increases and saturates the photosynthetic system, and as a result of stress. An increasing number of studies (e.g. Black and Guo, 2008; Guo and Trotter, 2004; Nichol *et al.*, 2000) have shown that *PRI* is a useful linear estimator of *LUE*. The approach can be particularly powerful when combined with multiangular sensing to derive sunlit and shaded fractions (Hall *et al.*, 2008) as outlined in Section 8.2.2 or when combined with *BRDF* modelling to correct *PRI* to a constant view and illumination angle (Hilker *et al.*, 2008). Unfortunately the relationship between *PRI* and *LUE* has been found to vary quite widely between different studies (see Fig. 7.10) so no single calibration appears valid (Grace *et al.*, 2007) and more development of the approach is needed.

When combined with modelling of the radiation environment, therefore, remote sensing of spatial variation of vegetation cover (or *fAPAR*) combined with classification techniques for the delineation of areas with different vegetation types provides a powerful tool for estimation of productivity of different ecosystems. Validation of such estimates is best achieved by comparison with data from eddy-covariance flux towers. These validated models can then be extrapolated using climatic data combined with water-balance models to derive seasonal changes in water stress.

9.7.3 Photosynthesis estimation from chlorophyll fluorescence

Only a fraction of the absorbed light energy is used to drive electron transport and photosynthetic carbon assimilation; a significant fraction is lost as heat and a rather small fraction (about 1–2%) is reradiated at a longer wavelength as chlorophyll fluorescence. As pointed out in Section 4.2.1 (eqn (4.1)), the amount of fluorescence can be used as an (inverse) indicator of electron transport and hence of photosynthesis, with Φ_F being approximately inversely related to the electron flow used in splitting water in photosystem II.

How can chlorophyll fluorescence be used remotely? In controlled environments it is possible to illuminate the leaves using only a narrow waveband, for example in the red, and to measure the chlorophyll fluorescence in the near infrared (>700 nm). In most environmental studies, however, one is primarily interested in studying chlorophyll fluorescence under normal solar illumination. In such cases there is a large component of reflected solar radiation that is at the same wavelength as the small amount of fluorescence. For laboratory studies, or for studies of leaves within a few metres in the field, the common way to achieve this is based on *active fluorescence systems* where one illuminates the leaf using a light source constrained by narrowband filters to a waveband shorter than that being sensed as fluorescence (fluorescence is always emitted at longer wavelength than the exciting signal because of the inevitable loss in energy) and to distinguish the fluorescence emitted from reflected ambient light by modulating the exciting signal (e.g. at 60 Hz) and detecting only the component of the emitted signal that fluctuates in phase with the exciting signal. In this way the much larger (c. two orders of magnitude) steady-state signals from reflected sunlight can be eliminated and hence ignored.

Although it is possible to derive much useful information on the properties of the photosynthetic system by analysis of the dynamics of the fluorescence signal after an illumination pulse, this generally requires a period of pre-equilibration in the dark (so is not convenient outdoors), and the interpretation tends to be rather empirical. Nevertheless, simple instruments that follow the induction kinetics (the 'Kautsky' effect) of chlorophyll fluorescence can provide useful discrimination between stressed and unstressed plants (Őquist and Wass, 1988). In such studies the fast induction kinetics (within a second or so) primarily reflect differences in photochemistry and the slower kinetics of increasing fluorescence quenching respond more to interactions with the 'dark' CO_2 fixation processes. Many studies have shown empirical relationships between parameters derived from induction curves and responses to stresses.

The introduction of fluorimeters using modulated chlorophyll fluorescence signals to probe photosynthetic functioning has greatly enhanced the power of chlorophyll fluorescence approaches (Schreiber *et al.*, 1986). The basics of fluorescence quenching analysis are outlined and explained in Box 9.1. Note that because of the sensitivity of the absolute signal

BOX 9.1 Chlorophyll fluorescence parameters and their measurement (terminology from Maxwell and Johnson, 2000)

A typical modulated chlorophyll fluorescence trace following illumination of a leaf after a period of acclimation in the dark (c. 30 min) is shown in the figure. When the very weak modulated light source is first switched on in the absence of any actinic light driving photosynthesis, one gets the initial basal F_o fluorescence, a flash of bright light is then used to saturate all the reaction centres in the chloroplast giving rise to peak fluorescence (F_m). Subsequent quenching of this signal results both from the induction of photosynthetic electron transport away from the reaction centre (photochemical quenching) and increased transfer to heat through non-photochemical quenching (NPQ). One can use repeated short flashes of saturating light to diagnose changes in electron transport rate (etr) and efficiency of photosynthesis. The key photosynthetic parameters that can be obtained from the various combinations of the steady-state fluorescence (F_t), and the maximum fluorescence obtained with a saturating flash, either on a dark-acclimated leaf (F_m) or on a leaf performing steady-state photosynthesis (F'_m) are listed below. Application in the field is limited first by the need to obtain a dark-adapted F_m, and secondly by the need to provide a saturating flash of high irradiance. This latter requirement, particularly, limits the range that the sensor can usually be from the leaf or canopy being studied and limits the application in remote sensing. Because it is usually not possible to follow fluorescence dynamics in remote sensing, interpretation often has to be based on variation in the sun-induced steady-state fluorescence alone (F_t) that tends to be inversely related to carbon assimilation because fluorescence competes with photosynthesis for electrons (eqn (4.1)).

$\Phi_{PSII} = (F'_m - F_t)/F'_m$ (quantum yield of photosystem II)

$F_v/F_m = (F_m - F_o)/F_m$ (maximum quantum yield of photosystem II)

$qP = (F'_m - F_t)/(F'_m - F'_o)$ (photochemical quenching, or the proportion of open PSII reaction centres)

$NPQ = (F_m - F'_m)/F'_m$ (non-photochemical quenching)

$etr \approx I_{PAR} \cdot \alpha \cdot \beta \cdot \Phi_{PSII}$ (where α is the leaf absorptance for PAR often assumed = 0.85, and β = the proportion of absorbed light delivered to photosystem II – often assumed = 0.5)

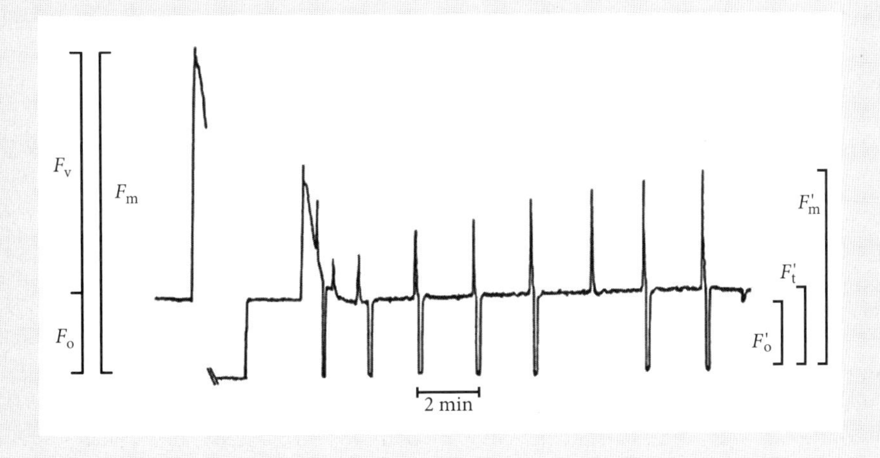

to small changes in geometry, it is usually necessary to normalize the signals, for example to the maximum fluorescence value (F_m). The most commonly used parameters are F_v/F_m that measures the maximum quantum yield of photosystem II and indicates any long-term damage to the photosystem, and the quantum yield of photosystem II (Φ_{PSII}), which is approximately proportional to the *electron transport rate* (*etr* – see Box 9.1, though note that both α and β can vary substantially). This estimated *etr* can also be used to estimate the rate of CO_2 fixation because a large proportion of the reducing power generated by electron transport is used to drive photosynthetic carbon fixation. Unfortunately, further approximations are required as the proportion of electrons used in photosynthesis varies. Collection and interpretation of chlorophyll fluorescence data is a specialist activity and there are many pitfalls for the unwary, even though the necessary instrumentation is now widely available (Logan *et al.*, 2007; Maxwell and Johnson, 2000).

Conventional chlorophyll fluorescence measurement is based on detection of the average signal from a given field of view. Recently, there has been increasing interest in chlorophyll fluorescence imaging. The latest imagers operate in a similar mode to conventional modulated fluorescence using pairs of sequential images to derive the spatial variation of the key photosynthetic parameters outlined in Box 9.1 (e.g. Nedbal *et al.*, 2000; Oxborough and Baker, 1997). It is even possible using modulated systems to get maps of photosynthetic activity for leaves exposed to sunlight in the field.

A major difficulty with scaling up the active fluorescence approach to airborne or satellite remote-sensing approaches is the difficulty of obtaining a high-enough-energy exciting signal when the illuminator is remote, combined with the problem that the fluorescence is emitted in all directions, not just towards the sensor, so that the signal received decreases with the square of the distance from the target. Nevertheless, Kolber *et al.* (2005) have demonstrated the principle of the use of laser-induced fluorescent transients (LIFT) operated at up to 50 m from the plants. The instrument involves the use of a highly collimated (100 mm diameter) excitation laser beam from a 1-W optical laser with detection through a 250-mm telescope. Using fast repetition stimulation these authors demonstrate that it is possible to derive a range of key photosynthetic parameters including F_t, F'_m, NPQ, and electron transport rate from such a system, while still remaining within standard guidelines for safe laser operation.

An alternative approach to the remote detection of chlorophyll fluorescence is to use passive *solar-induced fluorescence* (SIF): a very good review of available approaches to the use of SIF may be found in Meroni *et al.* (2009). Solar-induced fluorescence can be used to estimate the steady-state fluorescence (F_t) emitted by a photosynthesizing leaf as an estimate of photosynthetic light-use efficiency (*LUE*). The main approaches to estimation of F_t include radiance-based and reflectance-based techniques. The radiance-based methods exploit the *Fraunhofer line depth* (FLD) principle and make use of the narrow atmospheric absorption bands where molecular oxygen or hydrogen strongly absorb the incoming solar radiation (the Fraunhofer lines – see Box 9.2). The reflectance-based approaches, on the other hand, derive reflectance indices that are related to fluorescence and do not require Fraunhofer lines but compare wavelengths close to and remote from the fluorescence maxima.

Less precise estimates can be obtained by broader-band sensors, potentially even those at satellite level, with some success having been reported for MERIS band 11 (3.75 nm bandwidth centred on the Fraunhofer line at 760.6 nm) and band 10 that acts as a very near reference at 753.8 nm (Guanter *et al.*, 2007). Although this allows estimation of the 'steady-state' fluorescence it does not readily provide the information needed for quenching analysis (Box 9.1), so its application is currently limited to a rather simplistic inverse relationship with photosynthesis. When using line-filling methods remotely it can be useful to use a comparison between plant targets and non-fluorescing areas to help to correct for atmospheric interference (Moya *et al.*, 2004). Although development thus far of fluorescence sensing holds promise, and was the basis for a proposed ESA mission (FLEX), there is still much work to do before it can provide a reliable photosynthetic signal. Nevertheless, the use of sun-induced steady-state fluorescence has been used successfully in a data assimilation procedure in combination with a photosynthetic model and remotely sensed estimates of *fAPAR* to simulate successfully

BOX 9.2 The Fraunhofer line depth method for chlorophyll fluorescence measurement

Important bands for the detection of SIF include the very narrow Hα band at 656.28 nm, the O_2-B band at 687 nm, and the O_2-A band at 760.6 nm where incoming solar radiation is attenuated by c. 90%. The O_2-A atmospheric absorption band is shown in (Fig. a – data from Meroni *et al.*, 2009) as measured using either a 0.13-nm half-bandwidth spectrometer (fine line) or a 1-nm half-bandwidth (heavy line). In such bands, fluorescence can be detected by measuring the filling of such a 'well' as compared with the adjacent continuum. This requires a very narrowband sensor as the Fraunhofer lines are typically 0.5 to 2 nm in width. The outgoing radiances in the well wavelength (λ_{well}) and at the reference wavelength (λ_{ref}) are given by the sum of reflected radiation and the fluorescence (F) as shown in (Fig. c) according to

$$L_{\lambda\text{-well}} = \rho_{\lambda\text{-well}} \cdot I_{\lambda\text{-well}}/\pi + F_{\lambda\text{-well}} \tag{9.2.1}$$

$$L_{\lambda\text{-ref}} = \rho_{\lambda\text{-ref}} \cdot I_{\lambda\text{-ref}}/\pi + F_{\lambda\text{-ref}} \tag{9.2.2}$$

Assuming that leaf reflectance and fluorescence are constant over the range of the in-well and reference wavelengths, these equations can be solved to give

$$\rho = \pi\,(L_{\lambda\text{-ref}} - L_{\lambda\text{-well}})/(I_{\lambda\text{-ref}} - I_{\lambda\text{-well}}) \tag{9.2.3}$$

$$F = (I_{\lambda\text{-ref}} \cdot L_{\lambda\text{-well}} - L_{\lambda\text{-ref}} \cdot L_{\lambda\text{-well}})/(I_{\lambda\text{-ref}} - I_{\lambda\text{-well}}), \tag{9.2.4}$$

where the π converts radiances to irradiances.

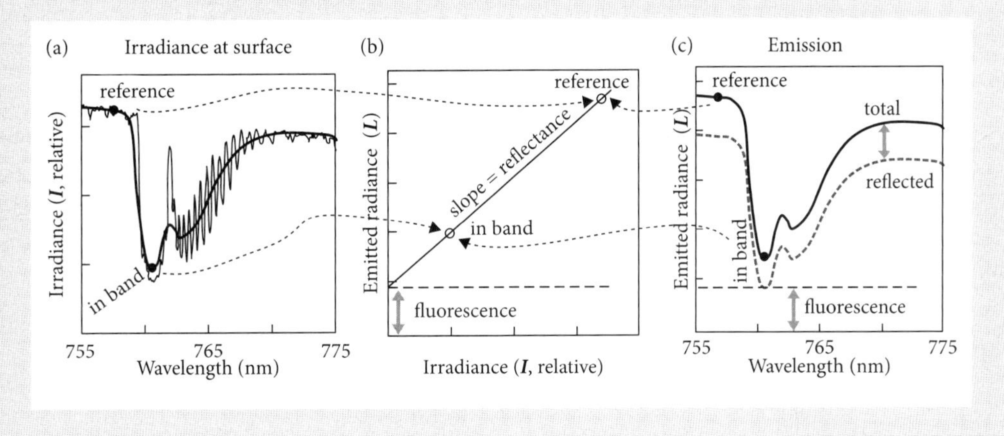

diurnal time courses of gross primary production (Damm *et al.*, 2009).

Carter *et al.* (1990) have demonstrated some success, at close range measurement of a Kautsky effect by using the Hα line at 656.28 nm. The principle is illustrated in Fig. 9.10 where the amount of fluorescence, F, was detected as

$$F = d - a(d-c)/(a-b), \tag{9.26}$$

where a is the radiance from a diffuse reflective surface over a region including the Hα line, b is the radiance from the same surface in the Hα line, d is the radiance from the target leaf in the same broad region, and c is the radiance from the target leaf in the Hα line. Potentially this approach could be extended to the diagnosis of plant stress responses where photosynthesis is affected by stress.

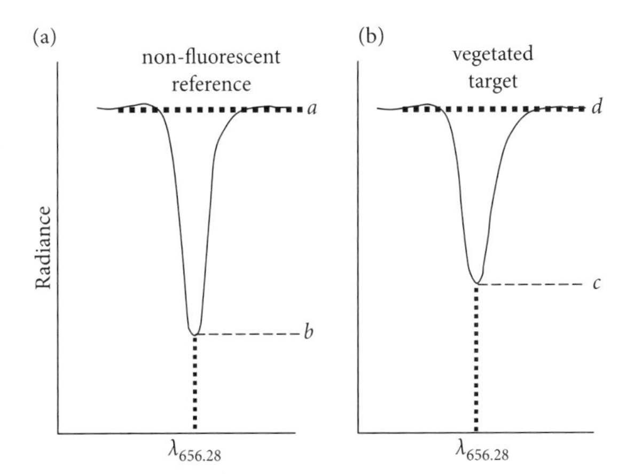

Fig. 9.10. Measurement of chlorophyll fluorescence using the Fraunhofer line depth method and the Hα line (Carter *et al.*, 1990). F is calculated from a comparison of measurements made over a non-fluorescing reference surface (a) and a vegetation surface (b), according to eqn (9.26).

9.8 Conclusion

The use of ecosystem and coupled soil–vegetation–atmosphere (SVAT) models is essential for the scaling up of smaller-scale measurements of fluxes using meteorological towers and remote-sensing data to the regional and global scales that are of particular interest in climate-change studies. For all the fluxes, including that of CO_2, data assimilation approaches are key (e.g. Dorigo *et al.*, 2007; Quaife *et al.*, 2008). As we have seen earlier, data assimilation involves the combination of remote-sensing data on reflectance (as a measure of *fAPAR*) and surface temperature (for estimation of evaporation) with climate data (or simulations) and with biosphere–atmosphere models such as the biosphere-energy transfer-hydrology model (BETHY) and used to estimate ecosystem fluxes including that of CO_2 (e.g. Rayner *et al.*, 2005). It should be noted that there are often substantial errors in the satellite-derived estimates of canopy parameters, even when using high-level composite products such as MODIS *LAI*/*fAPAR* products; these fluctuations need to be smoothed out to reflect the generally slowly changing *LAI*, and to improve the assimilation process.

Further reading

Further details on the relevant physiological processes and of heat- and mass-transfer processes may be found in texts such as Jones (1992), Campbell and Norman (1998) and Monteith and Unsworth (2008). There are many books on photosynthesis, but a useful description may be found in the text by Lawlor (2001). Much of the material on remote sensing of canopy function is not covered well in current remote-sensing texts and is best described in some of the research papers and reviews mentioned in the text.

Websites

References on surface energy balance: **http://badc.nerc.ac.uk/data/srb/references.html**

Sample problems

9.1 (a) Estimate the net radiation received by a grass canopy at midday with an $LAI = 6$ and at 20 °C, where shortwave irradiance is 650 W m^{-2} and the average apparent sky temperature is –5 °C. What is the soil heat flux expected in this case? (b) Assuming a Bowen ratio (β) of 0.1, estimate the boundary-layer conductance to heat transfer from the canopy, if the air temperature is 19 °C.

9.2 For a satellite image where the hottest pixels in an area average 32 °C and the coolest average 22 °C, estimate the evaporation rate from a cropped area where the average radiometric temperature = 26 °C, assuming an available energy ($R_n - G$) = 400 W m^{-2}. Indicate any other assumptions required.

9.3 A single MODIS pixel has an average temperature of 32 °C and a normalized *NDVI* of 0.40. Some adjacent pixels average 37 °C with normalized *NDVI*s of near zero. Estimate the average canopy temperature.

9.4 Expressed in conventional mass units, two contrasting canopies have boundary-layer conductances of (a) 300 mm s^{-1} and (b) 20 mm s^{-1}. Estimate the relative sensitivity of E to a change in stomatal conductance for each canopy when the stomatal conductance is either 10 mm s^{-1} or 40 mm s^{-1}.

10 Sampling, error, and scaling

10.1 Introduction

In Chapter 2 we traced the passage of electromagnetic radiation from the source to the detector and noted the requirement to correct for the effects of the intervening atmosphere on the radiation received at the detector. In Chapter 5 we considered some of the types of sensors and instruments for detecting and recording the radiation and their calibration so as to be able to relate the recorded data to the radiances emitted from the surface material in order that the geophysical and biological parameters may be extracted. In the majority of applications relevant to the study of vegetation, where quantitative measurements are made or when the data are used in biophysical models, it is essential to have accurate, reliable, and consistent radiance data that can be related to the variables being studied. Examples of such situations might be the measurement of temperature, the use of quantitative radiance measurements, when relating values relating to different places or times, or for relating values obtained using different instruments.

The validity of the information obtained from remote sensing depends both on accurate measurements of the relevant parameters and on the use of appropriate analytical and statistical techniques. The accuracy of the determination of a biophysical variable depends not only on the accuracy of the instruments themselves, but on the validity of subsequent steps relating these measurements to the variable in question. If you are not actually measuring the variables you think you are, no amount of sophistication in the analytical techniques used to process the data will provide an accurate measure of the phenomenon in question. The results of analysis relate to the data being processed and this may not relate to the reality of the situation. It is necessary to appreciate the errors that may enter at all stages and how these propagate in order to be able to assess the reliability of the final information.

Possibly the greatest accuracy can be achieved under laboratory conditions, but we have seen that such measurements on individual leaves or samples do not necessarily relate to those made on bulk samples or in the field under varying conditions. Similarly, many field measurements involve taking recordings at specific locations or at localized points. How do these point measurements relate to the extended variables measured by remote-sensing instruments? We consider the question of scale in measurements below. Geodynamics evolved from the application of spatial measurement techniques (remote sensing) to spatially distributed geographical processes that require spatial statistics in order to analyse their form and pattern, and that also evolve over time. It underlies all forms of dynamic modelling and addresses the question 'why' as opposed to the question 'what'. It is perhaps the link between the various topics discussed in this chapter and uses dynamic modelling such as *spatially distributed dynamic models* (SDDMs) (Darby *et al.*, 2005), as a means of predicting the response of complex, spatially distributed environmental systems to global change drivers.

In this chapter we consider a number of issues that relate to the optimal experimental design for the collection and use of field or *in situ* data and other ancillary data to aid in the analysis and interpretation of remotely sensed data and the factors that must be considered in order to produce accurate and reliable information. This includes aspects relating to the selection of test sites, choice of sampling strategies, and the impacts of errors and scale effects on the results. Finally, we will consider ways in which accuracy or precision can be assessed.

10.2 Basic sampling theory

A basic understanding of statistics and experimental design is crucial when one is planning any environmental study to ensure that the monitoring or experiments planned are in fact capable of providing the information that is required. Although a treatment of statistics is outside the scope of this book, it is appropriate here to remind ourselves about some of the key aspects of sampling, experimental design, and interpretation of results. Further details may be found in any introductory statistics text; some particularly useful texts are listed in the further reading section below. The approaches used for quantitative studies tend to be different from those appropriate for spatial data and classification, so they will largely be treated separately.

It is no use, indeed it is worse than useless, to conduct an experiment or to undertake a survey that is not capable of answering one's questions at the level of precision needed. The key initial step in any study, therefore, is to determine the sorts of differences that one needs to detect if one is going to be able to act on the results. For example, an agricultural researcher is trying to develop a method to assess the effects of common fertilizer treatments on the remotely sensed nitrogen content of cereal crops and the consequential effects on yield. Let us assume that some preliminary studies in the field showed that on average the extreme treatments give rise to leaf nitrogen concentrations that differ by about 10% and yields that differed by 20%. It follows that any remote-sensing method developed must be able at the very least to discriminate nitrogen contents differing by 10%, but even this would be of very limited value in any real situation as it would only discriminate the extremes. Though it is difficult to be precise, for the test to be of much use it would need to have a sensitivity of detection about an order of magnitude better. We now consider the information we need to determine whether we can achieve this sort of discrimination and to provide guidance on choosing experimental design and the amount of replication required.

To do this we need to consider the causes of variation that lead to uncertainty in remote sensing (or laboratory) estimates of a variable such as leaf nitrogen concentration. These sources of uncertainty include those arising (i) from instrument error (which may include noise in the detector, poor calibration, and thermal and temporal drift), (ii) from biological variability both within plants and between plants growing at different sites, and importantly (iii) from failure in any model linking the observation to the quantity of interest. In the case of leaf area monitoring (as we saw in Chapter 7) it is assumed that the spectral index used (e.g. *NDVI*) is related to *LAI* by a simple mathematical function (often assumed linear), but in practice the relationship is only an approximation to the true relationship, and furthermore the relationship itself changes as a function of many other factors such as background soil, so that the assumption of a constant relationship will itself lead to error.

The overall uncertainty or error includes both a *random component*, such as that resulting from instrument noise or biological variability, and a *systematic component* or bias, such as that resulting from calibration error and from some model failures. The random errors largely affect the *precision* of estimates while both random and systematic errors both affect the *accuracy*. These therefore need to be treated differently. The choice of an appropriate sampling regime and experimental design can be used to minimize the effect of the random errors and this is what is primarily addressed in this chapter.

10.2.1 Data description and variation

The variables describing a population of plants include *continuous variables* (e.g. height) that can take any value within a range), *discrete variables* (such as the number of shoots) that can only take whole number values, and *categorical variables* (e.g. species, or classification results) that bear no ordered relationship to each other. The most appropriate statistical treatment depends on the type of variable. In any population of plants one would expect some natural variation in properties such as height, leaf nitrogen content, and number of shoots, even if genetically identical plants were being grown as in a typical agricultural monoculture. This is because even small local differences in the environment resulting from soil or positional effects can have compound effects on plant growth. Even more variation would be expected in a typical natural ecosystem consisting of mixed genotypes and species. The distribution of values for any variable is frequently approximated by a theoretical distribution known as the *normal distribution* (Fig. 10.1).

We estimate the population mean (μ) for any variable and the variability in the population by randomly sampling individual plants from the population. The sample mean for any variable ($\bar{x}$) is determined, as usual, as the sum of the individual values divided by the number of observations. The important thing to note is that with small numbers of samples the mean is only an approximation of the true population mean, but with increasing numbers of samples the sample mean approaches μ; that is the precision of our estimate of the mean improves with a larger sample size, and our estimate of the distribution of values in the samples also approaches closer and closer to the true frequency distribution for the population. For many, but by no means all, situations it is found that the frequency distribution is close to that described by the normal curve as shown in Fig. 10.1. The shape of the normal distribution is defined by the standard deviation (σ) or the variance (σ^2), while the position is defined by the mean. The standard normal curve is a normal curve scaled to a mean of 0 and a standard deviation of 1. Although a range of different distributions is appropriate in different situations, the normal distribution is particularly useful. The standard deviation calculated for a sample taken from the population is usually indicated by s, which is as an estimate of σ, the true population standard deviation.

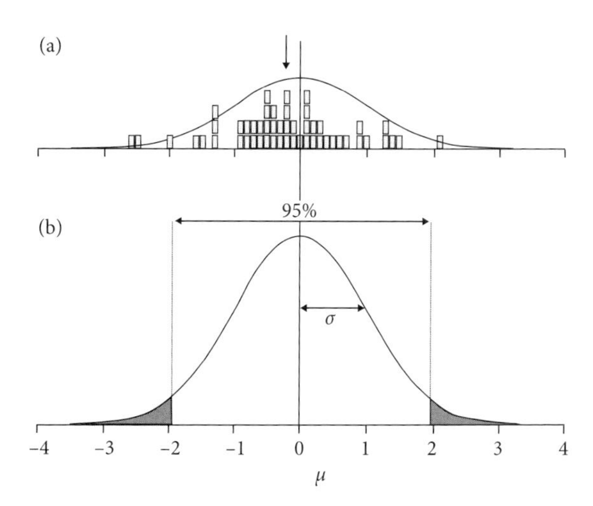

Fig. 10.1. (a) Random sample of 115 objects from a population having a mean (μ) of zero and a standard deviation (σ) of one, showing the scatter of different sizes obtained by chance and which had a mean of −0.201, rather than the expected zero. (b) The standard normal curve showing the range (−1.96 to 1.96) within which 95% of the observations fall.

The precision of our estimate of the population mean increases as we increase the number of samples by a simple law of diminishing returns. The precision is estimated by the *standard error of the mean* (*SE*) as $s/\sqrt{n}$. Although standard errors are widely quoted in experimental studies, it is generally more useful, especially when comparing different studies with differing numbers of samples, to derive *confidence intervals* (*CI*) that indicate the range of values which are likely, at a given percentage probability, to contain the population mean (given the observed sample mean)

$$CI = |\bar{x} - \mu| = t\,.s/\sqrt{n}, \qquad (10.1)$$

where t is the value of Student's t-distribution which may be found in standard statistical tables. The value of t depends on the number of degrees of freedom (one less than the number of samples, n); for a large sample, $t = 1.96$ at a probability level of 5% (i.e. $P = 0.05$) while the value of t required for the same probability level increases as the sample size decreases. This means that there is a 95% probability that μ falls in the range $\pm$ *CI*, or a 5% chance that it falls outside that range. Note that a high *precision* of estimate does not necessarily imply a high *accuracy* as a precise estimate may be very inaccurate if the instrument used has an incorrect calibration.

10.2.2 Tests and power

We are usually interested either in *testing for differences* between values or in *testing for relationships* between variables. Relationship tests such as correlation and regression can both be used to determine associations between variables, while regression is used to determine the relationships between variables and potentially to predict values of one on the basis of measured values of the other.

Much scientific investigation is based on *hypothesis testing* where one is usually comparing two contrasting hypotheses: a *null hypothesis* (H_0) that there is no difference, for example, between the means of two populations and an alternative hypothesis (H_1) that there is a difference. Our statistical test usually determines the probability (P) of obtaining a result at least as extreme as the one that was actually observed, assuming that the null hypothesis is true; this is called the P-value. A problem is that there are two types of possible error we can make as a result of the one in twenty chance of the wrong classification (Table 10.1). First, there is the possibility of a false positive result (i.e. declaring a difference when in fact there is no difference, called a *type I error*) or the possibility of a false negative where we incorrectly declare that there is no difference in spite of a true difference (a *type II error*). Strictly this applies only to the case where we fail to pick up a difference where the true difference is greater than the confidence interval. In practice, however, this type of error also applies where our test is not powerful enough to detect the difference because the confidence interval is larger than the real difference. False positives are potentially more serious than false negatives, because they can give rise to expenditure on an unnecessary intervention strategy.

When testing significance levels of differences between samples we need to bear in mind three aspects of the comparison: the significance level (α) that we decide is appropriate (often set at $\alpha = 0.0/5$), the required *power* of the test ($1-\beta$), and a minimum meaningful effect size (though the latter is often not explicitly considered). The values of α and β determine, respectively, the likelihood of rejecting H_0 when it is true and the likelihood of failing to reject H_0 when it is false – i.e. a true difference is not detected. The experimenter can set the values of α and β according to the relative costs of making these two types of error. For example, there may be cases, such as when one is surveying for invasive plants that the costs of failing to detect an outbreak are significantly greater than the costs associated with a few false positives.

The value of α is usually chosen to be small (0.05 or 0.01) so that the chance of false positives is kept small. The *power* for detecting a specified alternative hypothesis is equal to $1 - \beta$. In fact, power is broadly defined as the probability that a statistical test will reject the null hypothesis when it is false for a specified effect size. Many workers choose a value of 0.8 for the power, though this is rather arbitrary.

Most biologists are familiar with type I error and the need to ensure that few false positives are obtained; it follows that most effort has been put into ensuring that type I errors are kept small and tests are conventionally rather conservative (because of the frequently greater 'cost' of a false positive). However, the choice of too small a value for α means that the chance of detecting a difference when it exists may be too small. In some cases, therefore, it may be better to attempt to balance out the probability of false positives and false negatives. In any case it is important for an experimenter to consider the power of the experiment defined in terms of the probability of type II error. In general, the power increases with the size of the effect one wishes to detect, increases with the number of samples, and increases with both decreasing variance and decreasing α. Power analysis should be more commonly used at an early stage in any study to determine the sample size required to detect a given size of effect and readers are recommended to consult appropriate statistical texts.

Table 10.1 Errors in hypothesis testing.

Outcome of statistical test	Null hypothesis is	
	True	False
Reject H_0 (i.e. a difference exists)	False positive or Type I error with probability α chosen by experimenter	Correct conclusion (difference exists) Power $(1 - \beta)$
Accept H_0 (i.e. no difference exists)	Correct conclusion (there is no difference) Confidence level $(1 - \alpha)$	False negative or Type II error with probability β chosen by experimenter

Table 10.2 Propagation of errors for some commonly used formulae such as those used in the calculation of stress indices, where a and b are constants, x_1 and x_2 are variables, and $s^2_{x_1}$, $s^2_{x_2}$, and $s^2_{x_1x_2}$ are, respectively, the variances of x_1 and x_2, and their covariance. http://itl.nist.gov/div898/handbook/mpc/section5/mpc552.htm

	Standard error of the function y
$y = ax_1 \pm bx_2$	$s_y = \sqrt{a^2 s^2_{x_1} + b^2 s^2_{x_2} \pm 2ab s^2_{x_1x_2}}$
$y = x_1 \cdot x_2$	$s_y = x_1 x_2 \sqrt{\frac{s^2_{x_1}}{x_1^2} + \frac{s^2_{x_2}}{x_2^2} + 2\frac{s^2_{x_1x_2}}{x_1 x_2}}$
$y = x_1/x_2$	$s_y = \frac{x_1}{x_2}\sqrt{\frac{s^2_{x_1}}{x_1^2} + \frac{s^2_{x_2}}{x_2^2} - 2\frac{s^2_{x_1x_2}}{x_1 x_2}}$
$y = (x_1 - x_2)/(x_1 + x_2)$	$s_y = \frac{1}{(x_1 + x_2)^2}\sqrt{4x_2{}^2 s^2_{x_1} + 4x_1{}^2 s^2_{x_2} - 8x_1^2 x_2^2 s^2_{x_1x_2}}$

10.2.3 Propagation of error

Because almost any remote-sensing estimate of a biophysical variable involves use of a model or formula (e.g. $NDVI = (N-R)/(N+R)$) containing two or more independently measured variables, each of which is subject to error, it is important to understand how these errors combine in affecting the overall precision of estimate. Of course, in many cases we can simply estimate the errors in our prediction using observed uncertainty in the relationship between the calculated index and the output using a series of replicate observations. This is sometimes known as the 'top-down' approach to error analysis. Nevertheless it is often useful, particularly when designing new indices, to consider the contribution of the different measurements to the error as this helps us to identify particularly important terms in the formula for which improvements in accuracy will have the greatest impact on the outcome. This can allow us to optimize the index.

Details of error analysis are widely discussed in engineering and statistics texts and will not be discussed in detail here. The general formula for evaluating the error in y, where y is a defined function of a number of variables $(x_1, x_2, \ldots)$ is

$$s_y = \sqrt{\left(\frac{\partial y}{\partial x_1}\right)^2 s^2_{x_1} + \left(\frac{\partial y}{\partial x_2}\right)^2 s^2_{x_2} + \ldots + 2\left(\frac{\partial y}{\partial x_1}\right)\left(\frac{\partial y}{\partial x_2}\right) s^2_{x_1x_2} + \ldots}, \tag{10.2}$$

where s_y is the standard error of y, $s^2_{x_1}$ and $s^2_{x_2}$ are, respectively, the variances of the measured variables x_1, x_2 and so on, $s^2_{x_1x_2}$ is the covariance between x_1 and x_2, and the $(\partial y/\partial x)$ terms are the partial derivatives of the function y with respect to each of the x measurements. These partial derivatives are the sensitivity coefficients describing the sensitivity of y to changes in each of the x variates. Where the x variates are independent, the covariance terms are zero and in many cases it is assumed that they can be ignored in the calculation. Some typical examples of the application of this formula are given in Table 10.2.

10.3 Field measurements and collection of other reference data

Field work in remote sensing, whether it is to collect training data for classification (Section 7.4), accuracy assessment (Section 10.6), instrument calibration (Section 5.9), or parameter estimation or validation (Section 10.5), is essential to the success of any project and requires careful planning and preparation.

In situ measurements for the calibration or verification of data can be undertaken either in the laboratory or in the field and will be considered in the next section. Here, we consider the collection of data for use in the analysis and interpretation of remotely sensed images. Note that the term 'ground truth' is often used for this reference information, but this term should be avoided as it implies that it is absolutely correct; this is very rarely the case as the ground reference data are nearly always subject to some error. The error in the reference may even be substantial, as we shall see.

There are a number of essential steps in project planning. The location and size of the test site will depend on the project, of course, but also on the resources available, both financial and personnel. The level of detail must be adequate for the given project, not too much, which is wasteful of resources, nor too little. Also to be considered are the scale of the final product (often a map), the desired accuracy, the purpose of the work and the requirements of the end user, the field methods to be employed and the type of data required, and what collateral data are available. Power analysis and optimization of resource allocation is critical to ensure that what is proposed is likely to be able to answer the question(s) posed and give guidance on the optimal allocation of limited resources. The timing of the work will depend on the seasons, and whether it has to be coincident with image collection. An appreciation of the response of different features and their dynamics are important so that the relevant biophysical variables are measured to satisfy the project objectives, bearing in mind the parameters that may affect the response in the image such as wind speed, cloud cover and solar radiation, rainfall, humidity, and so on, which may need to be measured at the same time. Field work may also involve taking samples for subsequent laboratory analysis, such as soil conditions, moisture content of vegetation (for fire potential assessment – Section 11.5).

One of the main uses of field data is in conjunction with image classification, both to provide training data and against which to check the success of the procedure. Classification usually assumes that data points are randomly distributed over the study area, so it is essential that field measurements are representative of the distribution of landcover. Sampling strategies are discussed below where we also consider the optimum number of training points required for each class. The size of sample sites will depend on the size of the pixels, and to some extent on the spatial accuracy required of the project. The number of observations over a site depends on the variability; this also relates to scaling up, which will be discussed below. The assessment sites should be different from the training sites.

The accuracy of the training data depends on a number of factors. First, is the positional accuracy of the point at which a measurement is taken. Samples may be taken at arbitrary points and at different times. The spatial accuracy may be determined from a map or by use of GPS, which itself has an inherent error. There may also be ambiguity in the classification of different cover types on the ground due to personal bias and inconsistencies between surveyors, since such decisions are subjective. There may also be a time lapse between the ground recording and the acquisition of the remotely sensed data.

10.3.1 Sampling

Statistical tests generally assume a random sample of the population. Unfortunately, this is often not easy to achieve in practice and care is needed to avoid the many possible pitfalls in sampling. One approach is to subsample the area of interest by only studying in detail small randomly chosen areas, often known as quadrats after the sampling frames (frequently 1 m square) often used, though quadrats may be any size appropriate to the vegetation being studied. Rigorous use of random numbers to locate quadrats or sample points is usually substantially better than 'quasi-random' approaches such as throwing quadrats that inevitably introduce some observer bias. As a simple example of a non-obvious problem that can arise in sampling, one might expect that if we choose a point at random in a field and then take the nearest plant; we might assume that this would be a random sample. However, if the plants showed a very clumped or uneven distribution we might have ended up with a sample that is biased to the plants growing at the edges of the clumps. Choosing the sampling strategy is a critical step in the optimization of sampling; some of the options available are outlined in Box 10.1, and

BOX 10.1 Sampling strategies

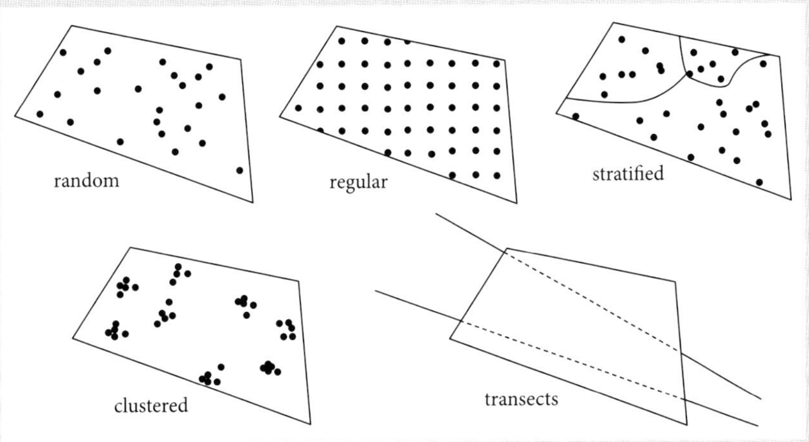

	Advantages	Disadvantages
Random	Statistically optimal	Smaller categories may be undersampled or missed
	Avoids operator bias	Possible inaccessibility of points in difficult terrain
Stratified random	Reduces chance of undersampled categories	Possible inaccessibility of points
	Largely avoids operator bias	
	Often the most efficient strategy	
Regular	Ease of sampling	Lacks true statistical randomness
	Good spread over area	Possible bias where linear features exist
Clustered	Reduces travel time in the field	If individual sites are too close to nodes, results can be subject to autocorrelation
		Can be subject to bias depending how nodal points selected
Transects	Ease of access and sampling	Generally non random
	Good where pre-existing gradients known (e.g. altitude)	Can be subject to bias depending on how transects chosen
		Incomplete coverage

In addition to the above, many workers use a 'subjective' or 'judgemental' sampling strategy on the basis that the operator has experience of the site and aims to select 'representative' sites. Unfortunately, this practice inevitably introduces bias, which is usually serious, and should therefore be avoided.

are discussed further in the text below. Although the use of true random samples is best in homogeneous habitats, many environments are heterogeneous and one might be interested in evaluating the variate of interest in each type of subarea independently; in such cases the choice of 'random' samples becomes more subjective. Although most statistical tests strictly require random samples, there are often arguments for the use of systematic sampling using some regular grid. Stratified random sampling has various applications, including the situation where one is interested, for example, in estimating the number of leaves per unit area. Rather than counting all the leaves in a quadrat, which may be several thousand even for small quadrats with trees, it is possible to count the number of leaves on a small sample of randomly selected branches and to multiply up to give a number of leaves per unit area of ground as the product of leaves/branch, branches/tree and trees/unit area. A stratified approach can be used to randomly select trees for sampling, and to randomly select branches on those trees.

10.3.2 Choice of sample units and sampling strategy

It is difficult to generalize about the choice of sampling strategy as the details need to be considered carefully for each situation, bearing in mind both statistical and practical considerations. In each case the strategy used to determine the position and number of sample points will depend on the reason for undertaking the field work and the method of analysis employed, the relative costs of different strategies, and the desired accuracy. Sampling strategies for validation of land-cover classifications will be discussed in the following section.

Note that laboratory measurements on soils, plant parts, or even micro-communities cannot replace *in situ* measurements. The reasons include the difficulty of relocating representative communities to the laboratory, the limited sample area and hence limited variability, and especially the difficulty of recreating the natural illumination field in the laboratory.

An important question will be the choice of sampling units. The sampling unit is sometimes a natural unit such as a plant or a leaf, but more often the sample unit is chosen somewhat arbitrarily, for example as a quadrat. In such cases there will nearly always be a trade-off between the number of sample units and their size, depending on the precision expected and the cost of sampling. The appropriate size of the sample area also depends on the scale of the vegetation of interest; for example for grassland, quadrats of <1 m^2 may be suitable, but for forests much larger quadrats will be suitable, though other sampling strategies, such as the random selection of individual trees may become more useful. In many cases a stratified sampling strategy is used where one samples, for example, a few replicate leaves from each of a number of plants or quadrats. This is an example of a typical two-stage sampling strategy with primary samples and subsamples. This can be extended to three-stage sampling and beyond.

Unfortunately, the question of how one should allocate limited sampling resources is rarely addressed adequately during the experimental design phase. In particular, the choice between sampling a few large quadrats or many small quadrats, or between taking many subsample measurements from each of a few sample units and few subsamples from many sample units is an important question in any calibration or validation study. The answer depends both on the balance between within-unit variation and between-unit variation and on the relative costs of individual replicate measurements in comparison with the costs of getting additional sample units. As an example we might wish to estimate the chlorophyll content per unit fresh weight for leaves in a field. Possibilities might include randomly sampling a large number of individual leaves from the whole field (which is not easy to do in a truly random manner), or sampling a smaller number of whole plants and randomly sampling a number of leaves from each of them, or even by sampling a number of leaves from within quadrats placed at random in the field. In order to decide on the optimal sampling strategy to use in terms of the number of quadrats and the number of leaves to sample per quadrat we need to investigate the *components of variation*, which can be obtained by means of an *analysis of variance* (ANOVA) using some preliminary sample data. A hypothetical dataset and analysis is shown in Table 10.3.

The 'between quadrats mean square' (QMS) includes a component due to the variation between parts of the field and a component resulting from the leaf to leaf variation within a sampling unit (LMS). Therefore,

Table 10.3 Abbreviated ANOVA table resulting from some simulated data on leaf chlorophyll concentration measured on extracts of n_2 (= 6) individual leaves from each of n_1 (= 4) different quadrats.

Source of variation	d.f.	Mean square	Parameters estimated
Total	23		
Between quadrats	5	0.152 (= QMS)	$\sigma^2 + n_2\,\sigma_q^{\,2}$
Leaf determinations (= between leaves within plots, s^2)	18	0.005 (= LMS)	σ^2

s^2 (= LMS) estimates σ^2; $s_q^{\,2}$ (=(QMS–LMS)/n_2 = 0.037) estimates $\sigma_q^{\,2}$

the 'between quadrats component of variance' ($\sigma_q^{\,2}$) is estimated by (QMS – LMS)/n_2 and the between leaves component (σ^2) is estimated by the residual (LMS). The variance of the mean, σ_y is given by

$$\sigma^2_y = \sigma_q^{\,2}/n_1 + \sigma^2/n_1 n_2. \tag{10.3}$$

Once estimates have been obtained for σ^2 and $\sigma_q^{\,2}$, one can then attempt to optimize the sampling design and specifically the relative effort put into increasing the number of replicate measurements within a sample unit as compared with increasing the number of sample units as shown in Box 10.2. In principle it is always best to obtain as many samples as possible at the higher level, with for example, two leaves from 10 plants generally being better than four leaves from five plants, with one leaf from each of 20 plants being better still. It is only when the costs of getting more plants outweighs the statistical advantage that we should change that rule.

10.3.3 Sampling for reference or training data

Ground data need to be collected for a wide range of remote-sensing activities. The purposes include (i) both the classification of training pixels for classification studies and the identification of vegetation types for reference pixels to be used for accuracy assessment of the output classification, and (ii) the need to get accurate and representative ground data for the derivation and validation of models for the extraction of canopy biophysical properties from RS data. In all cases the basic reference for any study must be data obtained on the ground at each reference site.

The sample needs to be truly representative of the whole image area, particularly because in large images (e.g. AVHRR) there is often a gradient of illumination or view angle across the image that can have substantial impact on the classification. Approaches to training site selection can be truly random, which has advantages, particularly when enough sample sites are chosen. This can, however, have less power than a stratified sampling making use of known underlying gradients in characteristics such as soil type or altitude. For example one part of an image may be known to be on an upland area of infertile sandy soil and another part of the image may cover peat-rich lowland soils. Geographical stratification is a useful tool for selection of reference pixels, though in extreme cases it may even be necessary to apply classifications to each area separately. If the aim is to provide labels for the classes already produced in an unsupervised classification, then a simple visual identification of the landcover will suffice.

For classification studies it is often possible to map the vegetation type accurately using ground survey and GPS to identify areas of pure vegetation or crop stands for use as training pixels. As we shall see, however, the collection of ground reference data for accuracy assessment should ideally be collected on areas corresponding to randomly chosen pixels because the choice of pure pixels only as references can lead to some bias. In most images there will frequently be problems in assigning a particular vegetation class to areas of mixed vegetation; therefore clear rules will be needed to deal with mixed or indeterminate pixels. A fuzzy classification system that uses more than presence or absence as a classifier can be valuable.

It is usually important to obtain as much ancillary data as possible, such as tree height for woodland or

BOX 10.2 Optimization of resource allocation in sampling

One can estimate the best allocation of resources in a quantitative study by introducing the cost of sampling into the variance equation. In principle, the total additional cost of sampling (C) can be given by

$$C = c_1 n_1 + c_2 n_1 n_2, \qquad \text{(B10.2.1)}$$

where n_1 and n_2, respectively, are the number of primary units and replicates, and c_1 is the 'cost' of collecting data for one primary unit (e.g. travelling time to the sample site) which is independent of the number of subsamples, and c_2 is the cost per sub-unit (e.g. the analytical time per sample). Whether one is aiming to achieve a specified confidence interval for the least cost or whether we want to achieve the best result for a given resource we need to minimize the product of variance and cost as in combining eqn (B10.2.1) with the equation for variance (eqn (10.3)) to give

$$s_y^2 C = \left(\frac{s_1^2}{n_1} + \frac{s_2^2}{n_1 n_2}\right)(c_1 n_1 + c_2 n_1 n_2). \qquad \text{(B10.2.2)}$$

It can be shown (see Snedecor and Cochran, 1999) that for a given total cost this product is minimized for

$$n_2 = \sqrt{\frac{c_1 s_2^2}{c_2 s_1^2}}. \qquad \text{(B10.2.3)}$$

This can be substituted into eqn (B10.2.1) to obtain the corresponding value for n_1. The effect of varying ratios between the between-plot and within-plot variances and of changing the number of plots or replicates are shown in the figure.

Fig. Box 10.2.1 Illustrates how the variance (error) increases as the ratio of the number (n_1) of plot samples to the number (n_2) of replicates decreases (for a total of 48 measurements). Each line represents a different ratio of between-plot variance to replicate variance (with the mean of the two variances maintained at 1). The different lines indicate decreasing ratio of plot variance to replicate variance going from 10 (solid line), 2, 1, 0.5, 0.2, to 0.1, showing that the sensitivity to the ratio of n_1 to n_2 is greatest where the plot variance is greater than the between replicate variance.

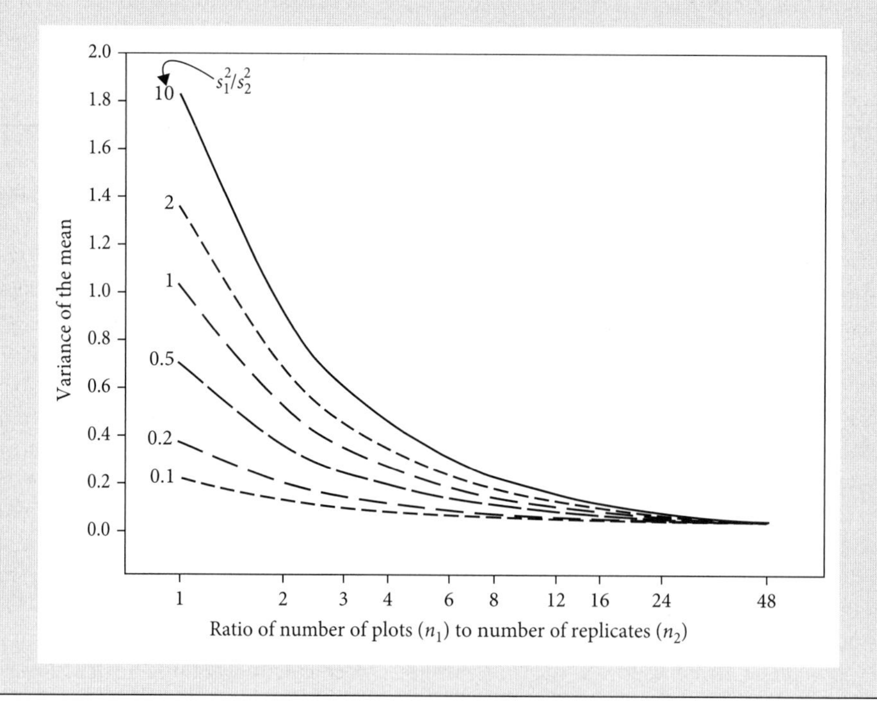

ground cover, leaf-area index or chlorophyll content for an agricultural crop. These detailed measurements are essential for calibration and validation studies and are often useful in classification studies. Many alternative sources of reference data are used in particular studies. These include aerial photographs, Google Earth images, soil maps, meteorological and climate data, together with other archive material. On occasions, field data are collected automatically using earth-based sensors that collect data and transmit them to a central location via telephone-based, telemetry, or satellite-based systems. Examples of data collection and relay satellites are the ARGOS system on NOAA, Meteosat, and other satellites. This technique is used particularly in meteorology where data from remote stations and floating buoys are continually being collected.

10.4 Scaling considerations

We have already alluded to the concept of scale, and some of the problems associated with differences in scale, at a number of points in the foregoing text. Examples include the spatial resolution of digital images (Chapter 5), the integration of point and spatial data (Section 6.5), the extension of leaf parameters to canopies (Chapter 9), texture, spatial dynamics and fractals (Chapter 6), data fusion (Chapter 6), and mixed pixels (Chapter 7). Here we consider further the concept of scale in remote sensing, including both the impacts of changing scale of observation on the calculated variables and the choice of an optimum scale for observations that will give meaningful information.

In most of the analytical techniques discussed above, the scale of an image refers to the pixel size, so it is important to select the optimum resolution for the particular study at hand. Physical and biophysical phenomena occur over a wide range of scales. The smallest with which we are concerned is at leaf level (although the underlying properties are governed by processes at much smaller scales), and the largest is probably at regional scales (although climate change may occur over global scales). Pixel size is inherently important because it limits the spatial precision with which measurements can be made. Pixels that are larger than the spatial variability of the features will average out any variations within that area and small variations will not be observable. An example of this could be the inability to discriminate between tree species within a general 'forest' area. Pixels that are smaller than the spatial variability will, on the other hand, pick up small, possibly irrelevant variations such as slight changes in the vigour of growth of a crop in a field due to variations in soil moisture or soil type. It may be tempting to select images with the highest spatial resolution available, but that may be unnecessarily expensive and may in fact mask the phenomena that are being studied. It is essential to select imagery of the appropriate spatial resolution and this is one of the first steps in project planning. Further discussion of scale in remote sensing may be found in Curran and Atkinson (1999) and Quattrochi and Goodchild (1997).

There is often a mismatch between the scale at which ground reference or calibration data are collected and the scale of the remote-sensing data or the scale of the required outputs. This mismatch can relate both to the spatial scale and the temporal scale of observations. For example instrumentation for measuring temperature and meteorological data usually provide point observations; nevertheless these relate to the surface–atmosphere interactions of a variable area of surface that is determined by the height of the sensors above the ground. Increasing height expands the footprint and fetch; this is reflected in the sensor reading, with the upwind fetch often being assumed to be c. 100-fold the height of the sensor (Monteith and Unsworth, 2008). For the same instrumentation there is often a temptation to use high-frequency data collection (minutes or hours) when the process of interest may operate over seasonal or longer timescales. Multiscale geophysical RS data need to be integrated with other types of data, such as point observations. What is the scale of such an observation network? It may be better to represent such phenomena as a mathematical field rather than as a set of points. For example, a meteorological measuring network can be considered as a density field representing the number of measuring stations per unit area, and forest cover as a field representing the density of trees.

Similarly, many physiological techniques such as the use of porometers, gas-exchange and fluorimeter systems for stomatal conductance, and photosynthesis measurements operate at the leaf or branch scale, and over short timescales (minutes to hours), while tissue mineral analysis may integrate effects over weeks to months. Similarly both canopy harvests for determination of *LAI* and below-canopy radiation transfer measurements for calibration of radiation-transfer models and estimation of *LAI* also are most suited to small plots of a few m^2, except perhaps in the case of forests.

Solutions to this mismatch problem depend on the application, but a usual approach is to adopt a stratified two- or three-step subsampling procedure as discussed above. Using such a nested strategy the distribution of sample sites must relate to the scale of the image. Modelling is often required to scale up (aggregate) small spatial- or temporal-scale field observations to the larger areas that are needed if we are to link effectively with 1–5 km satellite data or larger regional-scale studies.

10.4.1 Up-scaling (aggregation)

Up-scaling or aggregation (also sometimes known as degradation) is the process of combining high-resolution data into fewer lower resolution pixels; this process can involve a reduction of spatial or temporal resolution. Up-scaling inevitably involves the need to discard some detail, otherwise one either retains too much information to be of use, or else one simply obscures the relevant emergent phenomena that are only apparent at the larger spatial scales. A corollary of the loss of detail is usually a reduction in the number of degrees of freedom. Up-scaling may be regarded as a process of averaging that can be applied to any heterogeneous area, though some would argue that up-scaling is distinct from averaging in that it also involves the application of additional scaling models to describe how the simplification might work at larger scales (Wood, 2009). For example, the gas laws apply very precisely at the macromolecular scale where we can describe the properties of a mole of gas using one degree of freedom (e.g. P in the universal gas law), but would need an enormous number of degrees of freedom to describe the behaviour of all the individual molecules (position and motion for each of the 6.02×10^{23} molecules in the mole). For normal practical purposes the macromolecular description is adequate though it has lost some potentially important information.

In practice, up-scaling may be achieved either by aggregating input data (radiances) at the small scale and then calculating the output (e.g. $\boldsymbol{E}$) on the resulting average ('input' up-scaling), or else it may be achieved by calculating the output ($\boldsymbol{E}$) for each pixel at the small scale and averaging the result ('output' up-scaling). In principle, errors are minimized, as we shall see below, by up-scaling the output data where possible.

One general feature of aggregation is that it leads to a reduction in the range and variance of the data. Another, related, general feature is that the number of mixed pixels increases as the spatial resolution becomes coarser. Nevertheless, it may sometimes be better to use coarse data to avoid oversampling and to avoid the detection of unnecessary variation within a feature (e.g. forest stand not individual trees). As spatial scale increases we are usually more restricted in the numbers of classes that can sensibly be detected; at a regional scale it is usual to aggregate many classes (e.g. individual crops that may occur at a subpixel scale) into a few broader classes (agricultural land).

10.4.2 Down-scaling (disaggregation)

In many cases remote-sensing imagery is at too large a spatial scale to obtain the detail required for specific studies. Even 250 m pixels for a sensor like MODIS cannot give much information at a scale that might be suitable for precision agriculture. Therefore, there is much interest in disaggregation of imagery to attempt to improve spatial resolution, though of course disaggregation cannot add any more information. As was outlined in Section 6.4.6 the most effective approaches are based on data-fusion techniques where lower-resolution images are enhanced by using information from higher spatial resolution sensors. As an example of this approach Kustas *et al.* (2003) demonstrated the successful disaggregation of 1- to 5-km spatial resolution thermal data from meteorological satellites using optically derived *VI* data at up to one order of magnitude better resolution. This relies on the generally good relationship between *NDVI** or other *VI* and temperature, with high *NDVI* corresponding to rapidly transpiring crop (and low temperatures). Similarly some subpixel information can be extracted by spectral unmixing techniques (Section 7.4.3), though without providing subpixel localization.

10.4.3 Problems of non-linearity

The uncertainty involved in aggregating remote sensing data from smaller to larger spatial scales is related both to non-linearity in the response function being studied and to the heterogeneity of the site. Where the response is non-linear, conventional averaging of an independent input variable (such as reflected radiation) gives rise to a biased estimate of the output response (e.g. photosynthesis, temperature, *LAI*, or evaporation).

A good example is provided by photosynthesis, which illustrates two aspects of the scaling problem. The first relates to the different behaviour of single leaves and canopies. As we saw in Chapter 4 photosynthesis shows a saturation response to increasing light (Fig. 10.2). We see in this figure that individual leaves show sharply saturating responses to light, though the actual curves are somewhat different for the lower leaves in the canopy, which are adapted to low irradiances, and the upper leaves. For the whole canopy, however, the photosynthetic response to irradiance is more nearly linear with saturation occurring at higher irradiances and higher assimilation rates per unit area (of ground in this case). There are several reasons for the different behaviour of the canopy. These include: (i) not all radiation incident on a leaf is absorbed so that a canopy uses the incident radiation more efficiently with the scattered and transmitted radiation used by leaves lower in the canopy, (ii) many leaves avoid light saturation by being at an angle to the incident radiation; this means that the radiation is used more efficiently, and (iii) the differential biochemical responses of leaves in different parts of the canopy. Effective treatment of this type of scaling problem requires the use of a functional model that integrates the small-scale observations to the larger scale (e.g. Asner and Wessman, 1997).

As a second illustration of the scaling problem we can consider the impact of the non-linearity of the relationships between fractional ground cover and *NDVI* and especially *LAI* (see Table 10.4, also presented in graphical form in Fig. 10.3). For example, we can consider a hypothetical remote-sensing situation where we have a patchy canopy with, on average, equal areas of dense canopy patches (each c. 10 m^2 at 90% cover) and very sparse canopy patches (each c. 10 m^2 at 10% cover) giving an overall average ground cover of 50%. This can be compared with a homogeneous canopy having the same (50%) ground cover. We can see (Table 10.4 and Fig. 10.3(b)) that the average *LAI* for the homogeneous case is 1.02, while for the patchy case the average *LAI* is 2.41 (see the bottom line of Table 10.4), as a result of the proportionately much higher *LAI* in the dense patches.

Averaging input data – In each case we can calculate the appropriate R and N reflectances by assuming the linear mixing model (eqn (7.3)); these values are the same for the patchy mixtures and for the homogeneous case (top two lines of the lower part of Table 10.4). It is clear that such an 'input' aggregation based on reflectances of the component patches would give the same values whether one is viewing a homogeneous canopy or a patchy canopy with the same average fractional ground cover. It follows, therefore, that averaging input reflectance data from either a high spatial resolution sensor such as IKONOS (with about c. 1 m pixels) or from a medium-resolution sensor (such as MODIS with 250 m pixels) and then calculating the output, we estimate the same *LAI* (=1.02) for both homogeneous or patchy scenes, whichever sensor we use.

Averaging output values – We get very different results if, on the other hand, we use the input data to calculate 'output' values (i.e. *LAI* or *NDVI*) for each patch and then average these. For the homogeneous

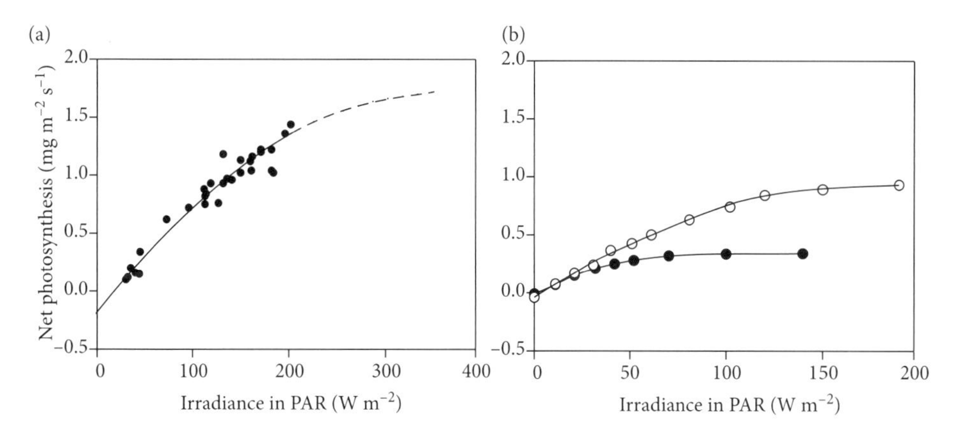

Fig. 10.2. Non-linearity effects on photosynthesis: (a) photosynthetic light response curve for a tomato canopy, (b) corresponding light response curves for lower (solid symbols) and upper (open symbols) leaves (modified from Jones, 1992; using data from Acock *et al.*, 1978). Note that this experiment was conducted in a controlled environment so light saturation occurs at substantially below that which it would in the field.

Table 10.4 Averaging at different scales illustrated for *NDVI* using the data from Table 7.1. The top halves of the columns give the calculated values of ρ_R, ρ_{NIR}, *NDVI*, *NDVI**, and *LAI* calculated for pure stands of different densities. Note the strong non-linearity between *LAI* and % cover. The lower half of the table gives the values of ρ_R and ρ_{NIR} estimated by averaging the values for the component patches (i.e. input aggregation) for patchworks with different combinations of patches giving 50% cover, together with estimates of *NDVI*, *NDVI**, and *LAI* calculated for these combinations by averaging the calculated values for each patch (i.e. output aggregation). The relationship between *LAI* and cover was taken from Asner and Wessman (1997) using data for 10% diffuse radiation.

	Per cent cover (%)						
	0	10	20	50	80	90	100
ρ_R	**0.27**	0.25	0.23	0.165	0.10	0.08	**0.06**
ρ_{NIR}	**0.31**	0.32	0.32	0.345	0.37	0.37	**0.38**
NDVI	0.07	0.12	0.17	0.35	0.56	0.64	0.73
*NDVI**	0	0.08	0.16	0.43	0.75	0.87	1.00
LAI	0	0.22	0.33	1.02	3.15	4.60	6.71
Averages	(0 + 100)/2	(10 + 90)/2	(20 + 80)/2	50			
ρ_R	0.165	0.165	0.165	0.165			
ρ_{NIR}	0.345	0.345	0.345	0.345			
NDVI	0.4	0.38	0.37	0.35			
*NDVI**	0.5	0.47	0.46	0.43			
LAI	3.35	**2.41**	1.74	**1.02**			

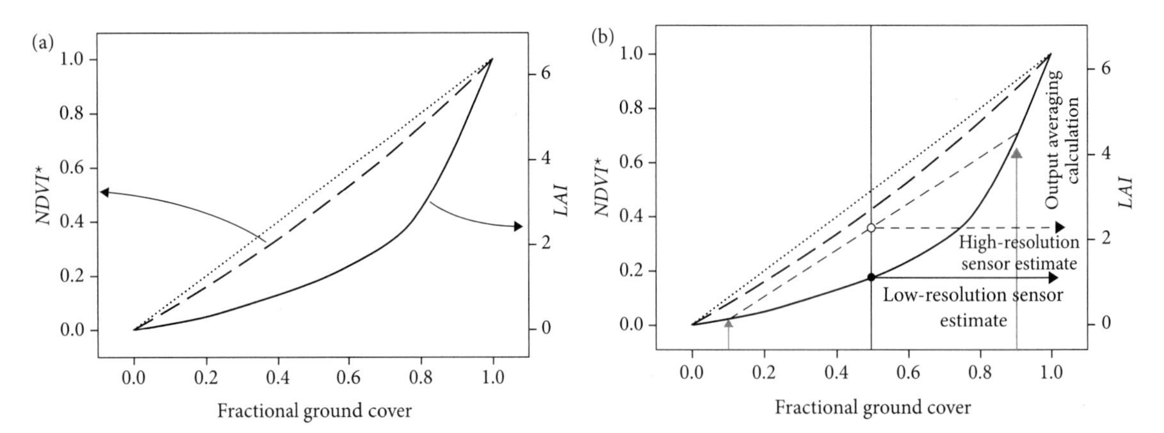

Fig. 10.3. An illustration of the effect of up-scaling on errors in the estimates of canopy biophysical parameters by remote sensing (data from Table 10.4). (a) Both *NDVI** (dashed line) and especially *LAI* (solid line) are non-linearly related to ground cover. (b) Shows the difference between the *LAI* calculated with low and high spatial resolution sensors for a heterogeneous mixed community 50% of which comprises patches with 90% ground cover and 50% with 10% ground cover. The open circle indicates the true overall average values of *LAI*, as would be derived using high-resolution sensors (with output averaging), while the closed circle indicates the value that would be derived from a low-resolution sensor on the basis of the average radiative properties, or using a high-resolution sensor with input averaging.

case both sensors estimate the correct *LAI* of 1.02, but for the patchy case we get very different answers. Averaging the values (e.g. of *LAI*) calculated from IKONOS reflectances gives the correct average *LAI*s of 2.41 for the patchy scene and 1.02 for the homogeneous scene, but MODIS will give a value of 1.02 in both cases because it uses the averaged reflectance over the area.

It is now easy to see that not only can the use of low spatial resolution sensors give poor estimates of biophysical variables such as *LAI*, but even incorrect scaling of high-resolution data can be problematic.

In practical situations with remote sensing data obtained at two different scales one can compare the area-average obtained by averaging the derived values at the small scale with the value calculated using the large-scale image. Although estimates of T and reflectance may be quite similar at the two scales, estimates of energy fluxes ($\mathbf{C}$ and $\lambda\mathbf{E}$) may be subject to considerable error (Moran *et al.*, 1997). Not only does non-linearity of the response function cause problems in up-scaling, but this effect can be compounded where there is radiative scattering between the components of the scene. The usual linear mixture modelling assumes that each component behaves independently, but in some cases, such as for patches of soil in a vegetated area that are illuminated by a mixture of skylight and scattered light, this can have significant impact.

In classification studies, spatial scale can also directly affect classification accuracy, with aggregation often leading to an increased proportion of unclassified pixels (Gupta *et al.*, 2000), though the effects are not necessarily consistent (Raptis *et al.*, 2003).

10.5 Calibration and validation

Data are now being acquired by more than 50 earth observation satellites, each using different methodologies, and so it is essential that globally recognized guidelines are put in place in order to ensure that these datasets are compatible. The Committee on Earth Observation Satellites (CEOS) is working to establish a consensus within the international community that calibration, validation, and quality assurance processes are incorporated into satellite programmes in a harmonized way. At present, these processes vary widely, and it is often up to the user of the data to undertake such steps as he may require.

In practice, there are several aspects to the extraction of useful information from remotely sensed data. First, the instrument itself needs to be *calibrated*, in order that there is a known and consistent relationship between the radiation received and the output value. Some further calibration is also required to relate this output quantitatively to the values of the biophysical parameters being studied. The remotely measured values must then be *validated* against known targets to check the accuracy of interpretation made from the data. If the measurements are being made and distributed commercially by an agency then the customer will also require some assurance as to the quality and consistency of the information being provided.

A commonly used estimator of the accuracy or validity of measurements involves the comparison of observed values with reference values and the calculation of a measure related to the standard error of the mean known as the *root mean square error* (RMSE). This is calculated according to

$$\mathrm{RMSE} = \sqrt{\frac{\sum_{i=1}^{n} (\mathrm{observed_i} - \mathrm{reference_i})^2}{n}}, \qquad (10.4)$$

where $\mathrm{observed_i}$ and $\mathrm{reference_i}$ refer, respectively, to the ith pair of observed or calculated values and the comparable reference values, and n is the number of pairs of data points. Advantages of such an error measure is that it ignores the sign of the difference and weights more strongly for large differences than for smaller differences.

Instrument calibration

Instruments that are hand-held or that are flown on aircraft can be calibrated fairly easily against known standards, either in the laboratory or in the field. Those carried on satellites are usually subject to both *pre-launch* and *post-launch* (or inflight) calibration. The pre-launch stage will provide relative calibration between all the detectors in any one band. It will also relate the output data to known input values (radiances or temperatures) and will take account of any non-linearities in the instrument. Radiometric calibration coefficients are usually provided in the header file of data obtained from a data centre. But aging of electronic components results in sensor degradation and a change in the calibration coefficients over time, and sometimes some form of on-board calibration is provided. This is most common for thermal sensors. The AVHRR, for example, uses an on-board black body maintained at a temperature that is measured using thermocouples. This black body is scanned as the mirror rotates, thus giving a reading of a known temperature once every scan line. This is incorporated into the data stream of this scan line when it is telemetered down to the receiving station, together with the readings of the thermocouples. This enables every scan line to be individually calibrated eliminating the effects of both long-term and short-term drifts. The optical channels on some environmental satellites have on-board calibration capabilities although it is more difficult to maintain standard light sources than black bodies. Landsat TM/ETM+ has inflight calibration at the start and end of each scan using solar and lamp-based approaches (three lamps) for the solar reflective bands and black bodies for the thermal band. The responses of the 3000 detectors in each band of SPOT's HRVIR scanner are balanced (called *detector response normalization*) while the instrument views a perfectly uniform landscape (see below), and absolute calibration to determine the instrument's dynamic response is achieved by establishing a precise relationship between a perfectly stable external source (the sun) and the instrument's output signal. The calibration is used at regular intervals to check and, if necessary, to adjust the instrument response.

Non-linearity in the response of the instrument can lead to several problems. The usual effect is that the change in output level for a given change in input gets less for larger values so that the contrast at high levels of illumination gets smaller. The result is that all bright areas appear to have similar brightness and such areas in the image are 'burnt out'. The effect is particularly striking in thermal measurements, and it becomes impossible to differentiate between temperatures above a certain level. This is a problem particularly with the thermal channel on TM that has a very much smaller dynamic range than say the AVHRR. It was designed to measure normal land-surface temperatures and cannot be used to monitor volcanic activity or wild fires, for example. It is not just the pixel(s) covering the hotspot that are affected. If the detector suffers severe saturation, it may take a while for the electronics to recover, leading to a loss of data for some number of pixels along the scan line following the point in question.

Calibration against biophysical parameters

Given accurate radiance data we now need to relate such radiances to the biophysical or other parameters we are studying. In statistical terms such calibration is the reverse of regression. In regression we estimate the relationship between a dependent and an independent variable, while in calibration we use this relationship and information on the errors to estimate (or *predict*) new values of the independent variable from observed values of the dependent variable (see Fig. 10.4).

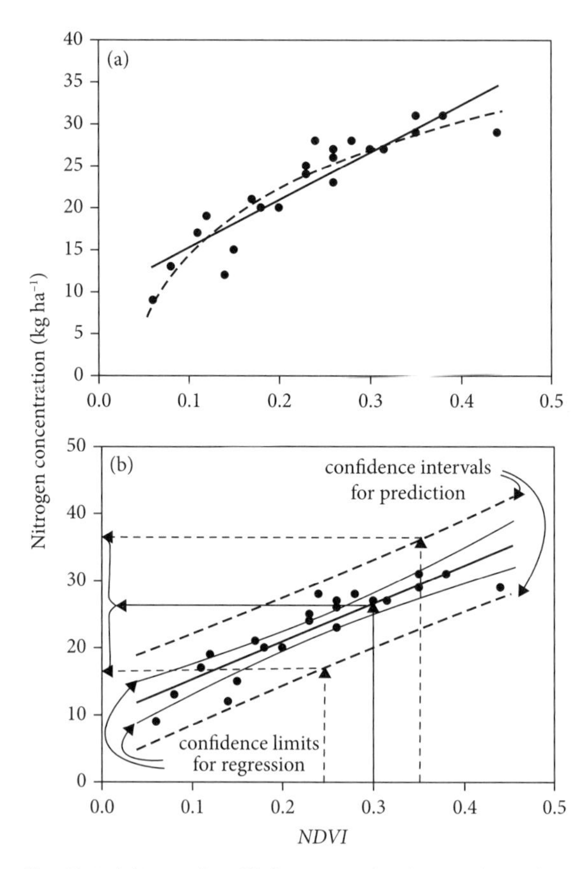

Fig. 10.4. (a) Regression of [*N*] on *NDVI*, showing experimental data points collected during calibration and fitted linear ([*N*] = 55.62 × *NDVI* + 9.64; R^2 = 0.785) and non-linear ([*N*] = 11.50*ln(*NDVI*) + 40.8; R^2 = 0.847) curves, (b) the same data being used with the linear model to estimate the value of [*N*] for a sample for which we have an observed *NDVI*. The confidence limits for the regression line are shown by the inner pair of curved lines, while the dashed lines show the confidence intervals for prediction. Note that the potential error in measurement of *NDVI* should also be taken into account, with the potential prediction error being indicated by the brace on the ordinate; this error depends on the number of replicates used.

For example, a farmer might be interested in estimating the nitrogen concentration in his crop and wishes to use a remote sensing measure such as a vegetation index. He might then measure the *VI*s for a range of crops grown in different areas and different years with as wide a variety as possible of input nitrogen fertilizer. At the same time he would sample the crops for chemical analysis of leaf nitrogen concentration ([*N*]). In principle, regressing [*N*] on *VI* gives the experimental relationship, together with estimates of the variance of [*N*] for any given *VI*. This allows the farmer to estimate [*N*] for a new field from any measured *VI* as shown in Fig. 10.4.

As illustrated with this dataset we often find that the calibration is non-linear, either as a result of a fundamentally non-linear relationship, or as a result of sensor saturation (as frequently occurs with *NDVI*). In such cases, estimation of the errors of prediction is more difficult, usually involving data transformation, and requires specialist statistical input.

A real example that illustrates some of the problems, both of scaling up field data and of using that data for prediction, is shown in Fig. 10.5. This is based on data where the researchers were attempting to determine the relationship between CO_2 assimilation and spectral indices with the implied objective of being able to derive rapid non-destructive RS approaches (e.g. using spectral measurements) to estimate rates of photosynthesis or other biophysical quantities.

Although a statistically significant regression was obtained (meaning that the slope was different from zero at $P = 0.012$) this does not really achieve the objective of being able to estimate photosynthesis from a spectral *VI*. This is because it does not have much (if any) power for estimating photosynthesis from observations of *PRI*, as is clear from the extremely wide prediction interval shown in Fig. 10.3. In fact this is as large as the range of the experimental data. This example illustrates some of the potential difficulties with field sampling and the impacts of mismatch of scaling with remote-sensing studies. These include:

(i) low site replication, even though large numbers of the primary measurements were made,

(ii) the mismatch between the scale of measurement and the scale of comparison; both photosynthesis and *PRI* were estimated on large numbers of individual leaves or small subplots (39 × 39 cm) for the spectral data, and averaged over the whole site, thus leading to potential scaling errors. Unfortunately, no use was made of the stratified sampling regime adopted (i.e. no association between individual photosynthesis and *PRI* estimates),

(iii) substantial scatter in the data, with one point (*PRI* = −0.064) having very large leverage in the regression.

Validation

The validation of remotely sensed estimates of surface variables is a critical component of any remote-sensing study. We have introduced above the considerations

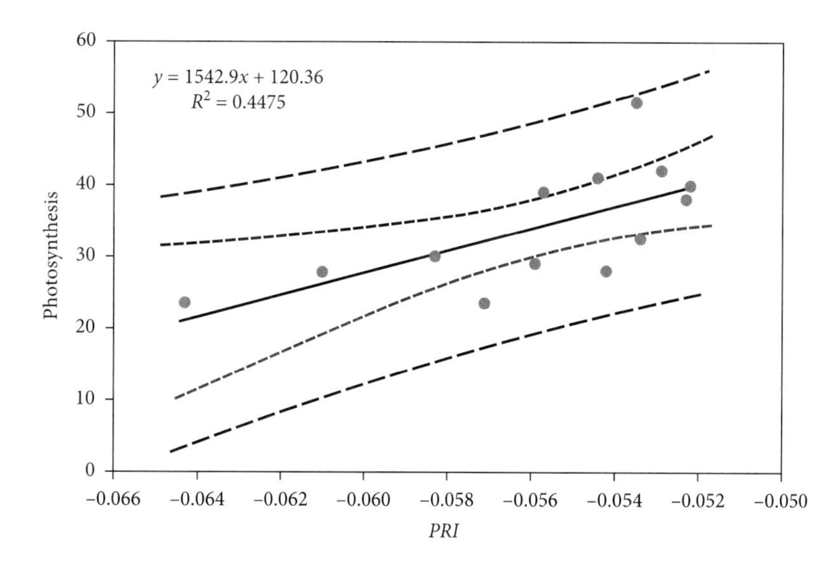

Fig. 10.5. Relationship between photosynthesis and *PRI* for thirteen 100 m × 100 m plots in a mixed-grass prairie ecosystem in Saskatchewan, Canada (data from Black and Guo, 2008). Estimates of both photosynthesis and *PRI* were obtained as averages of a large number of stratified subsamples. The fitted regression line was significant ($P = 0.012$). The inner long-dashed lines indicate the confidence limits for the regression and the outer, shorter-dashed lines indicate the prediction intervals.

that guide the choice of sampling strategy for the collection of high-quality *in situ* data. Here, it is just as important to reiterate the need to allow for the different scale of most field data collection and the satellite imagery. In many cases measurements are scale dependent and success often relies on the application of an appropriate scaling model to link the scales. In some cases, such as the use of eddy-correlation flux sites for validation, it is possible to scale the ground measurements to match the satellite pixels by adjusting the height of the sensors (and hence the footprint size), but in other cases, such as the validation of km-scale satellite estimates of surface temperature, there remain major difficulties in validation.

10.6 Uncertainty in spatial data classification accuracy

The user of thematic information derived from remote sensing needs to know how much credence he can put on the results, so it is necessary to consider ways in which the accuracy can be assessed.

Unfortunately, the simple question 'how accurate is the classification?' is actually rather difficult to answer with precision as there are several alternative measures of accuracy available, each of which may be appropriate in particular cases. Any classification will contain error that may result from inaccuracy in any of the following steps: errors in the basic remote sensing measurements, inaccuracies in any image correction procedure such as those used for geometric correction (where 'new' pixels are formed from admixtures of the original ones), and for radiometric correction (especially when images are heavily affected by contaminants such as smog), from inherent variability in the surface resulting in impure pixels, from scaling errors, as well as from inaccuracies or uncertainty in any training pixels used or failures in the classification algorithm itself. Most seriously there may also be problems in the classes used with those chosen not necessarily being exclusive or exhaustive. It is therefore essential to undertake an accuracy assessment that allows one to ascribe confidence limits to the thematic information derived.

These various sources of uncertainty in spatial data can usefully be separated into those relating to *positional uncertainty* (including both locational and temporal uncertainty), and *attribute uncertainty* (e.g. class). For example, the locational uncertainty leads one to need to sample a larger area than the pixel if one is to be sure of being representative. This area, A, is given by $A = (S_d{}^*(1 + 2S_g)^2)$, where S_d is the pixel size and the geometric accuracy of pixel location in pixel units is S_g (Curran, 1985). Further details of the treatment of uncertainty in spatial data can be found in Foody and Atkinson (2002) and Heuvelink (1998).

10.6.1 The error matrix

In order to assess classification accuracy we need to compare the output from the classification against ground reference information. Error analysis makes use of the *error matrix* (often referred to as a *confusion matrix* or a *classification matrix*) as illustrated in Table 10.5. For the case where we have a classification into k classes there will be k^2 cells in the matrix.

The most straightforward measure of accuracy is the *overall accuracy* defined as the total of correctly classified samples (values in the major diagonal) divided by the total number of samples, i.e. in this case 80/121 = 0.66, or 66%.

One can also derive measures of accuracy for each of the classes; this can be particularly useful where, for example, one is a forester who might be primarily interested in only one of the land classifications such as the broadleaved forest. In such cases one can count the number of correctly identified pixels for a class either as a fraction of the 'true' number of pixels of that class (i.e. the reference number obtained as the column total), or as a fraction of the number classified in that class (i.e. the row total). The first of these options (number of correct pixels for a class divided by the reference number in the class) is a measure of what is called *producer's accuracy*. The misclassifications in this case are termed *omission error* and indicate the number of known pixels

Table 10.5 A typical error matrix for assessing classification accuracy. In this case each of the n = 121 test pixels was classified into one of four vegetation classes by means of ground observation to produce ground reference data. These assignments were compared on a pixel-by-pixel basis with the results of the automatic classifier being tested. Shaded cells along the diagonal are the number of correctly classified reference samples.

		Ground reference data						
	Class	Evergreen	Deciduous	Grass	Scrub	Total	User accuracy	Commission error
Classification	Evergreen	25	1	6	0	32	25/32 = 0.69	7/32 = 0.22
	Deciduous	2	33	8	4	47	33/47 = 0.70	14/47 = 0.30
	Grass	1	4	17	2	24	17/24 = 0.71	7/24 = 0.51
	Scrub	2	0	1	5	8	5/8 = 0.625	3/8 = 0.38
		30	38	32	11	n = 121		
	Producer's accuracy	25/30 = 0.833	33/38 = 0.87	17/32 = 0.53	5/11 = 0.45		Overall accuracy: 80/121 = 0.66	
	Omission error	5/30 = 0.167	5/38 = 0.13	15/32 = 0.47	6/11 = 0.55			

for that class that were not correctly identified. For example for class 1, the producer's accuracy is 83% with a corresponding omission error of 17%. The second option, where one divides the number of correctly classified pixels in a category by the total number classified in that group, is a measure of reliability of the classification and indicates the *user's accuracy*. The complement of this is known as *commission error*. Reliability is essentially the degree of agreement between repeated measurements; this is not necessarily a measure of the validity of the measurements – which may all be wrong because of misclassification of the reference pixels. The user's accuracy indicates the probability that a classified pixel actually represents that class. It is generally appropriate to indicate all three types of error when describing the success of a classification.

In addition to the above descriptive evaluations of error there are a number of other approaches available for evaluating the success of a classification, often involving normalization of the data. A particularly widely used method of accuracy assessment or the assessment of agreement between different observers is *kappa analysis*. The kappa statistic is a general term for several similar measures of agreement that can be applied to categorical data. Kappa analysis estimates a coefficient of agreement as κ. The key difference from the overall accuracy is that it takes account of omission and commission errors in addition to the values in the major diagonal. κ may be calculated as

$$\kappa = \frac{n\sum_{i=1}^{k} x_{ii} - \sum_{i=1}^{k}(x_{i+} \times x_{+i})}{n^2 - \sum_{i=1}^{k}(x_{i+} \times x_{+i})}, \qquad (10.5)$$

where x_{ii} are the diagonal cells of the matrix, the x_{i+} are the row marginal totals and the x_{+i} are the column marginal totals and n is the total number of elements in the matrix. Kappa is a measure of agreement after chance agreement is removed, so it can be used to determine whether the classification is significantly better than chance. A value of zero indicates no agreement and a value of one indicates a perfect match between the classification output and the reference data. Values of kappa greater than about 0.75 indicate good to excellent classifier performance, while values less than 0.4 suggest rather poor performance, though the actual limits for any experiment depend on sample size and data type (Mather, 2004).

It is often of interest to decide whether one classification approach for a given set of data is statistically better than another. A useful comparison should take account of the non-diagonal cells as well as the diagonal cells. One approach to this is to apply a discrete multivariate analysis to the whole of the normalized matrices; this can be used to give a measure of overall agreement (Congalton *et al.*, 1983). A more commonly used approach is simply to compare the kappa values directly by deriving confidence intervals for kappa using the variance of kappa (σ_{κ}^2) estimated using the large sample variance (Congalton *et al.*, 1983). In this case the statistical significance of the difference between two κ values may be determined from

$$z = \frac{\kappa_1 - \kappa_2}{\sqrt{\sigma_{\kappa_1}^2 + \sigma_{\kappa_2}^2}}, \qquad (10.6)$$

where the difference is assumed to be significant at the 5% probability level ($\alpha = 0.05$) when $|z| > 1.96$ (which is the value cutting off the top 2.5% of the normal distribution, appropriate for a two-tailed test). As the use of the kappa statistic often adds rather little compared with the overall accuracy measure, some other approaches have been proposed. For example, a simple comparison of the proportions successfully classified in the two samples can be useful (Foody, 2009a), as in

$$z = \frac{|p_1 - p_2|}{\sqrt{\bar{p}(1-\bar{p})(1/n_1 + 1/n_2)}}, \qquad (10.7)$$

where p_1 and p_2, respectively, are the proportions successfully classified (the overall accuracy) for the two methods, n_1 and n_2 are the numbers of samples for each, and $\bar{p} = (x_1 + x_2)/(n_1 + n_2)$, where x_1 and x_2 represent the number of correctly classified cases.

It is worth noting that all of the above discussion implies that the reference classification is correct, yet in practice this will not usually be the case. Errors in the reference set can have quite substantial effects on results, both in terms of the apparent accuracy and in terms of the actual classification obtained (see Foody, 2009b).

10.6.2 Fuzzy classification and assessment

Unfortunately, it is rare that a hard classification is ideal because many mixed or indeterminate pixels are often found; this means that the conventional confusion matrix approach is not really appropriate. In such cases the error matrix approach may be extended to take account of this uncertainty by giving some weight to other possible classifications. One way to do this is to give each ground reference site a 'most likely' or 'correct' classification, together with any 'acceptable' alternatives. So, a particular area of mixed trees and shrubs might be classified as deciduous forest, but given the fact that it might also have perhaps 40% cover of shrubs, a class of shrubland could be considered an acceptable alternative for that site. This allows one to identify the number of acceptable alternatives in the non-diagonal cells of the error matrix; these can then be added to the totals of correct assignments to generate estimates of the 'fuzzy' producer's or user's accuracies (Congalton and Green, 2008). This type of fuzzy classification test gives higher estimates of accuracy than the corresponding deterministic test.

We have only discussed some of the many measures of accuracy here. There is no single measure that can be recommended for all situations; indeed it is normally suggested that users report a range of error measures with their results. The reader is referred to image-analysis texts such as Mather (2004) and Jensen (2005) or specific reviews such as Liu *et al.* (2007) and Foody (2009a) for further discussion and guidance on the choice of the most appropriate measure for any particular study.

10.6.3 Sources of error in reference data

Note that classification accuracy can only be assessed for those pixels for which we have a reference value – a subset of the total. We then assume that these pixels are randomly distributed in the image and that the same error rate applies to all pixels in the image. Strictly, therefore, the reference pixels should be chosen randomly. A common practice, however, is not to chose random pixels for collection of reference information, but to select clearly identified examples of each vegetation type as the reference pixels. This situation often arises where a similar selection procedure is used for the collection of both training pixels and reference pixels. This will almost inevitably lead to an overestimate of image-wide accuracy, particularly because there will be greater than expected error in the reference assignment for the less clearly defined (e.g. mixed) pixels in the image.

Note also that accuracy assessment should be undertaken using a completely independent set of reference pixels from those that were used as training pixels for the initial classification. Where there is overlap between data used for training the classification and for accuracy assessment this leads to statistical bias in the results favouring a rather overoptimistic estimate of true classification accuracy. This could result, for example, where the training dataset may have omitted to sample one of the vegetation types present. In such a case, use of the training set for accuracy assessment will also lead to an overestimate of true image-wide accuracy. Therefore these two processes need to be completely independent.

A wide range of other important considerations include: positional errors of reference and of image registration, while the minimum mapping unit size can also be important. These might lead to an underestimate of true accuracy. Problems also arise if the reference data were collected at a different time or date from the remotely sensed image.

Further reading

There are many suitable introductory texts on statistics, though specially recommended simple introductions to biological statistics are those by Fowler and Cohen (1990) and Clewer and Scarisbrick (2001). A particularly useful introductory text that provides simple formulae and clear approaches to performing the most important statistical texts using commonly available statistical software is that by Dytham (2003), while the book by Foody and Atkinson (2002) on 'Uncertainty in remote sensing and GIS' is a particularly useful compendium of articles relevant to the analysis of uncertainty in spatial data. For a more extensive treatment of statistical analysis techniques and software, the reader is referred to appropriate statistical textbooks (e.g. Snedecor and Cochrane, 1999; Sokal and Rohlf, 1995; Zar, 1999).

A useful introduction to field methods in remote sensing is that by McCoy (2004). The topic of geocomputation which addresses the interactions between remote sensing and spatially distributed modelling of land surfaces was the subject of a conference held in 2003. The contributions of the invited participants have been collected together in a book entitled *GeoDynamics*, edited by Atkinson *et al.* (2004).

Websites

http://itl.nist.gov/div898/handbook/mpc/section5/mpc552.htm

Sample problems

10.1 The confusion matrix shown below has been produced from an assessment of landcover classification derived from satellite imagery. Calculate the following from the information given in the matrix: (a) the overall accuracy of classification, (b) the class that shows the most errors of commission, and (c) the class (or classes) that was most correctly classified, justifying your answer.

Predicted class

Actual class	Class 1	Class 2	Class 3	Class 4	Class 5
Class 1	84	17	2	0	5
Class 2	5	26	8	1	1
Class 3	7	2	35	0	9
Class 4	0	3	6	102	1
Class 5	1	0	4	0	45

10.2 For a ground-validation exercise, estimates of *NDVI* were obtained for groups of 1 m^2 quadrats in different zones of a field. The between replicate variance was 0.06, while the between zone variance is 0.12. What is the variance of the mean for ten replicates in eight zones? If the cost of moving to different zones is 10x the cost of obtaining replicate samples, how does the optimum ratio between the number of replicates and the number of zones vary?

10.3 For a study of vegetation indices using a sample where the mean and standard deviation of ρ_R is 0.3 and 0.2, while the corresponding values for ρ_N are 0.5 and 0.1. Estimate the standard error of (a) the *DVI*, and (b) the *NDVI*, assuming that the variances of ρ_N and ρ_R are uncorrelated?

11 Integrated applications

11.1 Introduction

It would not be possible to cover fully the vast range of current and potential applications, or the many thousands of research articles published, where remote sensing is a significant tool in the study of plant canopies. We have therefore decided to concentrate on a small number of representative examples to illustrate the principles involved and the ways in which different remote sensing approaches can be combined with each other and with supplementary information to answer questions relating to vegetation structure and function. We include studies ranging from those applied at the leaf, plant or canopy level in the field, through airborne sensing to studies at field to regional scales where satellite sensing is appropriate. There is no intention to be comprehensive in the topics we cover, or the examples given; rather we aim to introduce a good cross-section of topics to illustrate the range of approaches that can be applied. Our selection of topics has been somewhat subjective and we recognize that several important areas have been omitted including for example, the many studies where remote sensing has been used as a tool in studies of global CO_2 balance and climate change, though the use of remote sensing in the estimation of mass and energy fluxes at large scales has been addressed in Chapter 9.

We start with a general section on methods for detection of plant 'stress' that may be applicable to most of the topics that follow, and then go on to specific example applications where each section starts with an introduction to the topic and its importance, the variables that need to be measured, and the available RS techniques. We then indicate the strengths and weaknesses of remote sensing for this application.

11.2 Detection and diagnosis of plant 'stress'

One of the most popular uses of remote sensing is in the diagnosis and monitoring of plant responses to environmental stress, with many hundreds of recent publications on the topic. It is worth noting that most techniques available monitor the plant *response* to environmental stress, rather than the environmental *stress* itself. For example, with water deficit, it is common to estimate changes in canopy cover (through use of a vegetation index) or stomatal closure (through some thermal stress index) as measures of 'stress', but both of these measures are more correctly indicators of the plant response, rather than of the stress itself. In only a few cases, such as the use of microwaves to estimate soil moisture or even the derivation of soil-moisture balances using combinations of meteorological and remote-sensing data with water-budget models, do we measure the actual stress being imposed on the vegetation.

Remote monitoring of stress responses of vegetation is particularly important in precision agriculture (Section 11.3) where managers frequently need to diagnose and quantify plant responses to stress with the aim of optimising management responses such as fertilizer application or irrigation. Both 'in-field' sensors, such as tractor-mounted instruments, and airborne or satellite-based instruments, have potential for such agricultural applications. Similarly, researchers concerned with natural ecosystems are frequently concerned to monitor the effects of both natural changes (e.g. climate change) and man-made changes (such as deforestation or pollution) on the health and performance of the vegetation in natural ecosystems (Section 11.4).

Unfortunately, many different environmental stresses can give rise to similar plant responses: for example water deficits, salinity, nutrient deficiency, and biotic stresses all tend to reduce leaf area, while many stresses lead to stomatal closure (Fig. 11.1). It is therefore frequently difficult to diagnose or to monitor the impact of a particular stress from only one observed response. Fortunately, we have a number of remote-sensing technologies, introduced in earlier chapters that respond to different features of the plant (e.g. temperature, structure, spectral reflectance, or water content), so it is theoretically possible to improve discrimination by the combined application of several such techniques. A range of other tools that are currently only applicable at a laboratory scale such as microscopy, thermo-luminescence, and magnetic resonance imaging will not be discussed here.

Figure 11.1 shows the complex relationships between some of the most important plant stresses, the primary plant responses, and the various remote sensing technologies available for detection. In contrast to the complexity of relationships between stresses and responses, most sensors study a limited range of responses. Unfortunately, however, the data from some sensors are frequently over-interpreted with a rather weak empirical linkage to the primary stress of interest. For example, changes in particular spectral *VI*s (often primarily related to changes in *fAPAR*) have been used to quantify stresses as different as drought, salinity, pollutant damage, disease, or pest infestation. Such empirical relationships must always be used with great care as the correlations obtained usually only hold over a very limited range of conditions, as we saw in Chapter 7.

Thus far, the vast majority of attempts to detect stress have concentrated on the relatively easy detection of changes in (i) *LAI* or (ii) chlorophyll content that are both characteristic of many stresses. These are both primarily estimated by the use of broadband,

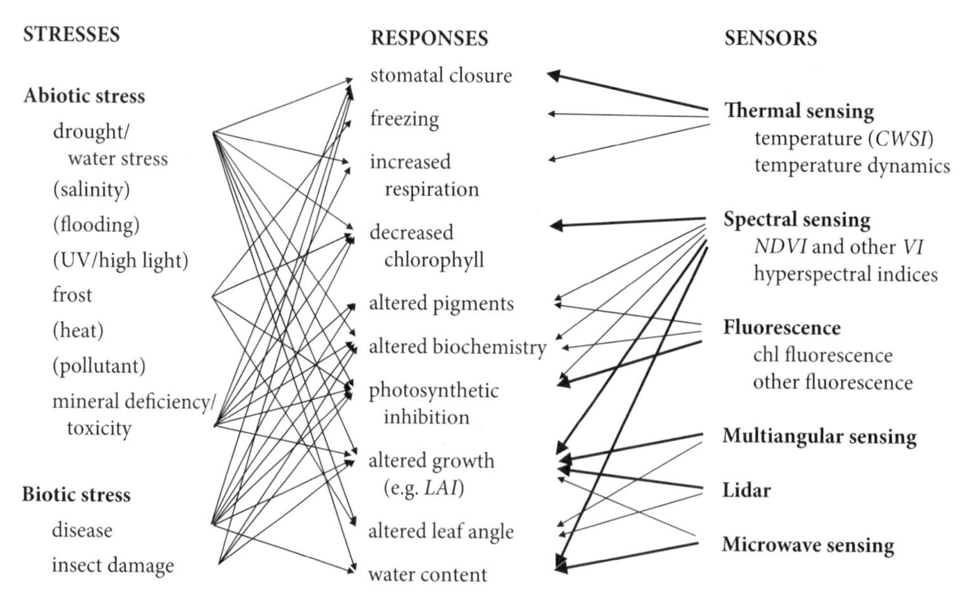

Fig. 11.1. Some of the relationships between primary stresses, the intermediate plant responses, and their detection by different remote sensing technologies. For clarity only some of the stress effects are shown (omitted ones are in parentheses). The most important sensing opportunities are indicated by thicker lines.

and sometimes hyperspectral, vegetation indices as outlined in Chapter 7, though thermally based stress indices as outlined in Chapter 9, which can detect both changes in *LAI* and stomatal closure, have also been widely used. For close-range studies in the field other approaches such as the use of fluorescence are increasingly being used. The complex response network indicated in Fig. 11.1 suggests that a robust tool for discriminating the causal stress in any situation might be developed by using the fact that each stress induces a characteristic suite of responses. Therefore a multisensor approach has particular potential for diagnosis and quantification of specific stresses. In what follows we outline the capability of each sensing technique individually before moving on to a discussion of multisensor approaches.

11.2.1 Spectral reflectance imaging and use of spectral vegetation indices

The main ways in which information on vegetation properties can be extracted from spectral data have been introduced in Chapter 7. As we have seen, spectral signatures can provide information not only on the pigment composition of the leaves but also on the leaf area, or even water content, of the canopy. All these properties can be of value for the detection and quantification of stress responses. The diagnostic power of broadband sensors, however, is rather limited, being most useful for following changes in leaf-area index; hyperspectral sensors, however, can potentially detect quite small changes in particular biochemicals and hence the rather subtle changes that are often characteristic of the early effects of stress. The development of new hyperspectral reflectance indices is a particularly active field of current research, and we will not attempt to cover all the indices that have been proposed.

Although hyperspectral data can provide quite powerful discrimination of both species and biochemical content when used in the laboratory on single leaves, the power of the approach tends to degrade on scaling up to the viewing of plants or canopies in the field, with even further degradation with more remote sensors. The reasons underlying this degradation include: interference by other components in the field of view (soil, other species, stems and branches, dead leaves and flowers), the effects of reflection and absorption with multiple scattering in the canopy, *BRDF* effects, and also atmospheric transmission and scattering effects. A particular problem is that many of the proposed spectral indices have been derived by empirical regression techniques for a specific experimental situation; it is not often that such empirically derived indices are capable of wide extrapolation For example the wavelengths selected as critical in one study (e.g. Carter, 1994), tend not to be the best when tried with another species or even the same species under slightly different conditions. It is always necessary to validate carefully any index that one plans to use on one's own experimental system.

In spite of these problems, spectral reflectance indices have been widely used for the study of abiotic and biotic stresses and a small subset of particularly useful indices have been listed in Box 7.1. The most successful studies have been at the leaf scale (where changes in the leaf-area index are not the main source of variation) and the main stress response is related to reductions in leaf chlorophyll content as the leaf yellows (Blackburn, 2007a; Carter and Knapp, 2001; Nilsson, 1995). This change can be detected using conventional broadband indices or by hyperspectral indices such as the red-edge position. Other particularly useful indices are based on physiological understanding (e.g. *PRI* was developed using an understanding of photosynthetic limitations). Interestingly the change of reflectance in the spectral range coinciding with the chlorophyll absorption maximum in the far red (around 680 nm), can only be observed at very low chlorophyll content (Buschmann and Nagel, 1993) because of the very quick saturation of the signal. In many studies, the highest increase in reflectance with decreasing chlorophyll as a result of stress occurred at wavelengths ranging from 696 to 718 nm. However, as this peak is very sharp it requires very narrowband data, so it is often more useful to use the broader peaks at around 600 nm.

There are relatively few studies that target potentially more specific pigment changes in response to stress, such as changes in carotenoids or xanthophylls (other than those that use *PRI* to study changes in photosynthetic activity – Section 9.7.2). Nevertheless, approaches such as wavelet decomposition of hyperspectral data (Blackburn, 2007b) appear to provide good opportunities for identification of specific pigments. If specific pigment changes can be associated with specific stresses this could become a useful approach.

Although average spectral responses for whole leaves or canopies can provide much useful information about stress responses, much greater power for discriminating between stress responses is available if we take account of the distribution of the colour across the leaf. Many diseases and mineral deficiencies or toxicities show characteristic colour patterns. For example, while nitrogen deficiency leads to a general loss of chlorophyll (and hence yellowing of the leaf), other mineral deficiencies lead to characteristic patterns: zinc deficiency results in interveinal browning, magnesium deficiency gives yellowing only in the interveinal areas, while sulphur deficiency leads to purpling of the veins (Fig. 11.2, Plate 11.1). A good summary of typical deficiency symptoms in tomato may be found in the web companion to Taiz and Zeiger (2006) while deficiency symptoms for citrus are well shown in Futch and Tucker (2001). Many other images may be found using a web search.

Similarly, many plant diseases are conventionally diagnosed from their characteristic patterns of lesions with often clear distribution of colour changes either near to, or remote from, the veins or the leaf edges, combined with colour changes that may be diagnostic for specific diseases (Fig. 11.3). Unfortunately, it requires a skilled observer to diagnose disorders from the visual patterning of leaves; it would be of particular interest to develop automated approaches to such diagnosis through image analysis or decision-support systems. The development of automated diagnosis or other expert systems is still at a rudimentary stage, though the first steps in automated segmentation of images of diseased plant tissue have been made (e.g. Moshou *et al.*, 2005); Fig. 11.3; Plate 11.2). It is still necessary to find ways to automate the identification and quantification of specific diseases from the patterning observed, while the scaling up to airborne or satellite remote sensing is not currently feasible.

In spite of the difficulties, the monitoring of the distribution of areas of poor canopy growth or canopy yellowing using airborne sensing of *NDVI* can often be useful as an indicator of pest or disease infestation, especially where the damage can readily be ascribed on the basis of ground knowledge to a specific cause (e.g. the spread of *Phytophthora* root rot in raspberry plots, or pest outbreaks in forest areas). For example, remote detection of defoliation using *NDVI* has been widely used since early studies on gypsy moth in

Fig. 11.2. Some examples of characteristic patterns of leaf colouration in response to different mineral deficiencies in tomato (from topic 5.1 of the web companion to Taiz and Zeiger (2006) at http://4e.plantphys.net/article.php?ch=t&id=289). (See Plate 11.1)

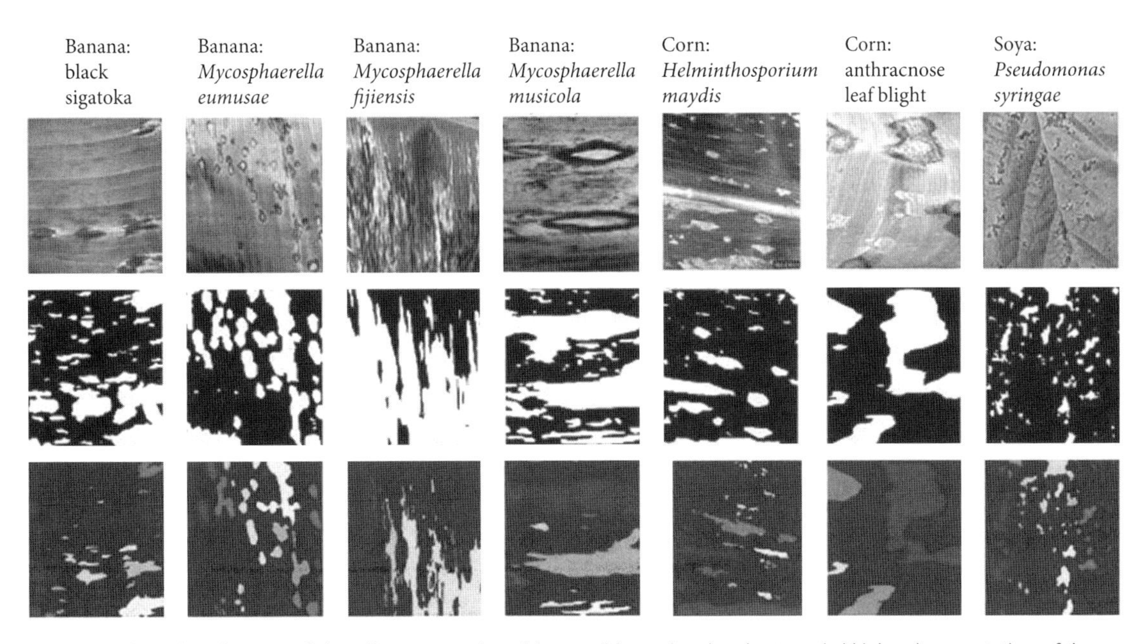

Fig. 11.3. Illustration of a range of plant diseases, together with manual (second row) and automatic (third row) segmentations of the images as a basis for automatic diagnosis systems (images from Camargo and Smith, 2009, with permission). (See Plate 11.2)

North America in the 1970s using Landsat MSS data (see Nelson, 1983) through to many recent studies such as those on outbreaks of cyclic geometrid moth on birch in Scandinavia (Jepsen *et al.*, 2009). In such cases the general mapping of infestation spread over large areas can be a useful tool for monitoring spread and aiding ground-based management or even for investigating impacts of climate change. In practice, the dominant use of spectral indices at the satellite scale has been based on the sensing of changes in either leaf chlorophyll content or *LAI* through a spectral vegetation index. Although many such indices have been tested, sometimes with success for a specific situation, they are still only detecting changes in chlorophyll or *LAI* rather than the specific disease. The potentially greater discriminant ability of hyperspectral indices, especially in combination with radiation-transfer models coupled with biochemical models (e.g. PROSAIL – see Chapter 7) offers increased scope for detection of biochemical changes associated with specific disease at a canopy scale.

Although many spectral indices have been correlated with water stress and the associated physiological responses, the relationship is usually empirical and rather indirect (acting largely through changes in *LAI*), except in the case of the various water indices (Box 7.1). Those indices that respond primarily to leaf area are less appropriate at the leaf-scale than when used at the canopy or airborne/satellite scales where the changing fraction of f_{veg} is more apparent. In general, one might expect mechanistically based approaches based on the inversion of RT models for quantification of canopy structure and biochemical content to be more robust than empirical stress indices, though they may be less easy to apply in practice.

11.2.2 Thermal sensing of water-deficit stress

As we saw in Chapter 9, thermal sensing is particularly useful for the study of plant–water relations and drought stress responses because of the key role of stomatal closure in the control of leaf temperature. In rare cases leaf temperature may be affected by other physiological processes: for example the heat generated (the exotherm) as water in a leaf freezes can be readily imaged (e.g. Wisniewski *et al.*, 1997), while in extreme cases of particularly high respiratory rates (e.g. in the *Arum* spadix) raised temperatures can be used as a measure of these increased respiration rates (Seymour, 1999). In most cases, however, the heat generated by respiration is too small in quantity to have a detectable effect on leaf temperature.

Practical aspects of the use of thermal stress indices

The basic theory underlying the use of thermal stress indices was outlined in Section 9.6; here we consider aspects relating to their practical application. Because of the sensitivity of absolute temperature to environmental conditions, the most robust approaches use reference surfaces. A particular advantage of using reference surfaces is that equations such as eqns (9.23) and (9.24) are not sensitive to absolute errors in temperature measurement, only to relative accuracy. This is important as many thermal cameras have a specified accuracy of only around ±2 °C (though they may have much better repeatability), and airborne or satellite data are subject to large absolute errors (resulting from emissivity errors and failures in atmospheric correction). In all cases where reference surfaces are used, it is essential that, (i) the radiative properties of the reference are similar to the real leaves, (ii) the boundary-layer properties are similar, and (iii) their orientation in relation to the incident solar radiation is similar to that of the canopy being studied. For canopies where a significant proportion of leaves are in shade, the average canopy illumination needs to be mimicked by the reference surface; otherwise calculated indices may go to unrealistic values. Further discussion on problems and precautions appropriate for the use of thermal stress sensing may be found elsewhere (Jones, 2004a), while applications to irrigation control are discussed below (Section 11.3.3).

The most critical aspect of using stress indices is the need to ensure that the temperature used in calculation of the index is that of the vegetation component rather than of the background soil (which is usually much hotter). Approaches to this depend on the scale of the images being used. Where the pixels are small in relation to the individual leaves, as in much close-range in-field imaging, it can be sufficient to select just those pixels that are pure vegetation, and exclude pixels that include soil. This can be achieved by coincident use of overlaid RGB or R/NIR images to identify plant, or else it is possible to use a thresholding approach to exclude the generally hotter soil pixels. Where pixels are large in relation to the units of vegetation, as is common with satellite images, approaches to pixel unmixing to estimate the temperatures of the soil and vegetation components separately are required. These use eqn (9.16) to estimate T_v from an estimated f_{veg} in the pixel. This requires either a separate estimate of the T_{soil} from reference or extreme pixels, or else T_v and T_{soil} can both be estimated from values for a number of pixels if they are assumed to be constant over a small area with only f_{veg} varying, by estimating the best fit values from eqn (9.16) (e.g. using 'Solver' in Microsoft Excel). Multiangular data can also be inverted to estimate the soil and vegetation temperatures separately (e.g. Jia *et al.*, 2003), or even the temperatures of shaded and sunlit areas of canopy and soil (Timmermans *et al.*, 2009) where data from enough view angles are available (see Box 11.1).

Crop phenotyping

An important application of thermal imaging is the identification of genotypes with advantageous characters for growing in dry conditions. The idea is that selection of lines with responsive stomata can be a route into the development of cultivars capable of tolerating drought environments where water deficits are short and sharp. On the other hand, genotypes that constitutively have low stomatal conductances or that are unresponsive to imposed drought, may be more appropriate for terminal stress environments such as Mediterranean climates. Thermal imaging has for a long time been used to select stomatal function mutants in the laboratory (Raskin and Ladyman, 1988); only recently has it become feasible to apply the technique at the field-plot scale, where it has proved possible to identify a number of genetic markers known as quantitative trait loci (QTLs) in rice germlasm (Fig. 11.4; Jones *et al.*, 2009).

11.2.3 Fluorescence and fluorescence imaging

A powerful tool for studying physiological responses to stress uses fluorescence emission (see Section 3.1.1). Section 9.6.3 outlined the uses that can be made of chlorophyll *a* fluorescence (primarily in the red at around 690 nm and the near infrared at around 740 nm); this provides a powerful probe for photosynthesis, and hence for stresses that affect photosynthesis. Other wavebands involved in fluorescence emission from a green leaf when excited by UV-A radiation are in the blue at 440 nm and in the green at 520 nm. These shorter-wavelength

BOX 11.1 Estimation of canopy and soil temperatures from multiangular thermal data

It is possible to estimate the temperatures of all the canopy components (sunlit leaves – T_{sl}; shaded leaves – T_{shl}; sunlit soil – T_{ss}; and shaded soil – T_{shs}) by combining information from a canopy radiation-transfer model, together with estimates made from different zenith view angles (θ) of both fractional vegetation cover as obtained from a conventional R/NIR vegetation index and observed scene temperature (T_θ).

This makes use of the following relationship:

where $f_{veg(\theta)}$ is the fraction of vegetation viewed at angle (θ) (estimated from a *VI*); $f_{sl(\theta)}$ is the fraction of viewed leaves that are sunlit (estimated from radiation-transfer model); and $f_{ss(\theta)}$ is the fraction of viewed soil that is sunlit (also estimated from the radiative-transfer model). Where one has data from multiple view angles it becomes possible to solve the series of equations for the component temperatures.

$$T_\theta = f_{veg(\theta)}\,[f_{sl(\theta)}\,T_{sl(\theta)} + (1-f_{sl(\theta)})T_{shl(\theta)}] + (1-f_{veg(\theta)})\,[f_{ss(\theta)}\,T_{ss(\theta)} + (1-f_{ss(\theta)})\,T_{shs(\theta)}],$$

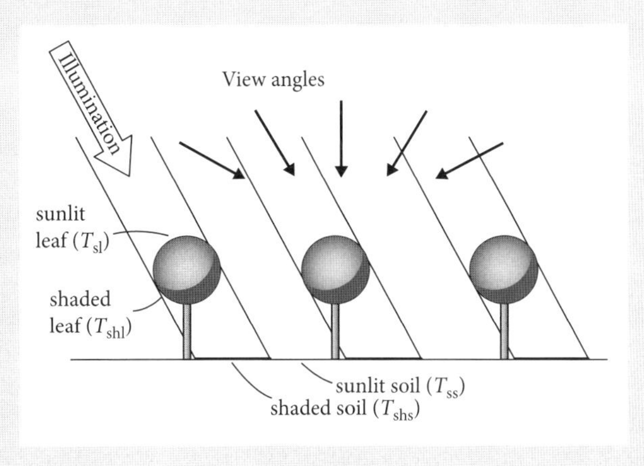

emissions are predominantly associated with fluorescence from a range of primarily phenolic compounds in the leaves, especially from ferulic and chlorogenic acid bound to the cell walls (e.g. Buschmann *et al.*, 2000; Lichtenthaler and Miehé, 1997). The fluorescence intensity at any wavelength is influenced not only by the illumination and detection geometry but also by factors such as the concentration of the emitting substance, the internal optics of the leaf (including factors that affect the partial reabsorption of the fluorescence), and especially for photosynthesis-related fluorescence, the energy partitioning between the photosystems and the energy-quenching processes in the chloroplast. It follows that the absolute intensity of fluorescence at any time is a relatively unhelpful signal, so that it is usual to do some normalization of the data, for example by deriving image analogues of the classic chlorophyll fluorescence parameters (Box 9.1) or by taking ratios between fluorescence signals at different wavelengths.

Many fluorescence imaging systems are now becoming available for work on single leaves or plants, not only for chlorophyll fluorescence (Nedbal *et al.*, 2000), but also for multicolour fluorescence (Buschmann *et al.*, 2000; Lichtenthaler and Miehé, 1997). Though most imaging systems are still confined to the laboratory, some fluorimeters such as the PAM (Heinz Walz GMBH, 91090 Effeltrich, Germany; http://www.walz.com/) and the Multiplex® (Force-A, Centre Universitaire Paris Sud, Orsay, France; www.force-a.fr) are particularly adapted to field measurements. Fluorescence systems have become particularly useful for the study of abiotic and biotic stresses, as both multicolour and chlorophyll fluorescence data have been used to demonstrate and distinguish the early stages of infection by fungi, viruses and bacteria before symptoms are visible in standard reflectance images (Chaerle *et al.*, 2007; Chaerle *et al.*, 2006; Rodríguez-Moreno *et al.*, 2008) and for the diagnosis of the nutrient status of

Fig. 11.4. Plant phenotyping at the plot scale showing possibilities of the approach when taking images from a height of c. 5 m at an angle to minimize detection of background soil. (a) Visible image, (b) corresponding thermal images (Jones *et al.*, 2009). (See Plate 11.3)

leaves (e.g. Langsdorf *et al.*, 2000). The most convenient method for normalization of multicolour fluorescence images is by the calculation of ratio images of the fluorescence intensities measured at different wavelengths (e.g. F_{440}/F_{690} or F_{440}/F_{520}). Empirical studies have shown that different stresses affect each of these fluorescence ratios in different ways (Table 11.1; Fig. 11.5 and Plate 11.4) so that fluorescence can be a useful component of a diagnostic system (Campbell *et al.*, 2007).

It is also possible to use fluorescence to estimate the content of many leaf compounds that are characteristic indicators of stress in plant leaves such as flavonols and anthocyanins as illustrated in Box 11.2. By using different illumination wavelengths in the UV and visible that are differentially absorbed by epidermal constituents such as flavonols and anthocyanins and measuring the resulting emitted chlorophyll fluorescence (e.g. at 740 nm) it is possible to quantify the epidermal content of such 'protective' compounds (Bilger *et al.*, 1997; Goulas *et al.*, 2004). The fluorescence emission ratio (FER), described in Box 11.2, therefore provides a good measure of flavonol concentration in the epidermis, which is a good indicator of the development

Table 11.1 Changes in the following fluorescence ratios blue/red (F_{440}/F_{690}), blue/far-red (F_{440}/F_{740}), red/far-red (F_{690}/F_{740}), and blue/green (F_{440}/F_{520}) in response to various developmental stages or environmental stresses (after Buschmann *et al.*, 2000). Responses range from a strong positive response (++), through no response (0), to a strong negative response (−−).

	F_{440}/F_{690}	F_{440}/F_{740}	F_{690}/F_{740}	F_{440}/F_{520}
Stages of leaf development				
Variegated vs. green leaf	++	++	++	0
Lower vs. upper surface	++	++	+	0
Yellow-green vs. green leaf	+	++	++	++
2nd flush vs. 1st flush leaf	−−	−−	++	−
Stress and strain				
Water deficit	++	++	0	0
N-deficiency	++	++	+	0
Sun exposure	++	++	+	−−
Mite attack	++	++	0	+
Inhibition of photosynthesis				
Heat treatment	−−	−−	0	−
UV-A treatment	−−	−−	0	−
Diuron treatment	−−	−−	+	0
Photoinhibition	++	++	−−	0

of photoprotection in response, for example, to high light stress.

Unfortunately, as we have seen, practical fluorescence techniques are still only available for stress detection at the leaf to plant scale, though extension to greater distances is the subject of active research, especially in relation to the analysis of photosynthetic responses to stress (Section 9.7.3).

11.2.4 Multiangular sensing, 3D imaging, and Lidar

Canopy structural information can be usefully determined by the use of multiangular imaging and data inversion (see Chapter 8) and from Lidar (both ground-based and airborne). Ground-based stereoscopic cameras and also the relatively new approach involving 'time-of-flight' cameras can also provide 3D canopy information. Not only can such data provide information on *LAI* and its height profile, but it can also potentially provide information on leaf angle variation, which itself may be an indicator of wilting as a result of water-deficit stress.

11.2.5 Multisensor imaging for diagnosis

Because the limited range of intermediary responses indicated by any individual imaging sensor can arise from multiple causes (Fig. 11.1), our ability to diagnose

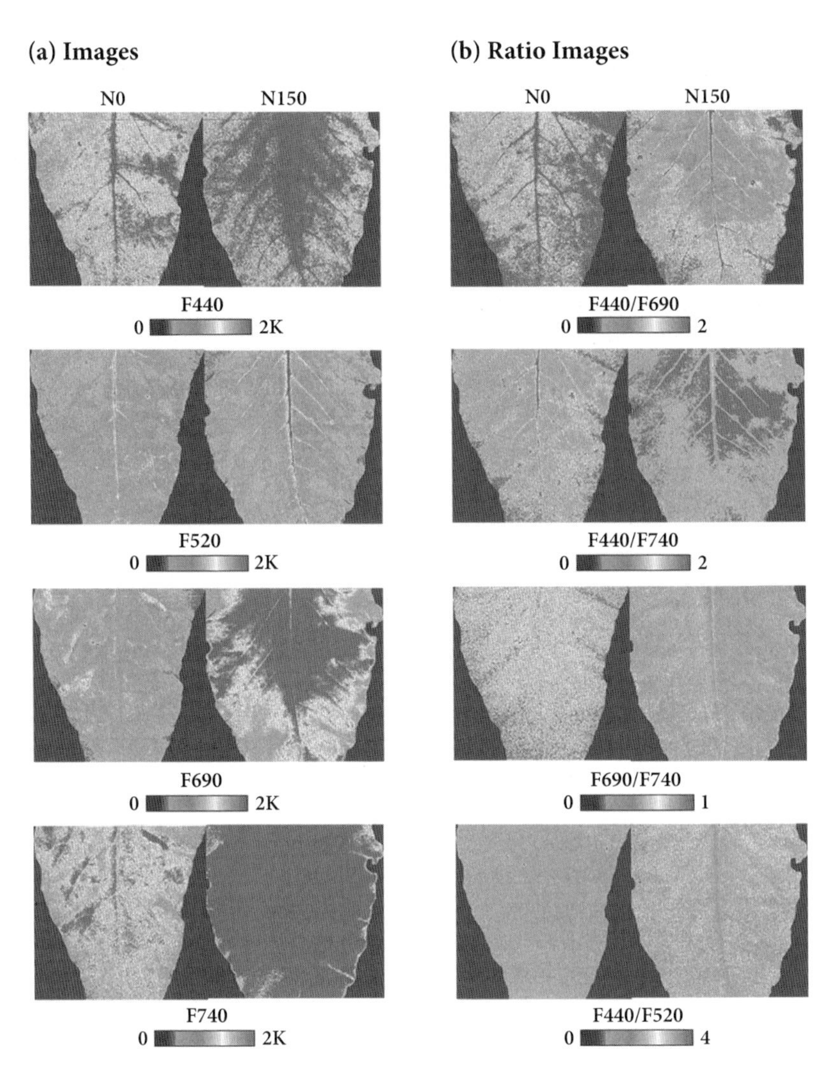

Fig. 11.5. Fluorescence imaging of single leaves from sugar beet plants grown with either no nitrogen supplied (N0) or a luxury supply (N150) showing the effects on UV-A stimulated fluorescence at four wavelengths (440, 520, 690, and 740 nm, and ratios between them presented in false colour with fluorescence intensity increasing from blue to red (with kind permission from Langsdorf *et al.*, 2000). (See Plate 11.4)

the particular primary stress is greatly enhanced by the combination of two or more sensor technologies. For example thermal imaging responds primarily to changes in evaporation rate, which are generally caused by changes in stomatal aperture, but stomatal closure can be a result of stresses as different as drought, flooding, salinity stress, fungal infection, or pollutants. In order to distinguish between these possible causes one needs further information. Multisensor approaches can range from a simple combination of, say, thermal and reflectance sensors, or visible reflectance and fluorescence sensors, through to combined fluorescence, reflectance, and thermal imaging sensors (Chaerle *et al.*, 2007). Possible relationships that operate at the leaf scale are summarized in Table 11.2.

There is increasing interest in the development of such diagnostic tools that include a range of sensors, with the Multiplex® system mentioned above, for example, routinely providing data on a number of stress-related fluorescence parameters. Not only is the multisensor approach appropriate at the leaf and in-field scale, but it is equally appropriate at coarser remote sensing scales. The widespread combination of thermal data with spectral vegetation indices for detection of vegetation stress (see below) is an example where two sensing systems can be combined with advantage. The approach can be taken further by incorporating radiative-transfer modelling and the use of multiangular sensing.

BOX 11.2 Use of chlorophyll fluorescence emission to quantify epidermal screening compounds

Accumulation of UV-absorbing phenolic compounds such as flavonols in the leaf epidermis screens out UV radiation and prevents its penetration into the leaf. The light that gets through to the chloroplasts then stimulates chlorophyll fluorescence, with the amount of fluorescence depending on the epidermal transmission of the exciting radiation. This effect provides a useful tool to quantify epidermal screening compounds by comparing the emitted fluorescence in response to modulated illumination at an absorbed wavelength (e.g. UV-A) with fluorescence emitted in response to illumination at a reference wavelength where the epidermis is transparent (e.g. blue-green or red). Figure (a) shows sample spectra for two epidermal phenolics (quercetin and coumarin) that differentially absorb in the UV, together with spectra for chlorophyll (data from Bidel *et al.*, 2007; Goulas *et al.*, 2004). Figure (b) illustrates the differential effect of UV and red light on chlorophyll fluorescence from leaves where the fluorescence emission ratio ($FER = F_R/F_{UV}$) is a measure of epidermal absorptance in the UV. Figure (c) shows some typical excitation wavelengths that can be used to distinguish compounds such as quercitin and coumarin that have different absorptances in the UV-A and UV-B regions; this also shows a typical chlorophyll fluorescence spectrum for an intact leaf. The log of the *FER* is expected from Beer's law (Chapter 2) to be linearly related to the absorbance and hence can often be closely related to the concentration of the main absorbing compounds (Fig. d) (data from Agati *et al.*, 2008).

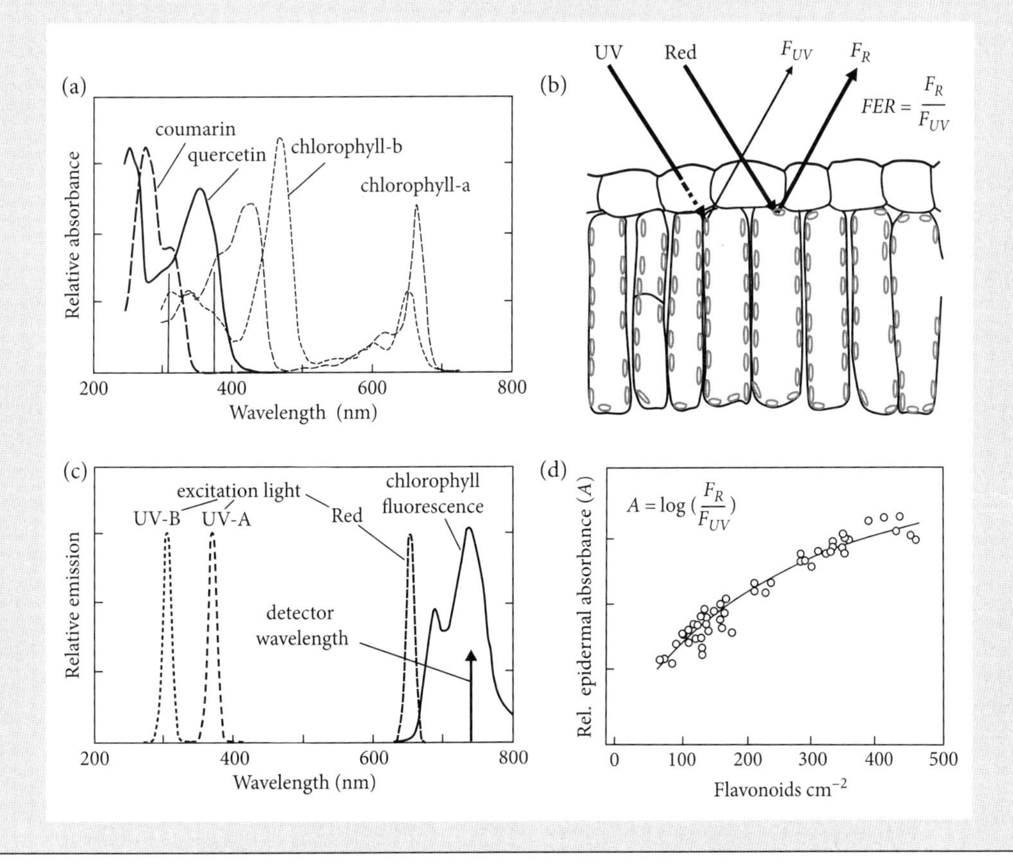

Table 11.2 An abstract of a table outlining the possible use of multisensor imaging for the diagnosis of different stresses, outlining the major responses to a number of stresses (contents from Chaerle, L., Leinonen, I., Lenk, S., Van Der Straeten, D., Jones, H.G. and Buschmann, C. unpublished).

Stress type	Thermography	Reflectance	Fluorescence
Abiotic stresses			
Water stress	Temperature rise (primary response – especially in 'isohydric plants')[1]	Leaf-angle distribution changes – sensed by multiangular sensing;[2] increase of reflectance[3]	Increase in blue-green fluorescence, decrease of Chl-F[4], decrease of variable Chl-F[5] or photochemical yield[5]
Nitrogen deficiency	Tendency to rise (but may relate to the reduced leaf area effect)[6]	Detectable by increasing yellow colour of leaves;[6] increasing reflectance at visible wavelengths, especially green and red;[7] specific patterns of colour change	Higher blue-green fluorescence and higher Chl-F at 690 nm[8]
Gaseous pollutants (NO_2, SO_2, O_3)	Rise (result of stomatal closure with often increased stomatal heterogeneity) [9]	Increase of reflectance in green and red regions[3]	Decrease of maximum quantum yield of PS II Fv/Fm[10]
Biotic stresses			
Fungal infection	Temperature decrease[12] temperature increase[13]	Increase of reflectance in red region and SWIR regions; specific patterns of colour change	Increase of Chl-F;[11,12] decrease of variable Chl-F[13]; decrease of PS II-efficiency
Viral infection	TMV: Initial temperature rise, followed by fall as cell death occurs[14]	Specific patterns of colour change	Increase of Chl-F and blue-green fluorescence[14] and variation in variable Chl-F[15]
Bacterial infection	Pectobacterium elicitor: pre-symptomatic temperature fall[16]	Specific patterns of colour change	Higher quantum yield, corresponding with a reduction in steady-state Chl-F[17]

[1] Jones (2004); [2] Casa and Jones (2005); [3] Carter (1994); [4] Lichtenthaler and Miehé (1997); [5] Meyer and Genty (1999); [6] Nilsson (1995); [7] Carter and Knapp (2001); [8] Langsdorf *et al.* (2000); [9] Omasa (1981); [10] Gielen *et al.* (2006); [11] Chaerle *et al.* (2006); [12] Chaerle *et al.* (2007); [13] Meyer *et al.* (2001); [14] Buschmann *et al.* (2000); [15] Osmond *et al.* (1998); [16] Boccara *et al.* (2001); [17] Berger *et al.* (2004).

11.3 Precision agriculture and crop management

Agricultural management decisions are conventionally made on the basis of crop walking and a limited number of sample measurements (often from random samples – e.g. of soil nitrogen or soil moisture content from a few sensors, or from sample measurements of leaf nitrogen or leaf water status). A particularly significant change in agriculture recently has been the shift towards increased precision in crop management, or *site-specific management* (SSM; Pinter *et al.*, 2003). This targets the substantial spatial variability in soils and in crop performance, aiming to identify the variability and then to optimize the management for specific parts of the field in an appropriate manner, for example by only applying pesticide to areas of crop showing disease symptoms. The use of high-precision global positioning systems (GPS) and geographical information systems (GIS) allows the farmer to map precisely variation in characters such as yield, soil type, soil water holding capacity and so on. This information can then be used in precision crop management, for example by using variable-rate application technology. Assimilation of the remotely derived measurements into a crop model provides the most powerful approach to the derivation of management recommendations, possibly by incorporation into decision support systems.

Information sources can include ground-based sampling including the use of an elevated platform such as a 'cherry picker' (Möller *et al.*, 2007), tractor-mounted sampling, and remote sensing from helicopter (Lee *et al.*, 2007), aircraft or satellite, with all data merged into a single GIS-based description of the field in question. Clearly there is more scope for real-time modulation of inputs of fertilizer or pesticide as the tractor traverses the field than there is for local modulation of inputs such as irrigation where treatments are usually applied on a field-wide scale. Nevertheless, at least for high value crops, establishment of local irrigation zones is feasible, while linear move and centre-pivot irrigators may potentially be regulated on the move. It is already common for crop 'nitrogen' sensors to be mounted on tractors, thus allowing precisely targeted application of N.

In-field measurements may usefully be combined with satellite remote sensing, which is particularly well suited to 'inventory-type' measurements – that is static mapping of variables such as cropping area and its spatial distribution, or soil mapping, since these latter measurements are not usually particularly time sensitive and are therefore readily obtained at any time during the cropping season (whenever suitable cloud-free images occur). High spatial resolution images are required, however, to provide information at the scale of individual fields or of parcels within them. In many cases the high-resolution data required for SSM can be obtained using airborne imagery where pixels of <1 m are readily obtained, and much useful information on characteristics such as *LAI* or disease spread can be obtained by simple multispectral or even RGB cameras without the need for the much more expensive hyperspectral data.

In recent years there has been increased interest in the potential of remote sensing to provide real-time information that can be of use for crop management during the course of a season; here, the requirement is for more frequent imagery. Although much research has been concentrated on the development of quite sophisticated systems, for example for predicting irrigation or nitrogen requirement, perhaps the greatest opportunities for the practical uptake occur with simple systems that permit rapid and timely provision of the data to farmers. An example of such an approach to irrigation scheduling was provided by the DEMETER (http://ec.europa.eu/research/dossier/do220307/pdf/demeter_sci_results_en.pdf) and PLEIADeS (http://www.pleiades.es) projects that combined ground-based estimates of a reference ***E*** that could then be continually adjusted for each farmer's land by remote estimates of local crop coefficients simply derived from remote estimates of *NDVI*.

An important limitation of RS data in precision agricultural systems is the general reliance on empirical relationships between spectral *VI* and variables of interest (e.g. *LAI*, canopy nitrogen, or biomass; see Fig. 11.6). Though good relationships with *LAI* and with chlorophyll concentration are common, the relationship to downstream variables such as biomass and yield as shown in this figure tend to be less good unless restricted to a limited set of conditions, so local and seasonal calibration is generally required.

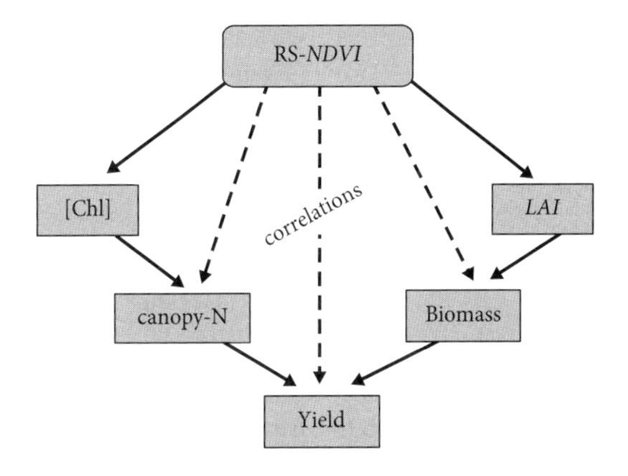

Fig. 11.6. An illustration of the uses of spectral indices such as *NDVI* in site-specific monitoring. The solid arrows indicate the more direct functional relationships, such as between the *VI* and *LAI* or chlorophyll content, while the dotted lines indicate the more indirect relationships used for the estimation of canopy nitrogen, biomass, or even yield from *VI*.

Ground-based or 'proximal' sensing

In spite of the opportunities for the use of satellite remote sensing in precision agriculture, most remote-sensing technology for SSM uses ground-based *proximal* sensing, with real-time monitoring using tractor-mounted sensors being particularly important. The main approaches to ground-based and remote (e.g. satellite) remote sensing of crop variables include (*a*) spectral information (Chapter 7), especially the use of simple or complex vegetation indices which can detect both variation in *LAI* and variation in crop status (e.g. chlorophyll or nitrogen content), (*b*) multi-angular sensing (Chapter 8) that provides further information on leaf area, leaf area distribution and leaf angle (the latter can be related to water status in those crops that wilt), though there are as yet few practical applications of this, (*c*) fluorescence sensing, for both photosynthetic function and biochemical constituents, (*d*) thermal sensing, which gives information on variation in evaporation rate and hence on crop water stress, (*e*) microwave sensing, which is particularly useful for detection of water availability, and (*f*) Lidar and even ultrasonic sensing which can potentially provide further information on canopy structure. A major difference between the 'in-field' and remote approaches, however, is the scope that in-field imaging provides for the analysis of variation within plants, avoiding the averaging that is inherent in larger pixels. Nevertheless, it should be noted that most current tractor-mounted sensors, such as the active *NDVI* sensors, and many hand-held sensors such as fluorescence sensors, are not imagers. Each reading just averages data over the field of view of the sensor, which is often several square m, then individual values over a whole field may be merged to provide whole-field maps or synthetic images.

11.3.1 Cropping inventories

Remote sensing is particularly well suited for estimating area and distribution of different crops and other components of land use and landcover. As well as its use for global mapping (Thenkabail *et al.*, 2009) there has been much interest in refining the landcover maps at smaller scales (see also Section 11.4 below). Although some information on landcover can be obtained by the use of standard vegetation indices, both spectrally based and texture-based classification techniques (Chapters 6 and 7) can be used to identify areas of each crop. Though it is generally easy to distinguish cropped and uncropped areas, when using only simple vegetation indices it is usually only possible to separate effectively broad vegetation types (e.g. bare soil, annual crops, and trees). It is also possible to develop a spectral library by the collection of many replicate samples of hyperspectral reflectance data for a range of likely crops and then using an appropriate classification method to assign observed data to the nearest member of the spectral library, though as spectral properties are time dependent, this is generally not a very reliable approach. Even when using such hyperspectral data it is still often difficult to distinguish the slight spectral differences between say wheat and barley, or between rice and sugar cane (Rao, 2008). Use of microwave sensing can complement spectral information; for example longer-wavelength microwaves that can penetrate the canopy are particularly useful for detecting flooded areas used for rice crops. Shorter-wavelength sensors (e.g. Ku-band scatterometer) that penetrate the crop only slightly can respond to surface characteristics such as appearance of the flowering head (heading). A further problem is that in general it has been found that the more complex the discriminator chosen, the more likely it is to be applicable only for the specific conditions under which it was derived.

Phenology

A particularly powerful tool for discriminating crops and even natural vegetation types is to use the characteristic differences in their seasonal development or phenology. Analysis of time series of satellite images provides a powerful tool for following phenological development of crops and of natural ecosystems. This is usually based on monitoring of the seasonal pattern of changes in *LAI* as measured by a vegetation index. The changes observed relate to the timing of germination and early growth in the spring (or leafing out in deciduous perennial crops) and senescence or leaf fall in the autumn. Because of the frequent data gaps (usually as a result of cloud cover) even for satellites such as MODIS that collect daily records it is usual to use 8- or 16-day composite products such as the Nadir *BRDF*-adjusted reflectance (NBARS) derived *EVI* for MODIS. Even when using a 16-day composite, it is possible to estimate key steps in phenology to within a few days by fitting a smoothed curve to the observed data (Zhang *et al.*, 2009). The time series of changing *NDVI* for even mixed scenes can potentially be decomposed into the component phenologies using Fourier analysis (Geerken, 2009) or wavelet analysis (Martínez and Gilabert, 2009); the latter having the advantage that it is not constrained to regular sinusoidal harmonics. The fact that different crops have different rates of phenological development during the season (e.g. different sowing or harvest dates), opens up a powerful approach to distinguish crop types based on a combination of spectral information with the use of multitemporal data for crop mapping. For example there have been a number of successful applications based on multi-pass *NDVI* or *EVI* from Landsat or MODIS (Fig. 11.7; Simmoneaux *et al.*, 2008; Wardlow *et al.*, 2007). A convenient way of using data on contrasting phenological cycles of *NDVI* for different vegetation types is simply to extract the *NDVIs* at key points in the annual cycle that emphasize differences; plotting three such *NDVIs* in the RGB channels will clearly allow one to distinguish crops of differing phenologies. More sophisticated classification can be achieved by classifying on the basis of the amplitudes and phases of the first few Fourier harmonics (Geerken, 2009). Classification of an area can then use the phases and amplitudes of the harmonics derived for separate pixels (Fig. 11.7(b)).

In addition to the use of characteristic phenological patterns of *VI* relating to sowing and growth to identify different species, it is also sometimes possible

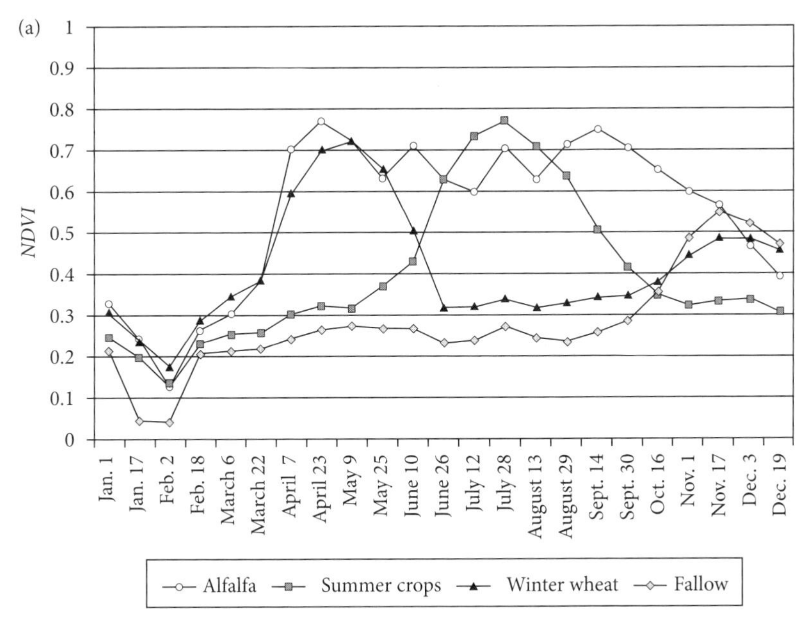

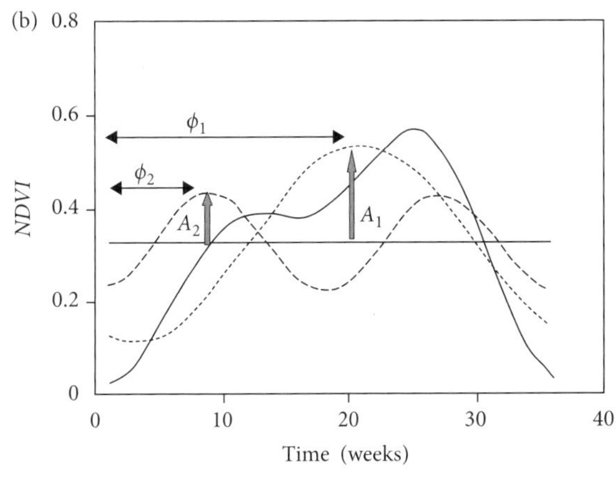

Fig. 11.7. (a) Crop patterns (from Wardlow *et al.*, 2007) showing that different crops can be distinguished by their characteristic temporal trends in *NDVI* without the need to make use of more detailed spectral differences. (b) An example of a smoothed seasonal phenological cycle of *NDVI* showing its decomposition into a mean and first- and second-order Fourier harmonics, each with an amplitude (A) and a phase (ϕ) (modified from Geerken, 2009).

to distinguish specific growth stages of a given crop even after the achievement of full ground cover. For example Pimstein *et al.* (2009) showed that heading in wheat was correlated with reflectance changes near the 1200 nm water absorption feature and developed a ratio-based *normalized heading index* ($NHI = (\rho_{1100} - \rho_{1200})/(\rho_{1100}+\rho_{1200})$) using the contrast with an adjacent insensitive band at 1100 nm. Exceeding a threshold value of $NHI >$ c. 0.18 was generally associated with heading; however, as crops senesced this rule became less reliable, so correction of the index by division by *NDVI* was proposed. This index was developed using ground data, but similar discrimination was reported using broadband satellite data, even on substituting $\rho_{845-890}$ for ρ_{1100} as the latter band was not available on the satellite used. In other cases, flowering can be detected as a change in spectral reflectance.

The number of studies where techniques such as subpixel classification, texture analysis, object detection, and time-series observations have been used for assessment of crop and vegetation distribution is increasing rapidly (Berberoglu and Akin, 2009; Vancutsem *et al.*, 2009; Verbeiren *et al.*, 2008).

11.3.2 Yield estimation/prediction

There has been extensive interest in the remote estimation of crop yields, and especially their prediction, using satellite remote sensing. For regional estimates of crop production, two steps are required: (i) the derivation of an area estimate for the appropriate crop from a landcover estimation routine, and (ii) the estimation of the yield per unit area of that crop. In the simplest cases only the land area information is used, but for accurate estimates of production the area needs to be modulated by an estimate of the yield per unit area. This can be achieved using a crop-growth model parameterized on the basis of characters such as leaf area and its seasonal development coupled with information on seasonal crop stresses and/or meteorological variables. Real-time refinement of the predictions can be achieved by updating key parameters such as *LAI* or water deficit on the basis of satellite-derived data.

Although the best results may be obtained with detailed crop models, the commonest and simplest approach to estimation of yield per unit area has been to use empirical relationships with instantaneous spectral vegetation indices (see Chapter 7). In general the *VI* tend to estimate ground cover as an indicator of *LAI*, which itself can indicate differences in productivity and hence yield. Particularly at higher *VI*s differences are also often assumed to reflect differences in chlorophyll content as a measure of crop nitrogen. Unfortunately, the relationships between *VI* and *LAI* can be quite variable depending on weather conditions, viewing geometry, and so on (see Section 7.2.1). Similarly, the relationships between *LAI* and biomass production, and between biomass production and final yield, are also subject to substantial error, varying with crop, location, year, and so on (Fig. 11.6). Therefore, these estimates tend to be rather unreliable, even though in a limited number of cases useful predictions have been obtained by this approach. It should be noted in particular that the relationship of *VI* to final yield tends to decline as the crop senesces and *NDVI* decreases. There have been attempts to improve estimates by use of an integrated *LAI* or green *LAI* from a series of measurements over the growing season, as this is fairly well associated with biomass productivity and hence crop yield.

As was pointed out in Section 9.7.1, the total biomass productivity of a crop (*NPP*) is closely related to the total amount of photosynthetically active radiation absorbed over the life cycle of that crop. We can therefore write eqn (9.25) in the following way

$$NPP = \int \boldsymbol{I}_S \, \varepsilon_c \, \varepsilon_f \, \varepsilon_A \, \varepsilon'_V \, dt, \quad (11.1)$$

where $\boldsymbol{I}_S$ is the global daily shortwave radiation at the surface, ε_c is the fraction of incident solar radiation that is in the photosynthetically active (PAR) wavelengths (≈ 0.48), ε_f is the average fraction of radiation intercepted by the canopy, ε_a is the fraction of intercepted radiation that is absorbed by the leaves, and ε'_V is the conversion efficiency of absorbed PAR into dry matter (DM). The value for ε_V' is often assumed to equal about 2.0 g MJ^{-1} for healthy crops. This value may be slightly lowered by water deficit or other stresses, but the major response of crops to stress is to reduce leaf area and hence the value of ε_f. Remote estimation of DM accumulation therefore depends on estimation of incident radiation (from knowledge of solar geometry and information on cloudiness and atmospheric transmissivity as may be obtained from remote sensing) and ε_f the fraction intercepted (from the *VI* as indicated in Chapter 7 with any necessary correction for the varying solar angle during the day). A standard value for ε_V' can usually be assumed with little error, though where direct estimates of water stress are available using thermal data, or of photosynthetic efficiency from *PRI*, it is possible to refine the estimates.

Improved estimates can be obtained by constraining the estimates using a priori knowledge, such as that obtained by the use of a crop growth model. An example of such an approach involves data assimilation is where attempts have been made to constrain a crop-growth model such as STICS by the use of RS data (Launay and Guérif, 2005) (see also Section 9.5.4).

11.3.3 Water management

Approaches to irrigation scheduling include control either (i) on the basis of soil-moisture measurement or estimation, or (ii) on the basis of monitoring the plant response to water deficits (e.g. Jones, 2004b). Although soil-monitoring methods are the more widely used, there is increasing interest in plant-based methods. RS can be used for both strategies. The application of RS in hydrological studies has been well reviewed (Schmugge *et al.*, 2002); here we just outline the main points. Some of the different direct and indirect measures of plant and soil water status and plant response (see Chapter 4) are listed in Table 11.3, together with the RS methods available for their quantification and their relationship to, or utility for, different types of application.

Soil-moisture-based methods

Water balance. The need for irrigation may by identified from changes in the amount of soil water available to the crop. This can be estimated indirectly from calculations of soil–water balance

$$W = Wo + Ppt - \boldsymbol{E}, \quad (11.2)$$

with irrigation being required when the available soil moisture, W (mm), falls below some threshold value appropriate for the crop or site, Wo is the initial amount (mm), Ppt is the amount of precipitation minus runoff (mm), and $\boldsymbol{E}$ is the evapotranspiration. One can use remote sensing to estimate $\boldsymbol{E}$ directly or via canopy cover estimates and through integration into GIS systems that address spatial variation (Chapter 9). One also needs to know the amount of rainfall, which is still difficult to estimate quantitatively by RS, though rain radar is becoming usable for this purpose.

Soil moisture. Alternatively, the amount of available soil moisture may be estimated directly using passive or active microwave sensing. As we have seen, microwave RS has particular advantages in that it is largely independent of time of day or cloud cover, and the longer wavelengths can partially penetrate the canopy, which optical wavelengths cannot. Passive sensing, with its low spatial resolution is particularly appropriate for mapping soil moisture at regional scales. Although C-band data retrieve soil moisture in only the top two to three cm of the soil, other wavelengths such as L-band penetrate deeper in the soil and provide soil-moisture information more relevant for crop and vegetation studies. The main problem with active microwave sensing of soil moisture is that the backscatter signal that relates to soil moisture is partially obscured by the volume scattering from the canopy signal and by attenuation by the canopy. This interference depends on canopy height, leaf and stem density, and on vegetation water content. These effects can be corrected for by using a canopy scattering model parameterized using optical or ground data, or by simple statistical or machine learning approaches using the same optical data. Other estimates of soil moisture may be obtained indirectly (with rather low precision) from any one of the range of indices that respond to the plant response to soil moisture (Chapter 7).

Plant response-based methods

Although plant response-based methods are often thought to be more sensitive than soil-based sensing, the fact that they do not directly indicate how much water to apply means that they usually are only suitable where it is possible to irrigate frequently (Jones, 2004b). Unfortunately direct estimation of plant water status or plant water deficit by remote techniques tends to be difficult, as the key variables of water potential (ψ) and cell turgor pressure (ψ_P) are not generally detectable remotely. Some of the spectral water indices in the MIR can provide some information on canopy water content, which can be a useful indicator of water status. More useful, however, especially for in-field measurements (e.g. tractor-mounted sensors) are the range of techniques that detect plant responses to water deficit, especially stomatal closure, but also the reduced leaf growth and reduced *LAI*. Both of these measures respond sensitively to water deficits and are therefore potentially early indicators of stress. As we saw in Chapter 9, stomatal closure can be detected

Table 11.3 The different measures of plant water status, methods for their remote detection, and a subjective assessment of their relevance for different purposes (expanded and modified from Jones, 2007). The value of each measure is indicated on a scale from +++ (very useful) through ++, and +, to (−) indicating little use; see text for further explanation.

	Remote sensing	Water transport	Drought adaptation	Plant breeding	Irrigation
Soil measures					
Water content	**μwave; MIR**	(−)	+	+	+++
Water potential (ψ)	[thermal]	+++	++	++	++
Plant measures					
Predawn ψ	[thermal]	+++	++	+	++
Leaf ψ	[thermal]	+++	+	(−)	(−)
Turgor P (ψ_P)	Multiangular; [thermal; *NDVI*]	+++	+++	(−)	+
RWC	MIR	(−)	++	++	+
Stomatal conductance	**Thermal (e.g. *CWSI*)**	(−)	++	+++	++
Leaf area (*LAI*)	***NDVI*** or other *VIs*	(−)	++	++	++
Notes	*Methods enclosed in [] are indirect, while those in* **bold** *are generally most effective*	*ψ and its components especially important for flow*	*RWC is often good surrogate for ψ_P*	*Measures of response such as g_s best*	Soil-based measures best, though plant responses can also be good

through thermal sensing (both in-field and from airborne or satellite platforms) using any of the various crop water-stress indices that have been proposed.

Similarly, leaf area can be readily assessed using *VI* approaches (Chapter 7). When combined with a crop-growth model that indicates the expected schedule of leaf-area development for the environment/crop combination of interest, one can use deviations below the expected *LAI*/crop cover trend as an indicator of water shortage and hence as a signal for scheduling irrigation. Of course, it is necessary to ensure that the poor crop growth is not due to some other limiting factor such as soil structure or nutrient availability. For some species, changes in canopy structure associated with leaf wilting can also be detected using especially multiangle sensing (Chapter 8) or Lidar, while spectral changes are characteristic of wilting in species such as sugar beet.

There is substantial current interest in the automation of thermal imaging, combined with visible (or red/infrared reflectance) sensing, for application to automated scheduling of irrigation. A particularly suitable target crop is the hardy nursery stock industry where it is important to ensure that all pots are well watered, while avoiding overirrigation with the associated problems of excess runoff and wastage of scarce water resources. In principle, such a system would combine the automated identification of plant/no-plant as achieved by Leinonen and Jones (2004) with an appropriate thermal-stress algorithm for detection of stomatal closure as an indicator of water need.

11.3.4 Nutrient management, pests, and diseases

Nutrient management. Field sampling for crop nitrogen status has historically been based on soil and leaf sampling and subsequent laboratory analysis, though field-usable colorimetric assays are now available. These analytical approaches are still essential for calibration of the spectral methods that are all more or less indirect. In recent years routine estimates of crop nitrogen status have been based on the use of hand-held instruments such as the Minolta SPAD chlorophyll meter (Spectrum Technologies, Inc., Plainfield, IL) which measures the differential transmission of light through a leaf at 650 nm (where the chlorophyll is strongly absorbing) and at 940 nm (where chlorophyll is non-absorbing), or from other wavelength combinations (Dualex®, Force-A, Centre Universitaire Paris Sud, Orsay, France). There are many comparable hand-held instruments that estimate chlorophyll content from reflectance, chlorophyll fluorescence ratios (Gitelson *et al.*, 1999), or from colour photography. Such instruments primarily detect changes in leaf chlorophyll content, which is usually closely related to leaf nitrogen content (because the largest proportion of nitrogen in the leaf is in the photosynthetic enzyme, rubisco). Nevertheless, SPAD readings need calibration if they are to be used successfully for estimation of nitrogen status or nitrogen demand (Wood *et al.*, 1993).

Precision application of nitrogen has been greatly helped by the development of tractor-mounted reflectance sensors that allow real-time monitoring of crop nitrogen status. The latest generation of commercial systems are 'active' sensors; these include the Greenseeker™ (NTech Industries, Ukiah. CA, USA) and the CropCircle™ (Holland Scientific, Lincoln, NE, USA). In these active systems light-emitting diodes generate modulated light in the visible and NIR regions and then detect the corresponding canopy reflectances. These two reflectances are used to derive a normalized vegetation index or chlorophyll index (see Box 7.1), which can then be used to determine nitrogen application recommendations in real-time. The use of a modulated light source (where only the signal that corresponds with an illumination pulse is recorded) means that these instruments can operate at night and in the presence of varying amounts of sunlight as the electronics discriminate the reflected sensor radiation from any ambient radiation. The precise visible and NIR wavelengths used depend on the system, with the CropCircle, for example, recommending the use of 590 nm and 880 nm as giving the best sensitivity to crop chlorophyll content. The calculated chlorophyll index ($CI_{590} = (\rho_{880} - \rho_{590})/\rho_{590}$) has been found in some cases to be more sensitive in assessing canopy nitrogen than is the conventional *NDVI*. More sophisticated systems include the Agricultural Irrigation Imaging System (AgIIS) that has nadir-pointing sensors that cover an area of about 1 × 1 m (see El-Shikha *et al.*, 2007) and measure five wavebands (reflectances in the green (550 nm), red (670 nm), far red (720 nm), and near infrared (790 nm)) and emission in thermal infrared (8–14 μm). The thermal channel provides information on crop water status (see Section 9.6).

There has also been substantial interest in the development of hyperspectral sensing approaches (Ferwerda and Skidmore, 2007) as well as the use of chlorophyll fluorescence (e.g. Adams *et al.*, 2000) for the diagnosis and monitoring of a wider range of nutrient deficiencies and toxicities (including major nutrients such as N, P, and K) and micronutrient deficiencies including those of Fe, Mg, Mn, Zn, and Cu. Some promising results have been obtained (Ferwerda and Skidmore, 2007), for example using both derivative spectra (after smoothing) and continuum-removed data (Kokaly and Clark, 1999). Such approaches can be reasonably successful at the leaf scale, but the results are generally not very robust or good at predicting between-species variation or in scaling up to canopies, though increasingly with the use of ANNs and similar approaches, useful remote estimates of canopy composition and even forage quality have been reported, at least from aerial imagery (Skidmore *et al.*, 2009).

Pests and diseases. As we have already seen, at the leaf or even plant scale, remote sensing potentially provides a powerful tool for diagnosing pests, diseases, and even nutrient deficiencies, because many diseases and deficiencies give rise to characteristic colour patterning on the leaves as well as changes in overall photosynthetic activity and stomatal conductance. There are many examples where infestations by specific pests or diseases have been successfully detected, though as pointed out in Section 11.2, the symptoms detected are often not truly diagnostic. Attribution of poor crop performance, such as yellowing or reduced growth, to a specific cause is usually problematic. Nevertheless, there have been many cases where useful correlations have been reported, particularly when hyperspectral data are used. For example, spectral indices have been used to detect spider mite in peach orchards where pairs of wavelengths were selected from partial least squares regression to generate normalized difference indices (Luedeling *et al.*, 2009), though, especially from airborne images the predictive ability is not good.

A key problem with all such studies based on correlation of spectral data with pest or disease abundance data, is that other characters that cause leaf-area reduction or senescence (e.g. drought) can also cause similar effects and it is hard to demonstrate that the differences are specific to the pest or disease of current concern.

Weed management. It is quite straightforward to use simple imagers based on *NDVI* or even RGB images to detect weeds on a bare soil background, and commercial systems are already available for this purpose (e.g. Weedseeker®, NTech Industries, Ukiah. CA, USA) that can automate localized spraying of weeds. Much more useful, however, would be techniques to identify and treat weeds growing in a crop, as precision application provides enormous scope to reduce herbicide applications. Weeds can sometimes be detected on the basis of characteristic spectral signatures, though machine-vision approaches seem to offer the most potential (Thorp and Tian, 2004). For example, where one is concerned with broad-leaved weeds, such as *Rumex*, in a narrow-leaved grass crop image analysis and object detection on the basis of the morphological differences between broad-leaved weeds and grass seems to offer much potential (van Evert *et al.*, 2009).

11.3.5 Practical considerations

The key aspects of the application of remote sensing and precision agriculture to farm management include (i) timeliness and availability of data and the interval between samples, (ii) a spatial resolution relevant for the management options available, (iii) accuracy of estimation of relevant parameters, and (iv) cost. The frequency of sensing, and the subsequent provision of data, must relate to the timescales in which farm-management operations need to be manipulated. For the control of irrigation and more particularly pesticide applications, measurements should ideally be available every day or so, while for the effective control of nutrient applications, weekly or even monthly data may be adequate. In relation to timeliness, tractor-mounted sensors offer scope for real-time response to sensor inputs (for example with crop N or water-status sensors); such immediacy is not possible with satellite sensors, as there is usually some delay in data provision after acquisition (though with efficient data provision chains this can now be within a day). The frequency of satellite sensing depends on the spatial resolution of the sensor, with higher resolution (and hence generally narrow swath widths) being a trade-off with the return frequency of the satellite (Chapter 5). For example, Landsat images (15/30 m resolution in the visible) are obtained every 16 days, while higher-frequency images tend to be available at the cost of lower resolution (e.g. MODIS samples approximately daily but only at 250 m spatial resolution) or at daily or more frequent intervals with meteorological satellites but at spatial resolutions of 1 km or more. Alternatively with pointable sensors it is possible to get more frequent high spatial resolution coverage but at high cost. A further problem with optical satellite sensing is the potential interference by clouds (partially overcome in microwave sensors) that can severely limit the number of seasonal images, especially in more humid regions. Airborne sensing can potentially overcome some of the limitations of satellite sensors as they can fly on demand and below clouds.

An important development for the handling of intermittent observational data is the interpolation between infrequent observations, for example by combination with a growth modelling approach that allows estimation of intermediate values. Further enhancement, for example for irrigation management, can be to combine growth models with satellite observations for scaling and monitoring crop growth with local meteorological data (e.g. rainfall) for the derivation of a continuous water balance.

Perhaps the most important consideration is the need to get robust estimates of the variables such as crop water status or nitrogen status that are needed as inputs to crop management decision making. Most remote sensing techniques estimate the variables such as *LAI* or chlorophyll content with substantial error, while the conversion to the more agronomically useful variables such as crop nitrogen or biomass is even more problematic. Therefore, thresholds for action usually need to be set in such a way that stress effects can be reliably distinguished from other environmental noise over a wide range of conditions. This inevitably results in rather coarse control. Although relative measures may be of use where there is a reference area within the imaged area of known optimal water or nitrogen status, this is not commonly available, so robust stress-detection algorithms are required.

Further details of specific applications and recent developments in sensing for precision agriculture applications may be found in recent editions of the journal *Precision Agriculture*.

11.4 Ecosystems management

Perhaps the greatest growth area in remote sensing is its use in ecosystem management and monitoring. It is now possible for anyone to see, often at spectacularly high spatial resolution, one's study sites using tools such as Google Earth (http://Earth.google.com/). The increasing availability of satellite imagery both in terms of the range of sensors available (e.g. hyperspectral, microwave, Lidar, and thermal) and also importantly in terms of the readily available higher level products such as regional and global maps of seasonal changes in variables such as landcover and net primary productivity, often at better than 1 km spatial resolution, has transformed the scope of ecology in recent years. It is now increasingly possible to move from the small-plot studies that have characterized many ecological studies to studies of whole reserves, regions, or even the world. This enhanced spatial capability, however, brings associated problems. In particular, as the scale of the study increases, it becomes more and more difficult to obtain the necessary ground data to validate the satellite outputs.

There are two key features of remote sensing as a tool for ecological applications. The first is the ability to acquire data for broad spatial extents that are difficult or impossible to obtain using field studies. The second, and perhaps most important strength of satellite remote sensing for the ecological community is its potential for collecting long series of consistent images that can be used for monitoring environmental and ecological change and also for identifying the causes of the observed changes. Also, the advent of what has been called hypertemporal imagery is starting to provide a range of new tools for classification and the study of vegetation and ecosystem dynamics. Satellite data are now being used to give reliable estimates of habitat loss, especially in the humid tropical forests (Achard *et al.*, 2002). The images archive of Landsat MSS and TM/ETM+, which operated at a spatial resolution of between 15 and 80 m and at 16-day intervals from 1972 for MSS or 1982 for TM until 2003 (with more intermittent data since), is the prime example where long time series have already proven invaluable. Since 1999, MODIS has been providing invaluable global data at 250 m resolution. The Advanced Land Imager (ALI) on EO-1 provides some Landsat continuity (at 10–30 m spatial resolution) since 2000 but does not include thermal bands. Unfortunately there is considerable uncertainty surrounding the long-term availability of this type of dataset, and although there are more and more high-resolution satellites coming on stream, these often do not have global coverage as a result of the fact that they may only record specific areas of the surface and data costs may be high or they may omit the crucial thermal bands (Loarie *et al.*, 2007). Although derivation of long-term environmental indices, even where there is one consistent set of data, is made difficult by the need for rigorous correction of atmospheric effects such as haze and effects of cloud shadows, the situation is even more difficult with mixed sets of satellites, where intercalibration of the different scanners is required. This is becoming increasingly necessary for many areas of the world,

There has been particular interest in the potential provided by hyperspectral imaging for ecosystem studies, though progress has largely been related to the estimation of non-pigment biochemical components of vegetation including water, nitrogen, cellulose, and lignin.

By the nature of the system, remote sensing is generally of more direct use for studies of ecosystem structure and especially the plant component of ecosystems than it is for the study of fauna and their movement. This is primarily a result of the limited spatial and temporal resolution of most satellite RS imagery. Nevertheless, useful indirect relationships can often be derived between observable characters and the distribution or abundance of specific animal species (Ngene *et al.*, 2009; Wang, 2009). In addition to the obvious limitations to ecological application that are set by spatial and temporal resolution, the lack of continuity is likely to be a major problem, especially with the plethora of smallsats that are now being used (Xue *et al.*, 2008). A good review of ecological applications of *NDVI* has been written by Pettorelli *et al.* (2005).

11.4.1 Landcover classification and mapping

The classification of landcover is a particularly important use of remote sensing technology – in principle it allows accurate mapping of land use and, when combined with long-term image sequences such as that from Landsat, the monitoring of changes with time. Essentially, the use of RS imagery in vegetation mapping involves image-classification procedures, often combining spectral, textural, and time-series information, together with ancillary information and the use of GIS systems, to delineate areas of each vegetation type. The development of landcover classification and mapping tools is an area of very active research, relying heavily on sophisticated classification techniques (see Sections 7.4 and 11.3.1), particularly involving the use of machine-learning algorithms, including those that use decision trees and other mixed techniques. Although substantial progress can be made at the pixel level, there can be real potential advantages of working at an object level for landcover-mapping applications. There are hundreds of methods that have been developed since the 1970s and many software packages that are commonly used to map landcover. We do not intend to cover all the applications in any detail; further details of the approaches available may be found in remote-sensing texts (e.g. Jensen, 2005), reviews (Xie *et al.*, 2008a) and in the extensive primary literature (Pignatti *et al.*, 2009; Price, 2003).

Although the use of optical imagery is the most usual, the use of SAR has particular advantages; these include relative independence of weather conditions and the ability to collect data at night. Single-band SAR data generally have limited ability to distinguish landcover types, so fusion of multifrequency and multitemporal SAR data has real advantages, while polarimetric SAR may be useful because of the characteristically different polarization properties of different surfaces, and interferometric SAR provides information on surface structure and complexity. Therefore, fusion of data from different SAR frequencies (L- and P-bands) and from different modes (polarimetric and polarimetric interferometric) can substantially improve landcover classification accuracy because of the complementarity of the various data sources (Shimoni *et al.*, 2009). Yet further improvements can in principle be achieved by combining such information with hyperspectral data.

Landcover classification and mapping is crucial for many aspects of conservation and land management, including provision of information on the geographical distribution of natural and cropped areas and the quality of the environment. Landcover data are very useful for predicting the distribution of both individual species and groups of species across large areas that would otherwise be difficult to survey. It can be done at different scales, with the EU CORINE Landcover project (http://www.eea.europa.eu/publications/COR0-landcover), for example, aiming to produce a European-wide landcover map at a 1:100,000 scale where vegetated land is partitioned into about 28 classes, with smallest units of 25 ha. Such a classification is rather coarse and though it has limited use for many studies provides a useful and consistent framework for more detailed landcover maps, such as those in individual countries. The approach to generation of the data involved use of existing topographic maps supplemented by satellite (initially Landsat MSS) and various other types of data and required substantial operator input for the photointerpretation. More detailed landcover maps, such as the UK 'Land Cover Map 2007' (http://www.countrysidesurvey.org.uk/land_cover_map.html), generally involve more substantial automation of the classification of the RS imagery. LCM2007 uses 19 vegetation classes, but these are provided at a spatial resolution of 25 × 25 m pixels. A number of other global landcover products are becoming available including the GLOBCOVER products calculated from MERIS full-resolution (300 m) composites (http://ionia1.esrin.esa.int/index.asp), and the GLC 2000 dataset derived from SPOT vegetation images (http://bioval.jrc.ec.europa.eu/products/glc2000/glc2000.php). The MODIS collection 5 landcover map (MOD12) provides a set of classifications at 500 m resolution (Friedl *et al.*, 2010). At a coarser scale there is the Global Inventory Modeling and Mapping Studies (GIMMS) product (Tucker *et al.*, 2004). This, as are many such products, is a global *NDVI* product at 8 km spatial resolution, so does not directly separate out different types of vegetation.

11.4.2 Landscape ecology

Landcover mapping provides the basic information for ecologists, but the remote sensing also has a lot to offer to the wider field of landscape ecology, which is largely based on the idea that environmental patterns influence ecological processes and the interactions between organisms and their environment. Landscape may be regarded as an area of land consisting of a mosaic of different habitats; clearly the scale relates to the organism of interest. Remote sensing is a powerful tool for identifying and describing the patterning of identifiable patches in landscape.

Although the textural metrics introduced in Chapter 6 provide some information about the scale and type of spatial nature (e.g. random or clumped distributions) of variation within an image or an area, there is frequently a need for more detailed descriptions of the spatial configuration of landscape features. *Fragmentation metrics* can provide information on the spatial configuration of landscape features. They have been used, for example, in studies of species distribution (e.g. Dufour *et al.*, 2006; Wang, 2009) and reserve management (Townsend *et al.*, 2008). A convenient software package to derive the necessary statistics is FRAGSTATS (McGarigal and Marks, 1995) (see http://www.umass.edu/landeco/research/fragstats/documents/fragstats_documents.html), which is a spatial pattern analysis program for categorical maps. This package calculates a wide range of possible *structural metrics* that describe the configuration of the patch mosaic without explicit reference to any ecological process. Calculations are performed on a moving window to give a continuous surface for each metric. The available landscape metrics (see examples in Table 11.4) include those that quantify the *composition* of the landscape map without reference to their spatial relationships (e.g. proportion of the landscape in each class, class richness, and class diversity). *Spatial configuration* is much more difficult to quantify and refers to the spatial character and arrangement, position, or orientation of patches within the class or landscape. Some aspects of configuration, such as patch isolation or patch contagion, are measures of the placement of patch types relative to other patches, other patch types, or other features of interest. Other aspects of configuration, such as shape and core area, isolation, and contrast are measures of the spatial character of the patches.

The difficulty in any experimental study, however, is the choice of appropriate measures for the biological problem of interest from among the many possibilities (Townsend *et al.*, 2009). One approach is to eliminate metrics that are highly correlated; this can then be followed by some regression analysis against ground validation data to derive a discriminatory set of metrics. In principle, however, it would seem better to start from metrics that would appear functionally relevant (e.g. patch size for the distribution of a large animal species).

A further step in the analysis of habitat fragmentation or connectivity and their impacts on wildlife can be to calculate *functional metrics* that explicitly measure landscape pattern in a manner that is functionally relevant to the organism or process under consideration. An example would be the use of 'habitat network analysis or connectivity analysis' that potentially goes one stage further as it incorporates species-specific functional permeabilities (Adriaensen *et al.*, 2003), allowing one to estimate the ability of a given species to move between core areas on the map (Fig. 11.8). In the example in Fig. 11.8, although a simple estimation of dispersal from the core habitats suggests that site 4 might be linked to others, once one takes account of the low permeability of the surrounding vegetation, we can see that that site is effectively isolated. An example of an application to a real landscape is shown in Fig. 11.9.

11.4.3 Biodiversity estimation

Conservation and enhancement of biodiversity is widely recognized as being a critical objective of sustainable ecosystem management. Landcover changes, especially those that lead to critical habitat loss and fragmentation, as well as a wide range of other processes of environmental degradation are associated with losses of biodiversity. Satellite remote sensing has much potential as a source of biodiversity information at the landscape level. In general, such estimation of biodiversity is almost always indirect, making use of known empirical or functional relationships between some aspect of biodiversity (usually only the number of species or *species richness*) and some readily identified feature of the landcover. Although biodiversity is commonly measured in terms of species richness, there are many more informative measures of diversity that also take account of the relative abundance of different species, otherwise known as *evenness* (see Magurran, 1988). Particularly well-known examples are the Shannon or Shannon–Weaver index (H), defined as

$$H = -\Sigma p_\mathrm{i} \log_\mathrm{e} p_\mathrm{i}, \tag{11.3}$$

where p_i is the proportion of individuals in the population that are of the ith species, and Simpson's diversity index (D), defined as

$$D = (1 - \frac{\Sigma n_i\,(n_i - 1)}{N(N-1)}), \tag{11.4}$$

where the summation is across all species and n_i is the number of specimens of the ith species and N is the total number of individuals. Both the Simpson and Shannon indices take some account of evenness between species and give less weight to the rare species. In some conservation situations, however, one is concerned not with the overall species richness of a site but with the preservation of specific, often iconic, species, in which case general biodiversity indices may be of little use.

Table 11.4 Some sample landscape metrics of use in ecological studies (see FRAGSTAT manual (http://www.umass.edu/landeco/research/fragstats/documents/fragstats_documents.html) and Townsend *et al.* (2009) for further information on the selection of key fragmentation parameters).

Composition	
Proportional abundance of each class	*Percentage of Landscape* (*PLAND*) The sum of the areas (m^2) of all patches of the corresponding patch type, divided by total landscape area (m^2).
Richness	*Number of different patch types.*
Diversity	*Shannon–Weaver diversity index* (*SHDI*) This equals minus the sum, across all patch types, of the proportional abundance of each patch type multiplied by the logarithm of that proportion.
Spatial configuration	
Patch-size distribution and density	*Number of patches* (*NP*) The number of patches in the landscape. *Mean patch area* (*AREA*) The sum of the areas (m^2) of all patches of the corresponding patch type, divided by the number of patches of the same type. *Patch density* (*PD*) The number of the corresponding patches divided by total landscape area (m^2). *Landscape shape index* (*LSI*) The total length of edge (or perimeter) involving the corresponding class, given in number of cell surfaces, divided by the minimum length of class edge.
Patch shape complexity	*Perimeter area ratio* (*PARA*) The ratio of the patch perimeter (m) to area (m^2). *Shape index* (*SHAPE*) Patch perimeter divided by the minimum perimeter possible for a maximally compact patch of the corresponding patch area. *Fractal dimension index* (*FRAC*) The sum of two times the logarithm of patch perimeter (*m*) divided by the log of patch area (m^2) for each patch of the corresponding patch type, divided by the number of patches of the same type.
Isolation/Proximity	*Proximity Index* (*PROX*) The sum of patch area divided by the nearest edge-to-edge distance squared (m^2) between the patch and the focal patch of all patches of the corresponding patch type whose edges are within a specified distance (m) of the focal patch. *Euclidean nearest-neighbour index* (*ENN*) The distance (m) to the nearest neighbouring patch of the same type, based on shortest edge-to-edge distance.
Contrast	*Total edge contrast index* (*TECI*) The sum of the lengths (m) of each edge segment involving the corresponding patch type multiplied by the corresponding contrast weight, divided by the sum of the lengths (m) of all edge segments involving the same type.
Connectivity	*Patch cohesion index* (*COHESION*) one minus the sum of patch perimeter divided by the sum of patch perimeter times the square root of patch area for patches of the corresponding patch type, divided by one minus one over the square root of the total area. *Connectance index* (*CONNECT*) The number of functional joinings between all corresponding patches, divided by the total number of possible joints between all patches of the corresponding patch type.

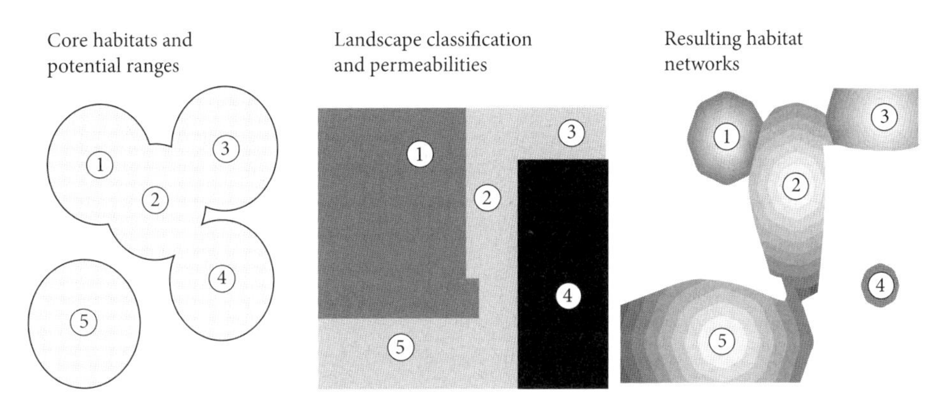

Fig. 11.8. Illustration of the principle of habitat networks showing (a) core areas and hypothetical dispersal ranges of a species, (b) the landscape patterning and the associated permeabilities (increasing in proportion to paleness), and (c) the calculated connectivity between sites (Watts *et al.*, 2005; with kind permission from Kevin Watts and the Forestry Commission). (See Plate 11.5)

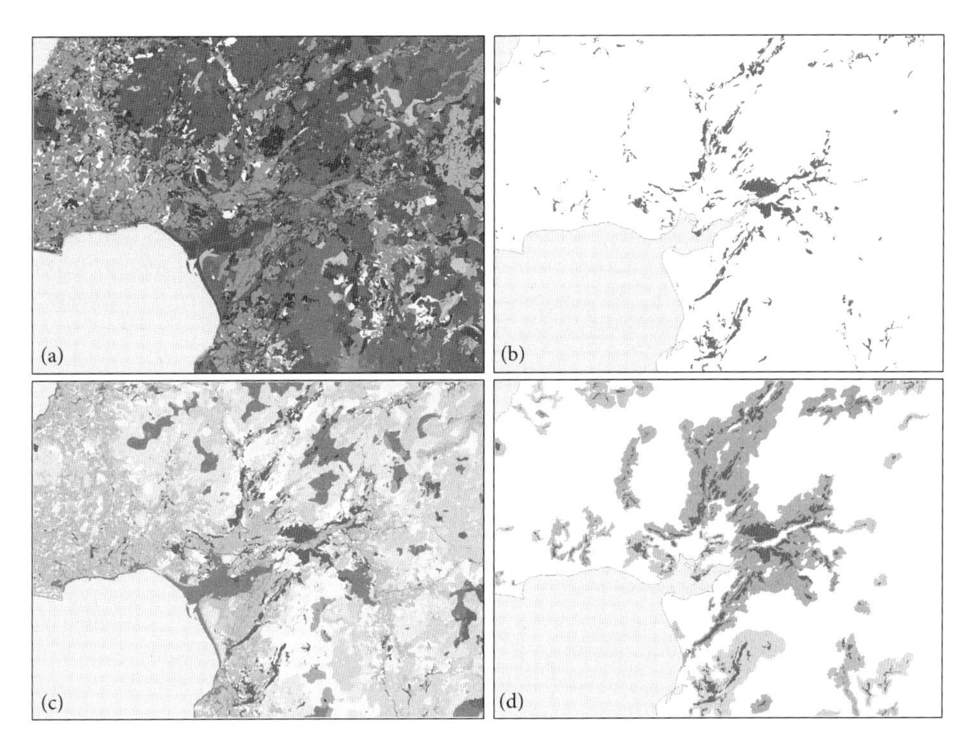

Fig. 11.9. Example of application of habitat networks to a landcover map of a portion of North Wales, showing (a) the landcover distribution with different cover types in different colours, (b) the area of core woodland habitat for a woodland specialist, (c) assumed permeabilities (higher permeability represented by lighter colours), and (d) the resulting habitat network (Watts *et al.*, 2005; with kind permission from Kevin Watts and the Forestry Commission). (See Plate 11.6)

Remote estimation of biodiversity

As diversity cannot usually be estimated directly from imagery, diversity estimation is generally dependent on the use of proxy measures or environmental descriptors that are more easily obtained by remote sensing. These proxy measures may include features such as primary productivity, standing biomass, or landcover type, all of which are known to be associated with biodiversity in certain situations, though there are many exceptions (Townsend *et al.*, 2008). Especially at smaller more local scales there are clear associations between landcover type and diversity with, for example, the differences between short grassland and scrub-land or farmland generally being quite obvious. Indeed, the question as to which factors drive variation in biodiversity has been the subject of much argument amongst ecologists for many years; there is little consensus with different factors being appropriate for different taxa and for different environments (Townsend *et al.*, 2008). Although there is a general trend to increasing biodiversity towards the tropics or to lower altitudes, at regional and local scales a combination of factors such as habitat extent, heterogeneity, topography, productivity, and water availability all may contribute, as will factors such as vegetation type.

Understanding the driving forces for biodiversity should allow the derivation of effective proxy measures that can indicate changes in diversity, but lack of knowledge often means that we have to rely on empirical or semi-empirical methods for selection of predictive variables (Foody and Cutler, 2003). The empirical approach generally operates by using a training set to select the best predictor(s) from all available remotely sensed variates (e.g. all the possible ratio indices from a hyperspectral dataset) by regression of ground-measured biodiversity for the biological group of interest (whether it be plants, birds, insects, etc.) on the possible remotely sensed variates. For example, Foody and Cutler (2003) showed that the use of reflectance data from the six optical bands of Landsat TM could be enhanced both by the incorporation of a local texture measure (standard deviation in a moving 3×3 window) and by analysis using a feedforward neural net. Other studies have attempted to relate species richness to various productivity, vegetation cover, and surface water indices (John *et al.*, 2008), while others have attempted to use hyperspectral sensing to predict richness over a range of habitat types (e.g. in Mississippi; Lucas and Carter, 2008). The latter workers found that none of the spectral indices related to richness when tested across all habitats, though data within habitats were sometimes closely related. This illustrates the problem that is commonly observed when attempting to use empirical associations. The selection of the most appropriate predictors can be enhanced by the use of various machine learning approaches such as decision trees (Coops *et al.*, 2009a) and neural networks (Foody and Cutler, 2003). Although this can be a powerful technique and has been widely used, unfortunately it is essentially correlative and hence is not usually capable of generalization to other situations, so it is usually essential to use a local calibration in any such study.

It has been found that particular types of vegetation may be associated with high biodiversity and others with low diversity; biodiversity theory (Whittaker *et al.*, 2001) provides some useful guidance on these relationships. Lidar is a particularly powerful tool for biodiversity studies as it provides directly information on key functionally relevant variables such as canopy height and vertical distribution. It has been noted that diversity often increases with canopy height (possibly through its association with biomass) while the pattern of canopy vertical distribution and canopy heterogeneity can be good predictors of bird diversity (Goetz *et al.*, 2007). Habitat heterogeneity, which is also a good proxy for species richness, is readily assessed by remote sensing using appropriate scale imagery and texture analysis (St. Louis *et al.*, 2006). It is worth noting that the relationships between species richness and measures of the canopy structure are dependent on the scale (resolution) of observation and on the taxonomic group of interest (Hawkins *et al.*, 2003). It is difficult, however, to obtain general relationships as there are clear geographic trends in diversity such that otherwise similar habitats show trends for increasing biodiversity as the equator is approached (e.g. Cox and Moore, 2000; Townsend *et al.*, 2008).

11.4.4 Other ecological applications

Invasive species

Invasive species are often not particularly spectrally distinct from other species so pure spectral approaches have their limitations. Nevertheless, for some species outstanding spectral features have been successfully used for their mapping (e.g. the water absorption features of the ice-plant; Underwood *et al.*, 2003). Because of the rather small spectral differences between some invasives and native species that might be present, the most powerful approaches to classification are based on artificial intelligence techniques such as neural networks, and more sophisticated classification and regression tree (CART) approaches (Andrew and Ustin, 2008). Spectral information can be supplemented by information on phenology if time-series data are available, or image texture or context may be used. The combination of conventional classification with object-based search routines (Xie *et al.*, 2008b) in a *geographic image retrieval* (GIR) approach can substantially enhance the conventional classification. GIR using Definiens or other object-oriented image-segmentation tools is particularly useful in situations where one is interested in a specific type of object (e.g. areas of a specific invasive species) that might occupy only a small but widely scattered part of the area being investigated.

Remote sensing is particularly suited to detection of canopy overstorey species, but in many cases the worst invasive species are hidden below the main canopy (Joshi *et al.*, 2006). In such cases indirect approaches are all that are accessible to remote sensing so that detection and mapping relies on understanding the key environmental determinants of invasion. Joshi *et al.* (2006) report the successful mapping of an important invasive cryptic shrub species, *Chromolaena odorata*, based on the fact that invasion depends on disturbance as it cannot apparently colonize dense forest. They trained an ANN to recognize areas of forest from Landsat TM data where light was reaching the forest floor and demonstrated a good correlation with *Chromolaena* invasion. In other cases understorey vegetation can be studied using images taken during the period of leaf-off in the overstorey (Wang *et al.*, 2009).

Parks and Protected Areas (PPA) monitoring

Parks and Protected Areas (PPA) managers can use RS in a number of ways to guide decision making. Example applications include monitoring of active fires (see Section 11.6 below), studies of landcover and land-use change, characterization of ecosystem processes and services such as primary productivity and water use, habitat connectivity mapping, and identification of disturbance and periodic events. The key requirement for these activities is the availability of repeated observations over time as can conveniently be provided by satellite imagery. Traditional monitoring was by means of sample plots, but this work is labour intensive and may not be representative of all the reserve, unless many sites are monitored, while airborne and satellite data can more easily provide information relating to the whole reserve. Spatial pattern analysis (Section 11.4.2 above) provides a particularly powerful tool because it allows the extraction of the key spatial relationships between habitat classes that are critical for survival and movement of individual species of interest (Townsend *et al.*, 2008). Widespread application of remote sensing to reserve management has for a long time been limited by poor availability of data at a suitable scale or frequency. Landsat acquisitions tend to be too infrequent (especially with cloud cover) while AVHRR and even MODIS and MERIS have rather too large pixels for many applications. Although airborne data can have high spectral and spatial resolution, the frequency of such images is often low and there tends to be a major requirement for image processing. Effective use of RS data in reserve management often requires integration of information from a range of sensors using specialized algorithms and models. All this information should be combined with ancillary data such as ground survey, local meteorological records, and so on, into an appropriate GIS. In addition to providing information at larger scales than observers on the ground can often achieve, RS often provides information on even larger scales about external events that may impact on specific reserves – e.g. climate change and drought or urbanization. A further limitation has been the cost, not only of the RS data, but more importantly of the essential ground validation and verification. RS addresses both spatial and temporal domains inaccessible to field observation.

Again this is a topic where there is much interest in application of RS techniques, some of which has been reported in a special issue of *Remote Sensing of Environment* (Issue 7, Vol. 113, 2009).

11.5 Forestry

Forests cover nearly one third of the earth's land area, and they play a very important role in carbon sequestration and global climate control. It is therefore important to map and to monitor for any significant changes that may take place whether due to natural causes or to human intervention such as illegal logging; remote sensing can play a vital part in this. Forestry is also particularly concerned with timber management, and again remote sensing is an invaluable tool for producing inventories and for estimating timber volume. Originally, forest inventory was considered to be the determination of the volume of logs, trees, and stands, and a calculation of increment and yield. Now, it has expanded to cover the assessment of various issues including wildlife, recreation, watershed management, and other aspects of multiple-use forestry (Hyyppä *et al.*, 2008) and is the subject of much active remote-sensing research.

The main operational methods for forest management for many decades have been the use of ground mensuration and the visual interpretation of aerial photographs. Maps can be produced of extent and species down to individual stand level and tree heights and canopy shapes can be obtained from stereoscopic pairs of photographs giving information about tree size and density. McLean (1982) proposed that the cross-sectional area of a forest canopy (or stand profile) obtained from photointerpretation is related to the gross marketable volume. Automated analysis of medium-resolution optical satellite data (e.g. from Landsat) gives two-dimensional information on a larger scale, but at lower spatial resolution, limited to coarse forest classes. Radar data can also be used for identifying and mapping forests, especially in tropical regions where cloud cover limits the use of optical satellite data. Three-dimensional information about trees and forests has until recently been obtained from fieldwork or terrestrial scanning; it is now also possible, as we shall see, to obtain such information using radar and airborne Lidar.

11.5.1 Mapping, monitoring, and management

Similar techniques are used for mapping forests, both at larger scales and for assessment of species composition and quality, as are used for agricultural systems (Section 11.3.1) and more general landcover mapping (Section 11.4.1). The use of aerial photographs and airborne scanner imagery provides good-quality, high spatial resolution maps, but is restricted to small areas. Satellite imagery, mostly from Landsat and SPOT, has been used extensively for many years for mapping the larger afforested areas, but at a much lower spatial resolution. High spatial resolution satellites, such as Quickbird, IKONOS, and WorldView provide higher-resolution data.

TM data, although widely used, does not allow detailed analysis of different forest species, not only because of its coarse spatial, but also its coarse spectral, resolution. Hyperspectral systems give much better discrimination because of their denser sampling of spectral signatures. As with agricultural systems, however, there has been a heavy reliance on simple *VI*s for the study of tree health and diseases, even though alone they have little diagnostic power. Again, both hyperspectral and multiangular approaches have much more ability to discriminate different causes of altered tree performance, particularly when used to extract relevant biophysical information through model inversion. Various workers have reported the use of image fusion of hyperspectral and Lidar data for classifying complex areas (Verrelst *et al.*, 2009); this information can be useful for generation of forest inventories. The two images can be coregistered and fed into the classifying system. Lidar can discriminate species having similar spectral signatures but different height properties or different preferences for growth at different elevations, so their joint use can increase separability. Buddenbaum *et al.* (2005) report the classification of forest stands regarding both tree species and age class using HyMap hyperspectral data using spectral angle mapper (SAM) and maximum-likelihood classification to an accuracy of 66–74%.

The extraction of forest features using texture analysis of high spatial resolution imagery provides a complementary source of data for those applications in which the spectral information is not sufficient for identification or classification of spectrally similar landscape features. Ouma *et al.* (2008) describe the results of grey-level co-occurrence matrix (GLCM) and wavelet transform (WT) texture analysis for the differentiation of forest and non-forest types in QuickBird imagery. They used semi-variogram fitting to find the optimal window size that was then used to produce eight GLCM texture measures (mean, variance, homogeneity, dissimilarity, contrast, entropy, angular second moment, and correlation). Up to five levels of macrotexture were calculated using wavelet transforms and these were tested in a classification process. The best accuracy obtained was nearly 78% as compared to 59% using conventional multispectral classification. Similar approaches are available with radar imagery where textural information provides a powerful additional dimension in studies of forest age and disturbance (Miles *et al.*, 2003), especially when combined with optical data.

Medium and high spatial resolution satellite imagery is increasingly being used to derive large-scale forest inventories and for forest monitoring purposes (e.g. Coops *et al.*, 2009b), with both regional and global maps becoming more widely available.

11.5.2 Canopy height, biomass, and 3D sensing

Whereas single optical images give radiometric information about a surface and allow landcover to be mapped (see Section 11.4.1 above), it is only by using indirect methods that information about the three-dimensional structure of the surface can be obtained from such images. Tree heights can be estimated from high spatial resolution imagery by measuring their shadows, the parallax displacement in stereo pairs, relief displacement, or from look-up tables of correlation between crown diameter and tree size. The estimation of tree height and volume from shadow areas is particularly feasible in open forests characteristic of semi-arid areas where appropriate high spatial resolution data are available. Indeed shadow area estimated either manually or by means of automated classification algorithms from pan-sharpened Quickbird high-resolution images has been found to be a better estimator of tree-stem volume than is the crown area for open juniper forest (Ozdemir, 2008). Early use of air photography was in *timber cruising*, the estimation of the volume of timber from individual trees or stands of trees for the production of inventories. Optical multidirectional images from digital cameras or line scanners provide radiometric information that can be used to generate automated *digital surface models* (DSM)[39] (or *canopy height models* (CHM)) of the upper part of the tree canopy (Baltsavias *et al.*, 2008), but Lidar is now extensively used in forestry for profiling and tree-height measurement. St-Onge *et al.* (2008) combined Lidar and digital photogrammetry to create hybrid photo-Lidar CHMs. Lidar was used to produce a DTM and a DSM was obtained using automatic stereomatching of aerial photographs. This opens up the possibility of using historical photos. Many studies have also shown that InSAR has the potential for estimating forest height and biomass (see, e.g., Askne *et al.*, 1997).

A special issue of the *International Journal of Remote Sensing* (Volume 29, number 5, 2008) was devoted to 3D remote sensing in forestry and provides reviews of a number of techniques and on-going work.

Lidar

The interaction of lasers and microwaves with vegetation was outlined in Chapter 5. The ability of lasers and especially of microwaves to penetrate vegetation to some extent, enables them to be used for three-dimensional measurements, and in particular the estimation of tree height. For laser-based systems, the first point of contact with the laser beam from an altimeter is the surface immediately below, and this can be used to produce a profile of the canopy surface. If the ground elevation at that point is known, this would enable the tree heights to be found. The ground level could be obtained from an accurate DEM from another source, such as InSAR, or from the last return of the laser pulse if the vegetation is not too dense. An excellent review of the extraction of forest inventory data from small footprint ALS data in boreal forests is given by Hyyppä *et al.* (2008). They claim that the retrieval of stem volume (and hence biomass) and mean tree height of a tree or at stand level from ALS is better than from photogrammetric methods. ALS data can also be used to detect the presence and height of understorey vegetation in deciduous woodland by comparing the profile of first returns in the leaf-on (summer) phase and the leaf-off (spring) phase when budburst had started in the understorey (Hill and Broughton, 2009).

Full-waveform laser scanners can be calibrated for measuring the scattering properties of vegetation and terrain surfaces in a quantitative way. As a result, a number of physical observables can be obtained (see Fig. 5.8), such as the width of the echo pulse and backscattering cross-section, which is a measure of the energy intercepted and reradiated by objects. Wagner *et al.* (2008) found that vegetation typically causes broadening of the backscattered pulse, while the backscatter cross-section is usually smaller for canopy echoes than for terrain echoes. These scattering properties enable classification of the 3D point cloud into vegetation and non-vegetation echoes with an overall accuracy of about 90%. Where full-waveform airborne Lidar is used (see Section 5.6), larger numbers of data points are retrieved than for conventional

[39]. It is essential to distinguish between a *digital surface model* (DSM) and a *digital terrain model* (DTM). It is necessary to have a good DTM in order to obtain meaningful vegetation height information. Airborne laser scanning has made possible the automated construction of high-quality DTMs, to an accuracy of 10 cm, in forested areas over regional-sized areas.

first and last pulse techniques, allowing the derivation of detailed spatial patterning of the points (Fig. 11.10) and hence the identification of individual tree positions and even species (Reitberger *et al.*, 2008). A particularly useful approach for identifying tree position is to segment the smoothed spatial canopy height model obtained from the Lidar data. This segmentation is often based on what is known as a *'watershed' algorithm*; this uses an analogy with hydrology where watersheds are defined as lines delimiting catchments. The approach is commonly applied to an inverted canopy height model that is notionally filled with water; areas that fill first are identified as the tops of individual trees and individual trees are delineated as the watersheds.

Some work has also been carried out using large-footprint Lidar systems (Section 5.6), such as GLAS on ICESat, SLICER, and SLA (Shuttle Laser Altimeter). The full-waveform data from GLAS was used by Duong *et al.* (2008) to distinguish between broad-leaved, mixed wood, and needle-leaved forests in Europe from summer and winter measurements. The actual tree heights barely changed in the six months, but the *height of median energy* (HOME, a measure of the effective height of the scattering centre) changed most in the broad-leaved (148%) and least for conifers (36%). The ratios of ground energy to canopy energy of the normalized waveforms also changed noticeably over time; 67% for broad-leaved and 47% for conifers. Attempts to classify forests on this basis produced a κ coefficient (see Chapter 10) of 0.57. More detailed estimates of size and volume and growth for smaller regions or individual stands can be obtained using the small footprint, full-waveform ALS systems that are also much more flexible and adaptable and hence more commercially attractive.

Lidar systems are also useful for derivation of fractional forest cover. Although standard optical vegetation indices have been widely used to determine fractional forest cover, especially in semi-arid areas where the background tends to bare soil, where there is a vegetated understorey that may also have a high *NDVI*, spectral methods lose discrimination. In such cases, small-footprint ALS systems can be useful. Even systems that only record first and last echoes can derive three classes of echo: first echo, last echo, and single echo. For vegetation, most single echoes will come from the ground. The fraction of cover (f_{veg}) for a forest can be estimated as $\Sigma E_{veg}/\Sigma E_{total}$ where the vegetation echoes (E_{veg}) are defined for forest as those above a threshold height (e.g. 1.25 m) above the ground. The calculated f_{veg} in this way has been found to be much

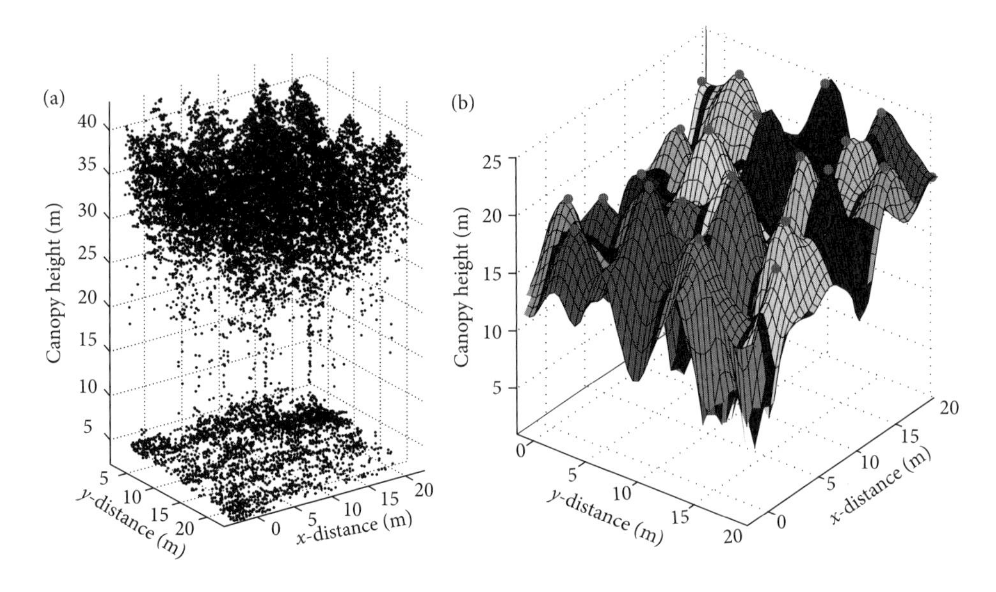

Fig. 11.10. (a) Illustration of a full-waveform Lidar point cloud for a tree stand, together with (b) a smoothed canopy height model derived from similar data and showing local maxima corresponding to individual trees (redrawn with permission from data in Reitberger *et al.*, 2008; Reitberger et *al.*, 2009).

higher when using first echoes than when using second or single echoes and approximations are needed to determine the true value for f_{veg} (Morsdorf *et al.*, 2006).

Microwaves

Numerous authors have demonstrated empirical relationships between the measured radar backscatter and biomass or other forest variables. There is evidence for a correlation between biomass and measured backscatter, though the signal tends to saturate rapidly as canopy biomass increases (Fig. 11.11) and volume scattering reaches a limit when microwaves no longer penetrate through the canopy. At this stage all the scattering takes place in the top layer of the canopy. Since longer wavelengths penetrate the canopy better than shorter wavelengths, the correlation of backscatter with biomass is usually better for longwave bands (L- and P-band) and for cross-polarization measurements where the response is dominated by canopy contributions rather than by the surface component.

Because the size of the scattered signal is influenced by the type of trees, the ground parameters, and the presence of water, results are not usually transferable between different areas and times. The scattering properties change with incidence angle as well as with the polarization of the transmitted and measured signal. Since the radar beam is directed obliquely to the side of the aircraft or satellite, the angle of incidence varies across the scene. As the incidence angle increases, the pathlength through the canopy increases so the microwaves penetrate less into the canopy. In this case, the backscatter from the canopy will increase, while that from the ground will decrease. Vegetation itself has a depolarizing effect with the H-V contrast decreasing as the amount of vegetation increases.

Radar can also be used for 3D measurements making use of the penetration of the signal through the canopy. The basis of the approach is rather similar to that used with Lidar. The depth of the canopy and the wavelength of the microwaves determine the extinction through the canopy: X-band and C-band (short wavelengths) are scattered mostly by leaves and twigs and tend not to penetrate dense canopy as well as the longer wavelengths. Therefore, shorter wavelengths estimate the location of the canopy surface.

The main use of interferometric SAR is for obtaining digital elevation models (DEMs). It is important, however, to remember that the measured elevations have a 'vegetation bias': that is they represent the ground-plus-canopy surface and not the ground surface alone. The scattering phase centre that determines the phase-difference geometry, and thus the height value obtained for the DEM, is some way above the actual ground surface, by an amount that depends on the density of the vegetation and the wavelength of the microwaves used, being nearer the ground for longer wavelengths. It is this vegetation bias that can be exploited to give information about the vegetation height.

As short-wavelength radiation, such as X-band, does not penetrate very far into vegetation, the SAR surface model will approximate the top-of-canopy surface. For accurate measurements, a suitable backscatter model could be used to take account of the underestimation of

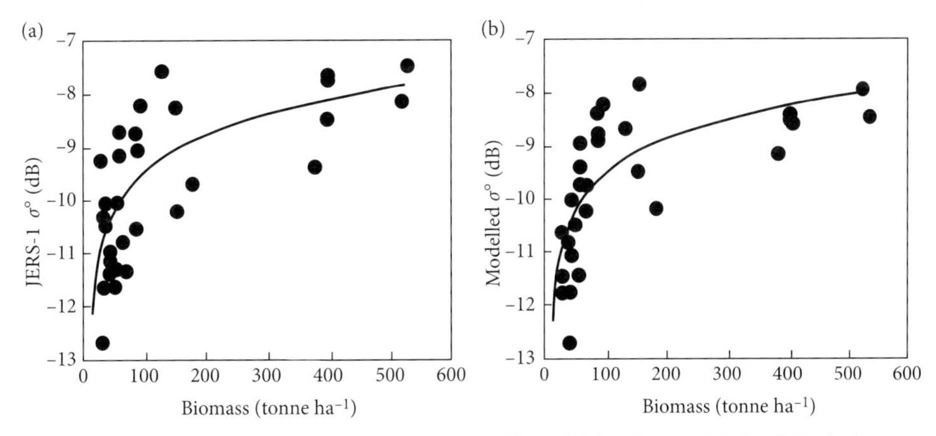

Fig. 11.11. Graph of backscatter against biomass for tropical forest: (a) data from JERS-1 L-band SAR backscatter at HH polarization and (b) modelled results after correcting for the leaf contribution. Biomass was estimated from tree height and stand density using a simple allometric relationship (data from Wang and Qi, 2008).

height due to the small amount of penetration. It is necessary to know the actual ground surface level to obtain the vegetation height. This can be found either from another DEM source, or, for small areas, to assume that the surface does not differ greatly from that of the surrounding non-vegetated areas. The ground surface could also be simultaneously measured using a longer wave P-band system that will penetrate the vegetation. A simple subtraction of the two values will give the vegetation height. Polarimetric interferometry that exploits the different scattering behaviour of the ground and canopy to different polarizations, can also be used. The canopy tends to be depolarizing, whereas the ground tends to have a particular polarimetric response that enables the returns from the two surfaces to be distinguished. The penetration and response from the different components of the trees will vary with the look direction (oblique viewing will tend to 'look' underneath the canopy and see more of the trunk), and so a *multibaseline* method could be used that uses the coherence measurements from data from up to four different passes that will view the vegetation from different angles.

11.5.3 Practical considerations

Forest inventory studies frequently acquire information on both continuous variables (e.g. volume, basal area, tree density) and categorical variables (forest type, soil class, etc.). Use of this information in mapping commonly involves the construction of a multivariate model of relationships between forest attributes and ancillary information from RS and other available data and their use to predict attributes for specific pixels. Because many variates are non-normally distributed it is often necessary to use non-parametric approaches, including the *k-NN* approach introduced in Section 7.4.2.

Although there are many sensors and analytical techniques that can provide both 2D and 3D information about trees and forests, the relationship between such measurements and real forest parameters such as species, tree heights, and trunk volume rely on a statistical relationship with field data from the same area, possibly down to the level of single trees. There are also unique factors for each acquisition, such as flying altitude and vegetation phenology. The collection of ground data for validation often limits the precision that can be achieved, though the use of terrestrial laser scanners can potentially improve on manual measurements for acquiring field data, especially over large areas (Hosoi and Omasa, 2007; Omasa *et al.*, 2007).

The accuracy of laser height measurements depends critically on the quality of both the DTM and the canopy-height model. Hyyppä *et al.* (2008) discuss the factors that influence these. The height accuracy of a laser point is a combination of several error components due to the integration of data from several different instruments (not only the laser itself but also the GPS and the inertial navigation system). The quality of the DTM is also influenced by the data characteristics (point density, first/last pulse, flying height, footprint size, and scan angle) as well as errors due to the complexity of the target (type and flatness of terrain, density and height of vegetation and understory) and the particular algorithm used to generate it. Errors ranging from about 10 to 50 cm have been reported using data from various systems in different types of terrain, but these are usually better than those for InSAR-derived DTMs.

Some of these measurement errors are also relevant when considering the accuracy of a CHM, but in addition one needs to consider the penetration of the laser beam into the canopy, which depends on the tree shape and species. This causes an underestimate of the tree heights (which can be up to 1.5 m) because the beam is mainly scattered from the tree's shoulder rather than the tree top. It will also depend on the density of the vegetation and of the crown itself as well as the size of the laser footprint relative to the crown diameter. Nevertheless, after taking account of penetration and DTM errors, height accuracy of around 0.5 m is possible.

Balzter *et al.* (2007) compared the accuracies of forest stand heights measured with L-band and X-band InSAR and Lidar. For X-band measurements, the height of the scattering phase centre is affected by both forest structural parameters (canopy density, understorey, and gaps) and sensor parameters (look angle and reduced coherence through temporal and volume decorrelation). L-band, which penetrates further into the canopy, had an RMSE of 3.1 m in the near range and 6.4 m in the far range, X-band had errors of 2.9 and 4.1 m, respectively, and Lidar of only 2.0 m. A critical difference between the InSAR and Lidar datasets is that the Lidar footprint (0.3 m^2) is considerably smaller than the (airborne) SAR footprints (2.9 m^2). The test site consisted of flat terrain and homogeneous single-species, even-aged plantations. Under less ideal conditions, terrain distortions, layover, and radar shadow, producing interferometric phase noise, would reduce the quality of the SAR data.

11.6 Wildfires and biomass burning

Wildfires occur in many parts of the world, often going unnoticed unless they have significant socio-economic or climatic impact. They are usually recognized as a hazard, especially in managed and populated regions where they can cause loss of life and property, and severe economic damage. But they are in fact important natural ecosystem processes with significant implications in the climate system. Their suppression is now recognized to have created far greater fire hazard than in the past and to have disrupted natural plant succession and wildlife habitat in many areas. Forest and brush ecosystems depend to some extent on frequent, natural, low intensity (surface) fires in order to maintain their typical structure and stability. In Australian forests, for example, season by season, sclerophyll, or hard-leaved woodlands, build up huge amounts of detritus by shedding leaves and bark and twigs. This will stifle new growth unless it is removed. Many Australian species, including most of the eucalypts, need fire in order to complete their reproductive cycle. Their seeds are encased in woody receptacles that need to be split by fire before they can be released to germinate. For millennia, the aboriginal peoples managed fire proactively, but more recently, particularly in developed areas, such practice has been discouraged, residents being prevented by law in some places from clearing native vegetation, contributing, for example, to the severe fire outbreaks in Victoria in early 2009.

Fires are particularly extensive in Africa (the 'fire centre' of the world), but occur in many tropical and temperate regions where several million square kilometres of forest and grassland are destroyed each year. They are endemic in the Mediterranean basin, South America, South-East Asia, and more locally in Australia and the USA (e.g. around California). It is said that humans may be responsible for over 50% of fires, being either accidentally or deliberately set, but many do occur in uninhabited parts of the world such as in the Arctic, where lightning strikes cause much damage to the tinder-dry vegetation. It was not until satellite imagery became available that huge burnt areas were discovered in Siberia – it is now estimated that an area the size of Belgium is destroyed annually in this region alone.

Since fires occur on scales ranging from local to regional, especially in inhospitable regions, and require to be monitored and mapped on a global scale, remote sensing is an ideal tool and plays an important role at four different levels:

(i) Prediction – the identification of conditions conducive to an outbreak.

(ii) Monitoring – the observation of the fire itself in order to check its extent and progress and to direct suppression activities.

(iii) Burnt-area mapping – to assess the extent of the damage in order to plan amelioration activities.

(iv) Regeneration – to monitor the extent of regrowth over periods of months or years.

We will consider each of these levels of activity in the following sections, looking at the parameters that need to be measured, the most suitable systems available, the best wavelengths and scales to achieve these ends, and the analysis/interpretation of the data. A good overview of the problem, including many references to work in the field of wildfires, is contained in the book by Emilio Chuvieco, *Wildland fire danger, estimation and mapping, the role of remote-sensing data* (2003).

11.6.1 Prediction

If areas susceptible to fires could be identified and monitored, it is possible that much damage could be prevented. The identification stage involves the mapping of the distribution of fuel types (species composition) and conditions (living or dead) in fire-prone areas and the monitoring stage involves the recognition of fuel-moisture levels and meteorological conditions (air temperatures, wind speeds, relative humidity, precipitation, etc.) conducive to fire occurrence. Danger variables can be combined to create *danger*, or *risk indices*. These numerical values can be produced on daily/weekly/monthly timescales, and, combined with the physical conditions that may lead to fire ignition and that support fire propagation, used to estimate danger levels from low to very high and to generate large-scale maps using GIS for distribution over the internet.

Fuel types include grass and scrub, undergrowth, litter and brush, and trees. Whether vegetation is living or dead will affect its fire potential and the volume (the fuel load) and distribution will affect the severity and extent of the fire. Fuel maps, volume, and moisture content can be produced from field sampling and from aircraft and satellite data. There are added difficulties for closed canopies as the conditions of the understorey cannot easily be detected by optical remote sensing. Susceptibility to fire is strongly influenced by canopy water availability, so various water indices (see Chapter 7) such as the *NDWI* are widely used as drought indicators, while many others such as *NDMI* have also been suggested (see Wang *et al.*, 2008). In principle, however, stress indices based on canopy temperature that are indicative of stomatal closure (such as *CWSI*) are also potentially useful; these can be enhanced further by combination with multispectral estimates of canopy cover (e.g. *NDVI*) to give trapezoidal or triangular indices (see Section 9.5).

Standard classification techniques can be used to map the extent of vegetation cover, and various parameters derived from remotely sensed data can give an indication of moisture content. Actual *fuel moisture content FMC*, defined as $[(FW - DW)/DW] \times 100\%$, where FW is the fresh weight of a sample measured in the field and DW is the oven dry weight of the same sample, can be obtained from field sampling. Either *NDVI* or the scaled *NDVI** (see eqn (7.2)) may be used as a measure of greenness, with the latter sometimes expressed as a *relative greenness index* (*RGI*), but unfortunately *NDVI* is only tenuously related to moisture content. Good agreement has been found between *FMC* and MIR reflectances using both spectro-radiometric measurements and satellite data, such as TM/ETM^+ band 7 (Chuvieco *et al.*, 2002; Cibula *et al.*, 1992). This is probably because MIR reflectance is more related to the *equivalent water thickness* (*EWT*). A *global vegetation moisture index*, *GVMI*, using NIR and MIR bands was found to be sensitive to the quantity of water per unit area of the canopy (the product of leaf-area index and *EWT*) (Ceccato *et al.*, 2002).

Dead fuels are very critical to the spread of fires, but their moisture content is difficult to estimate by remote sensing. Their moisture content is, however, basically determined by atmospheric conditions, and so good local weather information can be used as a proxy for estimating their risk potential. Since there is a correlation between σ^0 and weather variables such as rainfall amounts, microwaves have been used for fire prediction. These also provide complementary information for fuel mapping, being sensitive to the temporal and spatial variation of canopy biomass. Satellite SAR data and airborne profiling radar have also been used for measuring foliar biomass, tree volume, tree height, and canopy closure, but, because of their limitations, (accurate to only a few metres in height, not usable on steep slopes, and a tendency to saturate at high biomass levels – see Fig. 11.11), they tend to be used as complements to Lidar for interpolating between Lidar footprints or for extrapolating over larger areas.

Much work has been done (see, e.g., Luke and McArthur, 1978) on the development of algorithms suitable for predicting the possibility of fires occurring in high-risk areas in terms of a *fire prediction index* (*FPI*), or *fire danger rating index* (*FDRI*). These usually involve some combination of surface temperature and various vegetation indices. They can range from simply considering threshold values for T_s and *NDVI* (Slater and Vaughan, 1999), to the use of more sophisticated *classification and regression tree* (CART) analysis in which decisions are made on successive use of various thresholds for a number of derived parameters (Silva and Pereira, 2006). Frequent monitoring of the relevant parameters enables potential fire situations to be flagged up and the relevant local authorities to be alerted.

11.6.2 Monitoring

According to the Wien distribution law (eqn (2.6)), the energy from fires peaks in the wavelength region of 2–4 μm, which is in the MIR part of the spectrum. There are a number of sensors with channels suitable for observing the hotspots, such as Landsat TM/ETM+ band 7, MERIS, MODIS, etc. The fires can be detected in the thermal wavebands (10.5–12.5 μm) but these bands often saturate because of the intensity of the radiation.[40] The fires themselves are usually obscured by smoke in the visible channels, but the longer wavelengths can penetrate the plumes that consist of ash and other fine particles (they actually are absorbed and reradiated by the clouds).

Low spatial resolution satellite data, such as AVHRR or geostationary data (e.g. from GOES), can be used for continuous monitoring over large areas

[40] The radiation energy from a fire at 800–1000 K is more than a hundred times that from a normal surface at 300 K (see Fig 2.3) which most thermal scanners are designed to detect.

of high risk potential, where hotspots can be flagged up and their locations emailed to the relevant authorities in near-real time. The operational fire detection system used in Canada, Australia, and Finland uses AVHRR data. Since 2000, the Geostationary Wildfire Automated Biomass Burning Algorithm has been generating products for the western hemisphere in real time using data from GOES with a temporal resolution of 30 min. In Europe, Calle *et al.* (2008) used an algorithm, based on SEVIRI MSG 3.9-μm and 10.8-μm bands, that compares for any pixel, $T_{3.9}$ and $T_{3.9} - T_{10.8}$ with the corresponding means and standard deviations over a 9 × 9 window to eliminate false alarms. These have been validated against higher-resolution data from MODIS, which is also useful for detection and monitoring. Most detection algorithms are pixel based, but contextual information is increasingly used. In particular, MODIS active fire detection (Csiszar *et al.*, 2005) uses a contextural algorithm, making use of the characteristic signature of an active fire at 4 μm and the brightness temperature difference between 4 μm and 11 μm, looking for pixels where these are substantially different from the background pixels. The thresholds depend on the variability of the scene. Ground-based Doppler radar, in conjunction with satellite data, has been used in Alaska to monitor fire activity, and lightning detection networks in USA and Canada enable real-time detection of lightning strikes, allowing the dispatch of aircraft to verify fire start and deploy suppression forces.

Smoke plumes are difficult to differentiate from clouds, but have different dynamics due to their turbulence and dispersion over time. Multifractal analysis has been used by Yahia *et al.* (2008) to discriminate between plume and non-plume pixels.

11.6.3 Burnt-area mapping

Fires cause long-term changes in landcover, affecting soil erosion, biodiversity, and CO_2 budget, and mapping them is important both for environmental and economic reasons. Burnt areas are usually fairly obvious in a visible image and their extent can easily be mapped using data of suitable spatial resolution. Time is not now such a constraint, so high and medium spatial resolution polar-orbiting satellites can provide such data. They can often be identified by assiduous use of classification and vegetation indices, or by noting changes between pre- and post-fire images, but the success of identification depends on the severity of the fire and on the change in signature. Not all vegetation may be burned to the same extent, producing mixed pixels. Separability also depends on the region and on the pre-fire vegetation characteristics. Visible and NIR pixel values have been thresholded, and a number of vegetation indices have been used, including *NDVI*, *NDWI*, green *GNDVI*, GEMI, *SAVI*, as well as relative greenness, land-surface temperature, and single bands, either singly or in regression trees (Escuin *et al.*, 2008). The difference between pre- and post-*NDVI*s is quite useful. A number of indices have been used, such as the *normalized burn ratio*, *NBR* (the ratio of the sum of ETM+ bands 4 (0.76–0.9 μm) and 7 (2.08–2.35 μm) to their difference), which gives good correlation with damage but is not so good when the tree cover is sparse, such as in open woods, scrub and pasture. Various other *burned-area indices* (*BAI*) have been developed. Miettinen (2007) reports the use of MODIS, which has four suitable bands with resolutions of 250 or 500 m. He found that bands 1, 2, and 7, as indices or combinations of indices, were the most sensitive when green vegetation dominates the landcover, but when dry vegetation and soil dominate, band 5 alone showed good separation and could not be bettered by any multiband index. González-Alonso *et al.* (2007) applied linear mixing algorithms to the narrow spectral bands of post-fire MERIS data at 300 m resolution using a SPOT5-HRG image and ground data for end-member definition. Roy *et al* (2002) report an automatic burned-area mapping procedure using MODIS data and taking account of the directionality of reflection (*BRDF*) from such areas. This study forms part of the *global observation of forest cover* (GOFC) initiative of the IGBP. In it, a bidirectional reflectance model is inverted against multitemporal land surface reflectance observations, providing an expectation and uncertainty of subsequent observations with time. This allows the use of a statistical measure to detect change from a previously observed state.

11.6.4 Regeneration monitoring

Monitoring the regeneration of burned areas is slightly more difficult than just identifying them, mainly because of the small scale of the regrowth in relation to the total area, and also because the change in signature may be very subtle, at least initially. Regeneration in partly burned areas may be quite difficult to see. At least initially, high-resolution aircraft data or field observations may be the best way to see what is going on, followed by high-resolution and medium-resolution satellite data in the later stages. Again, thresholding of single bands and vegetation indices, supervised classification, multivariate analysis of the original bands, spectral unmixing and time-series analysis have been used for monitoring.

11.6.5 Practical considerations

Wildfires are multiscale events, which means that no one source of data is ideally suited for their study. Spotting of small events in high-risk areas when the weather conditions are conducive to outbreaks is often done visually from towers or from reconnaissance helicopters or aeroplanes. Larger outbreaks can be flagged up in low-resolution images. High temporal resolution is the major requirement here, whereas the other three aspects, fuel mapping, burnt-area mapping, and regeneration monitoring, have much less severe time restraints. Field work plays a significant part in both fuel mapping and moisture assessment and in regeneration monitoring. Imaging spectroscopy can be used to provide biophysical details. Lidar is useful in determining the 3D geometry of trees and vegetation layers with an accuracy of up to 5–15 cm, and full waveform Lidar can give information about the understorey if the vegetation is not too dense. Microwaves can give information about biomass, but is less sensitive at high density levels.

Xue *et al.* (2008) discuss the feasibility of using smallsats for hotspot and fire detection. The experimental BIRD (Biospectral InfraRed Detection) satellite (2001–2004), launched by Germany, was the first mission dedicated to the task of active fire detection. Most systems are not designed to measure hot events and easily saturate. The hotspot recognition sensor (HSRS) was specially designed for fire detection and the estimation of fire temperature, fire size and energy release in the subpixel domain. The *Disaster Monitoring Constellation* (DMC), a network of five smallsats, provides daily imaging capability at medium resolution of 30–40 m in three or four bands. These provide fast response for emergency environmental imaging for disaster relief under the International Charter for Space and Major Disasters.

Accuracy assessment: The use of coarse spatial resolution data can lead to a reduction in accuracy both in fuel mapping and burnt-area mapping, but can be checked using higher-resolution data and field surveys. Fuel type and condition can be checked using field data. The accuracy of identifying fire events is either 100% or 0% – they either did or did not occur and the evidence will be obvious. Regeneration accuracy is best checked with field work and observation.

Websites

http://www.umass.edu/landeco/research/fragstats/documents/fragstats_documents.html

http://ec.europa.eu/research/dossier/do220307/pdf/demeter_sci_results_en.pdf

http://www.pleiades.es/

http://Earth.google.com/

http://www.eea.europa.eu/publications/CORO-landcover

http://www.countrysidesurvey.org.uk/land_cover_map.html

http://gimms.gsfc.nasa.gov

http://4e.plantphys.net/article.php?ch=t&id=289

GLOBCOVER: **http://ionia1.esrin.esa.int/index.asp**

GLC 2000: **http://bioval.jrc.ec.europa.eu/products/glc2000/glc2000.php**

MODIS colection 5 land cover map: **http://www-modis.bu.edu/landcover/page1/page8/page8.html**

References

Abdou W.A., Helmlinger M.C., Conel J.E., Bruegge C.J., Pilorz S.H., Martonchik J.V. & Gaitley B.J. (2000) Ground measurements of surface BRF and HDRF using PARABOLA III. *Journal of Geophysical Research-Atmospheres*, **106**, 11967–11976.

Achard F., Eva H.D., Stibig H.-J., Mayaux P., Gallego J., Richards T. & Malingreau J.P. (2002) Determination of deforestation rates of the world's humid tropical forests. *Science*, **297**, 999–1002.

Acock B., Charles-Edwards D.A., Fitter D.J., Hand D.W., Ludwig L.J., Warren Wilson J. & Withers A.C. (1978) The contribution of leaves from different levels within a tomato crop to canopy net photosynthesis: an experimental examination of two canopy models. *Journal of Experimental Botany*, **29**, 815–827.

Adams J.B. & Gillespie A.R. (2006) *Remote sensing of landscapes with spectral images: A physical modelling approach*. Cambridge University Press, Cambridge. pp. 362, ISBN 978-0-521-66221-5.

Adams M.L., Norvell W.A., Philpot W.D. & Peverly J.H. (2000) Spectral detection of micronutrient deficiency in 'Bragg' soybean. *Agronomy Journal*, **92**, 261–268.

Adriaensen F., Chardon J.P., De Blust G., Swinnen E., Villalba S., Gulinck H. & Matthysen E. (2003) The application of 'least-cost' modelling as a functional landscape model. *Landscape and Urban Planning*, **64**, 233–247.

Agati G., Cerovic Z.G., Marta A.D., Di Stefano V., Pinelli P., Traversi M.L. & Orlandi S. (2008) Optically-assessed preformed flavonoids and susceptibility of grapevine to *Plasmopara viticola* under different light regimes. *Functional Plant Biology*, **35**, 77–84.

Allen R.G., Pereira L.S., Raes D. & Smith M. (1999) *Crop evapotranspiration – guidelines for computing crop water requirements. FAO Irrigation and Drainage Papers 56*. FAO Land and Water Division, Rome, Italy pp. 144.

Allen R.G., Tasumi M., Morse A., Trezza R., Wright J.L., Bastiaanssen W., Kramber W., Lorite I. & Robison C.W. (2007a) Satellite-Based Energy Balance for Mapping Evapotranspiration with Internalized Calibration (METRIC) – Applications. *Journal of Irrigation and Drainage Engineering*, **133**, 395–406.

Allen R.G., Tasumi M. & Trezza R. (2007b) Satellite-Based Energy Balance for Mapping Evapotranspiration with Internalized Calibration (METRIC) – Model. *Journal of Irrigation and Drainage Engineering*, **133**, 380–394.

Allen W.A., Gausman H.W., Richardson A.J. & Thomas J.R. (1969) Interaction of isotropic light with a compact plant leaf. *Journal of the Optical Society of America*, **59**, 1376–1379.

Amolins K., Zhang Y. & Dare P. (2007) Wavelet based image fusion techniques – An introduction, review and comparison. *ISPRS Journal of Photogrammetry and Remote Sensing*, **62**, 249–263.

Anandakumar K. (2009) Sensible heat flux over a wheat canopy: optical scintillometer measurements and surface renewal analysis estimations. *Agricultural and Forest Meteorology*, **96**, 145–156.

Anderson M.C. (1981) The geometry of leaf distributions in some South-eastern Australian forests. *Agricultural Meteorology*, **25**, 195–206.

Anderson M.C., Norman J.M., Diak G.R., Kustas W.P. & Mecikalski J.R. (1997) A two-source time-integrated model for estimating surface fluxes using thermal infrared remote sensing. *Remote Sensing of Environment*, **60**, 195–216.

Anderson M.C., Norman J.M., Kustas W.P., Houborg R., Starks P.J. & Agam N. (2008) A thermal-based remote sensing technique for routine mapping of land-surface carbon, water and energy fluxes from field to regional scales. *Remote Sensing of Environment*, **112**, 4227–4241.

Andrew M.E. & Ustin S.L. (2008) The role of environmental context in mapping invasive plants with hyperspectral image data. *Remote Sensing of Environment*, **112**, 4301–4317.

Askne J.I.H., Dammert P.B.G., Ulander L.M.H. & Smith G. (1997) C-band repeat pass interferometric SAR observations of the forest. *IEEE Transactions on Geoscience and Remote Sensing*, **35**, 25–35.

Asner G.P. & Wessman C.A. (1997) Scaling PAR absorption from the leaf to landscape level in spatially heterogeneous ecosystems. *Ecological Modelling*, **103**, 81–97.

ASTM (2000) American Society for Testing and Materials Standard Extraterrestrial Spectrum Reference E-490-00. [ASTM E490 - 00a(2006) ASTM E490 - 00a (2006) Standard Solar Constant and Zero Air Mass Solar Spectral Irradiance Tables. DOI: 10.1520/E0490-00AR06]

Aston A.R. & van Bavel C.H.M. (1972) Soil surface water depletion and leaf temperature. *Agronomy Journal*, **64**, 368–373.

Atkinson P.M., Foody G.M., Darby S.E. & Wu F., eds (2004) *GeoDynamics*. CRC Press, Boca Raton, FL., pp. 413, ISBN 0849328373.

Atzberger C. (2004) Object-based retrieval of biophysical canopy variables using artificial neural nets and radiative transfer models. *Remote Sensing of Environment*, **93**, 53–67.

Azam-Ali S.N., Crout N.M.J. & Bradley R.G. (1994) Perspectives in modelling resource capture by crops. In: *Resource Capture by Crops* (eds Monteith, J.L., Scott, R.K. & Unsworth, M.H.), pp. 125–148. Nottingham University Press, Nottingham. ISBN 1-897676-21-2.

Baldridge A.M., Hook S.J., Grove C.I. & Rivera G. (2009) The ASTER spectral library version 2.0. *Remote Sensing of Environment*, **113**, 711–715.

Baltsavias E., Gruen A., Eisenbeiss H., Zhang L. & Waser L.T. (2008) High-quality image matching and automated generation of 3D tree models. *International Journal of Remote Sensing*, **29**, 1243–1259.

Balzter H., Luckman A., Skinner L., Rowland C. & Dawson T.P. (2007) Observations of forest stand top height and mean height from interferometric SAR and LiDAR over a conifer plantation at Thetford Forest, UK. *International Journal of Remote Sensing*, **28**, 1173–1197.

Baret F. (1995) Use of spectral reflectance variation to retrieve canopy biophysical characteristics. In: *Advances in environmental remote sensing* (eds Danson, F.M. & Plummer, S.E.), pp. 33–51. John Wiley & Sons, Chichester.

Baret F. & Buis S. (2008) Estimating canopy characteristics from remote sensing observations: Review of methods and associated problems. In: *Advances in land remote sensing: system, modeling inversion and application*, pp. 173–201. Chinese Academy of Science, Beijing, People's Republic of China, 9th International Symposium on Physical Measurements and Signatures in Remote Sensing, Oct. 2005.

Baret F., Clevers J.G.P.W. & Steven M. (1995) The robustness of canopy gap fraction estimates from red and near-infrared reflectances: A comparison of approaches. *Remote Sensing of Environment*, **54**, 141–151.

Baret F. & Guyot G. (1991) Potentials and limits of vegetation indices for LAI and APAR assessment. *Remote Sensing of Environment*, **35**, 161–173.

Baret F., Guyot G. & Major D. (1989) *TSAVI: a vegetation index which minimises soil brightness effects on LAI and APAR estimation.* Paper presented at the 12th Canadian conference on Remote Sensing and IGARSS '90, Vancouver, BC.

Baret F., Jacquemoud S. & Hanocq J.F. (1993) The soil line concept in remote sensing. *Remote Sensing Reviews*, **7**, 65–82.

Barrett E.C. & Curtis L.F. (1999) *Introduction to Environmental Remote Sensing*. (4th ed.). Stanley Thornes (Publishers) Ltd, Cheltenham, UK. pp. 457, ISBN 0412371707.

Bartholic J., Namken L.N. & Wiegand C.L. (1972) Aerial thermal scanner to determine temperatures of soils and crop canopies differing in water stress. *Agronomy Journal*, **64**, 603–608.

Bastiaanssen W.G.M., Noordman E.J.M., Pelgrum H., Davids G., Thoreson B.P. & Allen R.G. (2005) SEBAL Model with Remotely Sensed Data to Improve Water-Resources Management under Actual Field Conditions. *Journal of Irrigation and Drainage Engineering*, **131**, 85–93.

Bastiaanssen W.G.M., Menenti M., Feddes R.A. & Holtslag A.A.M. (1998a) A remote sensing surface energy balance algorithm for land (SEBAL) – 1. Formulation. *Journal of Hydrology*, **212–213**, 198–212.

Bastiaanssen W.G.M., Pelgrum H., Wang J., Ma Y., Moreno J.F., Roerink G.J. & van der Wal T. (1998b) A remote sensing surface energy balance algorithm for land (SEBAL) – 2. Validation. *Journal of Hydrology*, **212–213**, 213–229.

Baumgardner M.F., Silva L.F., Biehl L.L. & Stoner E.R. (1985) Reflectance properties of soils. *Advances in Agronomy*, **38**, 1–44.

Becker F., Seguin B., Phulpin T. & Durpaire J.P. (1996) IRSUTE, a small satellite for water budget estimate with high resolution infrared imagery. *Acta Astronautica*, **39**, 883–897.

Beljaars A.C.M. & Bosveld F.C. (1997) Cabauw data for the validation of land surface parameterization schemes. *Journal of Climate*, **10**, 1172–1193.

Ben-Dor E. (2002) Quantitative remote sensing of soil properties. *Advances in Agronomy*, **75**, 173–243.

Berberoglu S. & Akin A. (2009) Assessing different remote sensing techniques to detect land use/cover changes in the eastern Mediterranean. *International Journal of Applied Earth Observation and Geoinformation*, **11**, 46–53.

Berk A., Bernstein L.S., Anderson G.P., Acharya P.K., Robertson D.C., Chetwynd J.H. & Adler-Golden S.M. (1998) MODTRAN cloud and multiple scattering upgrades with application to AVIRIS. *Remote Sensing of Environment*, **65**, 367–375.

Betts A.K. & Ball J.H. (1997) Albedo over the boreal forest. *Journal of Geophysical Research*, **102**, 28901–28909.

Bidel L.P.R., Meyer S., Goulas Y., Cadot Y. & Cerovic Z.G. (2007) Responses of epidermal phenolic compounds to light acclimation: In vivo qualitative and quantitative assessment using chlorophyll fluorescence excitation spectra in leaves of three woody species. *Journal of Photochemistry and Photobiology, B: Biology*, **88**, 163–179.

Bilger W., Björkman O. & Thayer S. (1989) Light-induced Spectral Absorbance Changes in Relation to Photosynthesis and the Epoxidation State of Xanthophyll Cycle Components in Cotton Leaves. *Plant Physiology*, **91**, 542–551.

Bilger W., Veit M., Schreiber L. & Schreiber U. (1997) Measurement of leaf epidermal transmittance of UV radiation by chlorophyll fluorescence. *Physiologia Plantarum*, **101**, 754–763.

Bird R. & Riordan C. (1984) *Simple spectral model for direct and diffuse irradiance on horizontal and tilted planes at the earth's surface for cloudless atmospheres* (TR-215-2436). Solar Energy Research Institute, Golden, CO.

Birth G.S. & McVey G.R. (1968) Measuring the color of growing turf with a reflectance spectrophotometer. *Agronomy Journal*, **60**, 640–643.

Bisht G., Venturini V., Islam S. & Jiang L. (2005) Estimation of the net radiation using MODIS (Moderate Resolution Imaging Spectroradiometer) data for clear sky days. *Remote Sensing of Environment*, **97**, 52–67.

Black S.C. & Guo X. (2008) Estimation of grassland CO_2 exchange rates using hyperspectral remote sensing techniques. *International Journal of Remote Sensing*, **29**, 145–155.

Blackburn G.A. (1998a) Quantifying chlorophylls and carotenoids at leaf and canopy scales. An evaluation of some hyperspectral approaches. *Remote Sensing of Environment*, **66**, 273–285.

Blackburn G.A. (1998b) Spectral indices for estimating photosynthetic pigment concentrations: a test using senescent tree leaves. *International Journal of Remote Sensing*, **19**, 657–675.

Blackburn G.A. (2007a) Hyperspectral remote sensing of plant pigments. *Journal of Experimental Botany*, **58**, 855–867.

Blackburn G.A. (2007b) Wavelet decomposition of hyperspectral data: a novel approach to quantifying pigment concentrations in vegetation. *International Journal of Remote Sensing*, **28**, 2831–2855.

Blackburn G.A. & Ferwerda J.G. (2008) Retrieval of chlorophyll concentration from leaf reflectance spectra using wavelet analysis. *Remote Sensing of Environment*, **112**, 1614–1632.

Blaschke T. (2010) Object based image analysis for remote sensing. *ISPRS Journal of Photogrammetry and Remote Sensing*, **65**, 2–16.

Boegh E., Soegaard H., Hanan N., Kabat P. & Lesch L. (1999) A remote sensing study of the NDVT-T_s relationship and the transpiration from sparse vegetation in the Sahel based on high-resolution satellite data. *Remote Sensing of Environment*, **69**, 224–240.

Borel C.C. (1996) Non-linear spectral mixing theory to model multi-spectral signatures. In: *Applied Geologic Remote Sensing Conference: Practical solutions for real world problems, February 1996*, Las Vegas, NV.

Borel C., Gerstl S. & Powers B.J. (1991) The radiosity method in optical remote sensing of structured 3-D surfaces. *Remote Sensing of Environment*, **36**, 13–44.

Bouttier F. & Courtier P. (2002) *Data assimilation concepts and methods March 1999*. European Centre for Medium-Range Weather Forecasts (ECMWF), Reading.

Bramson M.A. (1968) *Infrared Radiation – a handbook of applications.* (2nd ed.). Plenum Press, New York. pp. 623.

Bréda N.J.J. (2003) Ground-based measurements of leaf area index: a review of methods, instruments and current controversies. *Journal of Experimental Botany*, **54**, 2403–2417.

Briedenbach R.W., Saxton M.J., Hansen L.D. & Criddle R.S. (1997) Heat generation and dissipation in plants: can the alternative oxidative phosphorylation pathway serve a thermoregulatory role in plant tissues other than specialised organs? *Plant Physiology*, **114**, 1137–1140.

Broge N.H. & Leblanc E. (2001) Comparing prediction power and stability of broadband and hyperspectral vegetation indices for estimation of green leaf area index and canopy chlorophyll density. *Remote Sensing of Environment*, **76**, 156–172.

Brutsaert W.H. (1982) *Evaporation into the atmosphere: Theory, history and applications.* D. Reidel Publishing Co., Dordrecht, Netherlands. pp. 316, ISBN 9789027712479.

Brutsaert W. & Sugita M. (1996) Sensible heat transfer parameterization for surfaces with anisothermal dense vegetation. *Journal of The Atmospheric Sciences*, **53**, 209–216.

Buck A.L. (1981) New equations for computing vapor pressure and enhancement factor. *Journal of Applied Meteorology*, **20**, 1527–1532.

Buddenbaum H., Schlerf M. & Hill J. (2005) Classification of coniferous tree species and age classes using hyperspectral data and geostatistical methods. *International Journal of Remote Sensing*, **26**, 5453–5465.

Buschmann C., Langsdorf G. & Lichtenthaler H.K. (2000) Imaging of the blue, green, and red fluorescence emission of plants: An overview. *Photosynthetica*, **38**, 483–491.

Buschmann C. & Nagel E. (1993) In vivo spectroscopy and internal optics of leaves as basis for remote sensing of vegetation. *International Journal of Remote Sensing*, **14**, 711–722.

Calle A., González-Alonso F. & Merino de Miguel S. (2008) Validation of active forest fires detected by MSG-SEVIRI by means of MODIS hot spots and AWiFS images. *International Journal of Remote Sensing*, **29**, 3407–3415.

Camargo A. & Smith J.S. (2009) An image-processing based algorithm to automatically identify plant disease visual symptoms. *Biosystems Engineering*, **102**, 9–21.

Campbell G.S. (1986) Extinction coefficients for radiation in plant canopies calculated using an ellipsoidal inclination angle distribution. *Agricultural and Forest Meteorology*, **36**, 317–321.

Campbell G.S. (1990) Derivation of an angle-density function for canopies with ellipsoidal leaf angle distributions. *Agricultural and Forest Meteorology*, **49**, 173–176.

Campbell G.S. & Norman J.M. (1998) *An introduction to environmental biophysics.* (2nd ed.). Springer, New York. pp. 286, ISBN 0387949372.

Campbell G.S. & van Evert F.K. (1994) Light interception by plant canopies: efficiency and architecture. In: *Resource capture by crops* (eds Monteith, J.L., Scott, R.K. & Unsworth, M.H.), pp. 35–52. Nottingham University Press, Nottingham. ISBN 1897676212.

Campbell J.B. (2007) *Introduction to remote sensing.* (4th ed.). Taylor and Francis, London. pp. 546, ISBN 9780415416887.

Campbell P.K.E., Middleton E.M., McMurtrey J.E., Corp L.A. & Chappelle E.W. (2007) Assessment of vegetation stress using reflectance or fluorescence measurements. *Journal of Environmental Quality*, **36**, 832–845.

Carlson T.N., Capehart W.J. & Gillies R.R. (1995) A new look at the simplified method for remote-sensing of daily evapotranspiration. *Remote Sensing of Environment*, **54**, 161–167.

Carlson T.N. & Ripley D.A. (1997) On the relation between NDVI, fractional vegetation cover and leaf area index. *Remote Sensing of Environment*, **62**, 241–252.

Carter G.A. (1991) Primary and secondary effects of water content on the spectral reflectance of leaves. *American Journal of Botany*, **78**, 916–924.

Carter G.A. (1994) Ratios of leaf reflectances in narrow wavebands as indicators of plant stress. *International Journal of Remote Sensing*, **15**, 697–703.

Carter G.A. & Knapp A.K. (2001) Leaf optical properties in higher plants: linking spectral characteristics to stress and chlorophyll concentration. *American Journal of Botany*, **88**, 677–684.

Carter G.A., Theisen A.F. & Mitchell R.J. (1990) Chlorophyll fluorescence measured using the Fraunhofer line-depth principle and relationship to photosynthetic rate in the field. *Plant, Cell and Environment*, **13**, 79–83.

Casa R. (2003) Multiangular remote sensing of crop canopy structure for plant stress monitoring. Ph.D., University of Dundee, Dundee.

Casa R. & Jones H.G. (2005) LAI retrieval from multiangular image classification and inversion of a ray tracing model. *Remote Sensing of Environment*, **98**, 414–428.

Ceccato P., Flasse S. & Gregoire J.M. (2002) Designing a spectral index to estimate vegetation water content from remote sensing data – Part 2. Validation and applications. *Remote Sensing of Environment*, **82**, 198–207.

Čermák J., Gašpárek J., De Lorenzi F. & Jones H.G. (2007) Stand biometry and leaf area distribution in an old olive grove at Andria, southern Italy. *Annals of Forest Science*, **64**, 491–501.

Chaerle L., Hagenbeek D., Vanrobaeys X. & Van Der Straeten D. (2007) Early detection of nutrient and biotic stress in *Phaseolus vulgaris*. *International Journal of Remote Sensing*, **28**, 3479–3492.

Chaerle L., Pineda M., Romero-Aranda R., Van Der Straeten D. & Barón M. (2006) Robotized thermal and chlorophyll fluorescence imaging of Pepper Mild Mottle Virus infection in *Nicotiana benthamiana*. *Plant & Cell Physiology*, **47**, 1323–1336.

Chavez Jr P.S. (1996) Image-based atmospheric corrections – Revisited and improved. *Photogrammetric Engineering and Remote Sensing*, **62**, 1025–1036.

Chelle M., Andrieu B. & Bouatouch K. (1998) Nested radiosity for plant canopies. *The Visual Computer*, **14**, 109–125.

Chen J.M. (1996) Optically-based methods for measuring seasonal variation of leaf area index in boreal conifer stands. *Agricultural and Forest Meteorology*, **80**, 135–163.

Chen D.Y., Huang J.F. & Jackson T.J. (2005) Vegetation water content estimation for corn and soybeans using spectral indices derived from MODIS near- and short-wave infrared bands. *Remote Sensing of Environment*, **98**, 225–236.

Chen J.M. & Leblanc S.G. (1997) A four-scale bidirectional reflectance model based on canopy architecture. *IEEE Transactions on Geoscience and Remote Sensing*, **35**, 1316–1337.

Cho M.A. & Skidmore A.K. (2006) A new technique for extracting the red edge position from hyperspectral data: The linear extrapolation method. *Remote Sensing of Environment*, **101**, 181–193.

Cho M.A., Skidmore A.K. & Atzberger C. (2008) Towards red-edge positions less sensitive to canopy biophysical parameters for leaf chlorophyll estimation using properties optique spectrales des feuilles (PROSPECT) and scattering by arbitrarily inclined leaves (SAILH) simulated data. *International Journal of Remote Sensing*, **29**, 2241–2255.

Choudhury B.J. (1994) Synergism of multispectral satellite observations for estimating regional land surface evaporation. *Remote Sensing of Environment*, **49**, 264–274.

Choudhury B.J., Ahmed N.U., Idso S.B., Reginato R.J. & Daughtry C.S.T. (1994) Relations between evaporation coefficients and vegetation indexes studied by model simulations. *Remote Sensing of Environment*, **50**, 1–17.

Chuvieco E., ed. (2003) *Wildland fire danger estimation and mapping, the rôle of remote sensing data.* World Scientific, Singapore. pp. 280, ISBN 978-9812385697.

Chuvieco E., Riaño D., Aguado I. & Cocero D. (2002) Estimation of fuel moisture content from multitemporal analysis of Landsat Thematic Mapper reflectance data: applications in fire danger assessment. *International Journal of Remote Sensing*, **23**, 2145–2162.

Cibula W.G., Zetca E.F. & Rickman D.L. (1992) Response of Thematic Mapper bands to plant water stress. *International Journal of Remote Sensing*, **13**, 1869–1880.

Clark J.A. & Wigley G. (1975) Heat and mass transfer from real and model leaves. In: *Heat and mass transfer in the biosphere* (eds De Vries, D.A. & Afgan, N.H.), pp. 413–422. Scripta Book Co., New York.

Clark R.N. (1999) Spectroscopy of rocks and minerals, and principles of spectroscopy. In: *Manual of remote sensing, vol. 3: Remote sensing for the earth sciences* (ed. Rencz, A.N.), pp. 3–58. John Wiley and Sons, New York. ISBN 9780471294054.

Clark R.N., Swayze G.A., Wise R., Livo E., Hoefen T., Kokaly R. & Sutley S.J. (2007) *USGS digital spectral library splib06a* (Digital data series 231). USGS Spectroscopy Laboratory, Denver, CO, USA.

Clewer A.G. & Scarisbrick D.H. (2001) *Practical statistics and experimental design for plant and crop science.* John Wiley & Sons, Ltd., Chichester. pp. 332, ISBN 0 471 89909 7.

Cohen S. & Fuchs M. (1987) The distribution of leaf area, radiation, photosynthesis and transpiration in a Shamouti orange hedgerow orchard. I. Leaf area and radiation. *Agricultural and Forest Meteorology*, **40**, 123–144.

Colwell R.N. (1956) Determining the prevalence of certain cereal crop diseases by means of aerial photography. *Hilgardia*, **26**, 223–286.

Combal B., Baret F., Weiss M., Trubuil A., Macé D., Pragnère A., Myneni R., Knyazikhin Y. & Wang L. (2003) Retrieval of canopy biophysical variables from bidirectional reflectance – Using prior information to solve the ill-posed inverse problem. *Remote Sensing of Environment*, **84**, 1–15.

Congalton R.G. & Green K. (2008) *Assessing the accuracy of remotely sensed data: principles and practices.* (2nd ed.). CRC Press, Taylor & Francis Group, Boca Raton, FL. pp. 183, ISBN 978-1420055122.

Congalton R.G., Oderwald R. & Mead R.A. (1983) Landsat classification accuracy using discrete multivariate statistical techniques. *Photogrammetric Engineering and Remote Sensing*, **49**, 1671–1678.

Coops N.C., Wulder M.A. & Iwanicka D. (2009a) Exploring the relative importance of satellite-derived descriptors of production, topography and land cover for predicting breeding bird species richness over Ontario, Canada. *Remote Sensing of Environment*, **113**, 668–679.

Coops N.C., Wulder M.A. & Iwanicka D. (2009b) Large area monitoring with a MODIS-based disturbance index (DI) sensitive to annual and seasonal variations. *Remote Sensing of Environment*, **113**, 1250–1261.

Coudert B., Ottlé C. & Briottet X. (2008) Monitoring land surface processes with thermal infrared data: Calibration of SVAT parameters based on the optimisation of diurnal surface temperature cycling features. *Remote Sensing of Environment*, **112**, 872–887.

Courault D., Aloui B., Lagouarde J.-P., Clastre P., Nicolas H. & Walter C. (1996) Airborne thermal data for evaluating the spatial distribution of actual evapotranspiration over a watershed in oceanic climatic conditions – application of semi-empirical models. *International Journal of Remote Sensing*, **17**, 2281–2302.

Cox C.B. & Moore P.D. (2000) *Biogeography: An ecological and evolutionary approach.* (6th ed.). Blackwell Science, Oxford. pp. 298, ISBN 086542778X.

Cracknell A.P. (1997) *The advanced very high resolution radiometer (AVHRR).* Taylor and Francis, London. pp. 534, ISBN 0748402098.

Cracknell A.P. (2008) Synergy in remote sensing – what's in a pixel? *International Journal of Remote Sensing*, **19**, 2025–2047.

Cracknell A.P. & Hayes L.W.B. (2007) *Introduction to remote sensing.* (2nd ed.). CRC Press, Boca Raton, FL. pp. 293, ISBN 0850663350.

Crist E.P. & Cicone R.C. (1984) A physically-based transformation of Thematic Mapper data – the TM Tasseled Cap. *IEEE Transactions on Geoscience and Remote Sensing*, **GE-22**, 256–263.

Csiszar I., Denis L., Giglio L., Justice C.O. & Hewson J. (2005) Global fire activity from two years of MODIS data. *International Journal of Wildland Fire*, **14**, 117–130.

Curcio J.A. & Petty C.C. (1951) Extinction coefficients for pure liquid water. *Journal of Optical Society of America*, **41**, 302–304.

Curran P.J. (1985) *Principles of remote sensing.* Longman, London and New York.

Curran P.J. (1989) Remote sensing of foliar chemistry. *Remote Sensing of Environment*, **30**, 271–278.

Curran P.J. & Atkinson P.M. (1999) Issues of scale and optimal pixel size. In: *Spatial statistics for remote sensing* (eds Stein, A., van der Meer, F. & Gorte, B.), pp. 115–134. Kluwer Academic Publishers, Dordrecht.

Curran P.J. & Dash J. (2005) *Algorithm theoretical basis document ATBD 2.22: chlorophyll index.* MERIS ESL, Southampton.

Damm A., Elbers J., Erler A., Gioli B., Hamdi K., Hutjes R., Kosvancova M., Meroni M., Miglietta F., Moersch A., Moreno J., Schickling A., Sonnenschein R., Udelhoven T., van der Linden S., Hostert P. & Rascher U. (2010) Remote sensing of sun-induced fluorescence to improve modelling of diurnal courses of gross primary production (GPP). *Global Change Biology*, **16**, 171–186.

Darby S.E., Wu F., Atkinson P.M. & Foody G.M. (2005) Spatially distributed dynamic modelling. In: *GeoDynamics* (eds Atkinson, P.M., Foody, G.M., Darby, S.E. & Wu, F.), pp. 121–124. CRC Press, Boca Raton, FL.

Darvishzadeh R., Skidmore A., Atzberger C. & van Wieren S. (2008) Estimation of vegetation LAI from hyperspectral reflectance data: Effects of soil type and plant architecture. *International Journal of Applied Earth Observation and Geoinformation*, **10**, 358–373.

Dash P., Göttsche F.-M., Olesen F.-S. & Fischer H. (2002) Land surface temperature and emissivity estimation from passive sensor data: theory and practice - current trends. *International Journal of Remote Sensing*, **23**, 2563–2594.

Daughtry C.S.T., Walthall C.L., Kim M.S., Brown de Colstoun E. & McMurtrey III J.E. (2000) Estimating corn leaf chlorophyll concentration from leaf and canopy reflectance. *Remote Sensing of Environment*, **74**, 229–239.

Dawson T.P. & Curran P.J. (1998) A new technique for interpolating the reflectance red edge position. *International Journal of Remote Sensing*, **19**, 2133–2139.

de Smith M.J., Goodchild M.F. & Longley P.A. (2009) *Geospatial analysis: a comprehensive guide to principles, techniques and software tools*. (3rd ed.). Matador, 9 de Montfort Mews, Leicester, ISBN 9781848761582.

de Wit A.J.W. & van Diepen C.A. (2007) Crop model data assimilation with the Ensemble Kalman filter for improving regional crop yield forecasts. *Agricultural and Forest Meteorology*, **146**, 38–56.

Deering D.W., Rouse J.W., Haas R.H. & Schell J.A. (1975) Measuring "forage production" of grazing units from Landsat MSS data. In: *10th International Symposium on Remote Sensing of Environment, II*, pp. 1169–1178, University of Michigan, Ann Arbor.

Deneke H.M., Feijt A.J. & Roebeling R.A. (2008) Estimating surface solar irradiance from METEOSAT SEVIRI-derived cloud properties. *Remote Sensing of Environment*, **112**, 3131–3141.

Doan H.T.X. & Foody G.M. (2007) Increasing soft classification accuracy through the use of an ensemble of classifiers. *International Journal of Remote Sensing*, **28**, 4609–4623.

Dorigo W.A., Zurita-Milla R., de Wit A.J.W., Brazile J., Singh R. & Schaepman M.E. (2007) A review on reflective remote sensing and data assimilation techniques for enhanced agroecosystem modeling. *International Journal of Applied Earth Observation and Geoinformation*, **9**, 165–193.

Dufour A., Gadallah F., Wagner H.H., Guisan A. & Buttler A. (2006) Plant species richness and environmental heterogeneity in a mountain landscape: effects of variability and spatial configuration. *Ecography*, **29**, 573–584.

Duong V.H., Lindenbergh R., Pfeifer N. & Vosselman G. (2008) Single and two epoch analysis of ICESat full waveform data over forested areas. *International Journal of Remote Sensing*, **29**, 1453–1473.

Durbha S.S., King R.L. & Younan N.H. (2007) Support vector machines regression for retrieval of leaf area index from multiangle imaging spectroradiometer. *Remote Sensing of Environment*, **107**, 348–361.

Dytham C. (2003) *Choosing and using statistics*. (2nd ed.). Blackwell Scientific Publications, Oxford. pp. 264, ISBN 9781405102438.

Ehleringer J.R. & Forseth I. (1980) Solar tracking by plants. *Science*, **210**, 1094–1098.

Ehleringer J.R., Björkman O. & Mooney H.A. (1976) Leaf pubescence: Effects on absorptance and photosynthesis in a desert shrub. *Science*, **192**, 376–377.

Eller B.M. (1977) Leaf pubescence: the significance of lower surface hairs for the spectral properties of the upper surface. *Journal of Experimental Botany*, **28**, 1054–1059.

Ellingson R.G. (1995) Surface longwave fluxes from satellite observations - a critical review. *Remote Sensing of Environment*, **51**, 89–97.

El-Shikha D.M., Waller P., Hunsaker D., Clarke T. & Barnes E. (2007) Ground-based remote sensing for assessing water and nitrogen status of broccoli. *Agricultural Water Management*, **92**, 183–193.

Escuin S., Navarro R. & Fernández P. (2008) Fire severity assessment by using NBR (Normalized Burn Ratio) and NDVI (Normalized Difference Vegetation Index) derived from LANDSAT TM/ETM images. *International Journal of Remote Sensing*, **29**, 1053–1073.

Eshel G., Levy G.J. & Singer M.J. (2004) Spectral reflectance properties of crusted soils under solar illumination. *Soil Science Society of America Journal*, **68**, 1982–1991.

España M.L., Baret F. & Weiss M. (2008) Slope correction for LAI estimation from gap fraction measurements. *Agricultural and Forest Meteorology*, **148**, 1553–1562.

Evans G.C. & Coombe D.E. (1959) Hemispherical and woodland canopy photography and the light climate. *Journal of Ecology*, **47**, 103–113.

Evensen G. (2002) Sequential Data Assimilation for Nonlinear Dynamics: The Ensemble Kalman Filter. In: *Ocean Forecasting: Conceptual basis and applications* (eds Pinardi, N. & Woods, J.), pp. 101–120. Springer-Verlag, Berlin.

Fava F., Colombo R., Bocchi S., Meroni M., Sitzia M., Fois N. & Zucca C. (2009) Identification of hyperspectral vegetation indices for Mediterranean pasture characterization. *International Journal of Applied Earth Observation and Geoinformation*, **11**, 233–243.

Feret J.-B., François C., Asner G.P., Gitelson A.A., Martin R.E., Bidel L.P.R., Ustin S.L., le Maire G. & Jacquemoud S. (2008) PROSPECT-4 and 5: Advances in the leaf optical properties model separating photosynthetic pigments. *Remote Sensing of Environment*, **112**, 3030–3043.

Ferwerda J.G. & Skidmore A.K. (2007) Can nutrient status of four woody plant species be predicted using field spectrometry? *ISPRS Journal of Photogrammetry and Remote Sensing*, **62**, 406–414.

Fitter A.H. & Hay R.K.M. (2001) *Environmental physiology of plants*. (3rd ed.). Academic Press, London. pp. 1–367, ISBN 0122577663.

Foody G.M. (2008) RVM-based multi-class classification of remotely sensed data. *International Journal of Remote Sensing*, **29**, 1817–1823.

Foody G.M. (2009a) Classification accuracy comparison: Hypothesis tests and the use of confidence intervals in evaluations of difference, equivalence and non-inferiority. *Remote Sensing of Environment*, **113**, 1658–1663.

Foody G.M. (2009b) The impact of imperfect ground reference data on the accuracy of land cover change estimation. *International Journal of Remote Sensing*, **30**, 3275–3281.

Foody G.M. & Atkinson P.M., eds (2002) *Uncertainty in remote sensing and GIS*. John Wiley and Sons, London.

Foody G.M. & Cutler M.E.J. (2003) Tree biodiversity in protected and logged Bornean tropical rain forests and its measurement by satellite remote sensing. *Journal of Biogeography*, **30**, 1053–1066.

Foody G.M., Darby S. & Wu F. (2004) *GeoDynamics*. Taylor and Francis, Boca Raton. pp. 440, ISBN 9780849328374

Fowler J. & Cohen L. (1990) *Practical statistics for field biology*. John Wiley & Sons, Chichester, ISBN 0471932191.

Friedl M.A., Sulla-Menashe D., Tan B., Schneider A., Ramankutty N., Sibley A. & Huang X. (2010) MODIS Collection 5 global land cover: Algorithm refinements and characterization of new datasets. *Remote Sensing of Environment*, **114**, 168–182.

Fuchs M. (1990) Infrared measurement of canopy temperature and detection of plant water stress. *Theoretical and Applied Climatology*, **42**, 253–261.

Futch S.H. & Tucker D.P.H. (2001) *A Guide to Citrus Nutritional Deficiency and Toxicity Identification HS-797* (HS-797). Florida Cooperative Extension Service, Gainesville, FL.

Gamon J.A., Field C.B., Bilger W., Björkman O., Fredeen A.L. & Peñuelas J. (1990) Remote sensing of the xanthophyll cycle and chlorophyll fluorescence in sunflower leaves and canopies. *Oecologia*, **85**, 1–7.

Gamon J.A., Peñuelas J. & Field C.B. (1992) A narrow-waveband spectral index that tracks diurnal changes in photosynthetic efficiency. *Remote Sensing of Environment*, **41**, 35–44.

Ganguly S., Schull M.A., Samanta A., Shabanov N.V., Milesi C., Nemani R.R., Knyazikhin Y. & Myneni R.B. (2008) Generating vegetation leaf area index earth system data record from multiple sensors. Part 1: Theory. *Remote Sensing of Environment*, **112**, 4333–4343.

Gao B.-C. (1996) NDWI – A normalized difference water index for remote sensing of vegetation liquid water from space. *Remote Sensing of Environment*, **58**, 257–266.

Gardner B.R., Blad B.L. & Watts D.G. (1981) Plant and air temperatures in differentially irrigated corn. *Agricultural Meteorology*, **25**, 207–217.

Garrigues S., Shabanov N.V., Swanson K., Morisette J.T., Baret F. & Myneni R.B. (2008) Intercomparison and sensitivity analysis of Leaf Area Index retrievals from LAI-2000, AccuPAR, and digital hemispherical photography over croplands. *Agricultural and Forest Meteorology*, **148**, 1193–1209.

Gash J.H.C., Nobre C.A., Roberts J.M. & Victoria R.L., eds (1996) *Amazonian deforestation and climate*. John Wiley & Sons, Chichester, pp. 638, ISBN 978-0471967347.

Gastellu-Etchegorry J.P., Martin E. & Gascon F. (2004) DART: a 3D model for simulating satellite images and studying surface radiation budget. *International Journal of Remote Sensing*, **25**, 73–96.

Gates D.M. (1980) *Biophysical Ecology*. Springer Verlag, New York. pp. 1–611, ISBN 038790414X. [Dover Publications Inc. Reprint edition 2003]

Gausman H.W., Allen W.A. & Escobar D.C. (1974) Refractive index of plant cell walls. *Applied Optics*, **13**, 109–111.

Geerken R.A. (2009) An algorithm to classify and monitor seasonal variations in vegetation phenologies and their inter-annual change. *ISPRS Journal of Photogrammetry and Remote Sensing*, **64**, 422–431.

Ghulam A., Li Z.-L., Qin Q., Yimit H. & Wang J. (2008) Estimating crop water stress with ETM+ NIR and SWIR data. *Agricultural and Forest Meteorology*, **148**, 1679–1695.

Gillespie A., Rokugawa S., Matsunaga T., Cothern J.S., Hook S. & Kahle A.B. (1998) A temperature emissivity separation algorithm for advanced spaceborne thermal emission and reflection radiometer (ASTER) images. *IEEE Transactions on Geoscience and Remote Sensing*, **36**, 1113–1126.

Gillies R.R., Kustas W.P. & Humes K.S. (1997) A verification of the 'triangle' method for obtaining surface soil water content and energy fluxes from remote measurements of the Normalized Difference Vegetation Index (NDVI) and surface radiant temperature. *International Journal of Remote Sensing*, **18**, 3145–3166.

Gitelson A.A. (2004) Wide dynamic range vegetation index for remote quantification of biophysical characteristics of vegetation. *Journal of Plant Physiology*, **161**, 165–173.

Gitelson A.A., Buschmann C. & Lichtenthaler H.K. (1999) The Chlorophyll Fluorescence Ratio F_{735}/F_{700} as an Accurate Measure of the Chlorophyll Content in Plants. *Remote Sensing of Environment*, **69**, 296–302.

Gitelson A.A., Kaufman Y.J. & Merzlyak M.N. (1996) Use of a green channel in remote sensing of global vegetation from EOS-MODIS. *Remote Sensing of Environment*, **58**, 289–298.

Gitelson A.A. & Merzlyak M.N. (1997) Remote estimation of chlorophyll content in higher plant leaves. *International Journal of Remote Sensing*, **18**, 2691–2697.

Giunta F., Motzo R. & Pruneddu G. (2008) Has long-term selection for yield in durum wheat also induced changes in leaf and canopy traits? *Field Crops Research*, **106**, 68–76.

Gobron N., Pinty B., Verstraete M.M. & Widlowski J.-L. (2000) Advanced vegetation indices optimized for up-coming sensors: design, performance and applications. *IEEE Transactions on Geoscience and Remote Sensing*, **38**, 2489–2504.

Gobron N., Pinty B., Verstraete M.M. & Govaerts Y. (1997) A semidiscrete model for the scattering of light by vegetation. *Journal of Geophysical Research*, **102**, 9431–9446.

Goel N.S. (1989) Inversion of canopy reflectance models for estimation of biophysical parameters from reflectance data. In: *Theory and application of optical remote sensing* (ed. Asrar, G.), pp. 205–251. John Wiley and Sons, Chichester.

Goel N.S., Rozehnal I. & Thompson R.L. (1991) A computer graphics based model for scattering from objects of arbitrary shapes in the optical region. *Remote Sensing of Environment*, **36**, 73–104.

Goel N.S. & Strebel D.E. (1984) Simple beta distribution representation of leaf orientation in vegetation canopies. *Agronomy Journal*, **76**, 800–802.

Goel N.S. & Thompson R.L. (1984) Inversion of vegetation canopy reflectance models for estimating agronomic variables. V. Estimation of leaf area index and average leaf angle using measured canopy reflectances. *Remote Sensing of Environment*, **16**, 69–85.

Goetz S., Steinberg D., Dubayah R. & Blair B. (2007) Laser remote sensing of canopy habitat heterogeneity as a predictor of bird species richness in an eastern temperate forest, USA. *Remote Sensing of Environment*, **108**, 254–263.

González-Alonso F., Merino de Miguel S., Roldán-Zamarrón A., García-Gigorro S. & Cuevas J.M. (2007) MERIS Full Resolution data for mapping level-of-damage caused by forest fires: the Valencia de Alcántara event in August 2003. *International Journal of Remote Sensing*, **28**, 797–809.

Goudriaan J. (1977) *Crop micrometeorology: a simulation study*. PUDOC, Wageningen. pp. 249.

Goulas Y., Cerovic Z.G., Cartelat A. & Moya I. (2004) Dualex: a new instrument for field measurements of epidermal ultraviolet absorbance by chlorophyll fluorescence. *Applied Optics*, **43**, 4488–4496.

Govaerts Y.M., Jacquemoud S., Verstraete M.M. & Ustin S.L. (1996) Three-dimensional radiation transfer modeling in a dicotyledon leaf. *Applied Optics*, **35**, 6585–6598.

Govaerts Y.M. & Verstraete M.M. (1998) Raytran: A Monte Carlo ray-tracing model to compute light scattering in three-dimensional heterogeneous media. *IEEE Transactions on Geoscience and Remote Sensing*, **36**, 493–505.

Grace J., Nichol C., Disney M., Lewis P., Quaife T. & Bowyer P. (2007) Can we measure terrestrial photosynthesis from space directly, using spectral reflectance and fluorescence? *Global Change Biology*, **13**, 1484–1497.

Grant O.M., Tronina Ł., Jones H.G. & Chaves M.M. (2007) Exploring thermal imaging variables for the detection of stress responses in grapevine under different irrigation regimes. *Journal of Experimental Botany*, **58**, 815–825.

Gu J., Smith E.A., Hodges G. & Cooper H.J. (1997) Retrieval of daytime surface net longwave flux over BOREAS from GOES estimates of surface solar flux and surface temperature. *Canadian Journal of Remote Sensing*, **23**, 176–187.

Guanter L., Alonso L., Gómez-Chova L., Amorós J., Vila J. & Moreno J. (2007) A method for detection of solar-induced vegetation fluorescence from *MERIS* FR data. Paper presented at the Envisat Symposium 2007, Montreux, Switzerland.

Gueymard C.A. (1995) *SMARTS, a simple model of the atmospheric radiative transfer of sunshine: algorithms and performance assessment*. Technical Report FSEC-PF-270-95. Florida Solar Energy Centre, Cocoa, FL.

Gueymard C.A. (2001) Parameterized transmittance model for direct beam and circumsolar spectral irradiance. *Solar Energy*, **71**, 325–346.

Guilioni L., Jones H.G., Leinonen I. & Lhomme J.P. (2008) On the relationships between stomatal resistance and leaf temperatures in thermography. *Agricultural and Forest Meteorology*, **148**, 1908–1912.

Guo J. & Trotter C.M. (2004) Estimating photosynthetic light-use efficiency using the photochemical reflectance index: variations among species. *Functional Plant Biology*, **31**, 255–265.

Gupta R.K., Prasad T.S., Krishna Rao P.V. & Bala Manikavelu P.M. (2000) Problems in upscaling of high resolution remote sensing data to coarse spatial resolution over land surface. *Advances in Space Research*, **26**, 1111–1121.

Gupta S.K., Ritchey N.A., Wilber A.C., Whitlock C.H., Gibson G.G. & Stackhouse Jr P.W. (1999) A climatology of surface radiation budget derived from satellite data. *Journal of Climate*, **12**, 2691–2710.

Hadjimitsis D.G., Clayton C.R.I. & Retalis A. (2009) The use of selected pseudo-invariant targets for the application of atmospheric correction in multi-temporal studies using satellite remotely sensed imagery. *International Journal of Applied Earth Observation and Geoinformation*, **11**, 192–200.

Hall F.G., Hilker T., Coops N.C., Lyapustin A., Huemmrich K.F., Middleton E., Margolis H., Drolet G. & Black T.A. (2008) Multi-angle remote sensing of forest light use efficiency by observing PRI variation with canopy shadow fraction. *Remote Sensing of Environment*, **112**, 3201–3211.

Hall F.G., Shimabukuro Y.E. & Huemmrich K.F. (1995) Remote sensing of forest biophysical structure using mixture decomposition and geometric reflectance models. *Ecological Applications*, **5**, 993–1013.

Hansen M., Dubayah R. & Defries R. (1996) Classification trees: an alternative to traditional land cover classifiers. *International Journal of Remote Sensing*, **17**, 1075–1081.

Hapke B. (1993) *Theory of reflectance and emittance spectroscopy*. Cambridge University Press, Cambridge. pp. 455, ISBN 0521307899.

Haralick R.M., Shanmugam K. & Dinstein I. (1973) Texture features for image classification. *IEEE Transactions on Systems, Man and Cybernetics*, **3**, 610–621.

Hawkins B.A., Field R., Cornell H.V., Currie D.J., Guégan J.-F., Kaufman D.M., Kerr J.T., Mittelbach G.G., Oberdorff T., O'Brien E.M., Porter E.E. &

Turner J.R.G. (2003) Energy, water, and broad-scale geographic patterns of species richness. *Ecology*, **84**, 3105–3117.

Hemakumara H.M., Chandrapala L. & Moene A.F. (2003) Evapotranspiration fluxes over mixed vegetation areas measured from large aperture scintillometer. *Agricultural Water Management*, **58**, 109–122.

Heuvelink G.B.M. (1998) *Error propagation in environmental modelling with GIS*. Taylor and Francis, London.

Hilker T., Coops N.C., Hall F.G., Black T.A., Wulder M.A., Nesic Z. & Krishnan P. (2008) Separating physiologically and directionally induced changes in PRI using BRDF models. *Remote Sensing of Environment*, **112**, 2777–2788.

Hill R.A. & Broughton R.K. (2009) Mapping the understorey of deciduous woodland from leaf-on and leaf-off airborne LiDAR data: A case study in lowland Britain. *ISPRS Journal of Photogrammetry and Remote Sensing*, **64**, 223–233.

Hillel D. (1998) *Environmental soil physics*. Academic Press, London. pp. 771, ISBN 0123485258.

Hosgood B., Jacquemoud S., Andreoli G., Verdebout J., Pedrini A. & Schmuck G. (2005) *Leaf Optical Properties EXperiment 93 (LOPEX93)*, Ispra, Italy.

Hosoi F. & Omasa K. (2007) Factors contributing to accuracy in the estimation of the woody canopy leaf area density profile using 3D portable lidar imaging. *Journal of Experimental Botany*, **58**, 3463–3473.

Houborg R., Soegaard H. & Boegh E. (2007) Combining vegetation index and model inversion methods for the extraction of key vegetation biophysical parameters using Terra and Aqua MODIS reflectance data. *Remote Sensing of Environment*, **106**, 39–58.

Huang C., Wylie B., Homer C. & Zylstra G. (2002) Derivation of a tasselled cap transformation based on Landsat 7 at-satellite reflectance. *International Journal of Remote Sensing*, **23**, 1741–1748.

Huete A.R. (1988) A soil-adjusted vegetation index (SAVI). *Remote Sensing of Environment*, **25**, 295–309.

Huete A.R., Didan K., Miura T., Rodriguez E.P., Gao X. & Ferreira L.G. (2002) Overview of the radiometric and biophysical performance of the MODIS vegetation indices. *Remote Sensing of Environment*, **83**, 195–213.

Huete A.R., Liu H.Q., Batchily K. & van Leeuwen W. (1997) A Comparison of Vegetation Indices over a Global Set of TM Images for EOS-MODIS. *Remote Sensing of Environment*, **59**, 440–451.

Hyer E.J. & Goetz S.J. (2004) Comparison and sensitivity analysis of instruments and radiometric methods for LAI estimation: assessments from a boreal forest site. *Agricultural and Forest Meteorology*, **122**, 157–174.

Hyyppä J., Hyyppä H., Leckie D., Gougeon F., Yu X. & Maltamo M. (2008) Review of methods of small-footprint airborne laser scanning for extracting forest inventory data in boreal forests. *International Journal of Remote Sensing*, **29**, 1339–1366.

Idso S.B., Jackson R.D., Ehrler W.L. & Mitchell S.T. (1969) A method for determination of infrared emittance of leaves. *Ecology*, **50**, 899–902.

Idso S.B., Jackson R.D., Pinter Jr P.J., Reginato R.J. & Hatfield J.L. (1981) Normalizing the stress-degree-day parameter for environmental variability. *Agricultural Meteorology*, **24**, 45–55.

Irons J.R., Campbell G.S., Norman J.M., Graham D.W. & Kovalick W.M. (1992) Prediction and measurement of soil bidirectional reflectance. *IEEE Transactions on Geoscience and Remote Sensing*, **30**, 249–260.

Jackson R.D., Idso S.B., Reginato R.J. & Pinter Jr P.J. (1981) Canopy temperature as a crop water stress indicator. *Water Resources Research*, **17**, 1133–1138.

Jackson R.D., Reginato R.J. & Idso S.B. (1977) Wheat canopy temperature: a practical tool for evaluating water requirements. *Water Resources Research*, **13**, 651–656.

Jacquemoud S. (1993) Inversion of PROSPECT + SAIL canopy reflectance model from AVIRIS equivalent spectra: Theoretical study. *Remote Sensing of Environment*, **44**, 281–292.

Jacquemoud S. & Baret F. (1990) PROSPECT: A model of leaf optical properties spectra. *Remote Sensing of Environment*, **34**, 75–91.

Jacquemoud S., Ustin S.L., Verdebout J., Schmuck G., Andreoli G. & Hosgood B. (1996) Estimating leaf biochemistry using the PROSPECT leaf optical properties model. *Remote Sensing of Environment*, **56**, 194–202.

Jacquemoud S., Verdebout J., Schmuck G., Andreoli G. & Hosgood B. (1995) Investigation of leaf biochemistry by statistics. *Remote Sensing of Environment*, **54**, 180–188.

Jacquemoud S., Verhoef W., Baret F., Bacour C., Zarco-Tejada P.J., Asner G.P., François C. & Ustin S.L. (2009) PROSPECT + SAIL models: A review of use for vegetation characterization. *Remote Sensing of Environment*, **113**, S56-S66.

Jago R.A., Cutler M.E.J. & Curran P.J. (1999) Estimating Canopy Chlorophyll Concentration from Field and Airborne Spectra. *Remote Sensing of Environment*, **68**, 217–224.

Jarvis P.G. (1981) Stomatal conductance, gaseous exchange and transpiration. In: *Plants and their atmospheric environment* (eds Grace, J., Ford, E.D. & Jarvis, P.G.), pp. 175–204. Blackwell, Oxford. ISBN 06320005254.

Jarvis P.G. & McNaughton K.G. (1986) Stomatal control of transpiration: scaling up from leaf to region. *Advances in Ecological Research*, **15**, 1–49.

Jensen J.L.R., Humes K.S., Vierling L.A. & Hudak A.T. (2008) Discrete return lidar-based prediction of leaf area index in two conifer forests. *Remote Sensing of Environment*, **112**, 3947–3957.

Jensen J.R. (2005) *Introductory digital image processing: a remote sensing perspective.* (3rd ed.). Pearson Education Inc., Upper Saddle River, NJ 07458. pp. 526, ISBN 0131453610.

Jensen J.R. (2007) *Remote sensing of the environment: an Earth resource perspective* (2nd ed.). Pearson Prentice Hall, Upper Saddle River, NJ. pp. 592, ISBN 9780131889507.

Jepsen J.U., Hagen S.B., Høgda K.A., Ims R.A., Karlsen S.R., Tømmervik H. & Yoccoz N.G. (2009) Monitoring the spatio-temporal dynamics of geometrid moth outbreaks in birch forest using MODIS-NDVI data. *Remote Sensing of Environment*, **113**, 1939–1947.

Jia L., Li Z.-I., Menenti M., Su Z., Verhoef W. & Wan Z. (2003) A practical algorithm to infer soil and foliage component temperatures from bi-angular ATSR-2 data. *International Journal of Remote Sensing*, **24**, 4739–4760.

Jiang Z., Huete A.R., Chen J., Chen Y., Li J., Yan G. & Zhang X. (2006) Analysis of NDVI and scaled difference vegetation index retrievals of vegetation fraction. *Remote Sensing of Environment*, **101**, 366–378.

John R., Chen J., Lu N., Guo K., Liang, C., Wei Y., Noormets A., Ma K. & Han X. (2008) Predicting plant diversity based on remote sensing products in the semi-arid region of Inner Mongolia. *Remote Sensing of Environment*, **112**, 2018–2032.

Jonckheere I., Fleck S., Nackaerts K., Muys B., Coppin P., Weiss M. & Baret F. (2004) Review of methods for in situ leaf area index determination – Part I. Theories, sensors and hemispherical photography. *Agricultural and Forest Meteorology*, **121**, 19–35.

Jones H.G. (1973) Estimation of plant water status with the beta-gauge. *Agricultural Meteorology*, **11**, 345–355.

Jones H.G. (1992) *Plants and microclimate.* (2nd ed.). Cambridge University Press, Cambridge. pp. 428, ISBN 0521425247.

Jones H.G. (1999) Use of infrared thermometry for estimation of stomatal conductance as a possible aid to irrigation scheduling. *Agricultural and Forest Meteorology*, **95**, 139–149.

Jones H.G. (2004a) Application of thermal imaging and infrared sensing in plant physiology and ecophysiology. *Advances in Botanical Research*, **41**, 107–163.

Jones H.G. (2004b) Irrigation scheduling: advantages and pitfalls of plant-based methods. *Journal of Experimental Botany*, **55**, 2427–2436.

Jones H.G. (2007) Monitoring plant and soil water status: established and novel methods revisited and their relevance to studies of drought tolerance. *Journal of Experimental Botany*, **58**, 119–130.

Jones H.G., Archer N.A.L. & Rotenberg E. (2004) Thermal radiation, canopy temperature and evaporation from forest canopies. In: *Forests at the Land-Atmosphere Interface* (eds Mencuccini, M., Grace, J., Moncrieff, J. & McNaughton, K.G.), pp. 123–144. CAB International, Wallingford.

Jones H.G., Serraj R., Loveys B.R., Xiong L., Wheaton A. & Price A.H. (2009) Thermal infrared imaging of crop canopies for the remote diagnosis and quantification of plant responses to water stress in the field. *Functional Plant Biology*, **36**, 978–989.

Jones H.G., Stoll M., Santos T., de Sousa C., Chaves M.M. & Grant O.M. (2002) Use of infrared thermography for monitoring stomatal closure in the field: application to grapevine. *Journal of Experimental Botany*, **53**, 2249–2260.

Jördens C., Scheller M., Breitenstein B., Selmar D. & Koch M. (2009) Evaluation of leaf water status by means of permittivity at terahertz frequencies. *Journal of Biological Physics*, **35**, 255–264.

Joshi C., De Leeuw J., van Andel J., Skidmore A.K., Lekhak H.D., van Duren I.C. & Norbu N. (2006) Indirect remote sensing of a cryptic forest understorey invasive species. *Forest Ecology and Mangement*, **225**, 245–256.

Kalma J.D., Franks S.W. & van den Hurk B. (2001) On the representation of land surface fluxes for atmospheric modelling. *Meteorology and Atmospheric Physics*, **76**, 53–67.

Kalma J.D., McVicar T.R. & McCabe M.F. (2008) Estimating land surface evaporation: A review of methods using remotely sensed surface temperature data. *Surveys in Geophysics*, **29**, 421–469.

Karnieli A., Kaufman Y.J., Remer L. & Wald A. (2001) AFRI – aerosol free vegetation index. *Remote Sensing of Environment*, **77**, 10–21.

Kauth R.J. & Thomas G.S. (1976) *The tasseled cap – A graphic description of the spectral-temporal development of agricultural crops as seen by Landsat*. Paper presented at the LARS: Proceedings of the symposium on machine processing of remotely sensed data, West Lafayette, IN.

Khanna S., Palacios-Orueta A., Whiting, M.L., Ustin S.L., Riaño D. & Litago J. (2007) Development of angle indexes for soil moisture estimation, dry matter detection and land-cover discrimination. *Remote Sensing of Environment*, **109**, 154–165.

Kimes D.S. (1980) Effects of vegetation canopy structure on remotely sensed canopy temperatures. *Remote Sensing of Environment*, **10**, 165–174.

Kimes D.S., Idso S.B., Pinter Jr P.J., Reginato R.J. & Jackson R.D. (1980) View angle effects in the radiometric measurement of plant canopy temperatures. *Remote Sensing of Environment*, **10**, 273–284.

Kiniry J.R., Jones C.A., O'Toole J.C., Blanchet R., Cabelguenne M. & Spanel D.A. (1989) Radiation-use efficiency in biomass accumulation prior to grain-filling for five grain-crop species. *Field Crops Research*, **20**, 51–64.

Kiniry J.R., Tischler C.R. & Van Esbroeck G.A. (1999) Radiation use efficiency and leaf CO_2 exchange for diverse C_4 grasses. *Biomass & Bioenergy*, **17**, 95–112.

Kleinebecker T., Schmidt S.R., Fritz C., Smolders A.J.P. & Hölzel N. (2009) Prediction of $\delta^{13}C$ and $\delta^{15}N$ in plant tissues with near-infrared reflectance spectroscopy. *New Phytologist*, **184**, 732–739.

Knyazikhin Y., Martonchik J.V., Myneni R.B., Diner D.J. & Running S.W. (1998) Synergistic algorithm for estimating vegetation canopy leaf area index and fraction of absorbed photosynthetically active radiation from MODIS and MISR data. *Journal of Geophysical Research–Atmospheres*, **103**, 32257–32276.

Kokaly R.F. & Clark R.N. (1999) Spectroscopic Determination of Leaf Biochemistry Using Band-Depth Analysis of Absorption Features and Stepwise Multiple Linear Regression. *Remote Sensing of Environment*, **67**, 267–287.

Kolber Z., Klimov D., Ananyev G., Rascher U., Berry J. & Osmond B. (2005) Measuring photosynthetic parameters at a distance: laser induced fluorescence transient (LIFT) method for remote measurements of photosynthesis in terrestrial vegetation. *Photosynthesis Research*, **84**, 121–129.

Kramer H.J. (2002) *Observation of the Earth and Its Environment: Survey of Missions and Sensors*. Springer, Berlin. pp. 1510, ISBN 9783540423881.

Kramer H.J. & Cracknell A.P. (2008) An overview of small satellites in remote sensing. *International Journal of Remote Sensing*, **29**, 4285–4337.

Kubelka P. & Munk F. (1931) Ein Beitrag zur Optik der Farbanstriche. *Zeitschrift für Technische Physik*, **12**, 593–601.

Kucharik C.J., Norman J.M. & Gower S.T. (1998) Measurements of branch area and adjusting leaf area index indirect measurements. *Agricultural and Forest Meteorology*, **91**, 69–88.

Kucharik C.J., Norman J.M., Murdock L.M. & Gower S.T. (1997) Characterizing canopy nonrandomness with a multiband vegetation imager (MVI). *Journal of Geophysical Research–Atmospheres*, **102**, 29455–29473.

Kustas W.P., Choudhury B.J., Moran M.S., Reginato R.J., Jackson R.D., Gay L.W. & Weaver H.L. (1989) Determination of sensible heat-flux over sparse canopy using thermal infrared data. *Agricultural and Forest Meteorology*, **44**, 197–216.

Kustas W.P., Norman J.M., Anderson M.C. & French A.N. (2003) Estimating subpixel surface temperatures and energy fluxes from the vegetation index-radiometric temperature relationship. *Remote Sensing of Environment*, **85**, 429–440.

Kuusk A. (1995) A fast, invertible canopy reflectance model. *Remote Sensing of Environment*, **51**, 342–350.

Lagouarde J.P., Kerr Y.H. & Brunet Y. (1995) An experimental study of angular effects on surface-temperature for various plant canopies and bare soils. *Agricultural and Forest Meteorology*, **77**, 167–190.

Lang A.R.G. (1973) Leaf orientation of a cotton plant. *Agricultural Meteorology*, **11**, 37–51.

Langsdorf G., Buschmann C., Sowinska M., Banbani F., Mokry F., Timmermann F. & Lichtenthaler H.K. (2000) Multicolour Fluorescence Imaging of Sugar Beet Leaves with Different Nitrogen Status by Flash Lamp UV-Excitation. *Photosynthetica*, **38**, 539–551.

Launay M. & Guerif M. (2005) Assimilating remote sensing data into a crop model to improve predictive performance for spatial applications. *Agriculture, Ecosystems and Environment*, **111**, 321–339.

Lavergne T., Kaminski T., Pinty B., Taberner M., Gobron N., Verstraete M.M., Vossbeck M., Widlowski J.-L. & Giering R. (2007) Application to MISR land products of an RPV model inversion package using adjoint and Hessian codes. *Remote Sensing of Environment*, **107**, 362–375.

Lawlor D.W. (2001) *Photosynthesis*. (3rd ed.). Bios, Oxford. pp. 398, ISBN 978-0387916071.

Laymon C.A., Belise W., Cioleman T., Crosson W.L., Fahsi A., Jackson T., Manu A., O'Neill P., Senwo Z. & Tsegaye T. (1999) Huntsville '96: An experiment in ground-based microwave remote sensing of soil moisture. *International Journal of Remote Sensing*, **20**, 823–828.

Le Treut H., Somerville R., Cubasch U., Ding Y., Mauritzen C., Mokssit A., Peterson T. & Prather M. (2007) Historical overview of climate change. In: *Climate Change 2007: The Physical Science Basis. Contribution of Working Group I to the Fourth Assessment Report of the Intergovernmental Panel on Climate Change* (eds Solomon, S., Qin, D., Manning, M., Chen, Z., Marquis, M., Averyt, K.B., Tignor, M. & Miller, H.L.), pp. 93–127. Cambridge University Press, Cambridge.

Leblanc S.G., Chen J.M., Fernandes R., Deering D.W. & Conley A. (2005) Methodology comparison for canopy structure parameters extraction from digital hemispherical photography in boreal forests. *Agricultural and Forest Meteorology*, **129**, 187–207.

Leblon B. (1990) *Mise au point d'un modèle semi-empirique d'estimation de la biomasse et du rendement de cultures de riz irriguées* (Oryza sativa *L.) a partir du profil spectral dans le visible et le proche infra-rouge. Validation a partir de donnees SPOT.* Thesis, l'Ecole National Superieure Agronomique de Montpellier en Sciences Agronomiques, Montpellier.

Lee Y.J., Chang K.W., Shen Y., Huang T.M. & Tsay H.L. (2007) A handy imaging system for precision agriculture studies. *International Journal of Remote Sensing*, **28**, 4867–4876.

Lefsky M.A., Cohen W.B., Acker S.A., Parker G.G., Spies T.A. & Harding D. (1999) Lidar remote sensing of the canopy structure and biophysical properties of Douglas-fir Western hemlock forests. *Remote Sensing of Environment*, **70**, 339–361.

Leinonen I. & Jones H.G. (2004) Combining thermal and visible imagery for estimating canopy temperature and identifying plant stress. *Journal of Experimental Botany*, **55**, 1423–1431.

Leinonen I., Grant O.M., Tagliavia C.P.P., Chaves M.M. & Jones H.G. (2006) Estimating stomatal conductance with thermal imagery. *Plant, Cell and Environment*, **29**, 1508–1518.

Leroy M. (2001) Deviation from reciprocity in bidirectional reflectance. *Journal of Geophysical Research*, **106**, 11917–11923.

Lewis P. (1999) Three-dimensional plant modelling for remote sensing simulation studies using the Botanical Plant Modelling System. *Agronomie*, **19**, 185–210.

Lewis P. & Disney M. (2007) Spectral invariants and scattering across multiple scales from within-leaf to canopy. *Remote Sensing of Environment*, **109**, 196–206.

Li R., Min Q. & Lin B. (2009) Estimation of evapotranspiration in a mid-latitude forest using the Microwave Emissivity Difference Vegetation Index (EDVI). *Remote Sensing of Environment*, **113**, 2011–2018.

Li X.W., Gao F., Wang J.D. & Strahler A. (2001) A priori knowledge accumulation and its application to linear BRDF model inversion. *Journal of Geophysical Research–Atmospheres*, **106**, 11925–11935.

Li X.W. & Strahler A.H. (1985) Geometric-Optical Modeling of a Conifer Forest Canopy. *IEEE Transactions on Geoscience and Remote Sensing*, **23**, 705–721.

Li X.W. & Strahler A.H. (1992) Geometric-optical bidirectional reflectance modeling of the discrete crown vegetation canopy: effect of crown shape and mutual shadowing. *IEEE Transactions on Geoscience and Remote Sensing*, **30**, 276–292.

Liang S. (2004) *Quantitative remote sensing of land surfaces*. John Wiley and Sons, Inc., Hoboken, NJ. pp. 534, ISBN 0471281662.

Lichtenthaler H.K. & Miehé J. (1997) Fluorescence imaging as a diagnostic tool for plant stress. *Trends in Plant Science*, **2**, 316–320.

Lichtenthaler H.K. & Rinderle U. (1988) The role of chlorophyll fluorescence in the detection of stress conditions in plants. *CRC Critical Reviews in Analytical Chemistry*, **19, Supplement 1**, S29-S85.

Lillesand T.M., Kiefer R.W. & Chipman J.C. (2008) *Remote sensing and image interpretation*. (6th ed.). John Wiley & Sons, New York. pp. 768, ISBN 9780470052457.

Lindenmayer A. (1968) Mathematical models for cellular interaction in development. Parts I and II. *Journal of Theoretical Biology*, **18**, 280–315.

Liu C., Frazier P. & Kumar L. (2007) Comparative assessment of the measures of thematic classification accuracy. *Remote Sensing of Environment*, **107**, 606–616.

Loarie S.R., Joppa L.N. & Pimm S.L. (2007) Satellites miss environmental priorities. *Trends in Ecology and Evolution*, **22**, 630–632.

Logan B.A., Adams W.W.I. & Demmig-Adams B. (2007) Avoiding common pitfalls of chlorophyll fluorescence analysis under field conditions. *Functional Plant Biology*, **34**, 853–859.

Long M.W. (1983) *Radar reflectivity of land and sea*. (2nd ed.). Artech House, Inc., Dedham, Massachusetts. pp. 385, ISBN 0980061300.

Lu D. & Weng Q. (2007) A survey of image classification methods and techniques for improving classification performance. *International Journal of Remote Sensing*, **28**, 823–870.

Lucas K.L. & Carter G.A. (2008) The use of hyperspectral remote sensing to assess vascular plant species richness on Horn Island, Mississippi. *Remote Sensing of Environment*, **112**, 3908–3915.

Lucas R., Mitchell A. & Bunting P. (2008) Hyperspectral data for assessing carbon dynamics and biodiversity of forests. In: *Hyperspectral remote sensing of tropical and subtropical forests* (eds Kalacska, M. & Sanchez-Azofeifa, G.A.), pp. 47–86. CRC Press, Boca Raton, FL.

Luedeling E., Hale A., Zhang M., Bentley W.J. & Dharmasri L.C. (2009) Remote sensing of spider mite damage in California peach orchards. *International Journal of Applied Earth Observation and Geoinformation*, **11**, 244–255.

Luke R.H. & McArthur A.G. (1978) *Bushfires in Australia*. CSIRO, Canberra, Australia, pp. 359, ISBN 0642023417

Magurran A.E. (1988) *Ecological diversity and its measurement*. Croom Helm, London. pp. 179, ISBN 0709935404.

Makkink G.F. (1957) Testing the Penman formula by means of lysimeters. *Journal of the Institute of Water Engineers*, **11**, 277–288.

Martínez B. & Gilabert M.A. (2009) Vegetation dynamics from NDVI time series analysis using the wavelet transform. *Remote Sensing of Environment*, **113**, 1823–1842.

Mather P.M. (2004) *Computer processing of remotely-sensed images: an introduction*. (3rd ed.). Wiley, Chichester, UK. pp. 324, ISBN 0470849193.

Mather P.M. (2008) Editorial. *The Remote Sensing and Photogrammetry Society Newsletter*, October 2008, pp. 2.

Mätzler C. (1994) Passive microwave signatures of landscapes in winter. *Meteorology and Atmospheric Physics*, **54**, 241–260.

Maxwell K. & Johnson G.N. (2000) Chlorophyll fluorescence – a practical guide. *Journal of Experimental Botany*, **51**, 659–668.

McCoy R.M. (2004) *Field methods in remote sensing*. The Guilford Press, New York. pp. 159, ISBN 9781593850791.

McGarigal K. & Marks B.J. (1995) *FRAGSTATS: spatial pattern analysis program for quantifying landscape structure*. General Technical Report PNW-GTR-351. USDA Forest Service, Pacific Northwest Research Station, Portland, OR.

McGuire M.J., Balick L.K., Smith J.A. & Hutchison B.A. (1989) Modeling directional thermal radiance from a forest canopy. *Remote Sensing of Environment*, **27**, 169–186.

McLean G. (1982) *Timber volume estimation using cross-sectional photogrammetric and densitometric methods.* Masters thesis, University of Wisconsin, Madison, WI.

McNaughton K.G. & Jarvis P.G. (1983) Predicting effects of vegetation changes on transpiration and evaporation. In: *Water deficits and plant growth 7* (ed. Kozlowski, T.T.), pp. 1–47. Academic Press, New York.

McVicar T.R. & Jupp D.L.B. (2002) Using covariates to spatially interpolate moisture availability in the Murray–Darling Basin. A novel use of remotely sensed data. *Remote Sensing of Environment*, **79**, 199–212.

Meister G., Rothkirch A., Hosgood B., Spitzer H. & Bienlein J. (2000) Error analysis for BRDF measurements at the European Goniometric Facility. *Remote Sensing Reviews*, **19**, 111–131.

Meroni M., Rossini M., Guanter L., Alonso L., Rascher U., Colombo R. & Moreno J. (2009) Remote sensing of solar-induced chlorophyll fluorescence: Review of methods and applications. *Remote Sensing of Environment*, **113**, 2037–2051.

Meyer S. & Genty B. (1999) Heterogeneous inhibition of photosynthesis over the leaf surface of *Rosa rubiginosa* L. during water stress and abscisic acid treatment: induction of a metabolic component by limitation of CO_2 diffusion. *Planta*, **210**, 126–131.

Meyer S., Saccardy-Adji K., Rizza F. & Genty B. (2001) Inhibition of photosynthesis by *Colletotrichum lindemuthianum* in bean leaves determined by chlorophyll fluorescence imaging. *Plant, Cell and Environment*, **24**, 947–955.

Miettinen J. (2007) Variability of fire-induced changes in MODIS surface reflectance by land-cover type in Borneo. *International Journal of Remote Sensing*, **28**, 4967–4984.

Miles V.V., Bobylev L.P., Maximov S.V., Johannessen O.M. & Pitulko V.M. (2003) An approach for assessing boreal forest conditions based on combined use of satellite SAR and multispectral data. *International Journal of Remote Sensing*, **24**, 4447–4466.

Miller J.R., Hare E.W. & Wu J. (1990) Quantitative characterisation of the vegetation red edge reflectance. 1. An inverted Gaussian reflectance model. *International Journal of Remote Sensing*, **11**, 1755–1773.

Min Q. & Lin B. (2006) Remote sensing of evapotranspiration and carbon uptake at Harvard Forest. *Remote Sensing of Environment*, **100**, 379–387.

Moffiet T., Armston J.D. & Mengersoen K. (2010) Motivation, development and validation of a new spectral greenness index: A spectral dimension related to foliage projective cover. *ISPRS Journal of Photogrammetry and Remote Sensing*, **65**, 26–41.

Möller M., Alchanatis V., Cohen Y., Meron M., Tsipris J., Naor A., Ostrovsky V., Sprintsin M. & Cohen S. (2007) Use of thermal and visible imagery for estimating crop water status of irrigated grapevine. *Journal of Experimental Botany*, **58**, 827–838.

Monteith J.L. (1977) Climate and the efficiency of crop production in Britain. *Proceedings of the Royal Society of London, Series B*, **281**, 277–294.

Monteith J.L. & Unsworth M.H. (2008) *Principles of Environmental Physics.* (3rd ed.). Academic Press, Burlington, MA. pp. xxi + 418, ISBN 9780125051033.

Moran M.S., Clarke T.R., Inoue Y. & Vidal A. (1994) Estimating crop water deficit using the relation between surface-air temperature and spectral vegetation index. *Remote Sensing of Environment*, **49**, 246–263.

Moran M.S., Humes K.S. & Pinter Jr P.J. (1997) The scaling characteristics of remotely-sensed variables for sparsely-vegetated heterogeneous landscapes. *Journal of Hydrology*, **190**, 337–362.

Moran M.S., Jackson R.D., Slater P.N. & Teillet P.M. (1992) Evaluation of simplified procedures for retrieval of land surface reflectance factors from satellite sensor output. *Remote Sensing of Environment*, **41**, 169–184.

Morsdorf F., Kötz B., Meier E., Itten K.I. & Allgöwer B. (2006) Estimation of LAI and fractional cover from small footprint airborne laser scanning data based on gap fraction. *Remote Sensing of Environment*, **104**, 50–61.

Moshou D., Bravo C., Oberti R., West J., Bodria L., McCartney A. & Ramon H. (2005) Plant disease detection based on data fusion of hyper-spectral and multi-spectral fluorescence imaging using Kohonen maps. *Real-Time Imaging*, **11**, 75–83.

Moulia B. & Sinoquet H. (1993) Three-dimensional digitizing systems for plant canopy geometrical structure: a review. In: *Crop Structure and Light Microclimate: Characterization and Applications* (eds Varlet-Grancher, C., Bonhomme, R., & Sinoquet, H.), pp. 183–193. INRA, Paris.

Moya I., Camenen L., Evain S., Goulas Y., Cerovic Z.G., Latouche G., Flexas J. & Ounis A. (2004) A new instrument for passive remote sensing 1. Measurements of sunlight-induced chlorophyll fluorescence. *Remote Sensing of Environment*, **91**, 186–197.

Myneni R.B., Hall F.G., Sellers P.J. & Marshak A.L. (1995) The interpretation of spectral vegetation indexes. *IEEE Transactions on Geoscience and Remote Sensing*, **33**, 481–486.

Myneni R.B., Maggion S., Iaquinta J., Privette J.L., Gobron N., Pinty B., Kimes D.S., Verstraete M.M. & Williams D.L. (1995) Optical remote-sensing of vegetation – modeling, caveats, and algorithms. *Remote Sensing of Environment*, **51**, 169–188.

Myneni R. & Ross J., eds (1991) *Photon–vegetation interactions: applications in optical remote sensing and plant physiology*. Springer-Verlag, Berlin. pp. 560, ISBN 9783540521082.

Myneni R.B., Ross J. & Asrar G. (1989) A Review on the Theory of Photon Transport in Leaf Canopies. *Agricultural and Forest Meteorology*, **45**, 1–153.

Nagler P.L., Cleverly J., Glenn E., Lampkin D., Huete A. & Wan Z. (2005) Predicting riparian evapotranspiration from MODIS vegetation indices and meteorological data. *Remote Sensing of Environment*, **94**, 17–30.

Nedbal L., Soukupová J., Kaftan D., Whitmarsh J. & Trtilek M. (2000) Kinetic imaging of chlorophyll fluorescence using modulated light. *Photosynthesis Research*, **66**, 3–12.

Nelson R.F. (1983) Detecting forest canopy change due to insect activity using Landsat MSS. *Photogrammetric Engineering and Remote Sensing*, **49**, 1303–1314.

Ngene S.M., Skidmore A.K., van Gils H., Douglas-Hamilton I. & Omondi P. (2009) Elephant distribution around a volcanic shield dominated by a mosaic of forest and savanna (Marsabit, Kenya). *African Journal of Ecology*, **47**, 234–245.

Nichol C.J., Huemmrich K.F., Black T.A., Jarvis P.G., Walthall C.L., Grace J. & Hall F.G. (2000) Remote sensing of photosynthetic-light-use efficiency of boreal forest. *Agricultural and Forest Meteorology*, **101**, 131–142.

Nicodemus F.E., Richmond J.C., Hsia J.J., Ginsberg I.W. & Limperis T. (1977) *Geometrical considerations and nomenclature for reflectance*. NBS Monograph 160. National Bureau of Standards, Washington DC.

Nilson T. (1971) A theoretical analysis of the frequency of gaps in plant stands. *Agricultural Meteorology*, **8**, 25–38.

Nilson T. & Kuusk A. (1989) A reflectance model for the homogeneous plant canopy and its inversion. *Remote Sensing of Environment*, **27**, 157–167.

Nilsson H.-E. (1995) Remote sensing and image analysis in plant pathology. *Annual Review of Phytopathology*, **33**, 489–527.

Nobel P.S. (2009) *Physicochemical and Environmental Plant Physiology*. (4th ed.). Academic Press, Oxford. pp. 582, ISBN 9780123741431.

Norman J.M., Divakarla M. & Goel N.S. (1995a) Algorithms for extracting information from remote thermal-IR observations of the earth's surface. *Remote Sensing of Environment*, **51**, 157–168.

Norman J.M., Kustas W.P. & Humes K.S. (1995b) Source approach for estimating soil and vegetation energy fluxes in observations of directional radiometric surface temperature. *Agricultural and Forest Meteorology*, **77**, 263–293.

Ochsner T.E., Horton R. & Ren T. (2001) A new perspective on soil thermal properties. *Soil Science Society of America Journal*, **65**, 1641–1647.

Olioso A., Braud I., Chanzy A., Courault D., Demarty J., Kergoat L., Lewan E., Ottlé C., Prévot L., Zhao W.G.G., Calvet J.C., Cayrol P., Jongschaap R., Moulin S., Noilhan J. & Wigneron J.P. (2002) SVAT modeling over the Alpilles-ReSeDA experiment: comparing SVAT models over wheat fields. *Agronomie*, **22**, 651–668.

Omasa K., Hashimoto Y. & Aiga I. (1981) A quantitative analysis of the relationships between SO_2 or NO_2 sorption and their acute effects on plant leaves using image instrumentation. *Environmental Control in Biology*, **19**, 59–67.

Omasa K., Hosoi F. & Konishi A. (2007) 3D lidar imaging for detecting and understanding plant responses and canopy structure. *Journal of Experimental Botany*, **58**, 881–898.

Oosterhuis D.M., Walker S. & Eastham J. (1985) Soybean leaflet movements as an indicator of crop water stress. *Crop Science*, **25**, 1101–1106.

Őquist G. & Wass R. (1988) A portable, microprocessor operated instrument for measuring chlorophyll fluorescence kinetics in stress physiology. *Physiologia Plantarum*, **73**, 211–217.

Osmond C.B., Daley P.F., Badger M.R. & Lüttge U. (1998) Chlorophyll fluorescence quenching during photosynthetic induction in leaves of *Abutilon striatum* Dicks. infected with Abutilon Mosaic Virus, observed with a field-portable imaging system. *Botanica Acta*, **111**, 390–397.

Otterman J., Susskind J., Brakke T., Kimes D., Pielke R. & Lee T.J. (1995) Inferring the thermal-infrared hemispheric emission from a sparsely-vegetated surface by directional measurements. *Boundary-Layer Meteorology*, **74**, 163–180.

Otterman J. & Weiss G.H. (1984) Reflection from a field of randomly located vertical protrusions. *Applied Optics*, **23**, 1931–1936.

Ouma Y.O., Tetuko J. & Tateishi R. (2008) Analysis of co-occurrence and discrete wavelet transform textures for differentiation of forest and non-forest vegetation in very-high-resolution optical-sensor imagery. *International Journal of Remote Sensing*, **29**, 3417–3456.

Oxborough K. & Baker N.R. (1997) An instrument capable of imaging chlorophyll-*a* fluorescence from intact leaves at very low irradiance at cellular and subcellular levels of organisation. *Plant, Cell and Environment*, **20**, 1473–1483.

Ozdemir I. (2008) Estimating stem volume by tree crown area and tree shadow area extracted from pan-sharpened Quickbird imagery in open Crimean juniper forests. *International Journal of Remote Sensing*, **29**, 5643–5655.

Paloscia S. & Pampaloni P. (1992) Microwave vegetation indexes for detecting biomass and water conditions of agricultural crops. *Remote Sensing of Environment*, **40**, 15–26.

Parrinello T. & Vaughan R.A. (2002) Multifractal analysis and feature extraction in satellite imagery. *International Journal of Remote Sensing*, **23**, 1799–1825.

Parrinello T. & Vaughan R.A. (2006) On comparing multifractal and classical features in minimum distance classification of AVHRR imagery. *International Journal of Remote Sensing*, **27**, 3943–3959.

Paw U K.T. & Meyers T.P. (1989) Investigations with a higher-order canopy turbulence model into mean source–sink levels and bulk canopy resistances. *Agricultural and Forest Meteorology*, **47**, 259–271.

Paw U K.T., Ustin S.L. & Zhang C.A. (1989) Anisotropy of thermal infrared exitance in sunflower canopies. *Agricultural and Forest Meteorology*, **48**, 45–58.

Pearcy R.W. & Yang W. (1996) A three-dimensional crown architecture model for assessment of light capture by understory plants. *Oecologia*, **108**, 1–12.

Peddle D.R., Hall F.G. & LeDrew E.F. (1999) Spectral mixture analysis and geometric-optical reflectance modeling of boreal forest biophysical structure. *Remote Sensing of Environment*, **67**, 288–297.

Penman H.L. (1948) Natural evaporation from open water, bare soil and grass. *Proceedings of the Royal Society of London, Series A*, **193**, 120–145.

Peñuelas J., Filella I., Gamon J.A. (1995) Assessment of photosynthetic radiation-use efficiency with spectral reflectance. *New Phytologist*, **131**, 291–296.

Pettorelli N., Vik J.O., Mysterud A., Gaillard J.-M., Tucker C.J. & Stenseth N.C. (2005) Using satellite-derived NDVI to assess ecological responses to environmental change. *Trends in Ecology and Evolution*, **20**, 503–510.

Pignatti S., Cavalli R.M., Cuomo V., Fusilli L., Pascucci S., Poscolieri M. & Santini F. (2009) Evaluating Hyperion capability for land cover mapping in a fragmented ecosystem: Pollino National Park, Italy. *Remote Sensing of Environment*, **113**, 622–634.

Pimstein A., Eitel J.U.H., Long D.S., Mufradi I., Karnieli A. & Bonfil D.J. (2009) A spectral index to monitor the head-emergence of wheat in semi-arid conditions. *Field Crops Research*, **111**, 218–225.

Pinter Jr P.J., Hatfield J.L., Schepers J.S., Barnes E.M., Moran M.S., Daughtry C.S.T. & Upchurch D.R. (2003) Remote sensing for crop management. *Photogrammetric Engineering and Remote Sensing*, **69**, 647–664.

Pinty B., Gobron N., Widlowski J.-L., Gerstl S.A.W., Verstraete M.M., Antunes M., Bacour C., Gascon F., Gastellu J.-P., Goel N., Jacquemoud S., North P., Qin W.H. & Thompson R. (2001) Radiation transfer model intercomparison (RAMI) exercise. *Journal of Geophysical Research–Atmospheres*, **106**, 11937–11956.

Pinty B., Widlowski J.-L., Taberner M., Gobron N., Verstraete M.M., Disney M., Gascon F., Gastellu J.-P., Jiang L., Kuusk A., Lewis P., Li X., Ni-Meister W.,

Nilson T., North P., Qin W., Su L., Tang S., Thompson R., Verhoef W., Wang H., Wang J., Yan G. & Zang H. (2004) Radiation Transfer Model Intercomparison (RAMI) exercise: Results from the second phase. *Journal of Geophysical Research–Atmospheres*, **109(D6)**, D06210.

Pontius J., Martin M., Plourde L. & Hallett R. (2008) Ash decline assessment in emerald ash borer-infested regions: A test of tree-level, hyperspectral technologies. *Remote Sensing of Environment*, **112**, 2665–2676.

Price J.C. (2003) Comparing MODIS and ETM$^+$ for regional and global land classification. *Remote Sensing of Environment*, **86**, 491–499.

Priestley C.H.B. & Taylor R.J. (1972) On the assessment of surface heat flux and evaporation using large-scale parameters. *Monthly Weather Review*, **100**, 81–92.

Prusinkiewicz P. (1998) Modeling of spatial structure and development of plants: a review. *Scientia Horticulturae*, **74**, 113–149.

Prusinkiewicz P. (2004) Modeling plant growth and development. *Current Opinion in Plant Biology*, **7**, 79–83.

Purves R. & Jones C. (2006) Geographic information retrieval. *Computers, Environment and Urban Systems*, **30**, 375–377.

Qin Z. & Karnieli A. (1999) Progress in remote sensing of land surface temperature and ground emissivity using NOAA-AVHRR data. *International Journal of Remote Sensing*, **20**, 2367–2393.

Quaife T., Lewis P., De Kauwe M., Williams M., Law B.E., Disney M. & Bowyer P. (2008) Assimilating canopy reflectance data into an ecosystem model with an Ensemble Kalman Filter. *Remote Sensing of Environment*, **112**, 1347–1364.

Quattrochi D.A. & Goodchild M.F., eds (1997) *Scale in remote sensing*. CRC Press Inc., Boca Raton, FL., pp. 432, ISBN 978-1566701044.

Rahman H., Verstraete M.M. & Pinty B. (1993) Coupled surface–atmosphere reflectance (CSAR) model 1. Model description and inversion on synthetic data. *Journal of Geophysical Research–Atmospheres*, **98**, 20779–20789.

Rao N.R. (2008) Development of a crop-specific spectral library and discrimination of various agricultural crop varieties using hyperspectral imagery. *International Journal of Remote Sensing*, **29**, 131–144.

Raptis V.S., Vaughan R.A. & Wright G.G. (2003) The effect of scaling on land cover classification from satellite data. *Computers and Geosciences*, **29**, 705–714.

Raschke K. (1956) Über die physikalischen Beziehungen zwischen Wärmeübergangszahl, Strahlungsaustausch, Temperatur und Transpiration eines Blattes. [The physical relationships between heat-transfer coefficients, radiation exchange, temperature and transpiration of a leaf.] *Planta*, **48**, 200–238.

Raskin I. & Ladyman J.A.R. (1988) Isolation and characterisation of a barley mutant with abscisic-acid-insensitive stomata. *Planta*, **173**, 73–78.

Rauner J.L. (1976) Deciduous forest. In: *Vegetation and the atmosphere, Vol. 2: Case studies* (ed. Monteith, J.L.), pp. 241–264. Academic Press, London. ISBN 0125051026.

Raven P.H., Evert R.F. & Eichhorn S.E. (2005) *Biology of plants*. (7th ed.). W.H. Freeman & Company Ltd, San Francisco, CA. pp. 900, ISBN 9780716762843.

Rayner P.J., Scholze M., Knorr W., Kaminski T., Giering R. & Widmann H. (2005) Two decades of terrestrial carbon fluxes from a carbon cycle data assimilation system (CCDAS). *Global Biogeochemical Cycles*, **19**, GB2026.

Rees W.G. (1999) *The remote sensing data book*. CUP, Cambridge. pp. 262, ISBN 052148040X.

Rees W.G. (2001) *Physical principles of remote sensing*. (2nd ed.). CUP, Cambridge. pp. 343, ISBN 0521669480.

Reichle R.H. (2008) Data assimilation methods in the earth sciences. *Advances in Water Resources*, **31**, 1411–1418.

Reitberger J., Krzystek P. & Stilla U. (2008) Analysis of full waveform lidar data for the classification of deciduous and coniferous trees. *International Journal of Remote Sensing*, **29**, 1407–1431.

Reitberger J., Schnörr C., Krzystek P. & Stilla U. (2009) 3D segmentation of single trees exploiting full waveform LIDAR data. *ISPRS Journal of Photogrammetry and Remote Sensing*, **64**, 561–574.

Reusch S. (2009) Use of ultrasonic transducers for online biomass estimation in winter wheat. In: *Precision agriculture '09* (eds Lokhorst, C., Huijsmans, J.F.M., & de Louw, R.P.M.), pp. 169–175. Wageningen Academic Publishers, Wageningen.

Riaño D., Valladares F., Condés S. & Chuvieco E. (2004) Estimation of leaf area index and covered ground from airborne laser scanner (Lidar) in two contrasting forests. *Agricultural and Forest Meteorology*, **124**, 269–275.

Ribeiro da Luz B. (2006) Attenuated total reflectance spectroscopy of plant leaves: a tool for ecological and botanical studies. *New Phytologist*, **172**, 305–318.

Ribeiro da Luz B. & Crowley J.K. (2007) Spectral reflectance and emissivity features of broad leaf plants: Prospects for remote sensing in the thermal infrared (8.0–14.0 μm). *Remote Sensing of Environment*, **109**, 393–405.

Richards J.A. & Xia X. (2005) *Remote sensing digital image analysis*. (4th ed.). Springer Verlag, Berlin. pp. 439, ISBN 9783540251286.

Richardson A.J. & Wiegand C.L. (1977) Distinguishing vegetation from soil background. *Photogrammetric Engineering and Remote Sensing*, **43**, 1541–1552.

Richardson J.J., Moskal L.M. & Kim S.-H. (2009) Modeling approaches to estimate effective leaf area index from aerial discrete-return LIDAR. *Agricultural and Forest Meteorology*, **149**, 1152–1160.

Richter K., Atzberger C., Vuolo F., Weihs P. & D'Urso G. (2009) Experimental assessment of the Sentinel-2 band setting for RTM-based LAI retrieval of sugar beet and maize. *Canadian Journal of Remote Sensing*, **35**, 230–247.

Ripley E.A. & Redman R.E. (1976) Grassland. In: *Vegetation and the atmosphere, Vol. 2: Case studies* (ed. Monteith, J.L.), pp. 350–398. Academic Press, London. ISBN 0125051026.

Robinson I.S. (2003) *Measuring the oceans from space*. Springer Verlag, Berlin. pp. 714, ISBN 9783540426479.

Rodríguez-Moreno L., Pineda M., Soukupová S., Macho A.P., Beuzón C.R., Barón M. & Ramos C. (2008) Early detection of bean infection by *Pseudomonas syringae* in asymptomatic leaf areas using chlorophyll fluorescence imaging. *Photosynthesis Research*, **96**, 27–35.

Roerink G.J., Su Z. & Menenti M. (2000) S-SEBI: A simple remote sensing algorithm to estimate the surface energy balance. *Physics and Chemistry of the Earth. Part B: Hydrology, oceans and atmosphere*, **25**, 147–157.

Ross J. (1981) *The radiation regime and architecture of plant stands*. Dr W Junk, The Hague. pp. 391, ISBN 9061936071.

Rotenberg E., Mamane Y. & Joseph J.H. (1998) Long wave radiation regime in vegetation – parameterisations for climate research. *Environmental Modelling and Software*, **13**, 361–371.

Roujean J.-L., Leroy M. & Deschamps P.-Y. (1992) A bidirectional reflectance model of the earth's surface for the correction of remote sensing data. *Journal of Geophysical Research*, **97**, 20455–20468.

Rouse J.W., Haas R.H., Schell J.A., Deering D.W. & Harlan J.C. (1974) *Monitoring the vernal advancement and retrogradation (greenwave effect) of natural vegetation*. NASA/GSFC Final report, Greenbelt, MD, USA.

Roy D.P., Lewis P.E. & Justice C.O. (2002) Burned area mapping using multi-temporal moderate spatial resolution data – a bi-directional reflectance model-based expectation approach. *Remote Sensing of Environment*, **88**, 263–286.

Rubio E., Caselles V., Coll C., Valour E. & Sospedra F. (2003) Thermal-infrared emissivities of natural surfaces: improvements on the experimental set-up and new measurements. *International Journal of Remote Sensing*, **24**, 5379–5390.

Russ J.C. (2006) *The image processing handbook*. (5th ed.). Taylor and Francis Ltd., Boca Raton, Fl. pp. 832, ISBN 9780849370731

Rydberg A., Söderström M., Hagner O. & Börjesson T. (2007) Field specific overview of crops using UAV (Unmanned Aerial Vehicle). In: *Proceedings of the 6th European Conference on Precision Agriculture, Skiathos, Greece. Precision agriculture '07*. (ed. Stafford, J.V.), pp. 357–364. Wageningen Academic Publishers, Wageningen.

Sabins F.F. (1997) *Remote sensing – principles and interpretation*. W.H. Freeman and Company, New York. pp. 494, ISBN 0716724421.

Salisbury F.B. & Ross C.W. (1995) *Plant Physiology*. (5th ed.). Wadsworth, Belmont. pp. 682, ISBN 0534983901.

Salisbury J.W. & Milton N.M. (1988) Thermal infrared (2.5 to 13.5 μm) directional hemispherical reflectance of leaves. *Photogrammetric Engineering and Remote Sensing*, **54**, 1301–1304.

Schaepman-Strub G., Schaepman M.E., Painter T.H., Dangel S. & Martonchik J.V. (2006) Reflectance quantities in optical remote sensing—definitions and case studies. *Remote Sensing of Environment*, **103**, 27–42.

Schmugge T.J., Kustas W.P., Ritchie J.C., Jackson T.J. & Rango A. (2002) Remote sensing in hydrology. *Advances in Water Resources*, **25**, 1367–1385.

Schopfer J.T., Dangel S., Kneubühler M. & Itten K.I. (2008) The improved dual-view field goniometer system FIGOS. *Sensors*, **8**, 5120–5140.

Schreiber U., Schliwa U. & Bilger W. (1986) Continuous recording of photochemical and non-photochemical chlorophyll fluorescence quenching with a new type of modulation fluorometer. *Photosynthesis Research*, **10**, 51–62.

Sedano F., Lavergne T., Ibañez L.M. & Gong P. (2008) A neural network-based scheme coupled with the RPV model inversion package. *Remote Sensing of Environment*, **112**, 3271–3283.

Seelig H.-D., Hoehn A., Stodieck L.S., Klaus D.M., Adams III W.W. & Emery W.J. (2008a) Relations of remote sensing leaf water indices to leaf water thickness in cowpea, bean, and sugarbeet plants. *Remote Sensing of Environment*, **112**, 445–455.

Seelig H.-D., Hoehn A., Stodieck L.S., Klaus D.M., Adams III W.W.. & Emery W.J. (2008b) The assessment of leaf water content using leaf reflectance ratios in the visible, near-, and short-wave-infrared. *International Journal of Remote Sensing*, **29**, 3701–3713.

Seguin B. & Itier B. (1983) Using midday surface temperature to estimate daily evaporation from satellite thermal infrared data. *International Journal of Remote Sensing*, **4**, 371–383.

Sellers P.J., Rasool S.I. & Bolle H.-J. (1990) A review of satellite data algorithms for studies of the land surface. *Bulletin of the American Meteorological Society*, **71**, 1429–1447.

Seymour R.S. (1999) Pattern of respiration by intact inflorescences of the thermogenic arum lily *Philodendron selloum*. *Journal of Experimental Botany*, **50**, 845–852.

Shi J., Jackson T., Tao J., Du J., Bindlish R., Lu L. & Chen K.S. (2008) Microwave vegetation indices for short vegetation covers from satellite passive microwave sensor AMSR-E. *Remote Sensing of Environment*, **112**, 4285–4300.

Shimoni M., Borghys D., Heremans R., Perneel C. & Acheroy M. (2009) Fusion of PolSAR and PolInSAR data for land cover classification. *International Journal of Applied Earth Observation and Geoinformation*, **11**, 169–180.

Shuttleworth W.J. & Wallace J.S. (1985) Evaporation from sparse crops - an energy combination theory. *Quarterly Journal of The Royal Meteorological Society*, **111**, 839–855.

Silva J.M.N. & Pereira M.C. (2006) Burned area mapping in Africa with Spot-Vegetation imagery: accuracy assessment with Landsat ETM+ data, influence on spatial pattern and vegetation type. In: *25th Annual EARSel Symposium*, pp. 367–376. Millpress, Rotterdam, Rotterdam.

Simmoneaux V., Duchemin B., Helson D., Er-Raki S., Olioso A. & Chehbouni A.G. (2008) The use of high-resolution image time series for crop classification and evapotranspiration estimate over an irrigated area in central Morocco. *International Journal of Remote Sensing*, **29**, 95–116.

Singh S.K., Raman M., Dwivedi R.M. & Nayak S.R. (2008) An approach to compute Photosynthetically Active Radiation using IRS P4 OCM. *International Journal of Remote Sensing*, **29**, 211–220.

Sinoquet H., Thanisawanyangkura S., Mabrouk H. & Kasemsap P. (1998) Characterization of the Light Environment in Canopies Using 3D Digitising and Image Processing. *Annals of Botany*, **82**, 203–212.

Skidmore A.K., Ferwerda J.G., Mutanga O., Van Wieren S.E., Peel M., Grant R.C., Prins H.H.T., Balcik F.B. & Venus V. (2010) Forage quality of savannas — Simultaneously mapping foliar protein and polyphenols for trees and grass using hyperspectral imagery. *Remote Sensing of Environment*, **114**, 64–72.

Slater M.T. & Vaughan R.A. (1999) Mediterranean fire risk monitoring using AVHRR. In: *18th EARSel Annual Symposium*, pp. 463–469. EARSel,The Netherlands.

Smith K.L., Steven M.D. & Colls J.J. (2004) Use of hyperspectral derivative ratios in the red-edge region to identify plant stress responses to gas leaks. *Remote Sensing of Environment*, **92**, 207–217.

Smolander S. & Stenberg P. (2005) Simple parameterizations of the radiation budget of uniform broadleaved and coniferous canopies. *Remote Sensing of Environment*, **94**, 355–363.

Snedecor G.W. & Cochran W.G. (1999) *Statistical Methods*. (8th ed.). Iowa State University Press, Ames, Iowa. pp. 1–524, ISBN ISBN: 0813815614.

Snyder R.L., Spano D. & Paw U K.T. (1996) Surface renewal analysis for sensible and latent heat flux density. *Boundary-Layer Meteorology*, **77**, 249–266.

Snyder W.C., Wan Z., Zhang Y. & Feng Y.-Z. (1998) Classification-based emissivity for land surface temperature measurement from space. *International Journal of Remote Sensing*, **19**, 2753–2774.

Sobrino J.A., El Kharraz M.H., Cuenca J. & Raissouni N. (1998) Thermal inertia mapping from NOAA-AVHRR data. *Advances in Space Research*, **22**, 655–667.

Sobrino J.A., Gómez M., Jiménez-Muñoz J.C. & Olioso A. (2007) Application of a simple algorithm to estimate daily evapotranspiration from NOAA–AVHRR images for the Iberian Peninsula. *Remote Sensing of Environment*, **110**, 139–148.

Soille P. (2002) *Morphological image analysis: principles and applications*. (2nd ed.). Springer-Verlag, Berlin and Heidelberg. pp. 407, ISBN 9783540429883.

Sokal R.R. & Rohlf F.J. (1995) *Biometry*. (3rd ed.). W.H. Freeman & Co., New York, ISBN 0716724111.

Solomon S., Qin D., Manning M., Chen Z., Marquis M., Averyt K.B., Tignor M. & Miller H.L., eds (2007) *Climate Change 2007: The Physical Science Basis*. Cambridge University Press, Cambridge, pp. 996.

Song J. (1998) Diurnal asymmetry in surface albedo. *Agricultural and Forest Meteorology*, **92**, 181–189.

St. Louis V., Pidgeon A.M., Radeloff V.C., Hawbaker T.J. & Clayton M.K. (2006) High-resolution image texture as a predictor of bird species richness. *Remote Sensing of Environment*, **105**, 299–312.

St.-Onge B., Hu Y. & Vega C. (2008) Mapping the height and above-ground biomass of a mixed forest using lidar and stereo IKONOS images. *International Journal of Remote Sensing*, **2**, 1277–1294.

Stanhill G. (1981) The size and significance of differences in the radiation balance of plants and plant communities. In: *Plants and their atmospheric environment* (eds Grace, J., Ford, E.D., & Jarvis, P.G.), pp. 57–73. Blackwell Scientific Publishers, Oxford. ISBN 0632005254.

Sternberg P. (2007) Simple analytical formula for calculating average photon recollision probability in vegetation canopies. *Remote Sensing of Environment*, **109**, 221–224.

Steven M.D. (1998) The sensitivity of the OSAVI vegetation index to observational parameters. *Remote Sensing of Environment*, **63**, 49–60.

Stisen S., Sandholt I., Nørgaard A., Fensholt R. & Jensen K.H. (2008) Combining the triangle method with thermal inertia to estimate regional evapotranspiration — Applied to MSG-SEVIRI data in the Senegal River basin. *Remote Sensing of Environment*, **112**, 1242–1255.

Strebel D.E., Landis D.R., Huemmrich K.F., Newcomer J.A. & Meeson B.W. (1998) The FIFE data publication experiment. *Journal of the Atmospheric Sciences*, **55**, 1277–1283.

Suits G.H. (1972) The calculation of the directional reflectance of vegetative canopy. *Remote Sensing of Environment*, **2**, 117–125.

Sun X. & Anderson J.M. (1993) A Spatially Variable Light-Frequency-Selective Component-Based, Airborne Pushbroom Imaging Spectrometer for the Water Environment. *Photogrammetric Engineering and Remote Sensing*, **59**, 399–406.

Sutherland R.A. (1986) Broadband and spectral emissivities (2–18 μm) of some natural soils and vegetation. *Journal of Atmospheric and Oceanic Technology*, **3**, 199–202.

Taiz L. & Zeiger E. (2006) *Plant Physiology*. (4th ed.). Sinauer Associates, Sunderland, MA 01375, USA. pp. 650, ISBN 9780878938568.

ter Steeg H. (1993) *HEMIPHOT: a programme to analyze vegetation indices, light and light quality from hemispherical photographs*. Tropenbos Foundation, Wageningen.

Thenkabail P., Lyon J.G., Turral H. & Biradar C., eds (2009) *Remote sensing of global croplands for food security*. Taylor and Francis, Boca Raton, FL, pp. 556, ISBN 9781420090093.

Thom A.S. (1975) Momentum, mass and heat exchange of plant communities. In: *Vegetation and the atmosphere. 1. Principles* (ed. Monteith, J.L.), pp. 57–109. Academic Press, London, New York and San Francisco. ISBN 0125051018.

Thornthwaite C.W. (1948) An approach toward a rational classification of climate. *Geographical Review*, **38**, 55–94.

Thorp K.R. & Tian L.F. (2004) A review of remote sensing of weeds in agriculture. *Precision Agriculture*, **5**, 477–508.

Timmermans J., Van der Tol C., Verhoef W. & Su Z. (2008) Contact and directional radiative temperature measurements of sunlit and shaded land surface components during the SEN2FLEX 2005 campaign. *International Journal of Remote Sensing*, **29**, 5183–5192.

Timmermans J., Verhoef W., Van der Tol C. & Su Z. (2009) Retrieval of canopy temperature component temperatures through Bayesian inversion of directional thermal measurements. *Hydrology and Earth System Sciences Discussions*, **6**, 3007–3040.

Tomppo E.O., Gagliano C., De Natale F., Katila M. & McRoberts R.E. (2009) Predicting categorical forest variables using an improved k-Nearest Neighbour estimator and Landsat imagery. *Remote Sensing of Environment*, **113**, 500–517.

Townsend C.R., Begon M. & Harper J.L. (2008) *Essentials of ecology*. (3rd ed.). John Wiley and Sons Ltd, Chichester. pp. 532, ISBN 9781405156585.

Townsend P.A., Lookingbill T.R., Kingdon C.C. & Gardner R.H. (2009) Spatial pattern analysis for monitoring protected areas. *Remote Sensing of Environment*, **113**, 1410–1420.

Tso B. & Mather P.M. (2001) *Classification methods for remotely sensed data*. Taylor and Francis Inc., New York. pp. 332, ISBN 9780415259095

Tucker C.J. (1979) Red and photographic infrared linear combinations for monitoring vegetation. *Remote Sensing of Environment*, **8**, 127–150.

Tucker C.J., Pinzon J.E. & Brown M.E. (2004) *Global Inventory Modeling and Mapping Studies* (Report). Global Landcover Facility, University of Maryland, College Park, MD.

Uchijima Z. (1976) Maize and rice. In: *Vegetation and the atmosphere, Vol. 2: Case studies* (ed. Monteith, J.L.), pp. 33–64. Academic Press, London. ISBN 0125051026.

Ulaby F.T., Moore R.K. & Fung A.K. (1982) *Microwave remote sensing. Active and passive. Vol II. Radar remote sensing and surface scattering and emission theory.* Addison-Wesley, Reading, Massachusetts, ISBN 0201107600.

Underwood E.C., Ustin S.L. & DiPietro D. (2003) Mapping nonnative plants using hyperspectral imagery. *Remote Sensing of Environment*, **86**, 150–161.

Valladares F. & Pearcy R.W. (1999) The geometry of light interception by shoots of *Heteromeles arbutifolia*: morphological and physiological consequences for individual leaves. *Oecologia*, **121**, 171–182.

Valladares F. & Pugnaire F.I. (1999) Tradeoffs between irradiance capture and avoidance in semi-arid environments assessed with a crown architecture model. *Annals of Botany*, **83**, 459–469.

Valor E. & Caselles V. (1996) Mapping land surface emissivity from NDVI: application to European, African and South American areas. *Remote Sensing of Environment*, **57**, 167–184.

Van de Griend A.A. & Owe M. (1993) On the relationship between thermal emissivity and the normalized difference vegetation index for natural surfaces. *International Journal of Remote Sensing*, **14**, 1119–1131.

van Evert F.K., Smason J., Polder G., Vijn M., van Dooren H.-J., Lamaker E.J.J., van der Heijden G.W.A.M., Kempenaar C., van der Zalm A.J.A. & Lotz L.A.P. (2009) Robotic control of broad-leaved dock. In: *Precision agriculture '09* (eds van Henten, E.J., Goense, D., & Lokhorst, C.), pp. 725–732. Wageningen Academic Publishers, Wageningen.

Van Gaalen K.E., Flanagan L.B. & Peddle D.R. (2007) Photosynthesis, chlorophyll fluorescence and spectral reflectance in Sphagnum moss at varying water contents. *Oecologia*, **153**, 19–28.

van Gardingen P.R., Jackson G.E., Hernandez-Daumas S., Russell G. & Sharp L. (1999) Leaf area index estimates obtained for clumped canopies using hemispherical photography. *Agricultural and Forest Meteorology*, **94**, 243–257.

Vancutsem C., Pekel J.F., Evrard C., Malaisse F. & Defourny P. (2009) Mapping and characterizing the vegetation types of the Democratic Republic of Congo using SPOT VEGETATION time series. *International Journal of Applied Earth Observation and Geoinformation*, **11**, 62–76.

Vapnik V. (1995) *The nature of statistical learning theory*. Springer-Verlag, New York, ISBN 0471030031.

Venturini V., Islam S. & Rodriguez L. (2008) Estimation of evaporative fraction and evapotranspiration from MODIS products using a complementary based model. *Remote Sensing of Environment*, **112**, 132–141.

Verbeiren S., Eerens H., Piccard I., Bauwens I. & Van Orshoven J. (2008) Sub-pixel classification of SPOT-VEGETATION time series for the assessment of regional crop areas in Belgium. *International Journal of Applied Earth Observation and Geoinformation*, **10**, 486–497.

Verger F., Sourbès-Verger I., Ghirardi R. & Pasco X. (2003) *The Cambridge encyclopedia of space*. Cambridge University Press, Cambridge. pp. 423, ISBN 9780521773003.

Verhoef W. (1984) Light scattering by leaf layers with application to canopy reflectance modeling: the SAIL model. *Remote Sensing of Environment*, **16**, 125–141.

Verhoef W. & Bach H. (2007) Coupled soil–leaf-canopy and atmosphere radiative transfer modeling to simulate hyperspectral multi-angular surface reflectance and TOA radiance data. *Remote Sensing of Environment*, **109**, 166–182.

Vermote E.F., Tanre D., Deuze J.L., Herman M. & Morcette J.-J. (1997) Second simulation of the satellite signal in the solar spectrum, 6S: an overview. *IEEE Transactions on Geoscience and Remote Sensing*, **35**, 675–686.

Verrelst J., Geerling G.W., Sykora K.V. & Clevers J.G.P.W. (2009) Mapping of aggregated floodplain plant communities using image fusion of CASI and LiDAR data. *International Journal of Applied Earth Observation and Geoinformation*, **11**, 83–94.

Verstraete M.M. & Pinty B. (2001) Introduction to special section: Modeling, measurement, and exploitation of anisotropy in the radiation field. *Journal of Geophysical Research*, **106**, 11903–11907.

Verstraete M.M., Pinty B. & Dickinson R.E. (1990) A physical model of the bidirectional reflectance of vegetation canopies 1. Theory. *Journal of Geophysical Research-Atmospheres*, **95**, 11755–11765.

Wagner W., Hollaus M., Briese C. & Ducic V. (2008) 3D vegetation mapping using small-footprint full-waveform airborne laser scanners. *International Journal of Remote Sensing*, **29**, 1433–1452.

Wald L. (2002) *Data fusion, definitions and architectures – Fusion of images of different spatial resolutions*. Les Presses de l'École des Mines, Paris. pp. 200, ISBN 291176238X.

Walthall C.L., Norman J.M., Welles J.M., Campbell G. & Blad B.L. (1985) Simple equation to approximate the bidirectional reflectance from vegetative canopies and bare soil surfaces. *Applied Optics*, **24**, 383–387.

Wan Z. (1999) *MODIS Land-surface temperature algorithm theoretical basis document (LST ATBD) Version 3.3*. Institute for Computational Earth System Science, Santa Barbara, CA.

Wan Z. (2008) New refinements and validation of the MODIS Land-Surface Temperature/Emissivity products. *Remote Sensing of Environment*, **112**, 59–74.

Wang C. & Qi J. (2008) Biophysical estimation in tropical forests using JERS-1 SAR and VNIR imagery. II. Above ground woody biomass. *International Journal of Remote Sensing*, **29**, 6827–6849.

Wang L., Qu J.J. & Hao X. (2008) Forest fire detection using the normalized multi-band drought index (NMDI) with satellite measurements. *Agricultural and Forest Meteorology*, **148**, 1767–1776.

Wang S. & Davidson A. (2007) Impact of climate variations on surface albedo of a temperate grassland. *Agricultural and Forest Meteorology*, **142**, 133–142.

Wang T. (2009) *Observing giant panda habitat and forage abundance from space*. Dissertation, Wageningen University, ITC.

Wang T., Skidmore A.K., Toxopeus A.G. & Liu X. (2009) Understorey bamboo discrimination using a winter image. *Photogrammetric Engineering and Remote Sensing*, **75**, 37–47.

Wang Y.P. & Jarvis P.G. (1988) Mean leaf angles for the ellipsoidal inclination angle distribution. *Agricultural and Forest Meteorology*, **43**, 319–321.

Wanner W., Li X. & Strahler A.H. (1995) On the Derivation of Kernels for Kernel-Driven Models of Bidirectional Reflectance. *Journal of Geophysical Research-Atmospheres*, **100**, 21077–21089.

Wardlow B.D., Egbert S.L. & Kastens J.H. (2007) Analysis of time-series MODIS 250 m vegetation index data for crop classification in the U.S. Central Great Plains. *Remote Sensing of Environment*, **108**, 290–310.

Watts K., Humphrey J.W., Griffiths M., Quine C. & Ray D. (2005) *Evaluating Biodiversity in Fragmented Landscapes: Principles*. (Forestry Commission Information Note No. 073). Forestry Commission, Edinburgh.

Weiss M., Baret F., Myneni R.B., Pragnere A. & Knyazikhin Y. (2000) Investigation of a model inversion technique to estimate canopy biophysical variables from spectral and directional reflectance data. *Agronomie*, **20**, 3–22.

Weiss M., Baret F., Smith G.J., Jonckheere I. & Coppin P. (2004) Review of methods for in situ leaf area index (LAI) determination Part II. Estimation of LAI, errors and sampling. *Agricultural and Forest Meteorology*, **121**, 37–53.

Weng F., Yan B. & Grody N.C. (2001) A microwave land emissivity model. *Journal of Geophysical Research*, **106**, 20115–20123.

Whitlock C.H., Charlock T.P., Staylor W.F., Pinker R.T., Laszlo I., DiPasquale R.C. & Ritchie N.A. (1993) *WCRP surface radiation budget shortwave data product description – Version 1.1* (NASA Technical Memorandum 107747). NASA, Hampton, Virginia.

Whittaker R.J., Willis K.J. & Field R. (2001) Scale and species richness: towards a general, hierarchical theory of species diversity. *Journal of Biogeography*, **28**, 453–470.

Widlowski J.-L., Robustelli M., Disney M., Gastellu-Etchegorry J.-P., Lavergne T., Lewis P., North P.R.J., Pinty B., Thompson R. & Verstraete M.M. (2008) The RAMI On-line Model Checker (ROMC): A web-based benchmarking facility for canopy reflectance models. *Remote Sensing of Environment* **112**, 1144–1150.

Wigneron J.-P., Calvet J.-C., Pellarin T., Van de Griend A.A., Berger M. & Ferrazzoli P. (2003) Retrieving near-surface soil moisture from microwave radiometric observations: current status and future plans. *Remote Sensing of Environment*, **85**, 489–506.

Wisniewski M., Lindow S.E. & Ashworth E.N. (1997) Observations of ice nucleation and propagation in plants using infrared video thermography. *Plant Physiology*, **113**, 327–334.

Wood B.D. (2009) The role of scaling laws in upscaling. *Advances in Water Resources*, **32**, 723–736.

Wood C.W., Reeves D.W. & Himelrick D.G. (1993) Relationships between chlorophyll meter readings and leaf chlorophyll concentration, N status, and crop yield: A review. *Proceedings Agronomy Society of New Zealand*, **23**, 1–9.

Woodhouse I.H. (2006) *Introduction to microwave remote sensing*. Taylor and Francis, London. pp. 370, ISBN 9780415271233.

Workman Jr J.J. (2008) NIR spectroscopy calibration basics. In: *Handbook of near-infrared analysis* (eds Burns, D.A. & Ciurczak, E.W.), pp. 123–150. CRC Press, Boca Raton, FL., USA.

Wu J., Wang D. & Bauer M.E. (2005) Image-based atmospheric correction of QuickBird imagery of Minnesota cropland. *Remote Sensing of Environment*, **99**, 315–325.

Xie Y., Sha Z. & Yu M. (2008a) Remote sensing imagery in vegetation mapping: a review. *Journal of Plant Ecology*, **1**, 9–23.

Xie Z., Roberts C. & Johnson B. (2008b) Object-based target search using remotely sensed data: A case study in detecting invasive exotic Australian Pine in south Florida. *ISPRS Journal of Photogrammetry and Remote Sensing*, **63**, 647–660.

Xue Y. & Cracknell A.P. (1999) Advanced thermal inertia modelling. *International Journal of Remote Sensing*, **16**, 431–446.

Xue Y., Li Y., Guang J., Zhang X. & Guo J. (2008) Small satellite remote sensing and applications – history, current and future. *International Journal of Remote Sensing*, **29**, 4339–4372.

Yahia H., Turiel A., Chrysoulakis N., Grazzini J., Prastacos P. & Herlin I. (2008) Application of the multifractal microcanonical formalism to the detection of fire plumes in NOAA-AVHRR data. *International Journal of Remote Sensing*, **29**, 4189–4205.

Yi Y., Yang D., Huang J. & Chen D. (2008) Evaluation of MODIS surface reflectance products for wheat leaf area index (LAI) retrieval. *ISPRS Journal of Photogrammetry and Remote Sensing*, **63**, 661–677.

Zar J.H. (1999) *Biostatistical analysis*. (4th ed.). Prentice-Hall, Upper Saddle River, New Jersey, ISBN 013081542X.

Zhang X., Friedl M.A. & Schaaf C.B. (2009) Sensitivity of vegetation phenology detection to the temporal resolution of satellite data. *International Journal of Remote Sensing*, **30**, 2061–2074.

Zhang Y., Chen J.M., Miller J.R. & Noland T.L. (2008) Leaf chlorophyll content retrieval from airborne hyperspectral remote sensing imagery. *Remote Sensing of Environment*, **112**, 3234–3247.

Zhou X., Guan H., Xie H. & Wilson J.L. (2009) Analysis and optimization of NDVI definitions and areal fraction models in remote sensing of vegetation. *International Journal of Remote Sensing*, **30**, 721–751.

Appendix 1
Sundry useful conversions, constants, and properties

Units and conversions

Use International System of Units (SI):

	Symbol	SI base unit	SI derived units	Equivalents
Mass		1 kilogramme (kg)		= 2.2046 pounds
Length	l	1 metre (m)		= 3.2808 feet
Time	t	1 second (s)		
Temperature	T	1 kelvin (K)		= 1 °C
Amount of substance		1 mole (mol)		contains 6.022×10^{23} elementary entities (photons, molecules, etc.)
Energy	E	1 kg m^2 s^{-2}	joule (J) = N m	= 10^7 erg = 0.2388 calorie = $6.24150974 \times 10^{18}$ eV
Force		1 kg m s^{-2}	newton (N)	= 10^5 dyne
Pressure	P	1 kg m^{-1} s^{-2}	pascal (Pa) = N m^{-2}	= 10^{-5} bar = 0.9869×10^{-5} atmosphere = 7.5×10^{-3} mm Hg
Current		1 ampere (A)		
Charge		1 sA	coulomb (C)	= 6.242×10^{18} elementary charges
Power		1 kg m^2 s^{-3}	watt (W) = J s^{-1}	= 10^7 erg s^{-1}
Angle		radian (rad)		2π radian = 360°
Solid angle	Ω	steradian (sr)		
Frequency	f	1 cycle s^{-1}	hertz (Hz)	

Conversions between units

For evaporation:
The appropriate conversions at 20 °C are:
1 mm = 1000 cm^3 m^{-2} = 0.01 ML ha^{-1} = 1 kg m^{-2} = 55.55 mol m^{-2} ≅ 2.454 MJ m^{-2}.

Energy/wave number to Energy/wavelength:
Wave numbers are $1/\lambda$, but are conventionally given units of cm^{-1} (rather than the SI m^{-1}), therefore to convert from wavelength (λ, μm) to wave number (υ, cm^{-1}):

$$\upsilon = 10\,000/\lambda$$

Radiation units:
Note the incorrect use of Einstein in some literature to refer to photon fluxes as a mol (Avogadro's number) of photons

Conversions between molar and conventional units

The Penman–Monteith equation may be written using either molar units as in eqn (4.13) as promoted by Campbell and Norman (1998) or in the more commonly used mass units. The conversions are outlined below for this and for eqn (4.14):

Molar units		Conventional units
$\lambda \boldsymbol{E} = \dfrac{\hat{c}_p \hat{g}_H D_x + \hat{s} \boldsymbol{R}_n}{(\hat{\gamma} \hat{g}_H/\hat{g}_W) + \hat{s}}$	(4.13)	$\lambda \boldsymbol{E} = \dfrac{\rho c_p g_H D_a + s \boldsymbol{R}_n}{(\gamma g_H/g_W) + s}$
$T_l = T_a + \dfrac{\hat{\gamma} \boldsymbol{R}_n - \hat{g}_W \hat{c}_p D_x}{(\hat{\gamma} \hat{c}_p \hat{g}_H + \hat{c}_p \hat{s} \hat{g}_W)}$	(4.14)	$T_l = T_a + \dfrac{\gamma \boldsymbol{R}_n - \rho c_p g_W D_a}{\rho c_p (\gamma g_H + s g_W)}$

Where		***Where***	
$\boldsymbol{E}$	mol m^{-2} s^{-1}	$\boldsymbol{E}$	kg m^{-2} s^{-1}
λ	J mol^{-1}	λ	J kg^{-1}
$\hat{c}_p$	J mol^{-1} K^{-1}	c_p	J kg^{-1} K^{-1}
$\hat{s}$	K^{-1}	s	Pa K^{-1}
$\hat{g}_H$	mol m^{-2} s^{-1}	g_H	m s^{-1}
$\hat{g}_W$	mol m^{-2} s^{-1}	g_W	m s^{-1}
$\boldsymbol{R}_n$	J m^{-2} s^{-1}	$\boldsymbol{R}_n$	J m^{-2} s^{-1}
$\hat{\gamma}$	$(= \hat{c}_p/\lambda)$ K^{-1}	γ	$(= P c_p/0.622\,\lambda)$ Pa K^{-1}
D_x	mol mol^{-1} (air water vapour pressure deficit)	D_a	Pa
		ρ	kg m^{-3}
		P	Pa

Physical constants

h Planck's constant	6.6261×10^{-34} (J s)
c speed of light *in vacuo*	2.99792458×10^{8} (m s^{-1})
k Boltzmann constant	1.3807×10^{-23} (J K^{-1})
σ Stefan–Boltzmann constant	5.6703×10^{-8} (W m^{-2} K^{-4})
$\mathfrak{R}$ Universal gas constant	8.3143 (J mol^{-1} K^{-1})

Saturation vapour pressure of water vapour

A simple approach to the estimation of saturation water vapour pressure ($e_{s(T)}$) for any temperature (T) over water is the Magnus formula (equation from Buck Research Manual (1996) – http://www.hygrometers.com/buckrehome.html); modified from Buck (1981)

$$e_{s(T)} = f(a \exp (bT/(c + T))$$

where $a = 611.21$; $b = 18.678 - (T/234.5)$; $c = 257.14$; $f = 1.0007 + 3.46 \times 10^{-8}P$ (e_s and P are in Pa, T in °C)

Properties of air

Specific heat of dry air at 20 °C and 101.3 kPa (c_p) = 1010 J kg^{-1} K^{-1}
Molar specific heat of dry air at 20 °C and 101.3 kPa ($\hat{c}_p$) = 29.3 J mol^{-1} K^{-1}
Latent heat of vaporization of water (λ) = 2.454 MJ kg^{-1} = 44172 J mol^{-1}
Density of dry air at 20 °C and 101.325 kPa (ρ) = 1.205 kg m^{-3}
Psychrometer constant ($\gamma = Pc_p/0.622\ \lambda$) = 66.2 Pa K^{-1} at 20 °C and 100 kPa, or $\hat{\gamma} = \hat{c}_p/\lambda = 6.59 \times 10^{-4}$ K^{-1}.
Slope of curve relating saturation vapour pressure to temperature (s) = 145 Pa K^{-1} at 20 °C.
Slope of curve relating saturation vapour mole fraction to temperature ($\hat{s}$) = 0.001431 K^{-1}.

Table of molecular masses

$M_{water} = 0.018$ kg mol^{-1}
$M_{air} = 0.029$ kg mol^{-1}
$M_{CO2} = 0.044$ kg mol^{-1}

Diffusivities in air (at 20 °C)

$\mathbf{D}_{water} = 24.2$ mm^2 s^{-1}
$\mathbf{D}_{CO2} = 14.7$ mm^2 s^{-1}
$\mathbf{D}_{heat} = 21.5$ mm^2 s^{-1}

in laminar flow, the ratios of effective diffusivities vary approximately in proportion to $(D_1/D_2)^{0.67}$, while in turbulent flow, values approach equivalence (see Monteith and Unsworth, 2008).

Other useful data

Distance to sun (D_{sun}) (1.50×10^{11} m)
Radius of sun (r_{sun}) (6.96×10^{8} m)
Mean radius of earth (r_{earth}) (6.371×10^{6} m)
Solar angle subtended at earth ($=\pi\ r_{sun}^2\ D_{sun}^2 = 6.76 \times 10^{-5}$ sr)
Solar constant ($\bar{I}_S$, 1366 W m^{-2})

Appendix 2
Chronology of earth observation

1839:	Invention of photography
1858:	First aerial photograph taken by Tournachon from tethered balloon
1868:	Electromagnetic theory first published
1914–18:	Aeroplane proves its worth as platform during First World War
1930–40:	Development of RADAR
1939–45:	Infra-red film developed for camouflage detection
1940s:	Development of valve computers
1947:	Captured V2 rockets used by US military for cloud observations
1948:	Invention of the transistor
1950s:	Development of aerial colour and colour infrared film opened up possibility of vegetation classification
1955:	Development of transistorized computers
1957:	Launch of Sputnik by USSR
1958:	Launch of Explorer-1 by US Army
1959:	First low-resolution space photograph of the earth taken and transmitted by Explorer-6
1960:	The term *Remote Sensing* coined
1960s:	Use of integrated circuits in computers
1960–72:	Twelve US CORONA missions produced high spatial resolution space photography
1960:	Launch of TIROS-1 with TV cameras giving daily pictures of cloud cover
1960s:	Gemini missions produced some 1100 photographs of the earth's surface that stimulated interest in observing the earth
1960s:	Development of solid-state multispectral imaging
1966:	Planning of first earth observing system, Earth Resources Observation Satellite (EROS), started
1966:	First geostationary satellite (ATS) started to supply meteorological data
1967:	First colour image of earth taken by Molniya-1 of USSR
1967:	First colour image from near-synchronous orbit taken using vidicon camera by DODGE satellite (US)
1972:	Launch of first earth Resources Technology Satellite (ERTS-1), later renamed Landsat-1
1970s:	Digital image processing became available
1978:	First spaceborne imaging radar system (SAR) flown on SeaSat
1978:	TIROS-N with the first AVHRR sensor launched
1982:	First airborne hyperspectral imager (AIS) flown by JPL
1980s:	Personal computers became available bringing digital image processing to the desktop
1999:	Launch of first high-resolution satellite – IKONOS-2
1999:	Launch of first MODIS instrument on the Terra (EOS AM-1) spacecraft
2000:	Launch of EO-1 carrying the first space-borne hyperspectral instrument "Hyperion"
2005:	Google Earth launched
2008:	NASA makes its data free

Appendix 3
Current data suitable for vegetation monitoring

Many remote-sensing instruments have been flown on satellites over the years. Even though some of them may have been designed primarily for other scientific purposes, such as meteorology or oceanography, data from these (notably the AVHRR) have often found use in vegetation studies. Satellites are ephemeral creatures; they may operate for many years or they may cease to provide data tomorrow, while others may be launched at any time. So here we list only some of the more commonly used instruments, the criteria for selection being that they provide data suitable for vegetation studies, that they are currently operational (at the time of writing) and that the data are readily available. There are very many more instruments flown on aircraft, but few will satisfy these criteria – for example the data may not be publicly available since many will have been built or operated by individual laboratories for specific research programmes. We will, however, mention a few of the more relevant ones that are either commercial instruments or which are flown by public bodies such as NASA.

The individual entries are listed in order of increasing spatial resolution (as far as that is possible, since some instruments provide data at more than one resolution) and within those categories according to their wavebands (optical, thermal or microwave). We also indicate the platform on which they are flown and, where relevant, other information such as the type of scanning system employed, their pointability and programmability, etc. Some of these systems were mentioned in Chapter 5. Detailed specifications can be found in the literature (e.g. Kramer, 2002, Campbell, 2007, and Lillesand *et al.*, 2007), or on various web sites, which may be more up-to-date.

Optical and thermal

Sensor	Satellite	Pixel resolution (at nadir)	Swath	Spectral bands (approx. band centre) (μm)	Bandwidth (μm)	Revisit time (at equator)	Comments
VISSR	Meteosat	2.5 km 5.0 km	Full earth disc	0.7 (Vis) 6.4 (WV) 11.5 (TIR)	0.2 1.4 2.0	30 min	Geostationary. Meteorological. Spinscan.
SEVIRI	Meteosat Second Generation (MSG)	1 km 3 km	Full earth disc	HRV broadband 0.635, 0.81, 1.64 3.90, 6.25, 7.35 8.7, 9.66 0.8, 12.0, 13.4	 0.14–0.28 1.0–1.8 0.6–0.8 2.0	15 min	Geostationary. Meteorological. Spinscan.
MIR	GOES East/West	1 km 4 km 8 km	Full earth disc	0.65 3.9 10.7 12.0 6.75	0.2 0.2 0.5 0.5 0.25	30 min	Geostationary. Meteorological. 2-axis scanning mirror.
AVHRR	NOAA/Metop	1.1 km	2400 km	0.61 0.9 3.7 11.0, 12.0	0.13 0.28 0.38 1.0	12 h	Two satellites in orbit at any one time provide 6 hour revisit. Whiskbroom
AATSR	Envisat	1 km	500 km	0.55, 0.66, 0.86 1.61, 3.70, 10.85, 12.00	0.02 0.30 1.00	5 days	Conical scanning
MERIS	Envisat	Two selectable resolutions 1.04 × 1.20 km 260 × 300 m	1150 km	0.41, 0.44, 0.49, 0.51, 0.56, 0.62, 0.66, 0.70, 0.89, 0.90 0.68, 0.75 0.76 0.78 0.86	0.01 0.075 0.025 0.015 0.02	3 days	5 overlapping cameras. Pushbroom. Bands programmable in position and width.

Sensor	Satellite	Pixel resolution (at nadir)	Swath	Spectral bands (approx. band centre) (μm)	Bandwidth (μm)	Revisit time (at equator)	Comments
LISS-3	IRS-1D	24 m 70 m	142 km 148 km	0.55, 0.65 0.82 1.65	0.07 0.11 0.20	24 days nadir viewing (5 days pointable)	Pushbroom.
PAN	IRS-1D	6 m	70 km	0.5–0.75		24 days	Pushbroom.
WiFS	IRS-1D	188 m	810 km	0.65, 0.82	0.07	5 days	Pushbroom.
MODIS	Terra/Aqua	250 m (bands 1–2) 500 m (bands 3–7) 1 km (bands 8–36)	2330 km	36 bands in range 0.41–14.34	From 0.05 to 0.30	1–2 days	Whiskbroom.
ASTER	Terra/Aqua	15 m 30 m 90 m	60 km 60 km 60 km	0.56, 0.66 0.81 (N and B) 1.65, 2.16 2.21, 2.25 2.33, 2.39 8.32, 8.60, 9.15 10.6, 11.3	0.07 0.10 0.01 0.05 0.07 0.35 0.7	16 days (4 days pointable)	Along-track stereo viewing (Nadir and backward-pointing at 0.81 μm).
ETM+	Landsat 7	15 m 30 m 60 m	185 km	Pan 0.48, 0.57, 0.66 0.74 1.65, 2.22 11.5	0.40 0.70 0.12 0.20 2.00	16 days	Launched in 1999. Whiskbroom.
HRG	SPOT 5	5 m 10 m 20 m	120 km	2 pan images 0.55, 0.65, 0.83 1.66	 0.07–0.11 0.17	27 days	Pushbroom (2 pan images give 3 m spatial resolution in supermode).
HRS	SPOT 5	10 m × 5 m	120 km	0.51–0.73 pan		27 days	Along-track stereo

Vegetation	SPOT 5	2200 km	1.15 km	0.45, 0.65, 0.83 1.66	0.04–0.07 0.17	daily	Perfectly registered with HRG.
CHRIS	PROBA	25 m 50 m	19 km	19 spectral bands 62 spectral bands		16 days (2 days pointable)	Pushbroom. Wavebands can be reconfigured. For collecting BRDF data.
OSA	IKONOS 2	1 m 4 m	13 km	Pan (0.45–0.90) 0.49, 0.56, 0.68 0.82	 0.06–0.08 0.17	11 days (pointable)	Pushbroom. Tiltable along or across track. Stereo capability.
BGIS 2000	Quickbird 2	0.61 m 2.44 m	16.5 km	Pan (0.45–0.90) 0.48, 0.56, 0.66 0.72	 0.06–0.08 0.14	1–3.5 days	Pushbroom. Tiltable along or across track. Stereo capability.
OHRIS	Orbview 3	1 m 4 m	8 km	Pan (0.45–0.90) 0.48, 0.56, 0.66, 0.83	0.07–0.14	Up to 3 days	Pushbroom. Orbit does not repeat exactly.
ALI	EO-1	10 m 30 m	37 km	Pan 9 multispectral bands 0.43–2.35 (6 of which are almost identical to ETM+ bands)	0.48–0.68 0.02–0.27	16 days	Pushbroom. Experimental improved ETM+, but no thermal band.
Hyperion	EO-1	30 m	7.5 km	242 spectral bands 0.40–2.50			Hyperspectral.

Synthetic Aperture Radar (SAR)

Instrument	Satellite	Waveband	Modes	Polarization	Spatial resolution	Swath	Comments
ASAR	Envisat	C-band	Image	VV, HH	28 m	Up to 100 km	ScanSAR modes
			Wide swath	VV, HH	150 m	400 km	
			Alternating polarisation	VV/HH, HV/HH, VH/VV	30 m	up to 100 km	
						5 km vignettes	
			Wave	VV, HH	30 m	400 km	
			Global	VV, HH	950 m		
SAR	Radarsat 2	C-band	Ultrafine	Single, H or V	3 m	10 or 20 km	Orbit repeat 24 days
			Fine	HH, HV, VV	11 × 9 m	25 km	
			Standard		25 × 28 m	25 km	
XSAR	TerraSAR-X	X-band	Spotlight 1	H and V in all modes	1 × 1.2 m	10 km	
			Spotlight 2		2 × 1.2 m	10 km	
			StripMap		3 × 3.2 m	40 km	
			ScanSAR		15 × 15 m	100 km	
PALSAR	ALOS	L-band	High resolution	HH or VV	7–44 m	40–70 km	
				HH/HV or VV/VH	14–88 m		
			ScanSAR	HH or VV	<100 m	250–350 km	
			Polarimetry	HH/HV + VV/VH	24–88 m	30 km	

Airborne scanners

IFOV and swath widths are usually defined in terms of the angles subtended at the instrument. GIFOV and physical swath width depend on flying height (H) and can be calculated by multiplying H by the appropriate angle. Because of the low flying height of aeroplanes, the (across-track) pixel size may vary greatly across the swath. Some hyperspectral instruments can operate in more than one spatial or spectral mode, which are often programmable.

Instrument	Spectral range nm)	Spectral bands	Spectral resolution (nm)	IFOV (mrad)	FOV (degrees)	Comments
Airborne Thematic Mapper (ATM) (AADS1268)	420–13 000	8 (420–1050 nm) 1550–1750 nm 2080–1350 nm 8500–13000 nm	30–140 200 700 4500	1.25	+/–43	11 fixed bands. Whiskbroom scanner. Bands 2,3,5,6,9,10 and 11almost identical to Landsat TM. 716 spatial pixels.
Airborne Visible/Infra-Red Imaging Spectrometer (AVIRIS)	400–2500	224	10	0.85	31.5	17 m pixels and 11 km swath from 20 km height.
HyMap	450–1200	100–200	10–20 (VIS/SWIR) 100–200 (TIR)	1–10	60–70	Family of hyperspectral instruments.
Eagle	400–970	488 244 122 60	1.25 2.3 4.6 9.2	0.5–1.04	29.9–62.1	Up to 1024 spatial pixels.
Hawk	970–2450	254	8.5	0.95–1.9	17.8–35.5	320 spatial pixels.

Instrument	Spectral range (nm)	Spectral bands	Spectral resolution (nm)	IFOV (mrad)	FOV (degrees)	Comments
CASI-2*	405–950	288	1.8	1.2	54.5	512 spatial pixels in 18 programmable bands (Spatial Mode). 288 in full spectrum in up to 39 look directions (Spectral Mode). 288 in full spectrum in blocks of 101 adjacent pixels (enhanced spectral mode). Full frame mode – all spatial and spectral pixels, but 1–2 s integration time – for calibration or field use.
DAIS	360–12 000	32 (hyperspectral) 5 (TIR)	20 1000–2000	5	+/−45	Rotating mirror. Reference black bodies viewed every scan line.
ROSIS	430–830	115	4	0.56	8	512 spatial pixels can be tilted by +/−20° to avoid sunglint.
TIMS	820–12 600	6	400–1000	2.5	86	714 spatial pixels

* Casi-2 can operate in several default bandsets, such as VEGETATION (12 bands) and SeaWiFS/Ocean Colour (13 bands)

Answers to sample problems

2.1 Using Wein's displacement Law (eqn (2.6)) we get: (a) forest fires: at 800–1000 K, $\lambda_{max} \sim$ 3–4 μm, mid-IR; (b) vegetation: ~ 293 K, λ_{max} = 9.9 μm, thermal IR.

2.2 (a) Using Wein's Law we get the radiant temperature for dull red ($\lambda_{max} \cong$ 0.65 μm) $\cong$ 4500 K; (b) λ_{max} for incandescent bulb $\cong$ 410 nm.

2.3 (a) UV, 8.5×10^{14} Hz, 5.6×10^{-19} J, 3.5 eV; (b) mid-IR, 8.5×10^{13} Hz, 5.6×10^{-20} J, 0.35 eV; (c) microwave, 8.5×10^{9} Hz, 5.6×10^{-24} J, 3.5×10^{-5} eV.

2.4 Assuming an average surface temperature of 23 °C, and no change in emissivity, $\sigma\varepsilon$ cancels out so the percentage difference from eqn (2.4) will be $100 \times (300^4 - 296^4)/(100 \times 296^4)$ = 5.5%. Energy emitted is proportional to ε, so a 2% change in ε will produce a 2% change in E.

2.5 Using eqn (2.15), we see that $m \cong (\exp(-1000/8000))/\cos(30°) = 0.8825/0.866 = 1.019$. From eqn (2.13), optical thickness = ln(−0.7) = 0.35668, therefore $\boldsymbol{I} = \boldsymbol{I}_0.\exp(-0.35668 \times 1.019)$ and $\boldsymbol{I}/\boldsymbol{I}_0 = 0.6953$.

3.1 (a) The outgoing radiation is the sum of the emitted radiation ($\varepsilon \times \sigma \times 293^4$ = 397 W m^{-2}) and the reflected environmental radiation ($(1-\varepsilon) \times \sigma \times T_{back}{}^4 = 0.05 \times 5.67 \times 10^{-8} \times 283^4$ = 18.19 W m^{-2}) = 415.19 W m^{-2}. Ignoring the reflected radiation and solving this for $T_{leaf} = (425.3/(\sigma \times 0.95))^{0.25}$ = 296 K (a 3-K error). Subtracting the actual reflected radiation before calculating T_{leaf} gives the correct value. (b) If one assumes an ε = 0.94, this gives the estimated emitted radiation as 415.19–21.82 = 393.37 W m^{-2}; this equates to T = 293.1 K (an error of 0.1 K).

3.2 From eqn (2.4), $E_{forest} = 0.98 \times \sigma \times 273^4 = 0.98 \times 5.67 \times 10^{-8} \times 273^4$, $E_{snow} = 0.8 \times 5.67 \times 10^{-8} \times 268^4$, so percentage change = −100 × (234.01–308.66)/308.66 = −24.2%. $T_{est} = \sqrt[4]{(0.95/0.98)} \times 268$ = 265.9 K or a 2.1 K error. Even if the forest were entirely snow covered it behaves as a rough surface where ε is larger than the value expected for a smooth snow surface (often approaching 0.95).

3.3 (a) From Box 3.1 we see that k = 1, so fraction intercepted = $(1 - e^{-kL})$ = 0.699, and sunlit leaf area = $(1 - e^{-kL})/k$ = 0.699; (b) From Table 3.4, $k = 2/(\pi.\tan\beta)$ = 0.232, so the fraction intercepted = 0.453 and the sunlit area = 1.593, (c) $k = 1/(2\sin\beta)$ = 0.532, so fraction intercepted = 0.749, and sunlit leaf area = 1.408.

3.4 From Box 3.1, the fraction of radiation transmitted = e^{-kL}, and from Table 3.4, $k = (0.5^2 + \cot^2\beta)^{½}/(0.5 + 1.774 \times 0.5 + 1.182)^{-0.733}/0.5$. Substituting for β gives the fractions transmitted as 0.073, 0.330 and 0.484.

4.1 $\psi_P = \psi - \psi_\pi$ = 0.9 MPa. Turgor pressure reaches zero when $\psi = \psi_\pi$ so ψ will be −1.2 MPa if the wall is inelastic, but if the wall is elastic the cell will shrink as turgor pressure decreases, concentrating the solute so that equality will be reached at a lower value of ψ.

4.2 (a) Total conductance for one surface $(G1) = g_{1s} \times g_{1a}/(g_{1s} + g_{1a})$ = 45.5 mmol m^{-2} s^{-1} (conductances in series), while the total leaf conductance $(G) = G1 + G2$ (conductances in parallel for the two surfaces). Substituting gives G = 45.5 + 142.9 = 188.4 mmol m^{-2} s^{-1}. (b) The overall leaf resistance $(R) = 1/G$ = 5.31 m^2 s mol^{-1}. (c) From Section 4.3.5, G (mm s^{-1}) $\cong$ 188.4/44.1 = 4.27 mm s^{-1}. (d) Leaf boundary-layer conductance to water vapour = $g_{1a} + g_{1a}$ = 1000 mmol m^{-2} s^{-1}. In a laminar boundary layer, $G_H/G_W \cong (D_H/D_W)^{0.67} = 0.92 \times 1000$ = 920 mmol m^{-2} s^{-1}.

4.3 (a) $\delta x = x_{s(18)}$ (= 0.02024) − $x_{s(24)}/2$ (= 0.02945/2) = 0.005512. (b) δx would reach zero at 13 °C. (c) The Magnus formula gives the vapour pressure as 7.413 × 0.6 as 4.448 kPa (= 4.448/101.3 = 0.044 mol fraction at 101.3 kPa).

4.4 (a) From eqn (4.15), and assuming that ε = 1, the radiative resistance, $\hat{g}_R$, = $(4 \times 1 \times 5.6703 \times 10^{-8} \times 293^3)/29.2$ = 0.129. Therefore, the isothermal net radiation = 400 + 0.129 $\hat{c}_p$ (24–20) = 400 + 5.71 × 4 = 423 W m^{-2}. (b) With no evaporation $\boldsymbol{C} = \boldsymbol{R}_n = 400 = \hat{g}_H\,\hat{c}_p$ (24–20), therefore $\hat{g}_H$ = 400/(4 × 29.2) = 3.43 mol m^{-2} s^{-1}.

4.5 (a) $\tau = c_{area}/(\rho c_p[(1/r_{HR}) + s/\gamma(r_{aW} + r_{lW})]) = (800 \times 3500 \times 0.0002)/(1210 \times 0.141)$ = 3.3 s. (b) Substituting $r_{lW} = \infty$, this gives 4.4 s, while for r_{lW} = 0, this gives 1.4 s.

5.1 Since $t^2 \propto r^3$, and $r_{shuttle}$ = 315 + 6371 = 6686 km and r_{sat} = 360 + 6371 = 6731 km, $t_{sat}/t_{shuttle} = \sqrt{(r_{sat}{}^3/r_{shuttle}{}^3)}$ = 1.010113. Satellite first overhead when $n \times t_2 = (n + 0.5) \times t_1$, therefore n = 0.5/0.010113 = 49.4 orbits.

5.2 From Box 5.1, orbital velocity = $\sqrt{(GM_e/r)} = \sqrt{(398\,603/(6371 + 700))}$ = 7.5 km s^{-1}. Ground velocity = 7.5 × 6371/7071 = 6.77 km s^{-1}. Period at 1700 km = 2 × π × (1700 + 6370)/7.03 = 120.2 min.

5.3 720 km h^{-1} = 720 000/(60 × 60) = 200 m s^{-1}, therefore, the satellite travels 10 m in 10/200 = 0.05 s. Therefore, the mirror must rotate at 20 Hz to be at the same point for the next scan line. A 10-m pixel subtends 1 mrad at 10 km, therefore dwell time = 0.05 × 0.001/2π = 7.96×10^{-6} s.

5.4 Swath width = tan(angle) × height (assume 700 km). For small angles this is $\cong$ angle × height = 185 km. Circumference of earth = 2π × 6371 = 40 030 km, so number of orbits = 215, which will take 215 × 100/(60 × 24) = 14.93 days.

5.5 There will be 50 000/10 = 5000 pixels and hence detectors in a swath. Array size = 5000×10^{-5} = 0.05 m.

5.6 Assume ERS height = 900 km, therefore signal total transit time = 2 × 900 × 1000/(2.8998×10^8) = 0.06 s.

Velocity = $\sqrt{(GM_e/r)} = \sqrt{(398\,603/(6371 + 900))} = 7.4$ km s^{-1}, so it travels 7.4 × 0.06 = 0.44 km. Referring back to eqn (2.22), the footprint = height × λ/d = 900 × 6 × $10^{-5}/10^{-3}$ = 48 km.

6.1 There are 15 pairs of bands for six bands. There are 6 ways of allocating three bands to three output colours, the general formula is $n\text{Pr} = n!/(n - r)!$, which for 7 channels and 3 colours is 7!/4! = 210.

6.2 Trial and error indicates that use of bands 2 and 3 gives values that distinguish the two tree types irrespective of illumination.

6.3 A particularly useful filter for eliminating outliers is a median filter. This gives:

157	156	148	148
155	156	148	148
152	153	148	90
152	152	93	92

7.1 ρR = 0.18 × 0.4 + 0.08 × 0.2 + 0.36 × 0.4 = 0.232; ρN = 0.28 × 0.4 + 0.49 × 0.2 + 0.56 × 0.4 = 0.454. By substitution into the formulae from Box 7.1, the indices are, respectively, 0.303, 0.260, 0.202, 1.871, 0.172, and 0.248. After regreening, these become 0.506, 0.374, 0.246, 3.050, 0.576, and 0.532.

7.2 The reflected radiance = ρ × incident, therefore R = 23.2 and 81.2 W m^{-2}, respectively, and N = 43.4 and 151.9 W m^{-2}, respectively. '*NDVI*' = 0.303 for both irradiances, but '*DVI*' increases from 20.2 to 70.7 as irradiance increases, showing the sensitivity of *DVI* to irradiance where one uses radiance rather than reflectance.

7.3 The Euclidian distance from a class is $\sqrt{}$(difference in channel1^2 + difference in channel2^2) so differences from A, B and C are 24.3, 34.1 and 22.4, and the unknown is closest to class C.

7.4 f_{veg} is related to *LAI* by eqn (7.5). For a homogeneous horizontal leaved canopy k = 1. Since $(\rho_{\text{NIR-plant}} - \rho_{\text{NIR-soil}}) > (\rho_{\text{R-soil}} - \rho_{\text{R-plant}})$, as expected, the response is convex.

7.5 Calculating a conventional *NDVI* gives negative values in the top left-hand side of the square and low positive values in the bottom right. This suggests that the former represent water and the latter represent bare soil.

7.6 (a) The ratio between band 1 and band 3 is least sensitive to shading (a 0.4% change as compared with a 16% change for bands 3 and 2 and a 19% change for bands 2 and 1). (b) These same two bands also give the least sensitive *NDVI*.

8.1 From simple geometry, the linear dimension of the view area is given by $h[\tan(\theta + \text{fov}/2) - \tan(\theta)] + h[\tan(\theta) - \tan(\theta - \text{fov}/2)]$, where h is the height and θ is the nadir view angle and fov is the field of view, Substituting gives the linear dimension on the ground as 0.28 m, 0.374 m, 1.134 m. For a goniometer arrangement the height of the sensor varies with view angle according to radius × cos(θ) so for 0°, 30° and 60°, respectively, the corresponding heights are 2, 1.732 and 1 m, with viewed areas being 0.28, 0.323, and 0.57 m.

8.2 From eqn (8.11), the probability = $\exp(-(k(\theta).\,L_{\text{eff}})$, which for (a) = exp(−3.2) = 0.04 for all angles, for (b) $k = 1/(2\sin\beta)$, which gives values for k of 1, 0.707, 0.577, and 0.517, respectively, and P of 0.11, 0.21, 0.28, and 0.32, respectively. (c) $k = 2/(\pi \tan\beta)$, which gives values for k of 1.10, 0.637, 0.378, and 0.171 and P of 0.19, 0.38, 0.58, and 0.77. Increased clumping would increase the probability of viewing gaps, a value of λ_0 of 0.5 would decrease the effective leaf area index by 50% in each case.

8.3 Solar elevations can be calculated using tools such as SunAngle (http://susdesign.com/sunangle/) that gives values of 35.6°, 61.5° and 83° for the three times. Noting that the results for canopy (b) are similar at all angles, this suggests a horizontal leaved canopy, so using eqn (8.11), we get L = − (ln(1 − shade fract))/k ≅ − ln(0.59) = 0.9. By rearranging eqn (8.11) we get k = − (ln(1 − shade fract))/L, and assuming constant values for L we see that k changes approximately as expected for a random leaf distribution (see Fig. 3.22), therefore by substituting $k = 1/(2\sin\beta)$ we obtain L = 1.8.

9.1 (a) Assuming $\varepsilon_{sky} = 1$, $\varepsilon_{can} = 0.98$ and $\rho_{can} = 0.15$; $\boldsymbol{R}_n = (1 - \rho_{can}) \times \boldsymbol{I}_S) - \varepsilon_{can}\sigma T_{can}^4 + \varepsilon_{sky}\sigma T_{sky}^4 = 435$ W m^{-2}; from eqn (9.2), the soil heat flux ($\boldsymbol{G}$) ≅ $\boldsymbol{R}_n$ × 0.4 × exp($-k\,L$). If one assumes a random leaf distribution, $k = 1/(2\sin\beta) = 0.5$ (for the sun overhead), $\boldsymbol{G}$ = 8.7 W m^{-2}. (b) From Section 4.5.2, $\beta = \boldsymbol{C}/\lambda\boldsymbol{E} = 0.1$, and $\boldsymbol{R}_n = \boldsymbol{C} + \lambda\boldsymbol{E} + \boldsymbol{G}$. so $\boldsymbol{C} = (\boldsymbol{R}_n - \boldsymbol{G})/(1 + 1/\beta) = 426.8/11 = 38.8$ W m^{-2}, therefore from eqn (9.4), $\hat{g}_H = \boldsymbol{C}/(\hat{c}_p(T_s - T_a)) = 38.8/(29.3 \times 1) = 1.32$ mol m^{-2} s^{-1}.

9.2 We can assume that the hottest pixels are non-evaporating while the coolest pixels correspond to surfaces evaporating at the potential rate. If one assumes the available energy is the same for different surfaces, eqn (9.14) simplifies to $E = (\boldsymbol{R}_n - \boldsymbol{G})\,(T_d - T_{can})/(\lambda(T_d - T_a)) = 400 \times (6/10)\,/\,2.454 \times 10^6$ kg m^{-2} s^{-1} = 0.35 kg m^{-2} h^{-1} = 0.35 mm h^{-1}. One also needs to assume that radiometric temperature equates to aerodynamic temperature.

9.3 Assuming that the normalized *NDVI* is proportional to f_{veg}, we can calculate canopy temperature from $T_{canopy} = (T_{av} - (1 - f_{veg}) \times T_{soil})/f_{veg} = (32 - 0.4 \times 37)/0.6 = 28.7$ °C.

9.4 Substituting in eqn (9.19) (a) $(\text{d}\boldsymbol{E}/\boldsymbol{E})/(g_s/g_s) = 0.9$ or 0.7 for the two conductances, respectively. (b) The corresponding results are 0.39 and 0.14 for the relatively smooth canopy (showing a much smaller sensitivity to stomatal conductance).

10.1 Note that the confusion matrix is presented in a different form from Table 10.5. (a) Overall accuracy is given by diagonal element divided by the total = 292/364 = 80.2%;

(b) the most errors of commission occur for class 1 (24) (sum of incorrect values in row), but the largest percentage will be for class 2 with 36.6%; (c) the user's accuracy is highest for class 4 (102/112 = 91.1%), while the producer's accuracy is also highest for class 4 (= 102/103 = 99%). Note that class 4 also has the fewest incorrect pixels (one).

10.2 $\sigma^2_{mean} = \sigma^2_q/n_2 + \sigma^2/n_1n_2 = 0.12/10 + 0.06/80 = 0.01275$. If $c_1 = 10\ c_2$, eqn (B10.2.3) suggests that $n_2 = \sqrt{(10 \times 0.06/0.12)} = 2.24$. The value for n_1 is obtained from eqn (B10.2.1).

10.3 Substituting into appropriate equations in Table 10.2 gives (a) 0.224, and (b) 0.326.

Index

A

B

C

G

H

L

M

N

O

P

Q

R

S

T

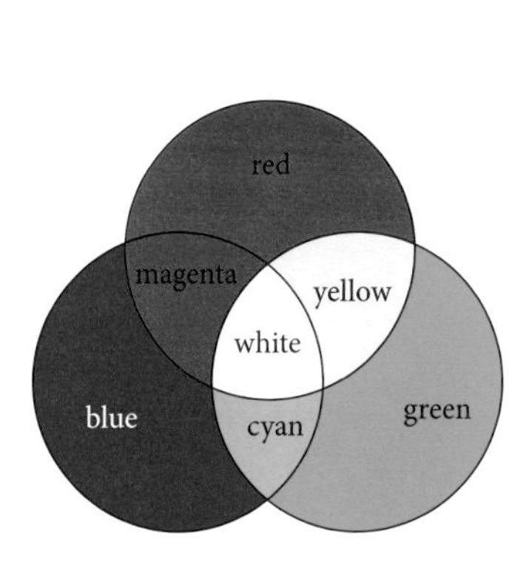

Plate 2.1. Illustration of colour addition and the three primary colours: red, green and blue. (See Box 2.3)

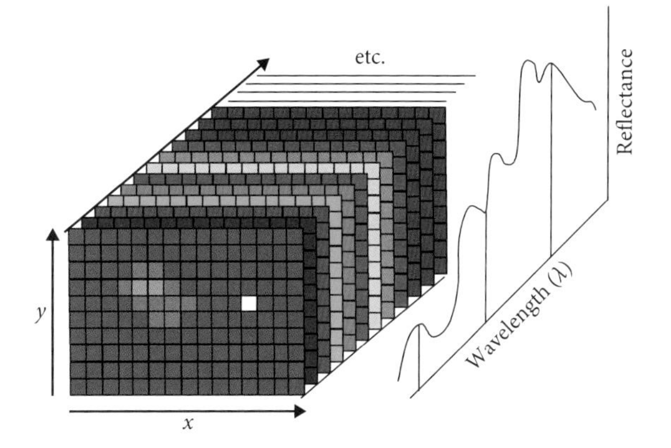

Plate 5.1. A hyperspectral image cube where many images at different wavelengths are stacked to form a cube with two spatial dimensions (x, y) and a spectral dimension (λ). (See Fig. 5.3)

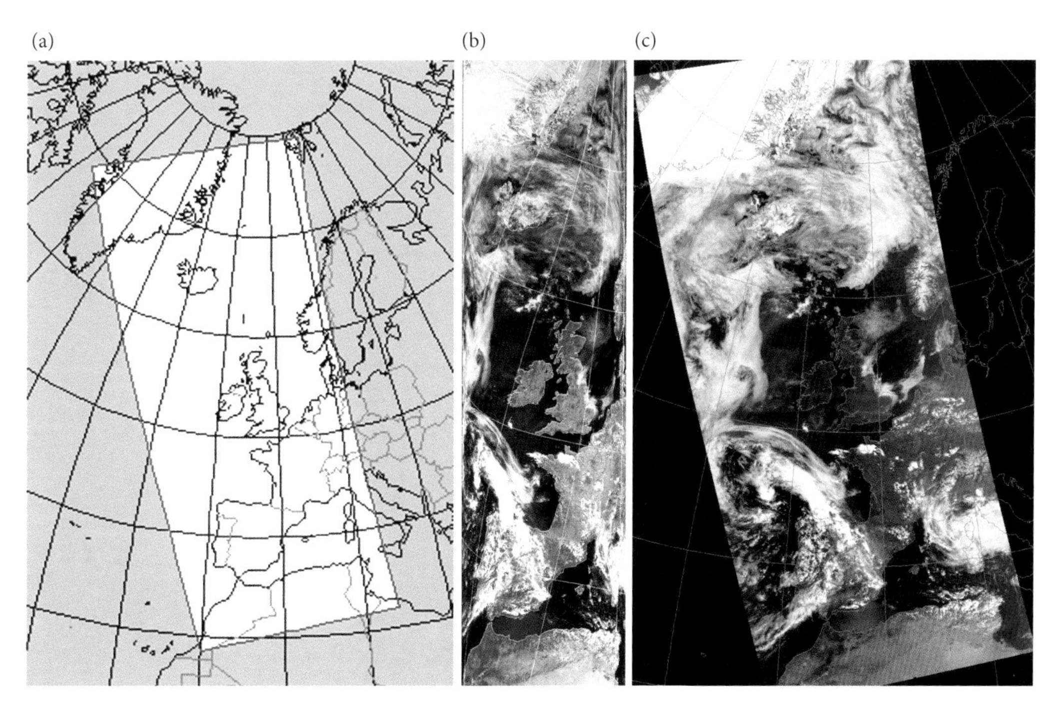

Plate 6.1. MODIS image over the UK (13.15 UTC) northbound over UK on 13 May, 2008, showing (a) the skewed path of the satellite (Aqua), (b) a raw image before geometric correction (near infrared (841–867 nm) showing geometric compression at the edges of the swath, and (c) a geocorrected RGB composite image (reproduced with permission from NEODAAS and University of Dundee). (See Fig. 6.1)

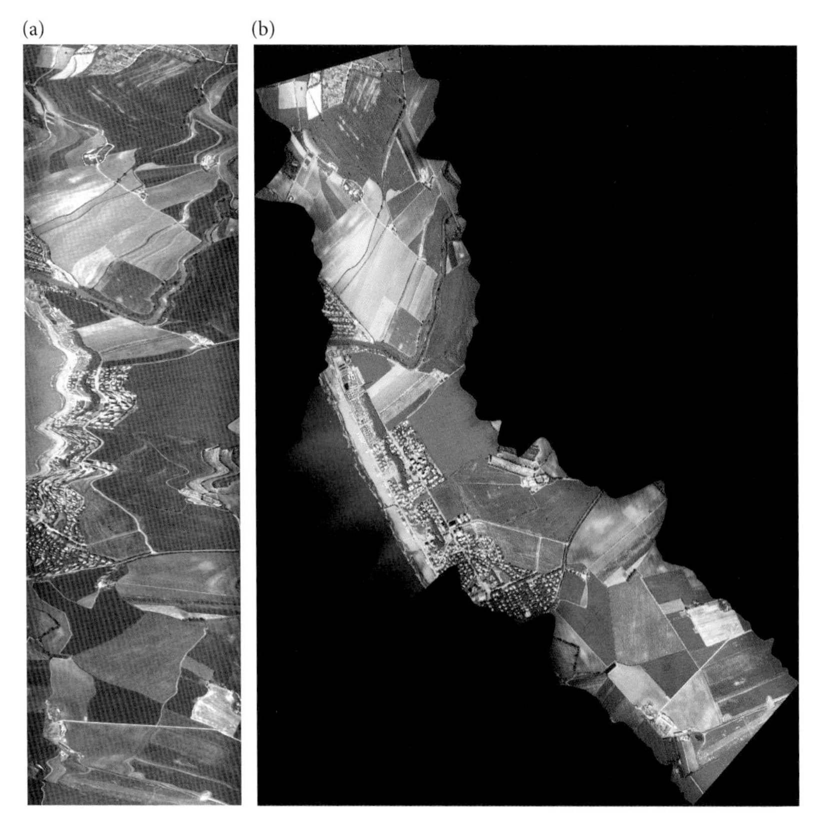

Plate 6.2. Consequences of aircraft instability on imagery: (a) original greyscale image as collected by an Airborne Thematic Mapper over Tarquinia in Italy, (b) the same image after geocorrection of distortions caused by aircraft instability (image from NERC ARSF project MC04/07). (See also Fig. 6.4 and the cover)

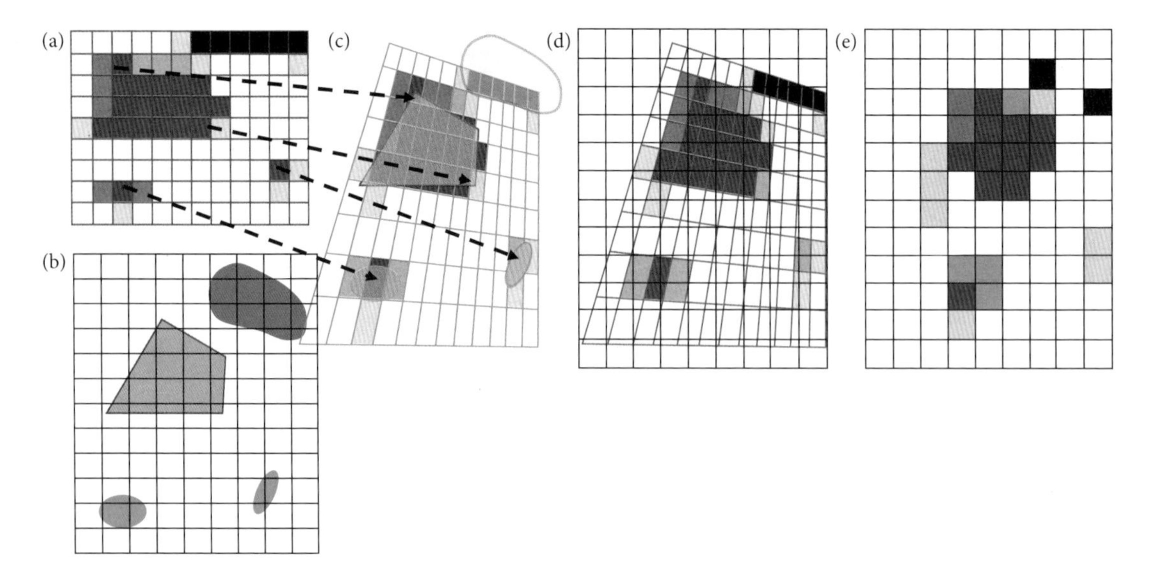

Plate 6.3. Reprojection and resampling of a simple image: (a) an original distorted greyscale image of an area, (b) a map of the corresponding real landscape showing some objects in blue, green, and brown, (c) the transformation (warping) of (a) so to overlay the map by using a series of ground control points (GCPs) in the image and on the map. The correspondence of GCPs in the image and map is shown using dashed lines. The warped image is then overlain by a rectangular grid in (d) and the new pixel values in the final georegistered image (e) are obtained using the nearest neighbour approach where the value for any pixel takes that of the nearest neighbour in (d). (See Fig. 6.5)

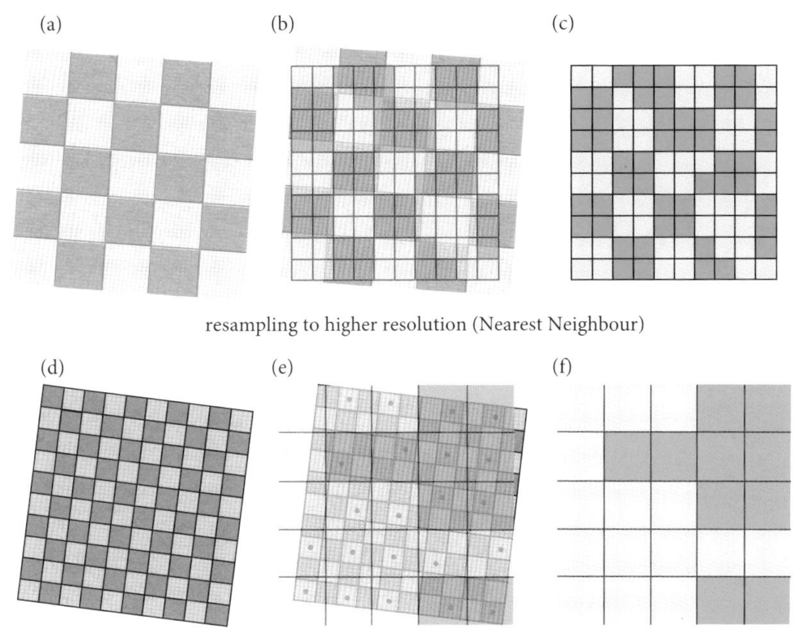

Plate 6.4. Illustration of nearest neighbour resampling (with reprojection) to a higher resolution (a, b, c) and resampling to a lower resolution (d, e, f). The dots represent the nearest original pixels to the new centres. The green grid shows the new projection and it is clear that resampling substantially alters the pattern of pixels obtained. (See Fig. 6.6)

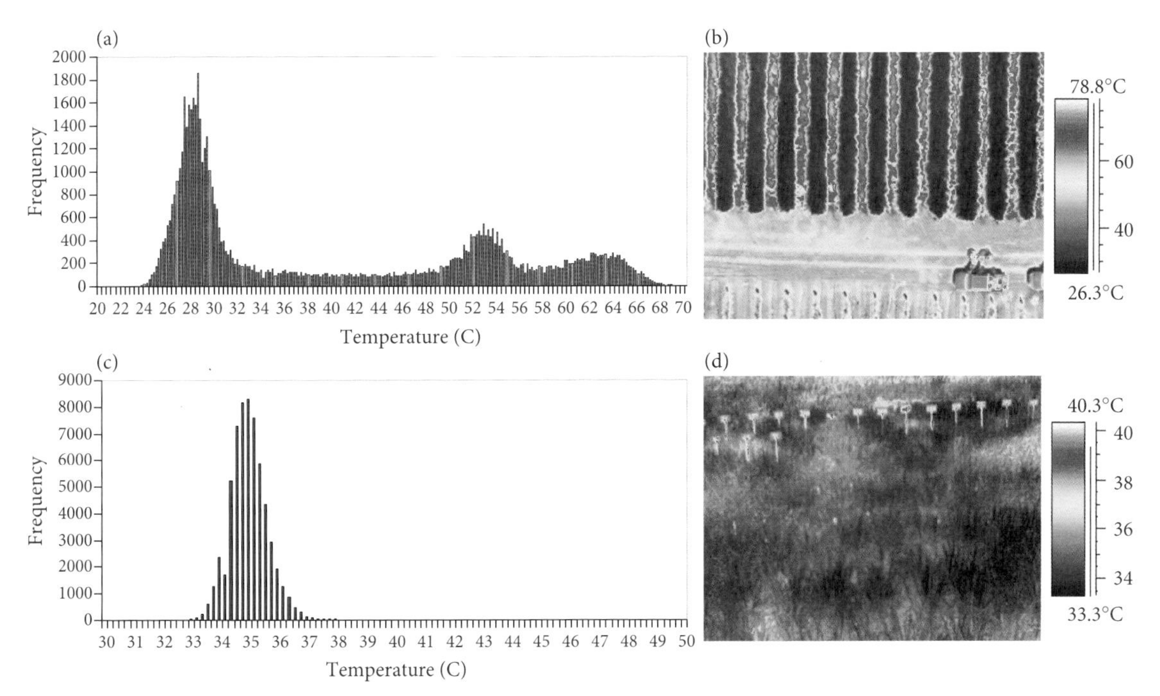

Plate 6.5. (a) Bimodal histogram of temperatures in the aerial image (b) of a vineyard in S. Australia (see also Fig. 8.1) showing the contrast between the cool canopy and the hot soil; (c) a relatively narrow and homogeneous temperature histogram for a thermal image of a rice field in Wuhan (d). (Photos A. Wheaton and H.G. Jones.) (See Fig. 6.7)

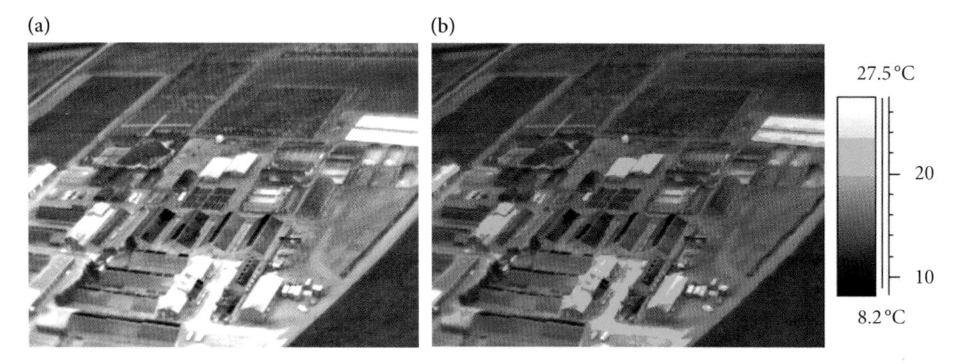

Plate 6.6. Illustration of density slicing, showing (a) greyscale thermal image of the farm buildings at the Scottish Crop Research Institute, Dundee (13 June, 2007), and (b) the same image with all pixels between 20 and 22 °C highlighted as a density slice in green. (See Fig. 6.11)

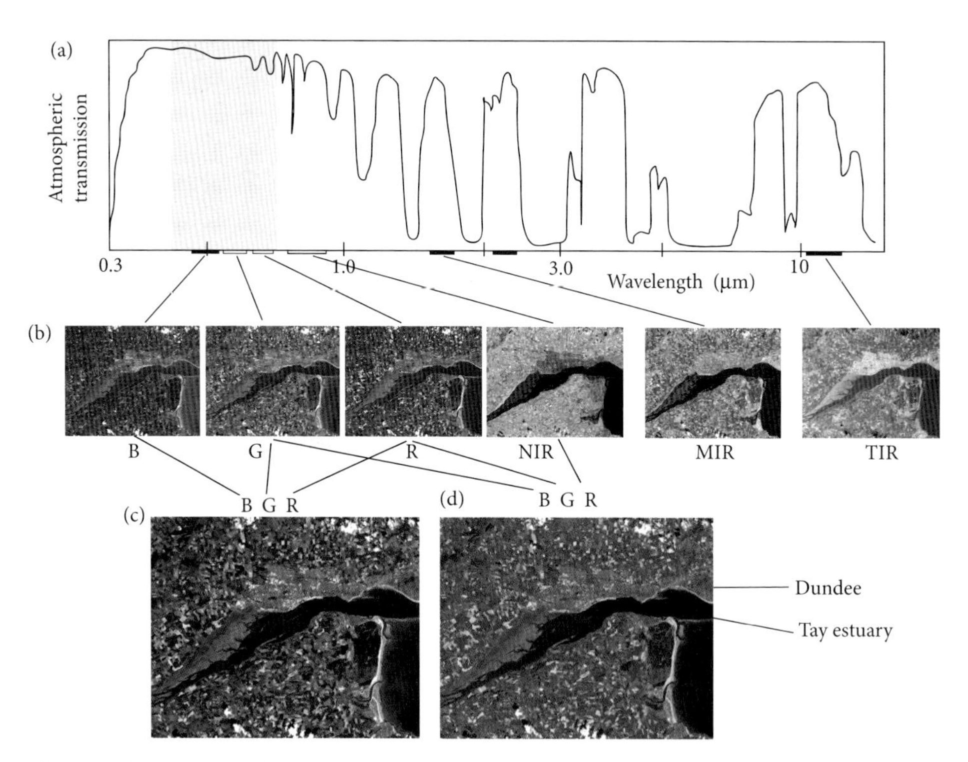

Plate 6.7. (a) An atmospheric transmission spectrum, showing the positions of the seven Landsat ETM+ bands, (b) corresponding greyscale single channel images for six of these bands for an image over part of eastern Scotland showing Dundee and the Tay estuary (path 205, row 21, for July 2000). These images can be combined in different combinations to highlight different aspects of the scene. The most obvious combination is to display the red, green, and blue channels in the R, G, and B screen colours as shown in (c) to give a natural colour composite. Alternatively different combinations may be combined to give one of several false colour composites, including, (d) a standard false composite where R = NIR, G = Red, and B = Green (Landsat image courtesy of USGS). (See Fig. 6.12)

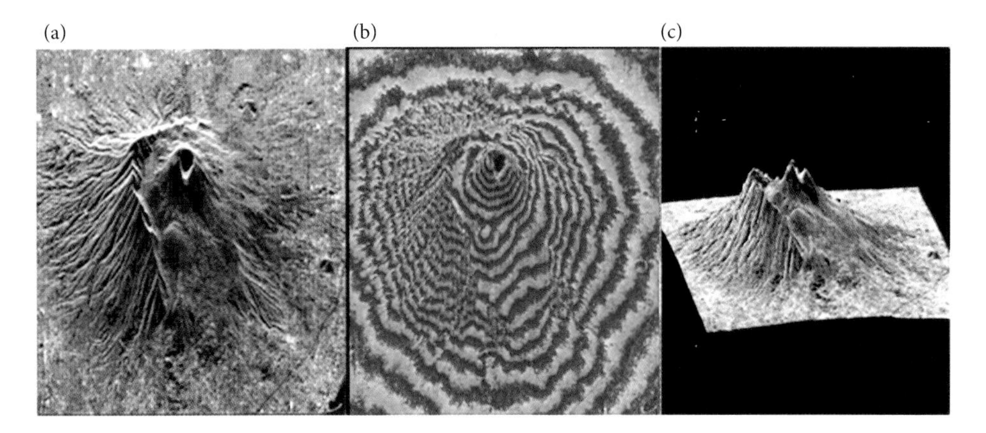

Plate 6.8. (a) A SAR image of Mt. Vesuvius from ERS-1, (b) a corresponding interferogram and (c) a 3D rendering of the derived Digital Elevation Model (images courtesy of ESA – European Space Agency). (See Fig. 6.20)

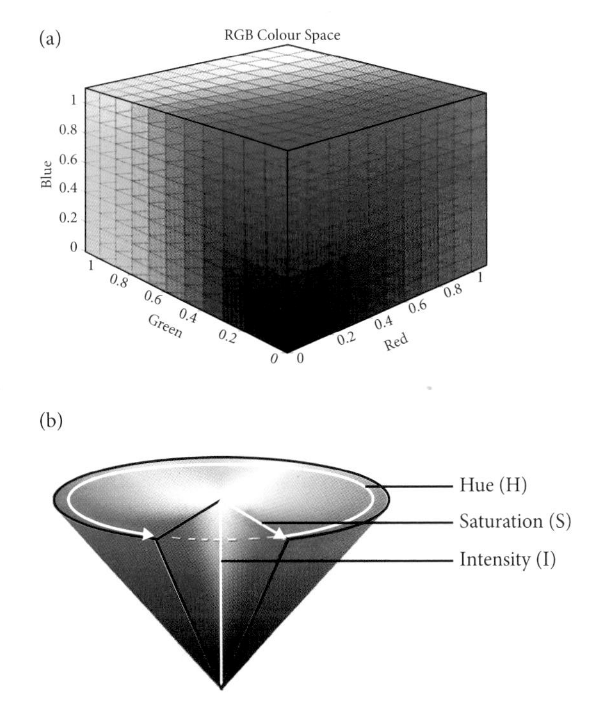

Plate 6.9. (a) Representation of RGB colour space as a colour cube, (b) representation of colour space in terms of Intensity (or Brightness), Hue, and Saturation. (See Box 6.2)

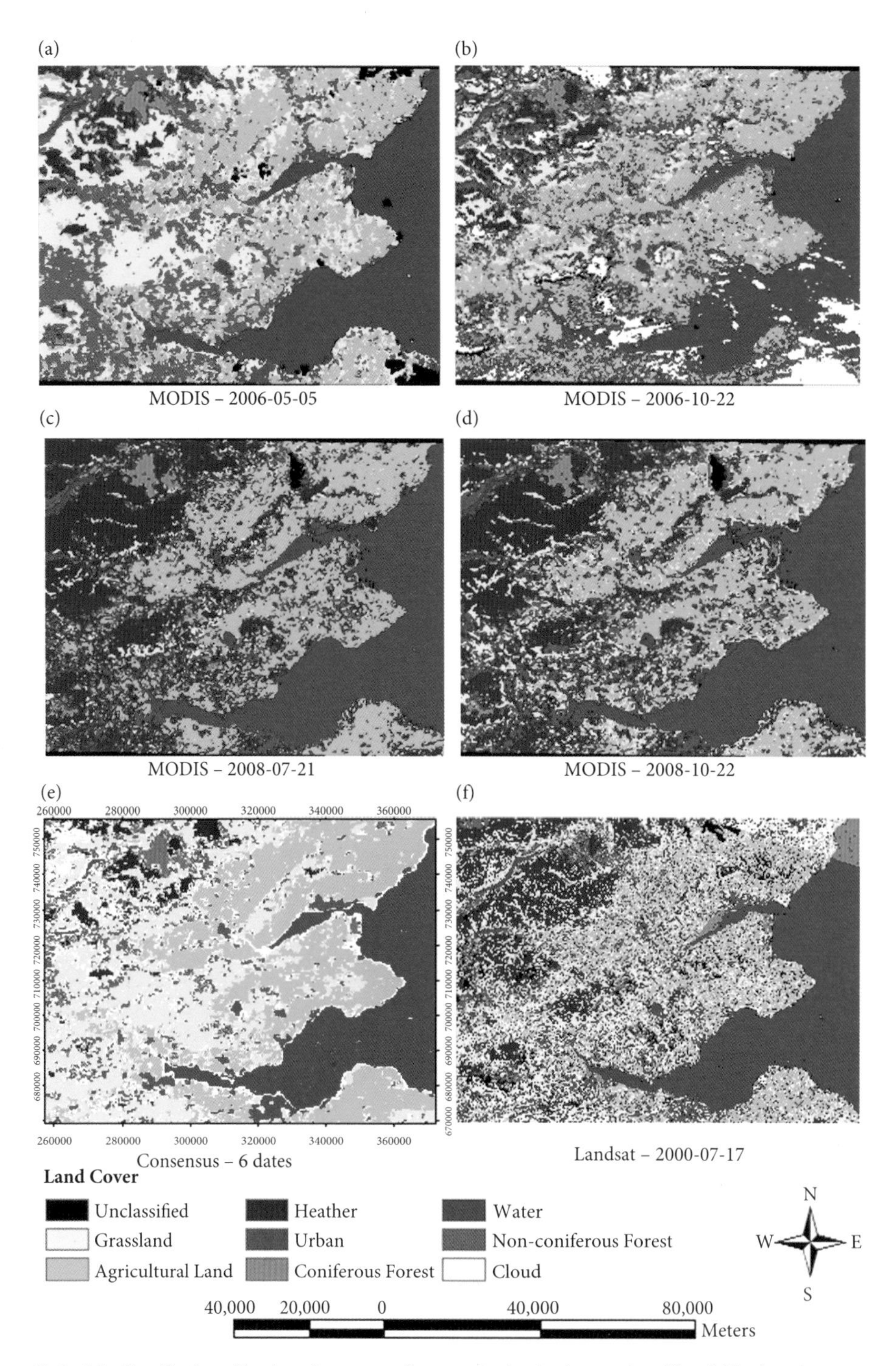

Plate 7.1. Classification of land use for an area of eastern Scotland using a series of four fairly clear sky images from MODIS (a, b, c, d) as compared with Landsat TM (f) using a maximum likelihood algorithm with the same parameters and the same training areas. In addition a consensus image (e) for MODIS is shown (where colours are shown for five or more similar classifications using 11 NBARs images over the period day 43 to day 289 in 2006 and 2008. (MODIS data provided by NEODAAS at University of Dundee and Plymouth; Landsat data for 17th July 2000, courtesy US Geological Survey and Landsat.org). (See Fig. 7.18)

(a)

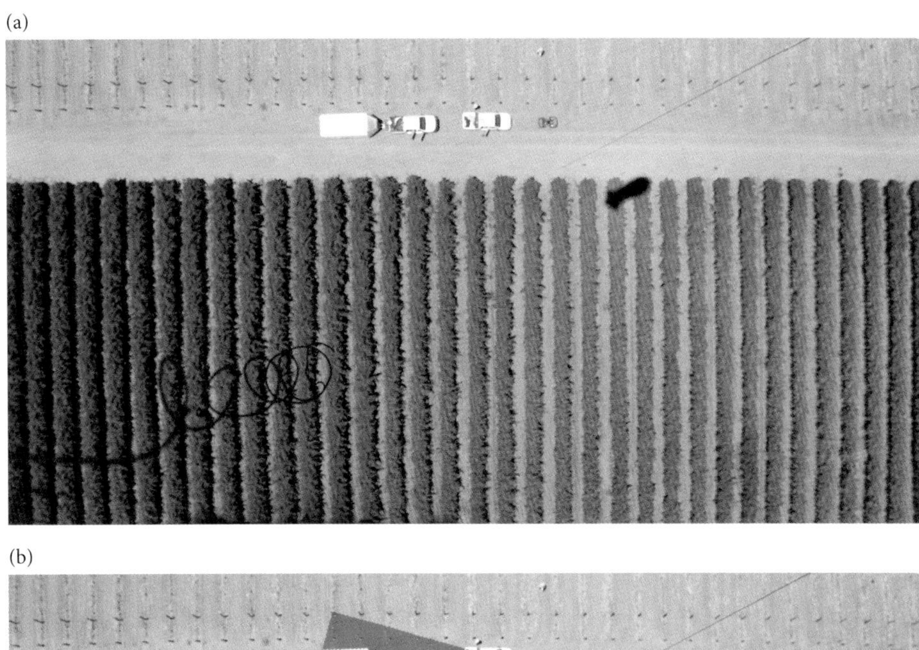

(b)

Plate 8.1. (a) Aerial image of a vineyard from a balloon at 90 m showing the decreasing brightness of the image as one moves from the hotspot (the area around the shadow of the balloon on which the cameras are mounted). The changes in brightness largely arise as a result of changes in the proportion of shadow visible, with the proportion increasing as one moves away from the hotspot. (b) same image with thermal image overlaid showing temperature increasing towards the hotspot. (See Fig. 8.1)

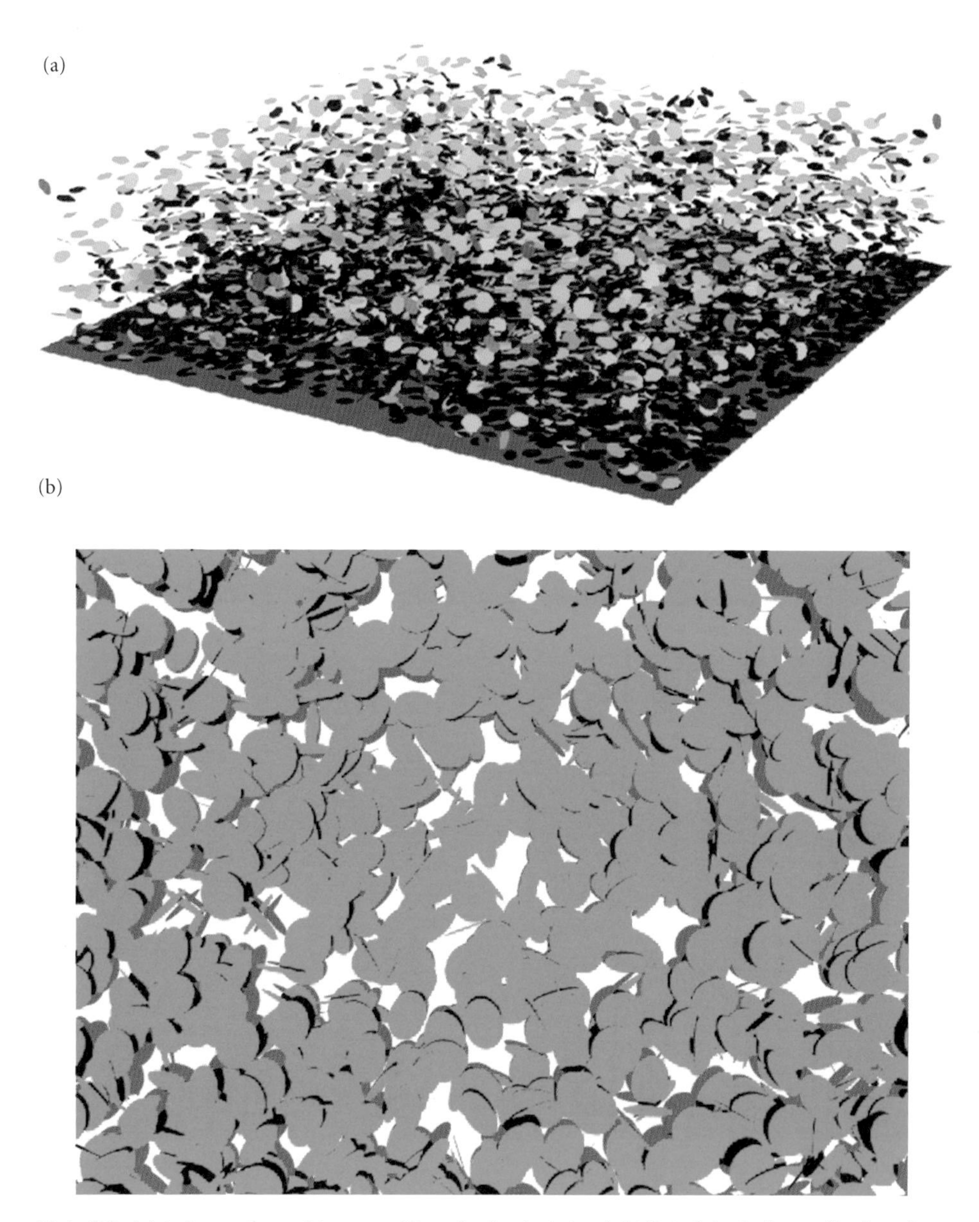

Plate 8.2. (a) An image of a model canopy with randomly oriented and distributed circular leaves, illuminated from the right, generated by POV-Ray. Shaded areas are indicated as darker areas on either the leaf or soil backgrounds (Fig. 8.16(a)). (b) View into the hotspot at +45°, showing increasing proportion of shadow as the angle of view moves away from the precise hotspot (Fig. 8.12(b)). (Images with permission from R. Casa.)

Plate 8.3. (a) A false colour (Red = red, NIR = green) image at +10° view angle of a potato plot using a two channel (Red/NIR) camera (ADC, Dycam Inc., Chatshereworth, CA), together with (b) the image classified into sunlit leaves, shaded leaves, sunlit soil and shaded soil using a supervised classification in the ENVI image processing software (Casa, 2003). (See Fig. 8.21)

Plate 11.1. Some examples of characteristic patterns of leaf colouration in response to different mineral deficiencies in tomato (from the web companion to Taiz and Zeiger (2006) at http://4e.plantphys.net/article.php?ch=t&id=289). (See Fig. 11.2)

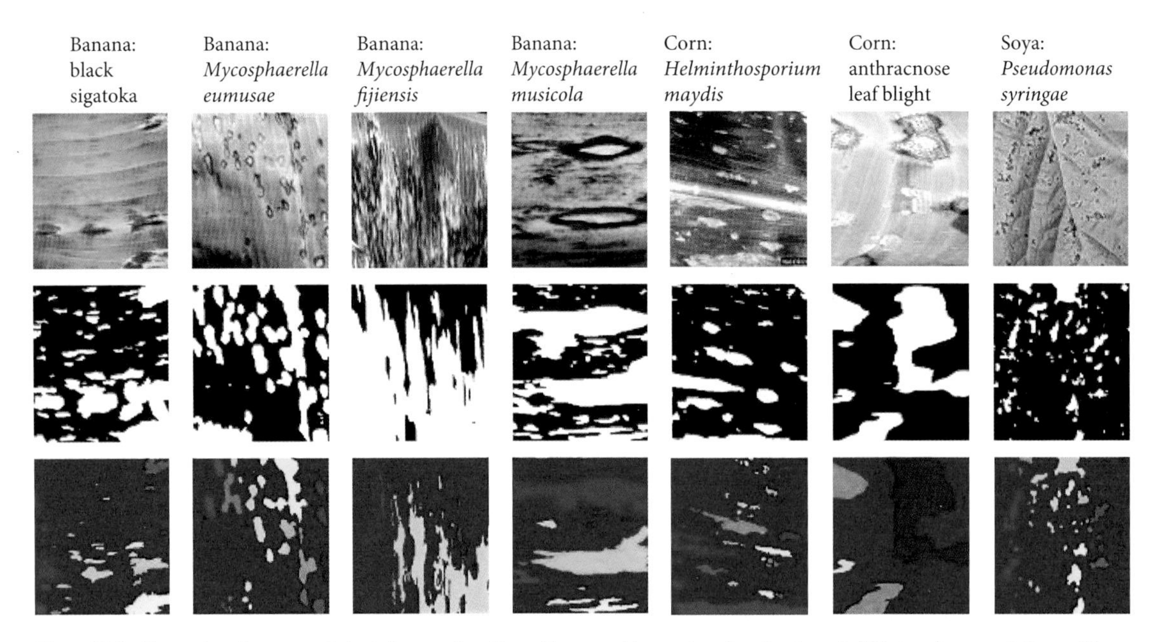

Plate 11.2. Illustration of a range of plant diseases, together with manual (second row) and automatic (third row) segmentations of the images as a basis for automatic diagnosis systems (from Camargo and Smith, 2009, with permission). (See Fig. 11.3)

Plate 11.3. Plant phenotyping at the plot scale showing a visible image overlaid by some thermal images to illustrate the possibilities of the approach when taking images from a height of c. 5 m and at an angle so as to minimise detection of background soil. (See also Fig. 11.4)

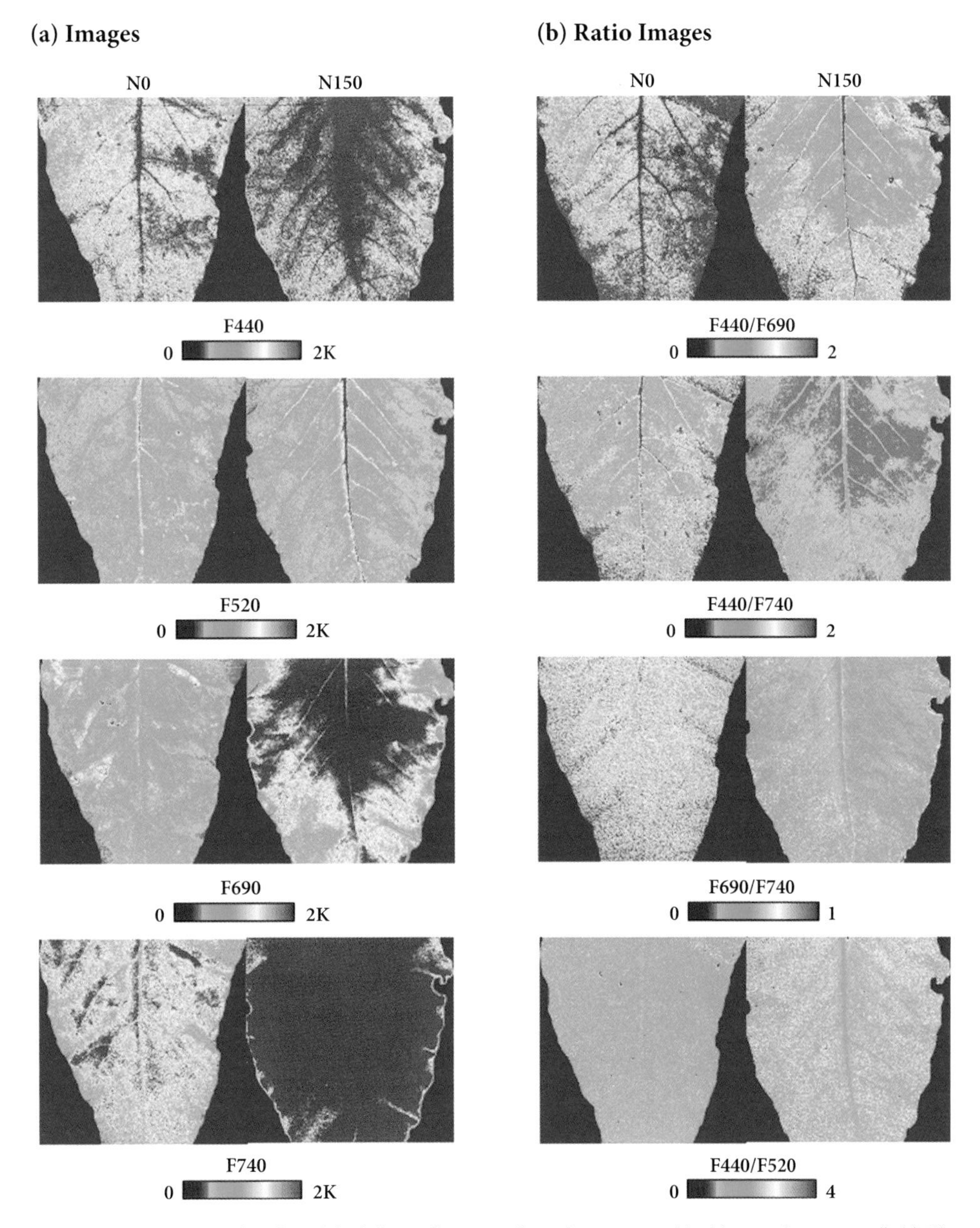

Plate 11.4. Fluorescence imaging of single leaves from sugar beet plants grown with either no nitrogen supplied (N0) or a luxury supply (N150) showing the effects on UV-A stimulated fluorescence at four wavelengths (440, 520, 690, and 740 nm, and ratios between them, presented in false colour with fluorescence intensity increasing from blue to red (with kind permission from Langsdorf *et al.*, 2000). (See Fig. 11.5)

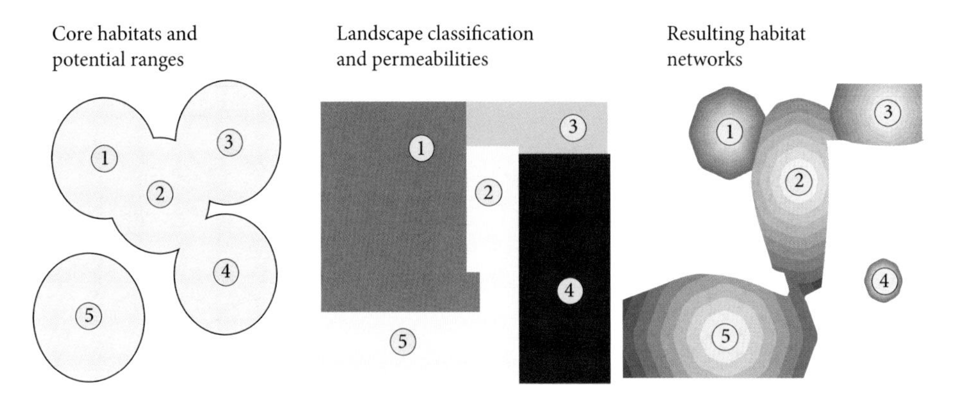

Plate 11.5. Illustration of the principle of habitat networks showing (a) core areas and hypothetical dispersal ranges of a species, (b) the landscape patterning and the associated permeabilities (increasing from yellow through brown and orange to brick), and (c) the calculated connectivity between sites (Watts *et al.*, 2005; with kind permission from Kevin Watts and the Forestry Commisssion). (See Fig. 11.8)

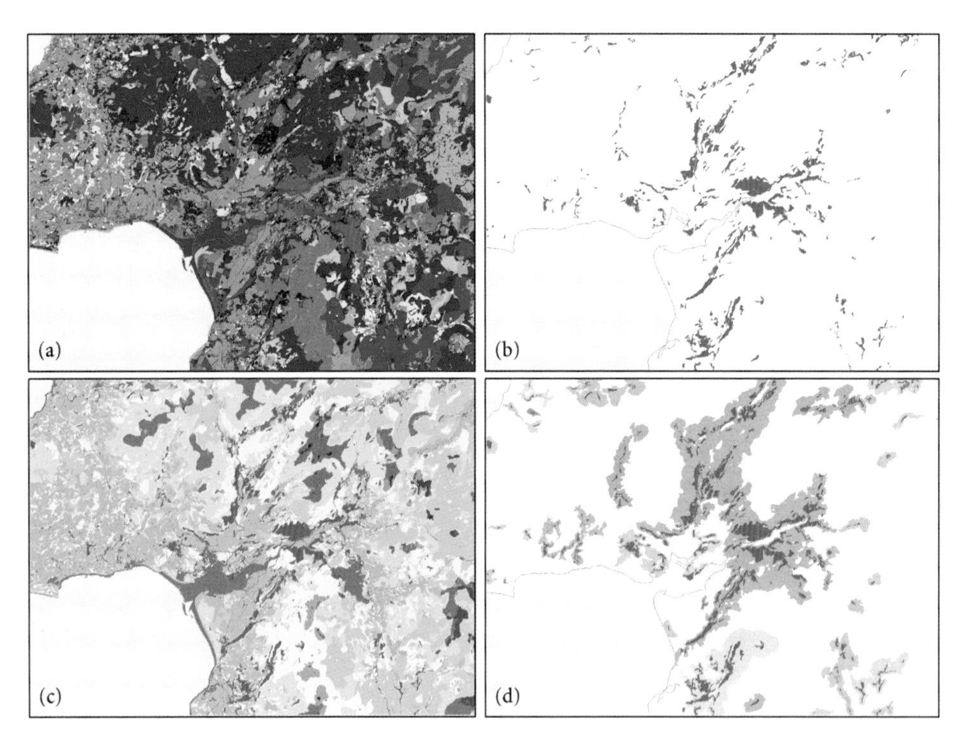

Plate 11.6. Example of application of habitat networks to a land cover map of a portion of North Wales, showing (a) the land cover distribution with different cover types in different colours, (b) the area of core woodland habitat for a woodland specialist, (c) assumed permeabilities (higher permeability represented by lighter colours), and (d) the resulting habitat networks, with non-connected networks illustrated in different colours (Watts *et al.*, 2005; with kind permission from Kevin Watts and the Forestry Commisssion).